D0329114

Risk Assessment of Chemicals in the Environment

Federation of European Chemical Societies
International Programme on Chemical Safety

Risk Assessment of Chemicals in the Environment

Edited by

Mervyn L. Richardson,
B.Sc., C.Chem., F.R.S.C., C. Biol., M.I. Biol.
Chairman, Chemical Information Group,
The Royal Society of Chemistry

ROYAL
SOCIETY OF
CHEMISTRY

British Library Cataloguing in Publication Data

Risk assessment of chemicals in the
environment.
1. Organisms. Toxic effects of environmental
pollutants
I. Richardson, Mervyn
574.5'222

ISBN 0-85186-118-0

Published by The Royal Society of Chemistry,
Burlington House, London W1V 0BN

Printed in Great Britain by
Whitstable Litho Printers Ltd., Whitstable, Kent

Foreword

These Proceedings cover the third European Conference on Chemistry and the Environment as formulated by the Federation of European Chemical Society's (FECS) Working Party on Chemistry and the Environment. The Working Party was formed in 1977 and in 1980 held its first Conference in Paris entitled 'Chemical Pathways in the Environment'. The second Conference in 1984 was concerned with the effects of acid rain on plants and soil and was held in Lindau. There are currently 20 countries represented on the Working Party, their representatives being G.J. Dickes (UK), J. Albaiges (Spain), A. Asenov (Bulgaria), L. Cagliotti (Italy), S. Cawley (Eire), M. Dular (Yugoslavia), S. Freyschuss (Sweden), A. Hackl (Austria), R. Hagemann (France), A.A. Jensen (Denmark), E. Lippert (Czechoslovakia), B. List (Switzerland), G. Mihalyi (Hungary), L. Pawlowski (Poland), M. Rabinowitz-Ravid (Israel), N. Schamp (Belgium), H.M. Seip (Norway), C. Troyanowsky (France), A. Valavanidis (Greece), E. Weise (West Germany), and Yuksel Inel (Turkey). Like all FECS bodies, the Working Party has no financial account and therefore relies on the generosity of individual Member Chemical Societies for its existence.

The third Conference therefore would not have taken place had it not been for the willingness of the Royal Society of Chemistry (RSC) to organise it on behalf of the Working Party. The usual professionalism of the RSC to organise such a Conference gave it a flying start at Guildford and I record my appreciation of the dedicated work of Dr. John F. Gibson and his staff, in particular Larraine Curzon and Penelope A. Yap.

In its remit from FECS, the Working Party has to choose Conference topics which have broad appeal to environmental chemists in order to attract not only European but also worldwide participation. Together with the hope that the subject matter would be topical, 'Risk Assessment of Chemicals in the Environment' was in my view, an ideal choice and this publication is based largely on the contributions of plenary and invited lecturers. In order to make a more complete book, contributions have been accepted from authors who did not participate in the Conference and I believe that this has resulted in a publication which is more balanced and better 'value-for-money'.

Having Proceedings of a Conference affords the opportunity on behalf of the FECS Working Party and of the Organizing Committee to thank the financial sponsors whose generosity meant that Conference registration fees were kept to a minimum. These sponsors included the World Health Organization, the executing agency of the International Programme for Chemical Safety (a co-operative programme with the International Labour Office and the United Nations Environment Programme), the Commission of the European Communities, the United States Air Force, and Schering Agrochemicals Limited. WHO also specifically sponsored the Indian and Chinese speakers. We are also grateful for the publicity by way of co-sponsorship given by the British Toxicological Society, the Institute of Biology, and the American Chemical Society.

No major Conference would ever take place without its Organizing Committee, in this case being R.H. Andrews, J.A. Deschamps, J.F. Gibson, P.A. Gilbert, R.M. Harrison, M.K. Johnson, D.V. Parke, M.L. Richardson, H.I. Shalgosky, and E. Smith. I thank them for their inspiratory ideas and hard work.

I also thank the Editorial Board of R.H. Andrews, A.G. Cubitt, J.A. Deschamps, J.F. Gibson, D.L. Miles, D.V. Parke, D.M. Sanderson, P.A. Sim, D. Taylor, and, in particular, Mervyn L. Richardson, the Editor, for maintaining such an impetus to ensure that this publication has become available so soon after the Conference. Mervyn's enthusiasm has been such that there was a real danger that the Proceedings would be published before the Conference took place!

Finally, I believe that in holding such a Conference, we in the scientific community not only gave our members the most recent hard data, but also provided a forum where people of different political and social backgrounds could meet to discuss each other's problems, not merely those associated with science.

G.J. Dickes
Chairman of the Organizing Committee
The Royal Society Chemistry

Preface

The word 'risk' probably originates from the Greek word ριψοκινδυνευσιζ, or the Italian word *rischio*. Originally it was related to the hazards of sailing near to dangerous rocks and cliffs, particularly in swirling tides. For the purpose of this book, 'hazard' is defined as the set of inherent properties of a chemical substance or mixture which makes it capable of causing adverse toxic effects on man or the environment, when a particular degree of exposure occurs. 'Risk' is defined as the predicted or actual frequency of occurrence of an adverse effect of a chemical substance or mixture from a given exposure to humans or the environment, and 'safety' is defined as the high probability that injury will not result from the use of a substance under specific conditions of quantity and manner of use. These definitions are consistent with those recommended by the Organisation for Economic Co-operation and Development and also by the Royal Society.

Risk management of chemicals in the environment is a complex matter with which we are all involved. The environment consists entirely of naturally occurring chemicals and all of us, and the rest of the animal kingdom, are continually exposed to both these and man-made substances. There are risks of exposure to both. Exposure, as was indicated in the Royal Society of Chemistry's publication **'Toxic Hazard Assessment of Chemicals'**, can arise from chemicals and materials handled at home, at work, and from food, water, or air.

Man-made chemicals are now essential to man's well-being, and effective means of minimizing their capacity to cause harm must be established. Risk assessment, which is the outcome of a series of processes involving identification, estimation, evaluation, and subsequent effective management, is a matter to be considered seriously by all those having responsibility for producing or handling chemicals.

Certain risks are inevitable, but when problems and uncertainties occur the greatest use must be made of information available to minimize any occurrence and also to understand the mechanisms involved.

The aim of this book is to review the current status of risk assessment procedures as they relate to the environment. It is not a dictionary of risks associated with any specific chemical. The book brings together some of the chemistry and other sciences which are necessary in the multi-disciplinary approach required in the risk assessment of chemicals. Over the past few years, the Royal Society of Chemistry has organized a number of conferences, workshops, and meetings dealing with this subject, and many of the chapters which follow have been previously discussed at these meetings, *i.e.*:

Risk Assessment of Chemicals in the Environment, Guildford, UK, 11-14th July 1988 - The 109th Federation of European Chemical Societies (FECS) Event and Third FECS Conference on Chemistry and the Environment.

Macro Effects from Micro Quantities - Environmental and Biochemical Effects of Micro Pollutants, part of the Royal Society of Chemistry Annual Chemical Congress, University of Kent at Canterbury 12-15th April 1988.

Pesticide Information, part of the Royal Society of Chemistry, Autumn Meeting, University of Nottingham 22-24th September 1987.

The Editorial Board has attempted to minimize overlap between chapters. However, in dealing with such important topics as hazard, risk, safety, environmental impact *etc.*, some duplication between chapters has been inevitable. Such repetition should enhance the contents of the book in view of the various and diverse experiences expressed by authors from such countries as Eastern and Western Europe, the United States of America, India and the Pacific basin, and representing the United Nations, Government Departments, Trans-National Industry, and Research Associations.

As Editor, and a member of the Royal Society of Chemistry's Chemical Information Group and Toxicology Group Committees, and a former committee member of the Environment Group and The Water Chemistry Forum, I am particularly fortunate in being able to draw on the advice and expertise of members of these committees. In particular, I have been well supported by members of my Editorial Board, J.A. Deschamps, G.J. Dickes, J.F. Gibson, D.L. Miles, D.V. Parke, D.M. Sanderson, E.

Smith and D. Taylor, and by friends and colleagues who have acted as additional referees who included F.S.H. Abram, T.V. Arden, S.K. Brown, S.S. Brown, J.A. Deschamps, G.J. Dickes, J.H. Duffus, P.C. Elwood, D.G. Evans, P.A. Gilbert, R.M. Harrison, M.K. Johnson, A.S. Kallend, D.P. Lovell, P.C. Merriman, D.L. Miles, D.V. Parke, R.J. Robinson, D.M. Sanderson, H.I. Shalgosky, D. Taylor, M. Tordoff, and W.B. Whalley.

The creation of this book was made possible by the unstinting secretarial support of Pauline A. Sim of the Gascoigne Secretarial Services who acted as secretary to the Editorial Board, undertook the typing of the international correspondence engendered in the preparation of this book, and retyped all of the manuscripts in camera ready format (using Vuwriter Professional, Scientific version 5.2E software by Vuman Computer Systems Limited). Furthermore, I am indebted to the members of the Books Department of the Royal Society of Chemistry for their support, R.H. Andrews and P.G. Gardam, and most particularly A.G. Cubitt for his unfailing assistance in the desk editing. Finally, my most sincere gratitude to my wife Beryl who so patiently accepted my working on this book and the enormous piles of papers for the past year.

Mervyn L. Richardson, Editor
Chemical Information Group
The Royal Society of Chemistry.

Contents

Section 2: Contribution of Toxicology to Risk Assessment

Section 4: Intentional Emissions

Contributors

M.E. Andersen, *Armstrong Aerospace Medical Research Laboratory*, Ohio, USA

R. Anliker, *Ecological and Toxicological Association of the Dyestuffs Manufacturing Industry*, Basel, Switzerland

S.C. Batt, *Monitoring and Assessment Research Centre*, London, UK

P. Benedek, Budapest, Hungary

F. Bro-Rasmussen, *Technical University of Denmark*, Lyngby, Denmark

D. Brown, *ICI plc, Brixham Laboratory*, Brixham, UK

C. Byrne, *Department of the Environment*, London, UK

H.J. Clewell, *Armstrong Aerospace Medical Research Laboratory*, Ohio, USA

N.E. Day, *MRC Biostatistics Unit*, Cambridge, UK

Sir Richard Doll, *Imperial Cancer Research Fund*, Oxford, UK

H.M. Donaldson, *ICI plc, Organics Division*, Manchester, UK

R.B. Drawbaugh, *Armstrong Aerospace Medical Research Laboratory*, Ohio, USA

F.R. Farmer, Warrington, UK

D.G. Ferry, *University of Otago Medical School*, Dunedin, New Zealand

J.S. Gow, *The Royal Society of Chemistry*, London, UK

L. Hordijk, *National Institute for Public Health and Environmental Protection*, Bilthoven, The Netherlands

C. Ioannides, *University of Surrey*, Guildford, UK

S. Jasanoff, *Cornell University*, Ithaca, New York, USA

C.R. Krishna Murti, *Scientific Commission for Continuing Studies of the Effects of the Bhopal Gas Leakage on Life Systems*, Madras, India

M.G. Lee, *Mersey Regional Health Authority*, Liverpool, UK

D.F.V. Lewis, *University of Surrey*, Guildford, UK

S. Li, *Chinese Academy of Preventive Medicine*, Beijing, China

R.L. Lipnick, *US Environmental Protection Agency*, Washington DC, USA

M. Mercier, *World Health Organization*, Geneva, Switzerland

R.J. Otter, *Department of the Environment*, London, UK

D.V. Parke, *University of Surrey*, Guildford, UK

P.J. Peterson, *Monitoring and Assessment Research Centre*, London, UK

I. Pollo, *Technical University of Lublin*, Poland

I.F.H. Purchase, *ICI plc*, Central Toxicology Laboratory, Alderley Park, UK

M.L. Richardson, *Chemical Information Group, The Royal Society of Chemistry*, London

J. Riha, *Czech Technical University*, Prague, Czechoslovakia

L.E.J. Roberts, *University of East Anglia*, Norwich, UK

H.M. Seip, *University of Oslo*, Norway

L. Somerville, *Schering Agrochemicals Limited*, Saffron Walden, UK

W.A. Temple, *University of Otago Medical School*, Dunedin, New Zealand

D.H. Trønnes, *Centre for Industrial Research*, Oslo, Norway

M.J. Van den Heuvel, *Department of Health and Social Security*, London, UK

Sir Frederick Warner, *University of Essex*, Colchester, UK

J.A.H. Waterhouse, *University of Birmingham*, UK

Editorial Board

Was ist das nit gifft ist? Alle ding find gifft und nichts ohn gifft. All ein die dosis macht das ein ding kein gifft ist.

Philippus Theophrastus Aureolus Bombastus von Hohenheim (Paracelsus)

1492-1541

Section 1: Introduction and Overview

1
Introduction

J.S. Gow

THE ROYAL SOCIETY OF CHEMISTRY, BURLINGTON HOUSE, PICCADILLY, LONDON W1V 0BN, UK

1 Introduction

The assessment of risk associated with chemicals which enter any part of the environment is a task which cannot be undertaken without an adequate knowledge of the chemistry of the substances involved. However, it is by no means simply a matter of the chemistry. Risk assessment is a multi-disciplinary subject with worldwide consequences. Obviously, poison gas clouds do not recognize national boundaries but it is important to remember also that risk assessment-based decisions taken in one country can have far-reaching effects in another part of the world. The scientific application of risk assessment techniques to the hazards of chemicals in the environment is the subject matter of this book. The probability that any given level of a chemical in the environment presents a hazard to the exposed population whether humans, animal, fish or other species, can be extremely difficult to determine, but it is essential that rigorous scientific methods are used in any such assessment. Unfortunately, in many cases the available data are not good enough to make precise predictions possible.

2 Federation of European Chemical Societies (FECS)

This book is largely based on the Conference held on 11th to 14th July 1988 at the University of Surrey, Guildford, UK, and which was organized by The Royal Society of Chemistry (RSC) on behalf of the FECS. The RSC welcomed the opportunity to make the arrangements for this, the Third FECS Conference on Chemistry in the Environment. Membership of the FECS now includes thirty-seven chemical societies from all over Europe, both East and West. The Federation carries out its detailed work via a series of specialist working parties of which the Working Party on Chemistry and the Environment is one of its most active and one of the most widely representative. The Conference brought together many speakers with international reputations and in addition to distinguished chemists included representatives of a number of disciplines, ranging from engineers and epidemiologists to toxicologists and zoologists.

3 Contents of the Book

The contents of the book fall into four major sections which are enumerated below. In addition, the book contains a very useful Glossary of Terms, and a list of Useful Addresses. The glossary is particularly important because, without a common definition of terms, meaningful international discussion of this complex subject area is impossible.

3.1 Section 1 - Introduction and Overview. This Section covers risk assessment and the acceptability of different risks. It deals with prediction and evaluation. International comparisons, control of industrial chemicals, standards of chemical safety and new procedures for the quantitative assessment of structure-activity relationships are all described.

3.2 Section 2 - Contributions of Toxicology to Risk Assessment. This Section highlights the techniques in risk assessment for carcinogenic compounds, *e.g.* use of animal data, mechanisms and epidemiology. It also discusses the effects of solvent abuse, and physiologically-based pharmacokinetic approaches to the problem of risk assessment.

3.3 Section 3 - Incidental Emissions - Air and Water.

3.3.1 Air - Topics include decision analysis in control of chemicals posing health risks, the modelling of the risks of acid rain, and problems related to nitrogen oxide emissions.

3.3.2 Water - The pollution of the River Danube and its risk assessment, total index environmental quality, structure-activity relationships as applied to fish and other aquatic species, risk assessment of dyestuffs in the environment, together with a novel means of selecting substances requiring priority action are described in detail.

3.4 Section 4 - Intentional Emission. This Section reviews the hazard and risk assessment and acceptability of chemicals in the environment. Examples cover pesticides, cytotoxic drugs, the acceptability of chemicals into the environment, including by-products of multi-purpose fine chemicals manufacture, *etc.* There is also a review of deterministic and probabilistic trends.

4 Perception of Risk

Although the risk of driving, or being driven. in a motor car is comparatively large, it is a widely accepted risk. On the other hand, the comparatively small risk associated with adding even minute quantities of a carcinogenic chemical to the environment is one which is not readily accepted by the public at large. The risk/benefit equation should be a major determinant but since both the risk and the benefit are frequently not fully understood, irrational fears develop. Whilst the benefits of the motor car are immediately obvious, the benefits of chemicals are not widely recognized and this makes it all the more important that objective methods of risk assessment are not only developed but are used on a common basis internationally. Pollution does not recognize national boundaries!

In assessing risk it is important to consider:

i The nature and degree of the potential health hazard to man or other species;

ii The models and techniques that are used for estimating the risk at low doses of substances that are believed to have genotoxic, epigenetic action or are proximate carcinogens;

iii The use of the notion of 'acceptable risk';

iv The assumptions that are made regarding human (or other species) exposure to organic compounds in air, drinking water, soil, food, *etc.* and exposures from other routes;

v How numerical values should be presented;

vi The criteria to be used for selecting chemicals for priority action; (since more than 100,000 industrial chemicals are in use, it is essential that a balanced scientific view be taken);

vii The importance which must be accorded to organ-specific toxicity, including reproductive, immunotoxic and behavioural toxicity;

viii The ways in which uncertainty factors are applied to guard against either excessively pessimistic or excessively optimistic projections.

It is important to stress that all risks are relative - there is no such thing as absolute safety. However, it is not sufficient to tell people that the risk of death from eating ten grams of peanut butter per day is ten times greater than the risk of drinking a glass of wine per day which, in turn, is one hundred times greater than the risk of drinking two litres of chlorinated water per day, since the first two may be seen as voluntary risks whereas the third is an imposed risk. However, without careful assessment of the relative risks, the debate on what is an acceptable risk cannot even be started.

5 Conclusions

The professional chemical societies of the world have an important role to play in encouraging the discussion and dissemination of the results of the assessments of the risks of chemicals in the environment. Chemicals are being produced all over the world in increasing numbers and quantities. Although many of these chemicals are toxic, very few of them have yet given rise to major health effects in the human population. The assessment of a chemical's health effects, and its fate in the environment, coupled with data on its production, distribution and usage must be carried out on an international basis so that agreements can be reached and the potential hazards appreciated. Increases in the use and production of chemicals, especially in Third World countries, is leading to greater storage, transportation and disposal problems and this emphasizes the need for international collaboration in risk assessment. At the present time, the benefits of chemistry and chemicals to society in general are not well appreciated. The presentation to the public of the assessment of the real level of risk posed by agrochemicals, pharmaceuticals, household chemicals, *etc.* is a matter of grave concern to chemical societies everywhere. This book, based on the Conference held at Guildford and attended by scientists from all over of the World, brings together a vast collection of expert knowledge. The Royal Society of Chemistry's Charter, states that the Society should:

i Serve the public interest in matters related to chemistry;

ii Advance the science and practice of chemistry;

iii Foster and encourage the growth and application of science by the dissemination of chemical knowledge.

It is hoped that by organizing the Guildford Conference and publishing this book, the Society will have made some contribution to all three of these objectives.

2
Risk Assessment and Risk Acceptance

L.E.J. Roberts

ENVIRONMENTAL RISK ASSESSMENT UNIT, SCHOOL OF ENVIRONMENTAL SCIENCES, UNIVERSITY OF EAST ANGLIA, NORWICH NR4 7TJ, UK

1 Introduction

The discipline of risk assessment should be seen as a branch of engineering. The results of risk analysis are an evaluation of risk in quantitative terms with a statement of the expected accuracy, but the evaluation will often depend on data which are not scientifically complete and therefore on decisions based on judgement, and perhaps of the resolution of conflicting data on the basis of experience or of prudence. Formal methodology and a system of standard practice is required. Increasingly those engaged in the assessment of risks have to be aware of the factors that influence the acceptance of risks by those possibly affected, both workers and the public. The perception of risks by concerned groups of the public will affect the criteria of safety insisted upon in different cases and therefore the priorities for remedial or avoiding action. This chapter is concerned with the difference between the technical discipline of risk assessment and the social factors that determine risk acceptance, *i.e.* with the interface between risk assessment and risk management. Actions to reduce risks cost money: in the extreme, the economic penalty associated with the loss of an activity deemed to be too risky; in practice, managing risks uses a mix of safety standards and the 'As low as Reasonably Practicable' approach.[1]

But who is to decide what is reasonable and on what evidence? Since risk assessment can never be an exact science, decisions about risk acceptance must always be taken in a situation of some residual uncertainty. Therefore, the procedures by which such decisions are made are of critical importance; the process of regulation will be seen as the ultimate guarantee of public safety and must inspire confidence, as must the management of an industry on whom rests the responsibility for implementing safety standards. Eiser[2] has argued in a recent paper that risk assessments are essentially made by an institution, not by an individual, and that certainly applies also to the formulation of regulations and to their application. So it must be confidence in institutions and in their internal procedures which matters, and such confidence can only be gained as a result of a long period of experience and exposition. However, in the final analysis, decisions may have to be taken on the basis of a political process with a timetable very different from that necessary for a deep scientific evaluation of a situation, and on the basis of considerations in which risk/benefit analysis plays only a small part.

It seems worthwhile to look at a few well documented cases where political considerations and processes have played a major part in decisions involving the management of environmental risk before attempting to summarize some general considerations concerning the way our institutions in the UK are working, and consequently some of the tasks that face the profession of risk assessment.

2 Lead in the Environment

The recent history and the sequence of events that led to the tightening of regulations concerning the use of lead compounds and the plans to phase out the use of lead in petrol have been described in many papers. Jasanoff[3] cited the course of the Lead in Petrol controversy in the USA and in the UK as one example of the effect that cultural predispositions may have; she suggests that these actually influence expert assessments of risk. However that may be, this case is a good example of the power of public campaigns to set priorities in environmental regulation and it is possible by now to look at the efficacy of the resulting regulation.

Lead and lead compounds are toxic materials which are very widely used in industry and in domestic applications. Lead has become widely disseminated in the environment, perhaps at levels two to three orders of magnitude higher than in pre-industrial times. The concentration in blood is a convenient measure of body burden, though this measure alone may be inadequate; the half-life of lead in blood is about 18 days, so blood levels should reach a new level in response to changed exposure in a matter of several weeks. Lead is stored in bone and the concentration in teeth is a measure of integrated dose. Lead is well known to be a neurological poison, with other clinical effects as well. Children are particularly at risk; any child with a blood lead level above 800 $\mu g\ l^{-1}$ is considered to be at serious risk, and those with a level above 250 $\mu g\ l^{-1}$ should be investigated. A Department of the Environment (DoE) circular in 1982 required follow-up investigations for any individual with a blood lead concentration above 250 $\mu g\ l^{-1}$.

In contrast to many elements which are toxic at high concentrations, lead is not an essential trace element. The question that became of the highest public concern was the possible effect of moderate levels of lead, below those that can cause acute clinical symptoms, on the intelligence and behaviour of children as a result of chronic neurological damage. Ratcliffe[4] lists 23 studies of groups of varying size, mainly in the USA. Some of these studies showed significant differences between low- and high-lead groups, but others did not.[4] It is arguable that these surveys did not establish an unequivocal effect or a dose-response relationship at moderate levels of exposure, but the majority of studies did report lower IQ scores (3-4 points) from a high-lead group. Such measurements are, of course, extremely difficult to make with precision. Intelligence tests are not universally accepted and behaviour patterns are difficult to establish. Such characteristics are known to depend on a number of confounding variables, such as age, sex, location, family circumstances, socio-economic status, and so on; the matching of control groups is difficult. The evidence for psychological damage, at least at blood content levels below 400 $\mu g\ l^{-1}$, seems inconclusive. For example, three studies in the UK were published in 1983; two showed no significant link between lead levels and behaviour, but one did show some correlation, though only in the case of children from poorer homes.[5] However, the earlier evidence was enough to form a strong association in the public mind, backed as it was by much attention in the media and a very well directed campaign by CLEAR.

If some link with behavioural symptoms is assumed at low levels, the next question is the major source and the pathway. Lead is so common in the environment that it can reach man via air, food, water, dust, soil, paint, or solder, and from industrial sources such as smelters. Lead in household lead paint has probably been responsible for more cases of overt childhood poisoning than any other single source, but the major contamination of the environment arises from atmospheric emissions, which can be inhaled or deposited and then ingested by various pathways. Ratcliffe decided that the inhalation of airborne lead is a comparatively minor pathway in most circumstances,[4] but the indirect uptake of petrol lead may be relatively larger than that from direct inhalation, especially in the case of children, because of the possibility of ingestion of dust and of contaminated foodstuffs.

Concern about the growing amount of lead in the environment due to the increasing volume of traffic led to reductions in the permitted lead content of petrol during the 1970s from the level which held in 1970, which was 0.84 g l^{-1} in the UK. The first reduction was to 0.64 g l^{-1} in 1973 and progressive reductions led to a level of 0.4 g l^{-1} in 1981. In its response to the Lawther report on Lead and Health,[6] the UK government agreed to reduce petrol lead to 0.15 g l^{-1} by 1986, the lowest figure allowed under the 1978 EEC Directive. However, the Campaign for Lead-Free Air, CLEAR, mounted a very effective campaign at the public, political, and scientific level during 1982 to eliminate lead from petrol altogether. The course of this campaign and the factors which led to its eventual success have been described by Jasanoff[3] and by Holmes.[7]

The Royal Commission on Environmental Pollution (RCEP) published its Ninth Report on 'Lead in the Environment' in April 1983.[8] The Commission decided that there had to be urgent and continuing actions to reduce the amount of lead dispersed in the environment by man. They recommended actions to reduce the lead content of water, with the incentive of grants for the replacement of lead plumbing in domestic and publicly owned buildings, and also the reduction of the allowable lead content of paints. Further, they recommended that the reduction of the permitted level of lead in petrol to 0.15 g l^{-1} should be regarded as an itermediate stage towards phasing out lead additives altogether, and that there should be no financial penalty inhibiting the use of unleaded petrol.

Largely owing to the CLEAR campaign, the question of lead in petrol had been put on the political agenda; public opinion polls showed that the public were in favour of abolition and there was to be a general election in the near future (1983). It is thus not surprising that the government accepted the recommendation of the RCEP to phase out petrol lead completely on the day of publication of the report, and to take other measures. The history may be seen as a good example of the influence a well-led pressure group can exert, given a measure of scientific support to establish a case and a favourable combination of political circumstances. It was a popular thing to do; it was bound to lead to a cleaner environment and, furthermore, it was an easy matter to introduce legislation and let powerful interests such as the oil companies and motor car manufacturers face up to the consequences, find the necessary investment, and eventually pass on the costs to the public.

It is now possible to begin to speculate about the effects of this policy, because of the reduction in lead levels from 0.4 to 0.15 g l^{-1} from 1st January 1986. As part of their examination of the problems, the RCEP published an estimation of the intake of lead by man, with a strong reservation that the calculations permitted only an estimate of the range and relative order of magnitude of the intake and uptake of lead from different sources and different pathways. Part of the summary of the Table in Appendix 4 to the Ninth Report is reproduced in Table 1. The very large figures for children compared to adults must be regarded as tentative; they are presumably mainly due to estimates of ingestion from dust, and the very variable figures included for the uptake of lead by children from petrol are presumably due to the wide range of possibilities for the source of lead in this dust.

The Department of the Environment (DoE) has undertaken an extensive programme to monitor blood lead concentrations from 1984 to 1987 in the wake of the reduction in lead content of petrol from 0.4 to 0.15 g l^{-1}, a reduction which was completed by 1st January 1986. Results for 1984 and 1985 and preliminary results for 1986 are available.[9,10] These studies include groups of adults in areas with heavy traffic, groups occupationally exposed to petrol lead, and children aged 6-7 years attending schools in urban areas; for comparison groups of adults and children in rural areas are included. The surveys are planned to cover 1500 adults and 1000 children in each year, and the groups are large enough to cover variations in age, sex, and smoking and drinking habits, which are known to affect blood levels of lead. Air lead concentrations were also measured at over 100 sites, and preliminary results are available for a long-term programme of monitoring air lead concentrations at 21 sites. In the first quarter of 1986, air lead concentrations fell by about 50% compared with the same period in 1985, before the restricted levels were introduced. It appears that the air lead concentrations fell by around 55% at roadside sites, rather less than the 65% reduction in the lead content of petrol, and by around 35% at rural, background sites.

As far as the results go to date, the effect on blood lead levels is not nearly so dramatic. The changes in blood lead levels of those sampled in each year from 1984 to 1985 and from 1985 to 1986 are recorded in Table 2, taken from Quinn and Delves.[10] The measurements for 1985 were completed before November, and hence do not depend upon any change in petrol lead concentrations. It can be seen that the average falls are about 10 $\mu g\ l^{-1}$ in both years in those living in exposed groups, and only slightly lower changes for the controls. The only groups showing a

Table 1 *Model balance schemes for lead intake and uptake*[8]

	Rural/small town			*Inner city*			*Extreme cases*	
	Adult	*Adult smoking/ drinking*	*Children*	*Adult*	*Adult smoking/ drinking*	*Children*	*High water*	*High air*
Total lead uptake/$\mu g\ d^{-1}$	12	18	70	18	23	105	44	26
% from inhalation	8	23	1	36	41	3	14	57
% from ingestion	92	77	99	64	59	97	96	43
Lead uptake from petrol/ $\mu g\ d^{-1}$	2	2	3–45	7	7	6–80	7	15
% of daily uptake from petrol	16	11	5–64	39	30	6–76	15	56
Predicted blood level concentration/$\mu g\ l^{-1}$	70	110	–	110	130	–	280	140

Table 2 *Changes in blood lead levels 1984-85 and 1985-86*[9]

	Group		N	*1984–1985* PbB*	%	N	*1985–1986* PbB	%
Adults	Exposed	Men	298	-8	-6.3	306	-10	-8.9
		Women	336	-10	-9.1	346	-9	-9.5
	Police	Men	104	+1	+0.5	147	-20	-18.1
	Controls	Men	157	-6	-5.3	144	-10	-9.1
		Women	171	-2	-2.4ns	172	-8	-10.4
Children	(exposed)			-8	-8.5		-14	-15.7

*For adults, the change in the geometric mean blood lead concentration (in $\mu g\ l^{-1}$) for all those sampled in *both* years. For children, the change in the geometric mean for all those sampled in each year, adjusted for age of dwelling, *etc.*; *N* = numbers in sample: 850, 870, and 903 in 1984, 1985, and 1986 respectively.

ns - not statistically significant at the 5% level.

greater change in 1985-86 than in 1984-85 are the police on traffic duties and the exposed children. Both of these groups would be expected to be more sensitive to petrol lead. However, the fall between 1984 and 1985 is not an isolated phenomenon; it seems that levels of blood lead have been falling for several years, due to a variety of causes.[11] If one assumes that the fall from 1984-85 represents an annual decrease due to other factors, it may be reasonable to conclude that the children's levels fell by ~6 $\mu g\ l^{-1}$ owing directly to the fall in petrol lead, a reduction of about 7%.

These results may be compared with predictions that might be based on the figures for uptake from the 'models' which are summarized in Table 1. A reduction of 55% in the lead uptake from petrol might have been expected to lead to a reduction of 20 $\mu g\ l^{-1}$ in the blood lead content of exposed adults and of 40 $\mu g\ l^{-1}$ in the case of extreme exposure, such as suffered by police officers on traffic duty. It is clear that such large falls have not happened, The fall in blood lead level for children of only 14 $\mu g\ l^{-1}$ in 1986, only a half of which may be due to petrol lead reductions, would be consistent with figures towards the lower end of the range for children quoted in Table 1.

It may be that the full effect of the reductions in lead in petrol have not yet been felt. Particularly in the case of children, whose intake depends so much on ingestion of dust *etc.*, there may be a time lag between the lead content of material deposited daily and the material ingested by the child. Further, the lead content of dusts and of contaminated foodstuffs may be more closely related to the total lead emitted into the environment, which is not necessarily proportional to the concentration in air and may be due to other factors, such as decaying lead paint. It seems, however, that a dramatic improvement is not to be expected as a result of the further decrease in petrol lead from 0.15 $g\ l^{-1}$ to zero in the 1990s, a reduction 60% as large as that which has occurred already from November 1984 to 1st January 1986.

Inspection of the estimated figures in Table 1 shows that the maximum blood lead levels are found in people exposed to high levels of lead in water; elimination of this source would seem to be the most beneficial single action to be taken, and indeed action to this end was recommended by the RCEP. Water as a source of lead was seen as an important factor in explaining regional variations,[12] and a study of various groups of women in Wales came to the same conclusion.[13] Water may be an important source, even in hard water districts. However, although action to reduce lead levels in water has been taken, it seems that high levels of lead will not be eliminated from all public drinking water supplies, at least in Scotland, by the end of 1989. Only 18 of the 103 public water supplies in Scotland which, in 1985, contained lead in excess of the levels prescribed by an EEC Directive in 1980 have been improved to date, and another 47 will be improved by 1989. Perhaps 3% of the population of Scotland will actually receive water in excess of the EEC limit after 1989.[14] Liming of a water supply is known to be a cheap and effective method of reducing lead levels in many instances, though the costs of replacing lead piping and tanks do not seem excessive. At a very rough estimate, 3% of the population of Scotland live in 60,000 dwellings. Allowing, say, £500 per house, the total cost of replacing the lead piping and lead-lined tanks in these houses might be of the order of £30 million.

Thirty million pounds sounds a large sum, and that deals only with Scotland. The total costs of water treatment and lead replacement in the UK would be much larger, and no figures are known for the EEC. But it is doubtful whether a determined attack on reducing lead in water would have cost more than the introduction of unleaded petrol.

The change from 1.5 $g\ l^{-1}$ of lead in petrol to zero will necessitate very considerable changes in refinery practice and in motor design to preserve octane levels and performance at something like the present levels. To attain the optimum octane number to minimize energy consumption with unleaded fuel will require another 44 tonnes of crude oil for every 1000 tonnes of petrol at a cost of about $19,000. The total industry investment might be of the order of $6000 million with $2.5-3.5 billion for fitting petrol stations in the EEC to take both leaded and unleaded grades during the transitional period when both will be required for the vehicles on the road.[7] The preservation of octane quality may also require the addition to petrol of a number of organic additives, such as alcohols and complex ethers. The total quality of exhaust emissions in the future therefore awaits evaluation.

No doubt the extra costs in moving to unleaded petrol will be absorbed by the oil industry and the motorist, and the slightly disappointing results in terms of blood lead levels do not mean that the advice of the Royal Commission on Environmental Pollution to move towards zero levels was wrong. But one can question whether this one of all their recommendations was the optimum choice for priority action. It may be so, but it has not been established, either by an analysis of different options for reducing the total environmental effects of exhaust emissions or by a costed study of the different options for reducing lead in the environment. One negative result seems to be a trend towards using more of our oil supply to run our transport during a period when we surely should be moving to reduce our hydrocarbon demand. However, there may be one major gain: the absence of lead in petrol may ease the transition to a technology that controls NO_x and hydrocarbons in exhaust gases, which is already mandatory in the USA, and which may be a much more significant environmental and health gain than a marginal reduction in blood lead levels.

3 Radioactive Chemicals

The second example is the assessment of risks arising from the release of radioactive chemicals into the environment, either deliberately or accidentally following the disposal of radioactive materials. There is one major distinction between this and the last example: lead is a very ubiquitous pollutant but almost all arises from human activities: radioactivity is also ubiquitous, but overwhelmingly of natural origin. Unfortunately the style of regulations adopted to control man-made radioactivity obscures this point, leading to or at least assisting some distortion of perceptions. The philosophy behind these regulations is well known; it is a good example of an explicit assumption of a dose-response relationship to allow of extrapolation from epidemiological data at high doses to a low-dose regime of practical interest. Apart from effects at very high doses which would only be experienced in a nuclear war or in major accidents like the one at Chernobyl, the only diseases proved to have been caused by radiation are leukaemias and cancers of the same types known from other causes, and there is also an assumption, based on animal experiments, that radiation can cause some increase in hereditary defects. The assumption made by the International Commission on Radiological Protection (ICRP) is that the dose-response curve is linear at the doses of interest, with no threshold, with a risk of cancer of 1.25×10^{-2} per Sievert and a smaller risk of genetic damage. The most recent advice from the National Radiological Protection Board (NRPB), based on a re-evaluation of the data from the Hiroshima and Nagasaki survivors, is that this risk coefficient should be multiplied by a factor of between 2 and 3.[15] This will not be enough to satisfy all the critics; risk coefficients more than ten times higher than the present ICRP figure have been proposed, whereas on the other hand, some argue for a zero or even slight beneficial effect at very low doses.[16] It is unlikely that this controversy will ever be settled by direct observation, since the effects due to radiation constitute such a small fraction of the total incidence of cancer.

Using the linear, no threshold hypothesis, maximum doses for workers with radiation have been fixed, so that the estimated risk of eventual death due to radiation is no greater than the actual risks run in comparable industries, with the additional legal requirement to reduce doses below these limits to values As Low As Reasonably Achievable (ALARA). The current limits set for workers are 50 mSv year^{-1} with an investigation level of 15 mSv year^{-1}, and for the general public 5 mSv year^{-1}, reducing to 1 mSv year^{-1} if the exposure is liable to be prolonged over many years. The NRPB recommends reducing the limits for workers to 15 mSv year^{-1} and to 0.5 mSv year^{-1} for doses to members of the public originating from a single source. Average exposures are indeed much lower, being between 1 and 5 mSv year^{-1} for workers, while only small groups of the public receive more than 0.1 mSv year^{-1} from industrial sources.[17]

The ALARA principle is then a very powerful regulatory tool; management is expected to do everything possible to reduce radiation risks up to a limit they can justify, and justify to the regulatory authorities: the aim is to reduce radiation doses until the cost of doing so is disproportionate to the benefit. To gauge benefit, in the absolute sense, you have to put a value on the risk avoided. The NRPB has published some guidance on this, based on the philosophy that the larger the risk, the greater the incentive to reduce it. On the assumption that some health effects will ensue, no matter how small the individual dose may be, the Board assigned a value of £3000 to the man-Sv at doses below 0.1 mSv with values up to sixteen times higher at doses near the maximum for workers. The Board quotes figures of £150,000 to £1.5 million for the value of a 'statistical life' compatible with these guidelines;[18] the United States Nuclear Regulatory Commission (USNRC) has recently used a figure of $10 million.[19]

The application of cost/benefit analysis to decisions in investment in radiological protection has proved to be useful in many contexts, certainly in ranking investments in an order of priority and effectiveness. In practice, such an approach can only be a guide, one input into a decision. Anderson and Mummery calculate that the investment in the most recent treatment plant to reduce the activity in the effluent from Sellafield led to an expenditure of £250 million for a reduction which is estimated to save two lives in the next 10,000 years.[20] This is far in excess of anything that would be expected on the approach summarized in the preceding paragraph; the reduction was necessary to reduce the radiation dose to the critical group, and to reduce effluent levels from Sellafield to those in other installations. The authors comment 'The Company and Radioactive Waste Management Advisory Committee (RWMAC) targets of 0.5 mSv year^{-1} public exposure and the general political/public situation at the time were of more importance'. Anderson and Mummery also draw attention to the difficulty of applying the ALARA principle without firm guidance on a minimum negligible level of risk; the principle could lead to an unnecessary and never-ending escalation of costs. In particular, the real meaning of risk equated to collective doses at extremely low levels of individual dose can be questioned.[21]

In part, the public disquiet about radiation at any level, however low, may be due to the convention that reporting and control is applied exclusively to radiation risks arising from industry. Radiation doses arising from medical uses or from natural background are excluded. Medical uses are a special case, in that the small risks inherent in the use of radiation to aid diagnosis or for therapy must be set against the positive or putative health benefit to each individual patient. But a better perspective might be inculcated if the whole truth and not a part of it were commonly reported. What the RWMAC has defined as the 'basic background dose' varies from 0.8 to 1.7 mSv year^{-1} across the country, with an average of about 1 mSv year^{-1}. When someone is exposed to an additional 0.1 mSv year^{-1} from industrial causes, the actual result is to raise their minimum annual radiation dose from 1.0 to 1.1 mSv year^{-1}, and the discharge of 20,000 man-Sv of ^{14}C from Sellafield in the past 20 years will have raised the collective dose to the world's population by a dose equivalent to natural background radiation lasting for less than 1 day.

Ignoring natural background levels may seem logical enough if nothing effective can be done to reduce them and if their range is small, but work in the past few years has demonstrated that neither of these assumptions is true. It has been realized since the early 1980s that natural radiation causes the highest doses received by groups of people as well as being the largest component of the collective dose. The highest exposures are due to the presence of the gases radon and thoron in buildings. The gases decay to short-lived α-emitters which become attached to indoor aerosols and are ingested. This is the phenomenon thought to account for an excess of lung cancers among uranium miners at high exposures. The effect may have grown worse as houses were better sealed with reduced ventilation, in order to conserve heat. Natural radiation has therefore emerged as a prominent matter in radiological protection.[22] Table 3 shows the distribution of doses due to radon in the UK population, as deduced from a preliminary survey of 2300 houses carried out by the NRPB. The distribution

Table 3 *Approximate sizes of populations in different bands of radiation dose*[23]

Artificial sources					
Annual radiation dose/mSv	0–5	5–15	15–50		>50
Workers in the nuclear power industry and research	39,677	3916	1200		2
Workers routinely monitored in other industries and in medicine	94,954	3142	747		10
Critical groups near nuclear installations receiving more than 0.1 mSv a^{-1} from discharges	~200				
Natural sources					
Annual radiation dose/mSv	2–5	5–10	10–20	20–50	>50
Population exposed to radon	4million	1million	85,000	50,000	5000

of radiation doses for workers in the nuclear industry reported in 1984 is included for comparison, as are figures for the critical groups of the public exposed to effluents from nuclear installations; since these are now less than 1 mSv year^{-1}, their total exposure will lie in the range 2-3 mSv year^{-1}.[23]

It is quite clear that some millions of people in the UK receive radiation doses far in excess of the maximum levels set for the general public arising from industrial sources, and many tens of thousands receive doses at levels above the maximum now recommended by the NRPB for radiation workers.

Recognizing that natural radiation levels pose the highest health risk to the public arising from ionizing radiation, both the ICRP and national authorities have attempted to set levels above which some remedial action should be taken. In this country, the NRPB has recommended an action level for existing dwellings in which the annual dose may be greater than 20 mSv year^{-1}, and that new dwellings should be built so that the maximum dose is less than 5 mSv year^{-1} These levels are respectively 40 and 10 times the maximum levels regarded as acceptable if the source is industrial. The levels set for radon exposure were set by comparison with the ordinary risks of living - the risk of cancer, or of being killed in a traffic accident. These are much higher than the safety standards set for industry and this comparison is particularly striking in this case since the risks are so similar.[24]

The ICRP has recommended that the cost/benefit approach be applied to natural as to man-made sources of radiation.[25] The costs of remedial measures applied to houses have been roughly estimated as £100-£10,000, with £1000 being a typical sum.[24] To reduce all houses in the UK to the 'action level' of 20 mSv year^{-1} would cost £20 million. On the reasonable assumption that treatment of a house in the high dose group would reduce doses by 20 mSv year^{-1}, and that an average family might occupy such a house for 20 years, then their collective dose over that time would be reduced by about 1 man-Sv at a cost of about £1000 (see also Chapter 16). This may be compared with the NRPB guidelines of £30,000/man-Sv at that dose level, and with the much higher investment per man-Sv in a further treatment plant already mentioned at Sellafield.

Hence, here is an excellent example of environmental investment being focused on an industry which is unpopular and distrusted in some circles rather than on the protection of the most vulnerable section of the public. Apparently it is radiation that originates in the nuclear industry that is feared. But this fear has had very serious consequences, and nowhere more obviously than in the long and continuing saga of the search for sites for the disposal of radioactive waste.

3.1 The Radioactive Waste Saga. This sad history is an example of a policy which was enunciated on ethical and environmental grounds by the Royal Commission, and accepted both by government and by the industry concerned, being opposed by organizations usually concerned to protect the environment and by the public. The Royal Commission, in its Sixth Report in 1976,[26] advocated strongly that more effort should be devoted to developing and proving technologies for the permanent disposal of radioactive wastes, so that it could be demonstrated that 'a method exists to ensure the safe containment of long-lived, highly radioactive waste beyond reasonable doubt for the indefinite future'. The volumes of radioactive wastes are small compared with the volumes of other toxic wastes - the UK generates about 30,000 tonnes a year, mostly of low radioactive content, compared with over 3 million tonnes of toxic wastes - and they are not difficult to store. But the principle that permanent disposal is preferable to indefinite storage on environmental and ethical grounds has been reiterated several times since 1976, most recently by the Environment Committee of the House of Commons,[27] and all countries with nuclear power programmes are in agreement on this.

There is also a broad degree of agreement on the means. Apart from the dilute streams of gases or liquids which are discharged to the environment, different categories of wastes are packaged so as to convert them in to a durable form in which they can be consigned to a permanent repository from which any escape of radionuclides will be sufficiently slow so that no appreciable radiation dose will accrue to people. The geological conditions that can meet these requirements have been defined; essentially, one needs a stable environment through which the movement of water is slow, since it is only by solution in water that the radioactive species can get back to the biosphere.

The major difficulty in the implementation of this policy arises because the geological properties of particular places can be proved only by field experiments carried out on site, and it is access to sites that has been vigorously opposed by the public and by the politicians who represent them. The environmental interest groups have played an ambiguous role. While basing their opposition to nuclear power partly on the assertion that there are no demonstrated ways of disposing of all nuclear wastes, they have consistently opposed any proposal to carry out the necessary geological research. The history of retreat from one proposal after another has been criticized in trenchant terms by the Advisory Committee.[21]

The last switch of policy occurred in May 1987, when the Secretary of State for the Environment cancelled the programme of evaluation of four sites as candidates for a second site for the near-surface disposal of low-level radioactive waste. The basis of this decision was advice from UK Nirex Ltd. that the costs of proving and engineering a site for near-surface disposal had risen to the point where there would be little economic advantage in opening a second site compared with the alternative of extending a deep disposal site, which would be needed for higher categories of waste, to take the low-level waste as well; the ranges of costs were given as £500-£1000 m^{-3} for the shallow repository compared with the marginal cost of £750-£1200 m^{-3} for deep disposal. This advice was accepted within one day and the programme cancelled without reference to the Advisory Committee. Again, there was an election in the offing; it will be assumed that an environmental decision was taken because of an atmosphere of public disquiet around the sites concerned, whether in fact it was so or not.

This decision presumably means that the major conclusions of an exercise launched by the Department of the Environment to determine the Best Practicable Environmental Option for waste disposal must be abandoned. But it is worth returning to that report[28] simply to see the orders of magnitude involved. Table 4 records the maximum individual risk in a year calculated to result from the disposal of one category of decomissioning wastes in different ways. The highest value for the annual risk was 6×10^{-9} $year^{-1}$, for disposal in an engineered trench; the outlook for the industry is bleak if that level of risk from waste disposal is not considered to be adequate, and the future of the chemical industry may be at risk if such a standard were to be applied to all toxic wastes. Further, disposal in the deep sea emerges as the cheapest option which is as safe as any other, with a maximum individual risk of 5×10^{-14} $year^{-1}$, and that option also seems to have been blocked by international political action, despite very favourable risk assessments.

The underlying reasons for the failure of the government, the regulatory authorities, the advisory mechanisms, and the industry to achieve a resolution of this problem which is acceptable to the public are complex. Risks arising from radioactive materials in the environment are not seen in the perspective of the risk from the natural radioactivity already there.[29] It seems that neither the industry nor the regulatory

Table 4 *Economic and radiological impacts of different management options for WAGR* decommissioning wastes*[28]

	Sea disposal	*Engineered trench disposal (10 years storage)*	*Deep cavity disposal (15 years storage)*
Cost (£M 1985)			
Storage cost	0	1.3	1.3
Disposal cost	1.0	1.3	5.5
Maximum individual risk in a year (a^{-1})			
Radionuclide migration	$4.6.10^{-14}$		
Inland site		$5.6.10^{-9}$	$1.1.10^{-15}$
Coastal site		$1.5.10^{-9}$	$4.7.10^{-18}$
Intrusion	$2.6.10^{-15}$	$4.4.10^{-13}$	$6.6.10^{-14}$

*WAGR: Windscale Advanced Gas-cooled Reactor

authorities are trusted to achieve the safety standards which they claim; there is a need for new and open procedures to be followed to establish confidence in the objectivity of the regulatory process.[30] In this respect, the recent attempt by UK Nirex Ltd. to canvass opinion among responsible public groups by circulating for discussion a document detailing the choices facing them now is a good new initiative and the results will be important new evidence.[31]

The results of this long controversy have been serious in economic and in safety terms. Major damage has been done to prospects for the nuclear industry by this perception of a long-term problem which is intractable, and which may be responsible for some future disaster. To meet this worry, large resources have been committed to research and development, with considerable diversion of effort and of talent. The industry has been urged towards 'Rolls Royce' solutions, almost without consideration of cost. The latest figures of estimated costs of disposal given by UK Nirex Ltd. would indicate total costs up to the year 2030 of £2 x 10^9 for low-level waste, and £1.5 x 10^9 for intermediate-level waste. Sums of up to a billion dollars have been mentioned for proving a site for high-level waste disposal in the USA. The effects of delay and of indecision can also be serious in operational terms, leading to a real decrease in safety levels: the Health and Safety Executive, in its recent audit of the Sellafield plant, drew attention to the burden laid on senior management, distracting them from their duties, by the lack of a firm forward plan for waste disposal and the need to investigate and report on any suspected leak of radioactivity, however trivial.

Perhaps the main lesson to be learned from this example is the need for wide and thorough explanation of an environmental policy in the very early stages before implementation is attempted;[32] once a connection with hazard and disaster has been established in the public mind, it is extremely difficult to eradicate or to counter. Admittedly, in the case of any aspect of the nuclear energy industry there is another factor: outright opposition to the industry by groups who regard it as a symbol of an industrial society which they oppose on general political grounds.[7] From the point of view of outright opposition to the industry, the waste problem has been a very convenient one; people may see a need for power, but they see no need for waste, and the long time-scales over which safety must be proved can be presented as a major difficulty, if it is forgotten that the risks decrease with time, due to radioactive decay, and that they are in any case extremely low. But the image of long-term threat has taken hold, to the point where recent experience has shown that a community close to an operating nuclear reactor would not tolerate the prospect of having close to it a repository for low-level waste, while any risk analyst would conclude rapidly that a reactor is a more dangerous neighbour than a repository. Despite these particular features, however, it would be a mistake to see this example as a unique aberration, and therefore to ignore the lessons that can be drawn from it concerning the presentation of estimates of risk and the institutional framework in which we have to consider them.

4 Pesticides and Herbicides: The 2,4,6-T Controversy

The third example discussed in this chapter is the controversy over the use of 2,4,5-trichlorophenoxy-acetic acid (2,4,5-T) as a herbicide to destroy woody weeds, where again there was a sharp difference between perceptions of safety and technical appraisals of risk, but where the outcome, in this country at least, was different.

2,4,5-T was a component of the notorious Agent Orange which was used to destroy foliage in the Vietnam war. There seems to be little evidence that 'pure' 2,4,5-T - if it exists - is a threat to human health although there may potentially be some risk associated with the use of any material as biologically active as a herbicide. But 2,4,5-T is commonly contaminated with one of the dioxins, particularly with 2,3,7,8-tetrachlorodibenzo-*p*-dioxin, known as TCDD. This compound is known to be highly toxic to some animals and to be a carcinogen in rats and mice, and it certainly causes chloracne in humans. The accident at Seveso in 1976, in which about 2 kg of TCDD was released from a plant manufacturing 2,4,5-trichloro-phenol caused 187 cases of chloracne, and perhaps some effects on the liver and nervous system of contaminated people, but no significant increase in mortality or of spontaneous abortions or congenital malformations in infants.

Nevertheless, concern grew in the 1970s that TCDD might cause some chronic damage to human beings, though it was clear that the toxic effects in humans were not nearly as severe as they were in some animals. The National Union of Agricultural and Allied Workers argued in 1978 for a complete ban on the use of 2,4,5-T to protect the health of workers, and the Trade Union General Congress called for a ban on its use in February 1980.[33] The Advisory Committee on Pesticides issued a second review of the safety of 2,4,5-T in December 1980.[34] They reviewed more than 20 case histories that had been referred to them, and all the international evidence to date. They stressed that there can never be total proof of the safety of any product, and that recommended safety precautions must be obeyed. However, their overall conclusion was that 2,4,5-T herbicides can safety be used 'in the recommended way and for the recommended purposes'. They remarked upon the resemblance between the controversy in 1979-80 and events in the USA, which culminated in the restriction of the use of 2,4,5-T in that country in 1970 and a temporary suspension by the Forestry Commission in the UK. The evidence was later found lacking and the suspended uses were restored in both countries.

Tschirley has recently reviewed all the evidence pertaining to the human toxicity of TCDD; it has been recognized as potentially toxic since 1949 and there were nine cases of multiple exposure before 1976. No ill effects in humans beyond chloracne have been proved.[35] Tschirley quotes a source in the Office of Science and Technology who has calculated that more than a billion dollars will have been spent by the Federal Government for research on dioxin matters, with more expenditure by chemical companies and other organizations; he concludes 'the total outlay is a tremendous amount for an issue of questionable importance.... the nation's limited scientific resources should be devoted to the issues posing a greater threat'.

However, the great difficulty facing any regulatory agency or advisory body in subjects such as this must be acknowledged; proof of the absence of chronic effects, which may take a long time to appear, from sub-acute doses of chemicals known to have some toxic effects on humans or animals at high concentrations is a classical problem. Despite the individual distinction of its members, the Advisory Committee on Pesticides (ACP) has been criticized for conducting most of its deliberations in private and for not consulting adequately the groups most affected by its work before issuing its conclusions. The TUC has not accepted the views carefully argued by the ACP, so far as is known, and 2,4,5-T may still be banned for use by some local authorities and other organizations; here again, a careful technical assessment of risks is not enough to change perceptions.

5 Conclusions

The three examples discussed were chosen because, in each case, resolution of an environmental problem was preceded and in some sense conditioned by much public airing of the issues. The pattern of events and the style of outcome was very different in each case and there are dozens of other examples which might be examined. But it seems possible to draw some general conclusions, and perhaps to pose some questions of importance to the future and health of the discipline of risk assessment as applied to environmental issues.

First of all, we are dealing in problems of this type with risks that are low and which can only be quantified by some assumed risk-dose relationship at very low doses. Particularly when the detrimental effects of concern are not specific, the major difficulty in any epidemiological study is to assign some fraction of, for example, cancer cases to a specific cause. We have to be quite honest with the public about these difficulties. The 'No observable effect level' does not guarantee that there are no effects. It will always be impossible to give an unequivocal answer to the question 'are you **sure** that the release of some chemical to the environment will harm **no-one**?', given the extreme variability of response of the human system to any stimulus, the possibilities of remote coincidence and synergies, and of particular weaknesses which may render some individuals particularly vulnerable.

Secondly, all the examples examined illustrate the difficulty of distinguishing sharply between the effects of one man-made component against a general background that will be variable and which may impose local risks of some kind. Lead, radioactive substances, and TCDD are all now part of our natural environment, as are a multitude of chemicals that may be harmful in high concentrations. This also should be better understood than it is, and the most promising approach to the protection of the public is to estimate total risks and the marginal change caused by some operation.

Thirdly, these examples illustrate well the effect that intense media interest can have on environmental management. Lord Ashby[36] described the three stages of environmental debate as 'ignition - examination - resolution', and my reading of the history is that the very essential second stage - rational examination of the evidence, balancing of risks against benefits, balancing of action against inaction or against alternatives - can be inhibited or distorted if public alarm is raised to a high level in stage one, and maintained at that level during stage two by continuous media attention. Sir Robin Day has been quoted as saying 'Television is a medium for shock rather than explanation. It is a crude medium which strikes at the emotions rather than the intellect. And because of television's insatiable appetite for visual action, for violence very often, it tends to distort and trivialize'. Of course, one cannot deny that excellent programmes about science and technology are screened in this country, and it is useless in a democratic society to imagine that this medium - any more than the written word - can simply be ignored. Professionals must learn to work more with the grain of this development and very serious thought must be given to the question of proper use of TV in particular in illustrating the balance of argument in contentious environmental issues. One must sympathize with the individual scientist or TV journalist who has unearthed some facts that appear to bear on public safety and which may challenge some orthodox consensus or regulation. And it will not be possible to insist that new results should first be published in the scientific literature and therefore be subject to peer review and critical scrutiny before they are launched at the public - though it is wished that this procedure were seen as the norm. So some producer will be faced with the problem of fair presentation of a complex subject in a very limited time and, however fair-minded he or she may be, there is no easy solution to this difficulty. The only action that can be taken is to see to it that the ground is adequately prepared. This means that good and interesting material on environmental questions and on the approach taken to safeguard the interests of the public should be available long before some crisis blows up and be repeated frequently enough to be absorbed. It is a task in which scientific societies and regulatory authorities should be engaged and it is never ending; the Royal Society Group under Professor Bodmer was right to recommend more continuous involvement of the scientific community in briefing journalists, politicians, and Parliamentarians.[37]

These considerations lead on to the fourth general point concerning public participation and consultations. The principle that sections of the public should be consulted and their views heard and taken account of before decisions which might affect their safety are made is unexceptionable, but it is a principle which is difficult to put into practice. Perhaps the ultimate expression in this country is the formal Public Inquiry, and another form is the 'tripartite' type of organization like the Health and Safety Commission, which allows for bargaining between Trade Union and management representatives. Both have their uses and their limitations, and both are truly concerned with the political balance of benefit and risk, that is, with risk **management**. There may be a need for a different form of interaction with the public or their representatives on technical questions of risk **assessment**. The excellent and dedicated work of the technical advisory committees such as the RWMAC and the ACP should be better known and appreciated; means to that end are certainly worthy of study since it appears that more is needed than the publication of occasional or even annual reports, however well written they may be. One can appreciate the reasons why the actual working data and the methods are regarded as confidential in many cases but yet argue that the advice from advisory bodies such as the ACP and the RWMAC might be more effective if the public profile of these bodies was higher than it is.

An attempt to separate risk management from risk assessment would be criticized by many; Jasanoff's work on the influence of cultural preconceptions is important here.[3] One must admit that the risk analysis of a complicated plant, for example, includes many points at which judgement in the estimation of some parameter is required, and the result is necessarily expressed in probabilistic terms, with a range of values. Different scientists or groups of scientists will differ in the conclusions they draw from the same set of data, particularly if some value judgement is implicit in the choice of the most significant. Indeed, Collingridge and Reeve have argued that an objective risk analysis is impossible; they aver that it will always be possible to point to conflicting data, and scientists engaged in risk analysis in environmental questions will choose data and results which are respectable enough to bolster one or other side in a political debate: the outcome will always be decided on political grounds, not on risk analysis.[38] It is argued that this is a dangerous doctrine for the profession to accept, involving in the end a diminution of function and of status. Scientists, and learned societies, are well used to controversy. But it is our bounden duty to seek objective truth, and, at the least, to separate out the subjective elements in an analysis so that the assumptions can be fairly debated. What is vital is to ensure that we have the organizations and procedures in place so that these arguments can take place and be seen to take place.

Learned societies can expose the basic causes for different interpretations of data, but it is the function of technical advisory committees to resolve conflicts of scientific opinion if possible and to report clearly how they come to their conclusions and what assumptions they have made. It is not their function to resolve conflicts of interest between different groups of people; that is a social and political judgement and truly demands a different sort of body - unless the questions are so broad that they involve Parliament itself. I would not be in favour of so broadening the membership of each and every advisory body that it tried to take the management as well as the assessment function. It is important to preserve a distinction between risk assessment and risk management, so that the responsibilities are clear.

The thought that serious environmental questions are decided by the accident that some aspect becomes politically important at a certain time has disturbing implications. While it is undoubtedly the judgement of a broad spectrum of people that must in the end decide on the acceptability of risks in a democratic society, this debate should not descend to the level of slogan and vituperation. To quote Ian Smart's review of the books by Holmes and Brackley 'The road hedged by such political realities may conceivably begin with burning cleaner coal, but it ends by burning books and witches'. Emotive language, perhaps, but the plea to sustain the element of rational argument in a political situation is clear enough.[39] That is our professional duty and we must expand the means of public presentation and education and the institutional framework that will enable us to pursue it. Beyond that, it is our duty to society in general to encourage a debate about safety perspectives, since they will determine priorities.[40] The huge expenditures devoted to some aspects of risk rather than others cannot just be dismissed as a charge on some sections of industry big enough to bear it; in the end, all societies have limited resources, particularly of ability, and distortions do matter. We should be guided more by the need to set sensible priorities which will actually benefit large sections of the population, as Peto has argued in the case of the prevention of cancer;[40] concentration on the avoidance of small risks may blind us to the possibility of mitigating much larger ones.

6 Acknowledgements

Assistance from Dr. David Everest, Visiting Fellow at UEA, in preparing this paper is gratefully acknowledged.

7 References

1. D.P.Lovell, 'Risk Assessment - General Principles' in 'Toxic Hazard Assessment of Chemicals', ed. M.L. Richardson, Royal Society of Chemistry, London, 1986, p.207.

2. J.R. Eiser, 'The Social Perception of Risk', in 'Risk Perception and the Safety Targets for Major Accidents', Environmental Risk Assessment Unit, Research Report No.4, UEA, Norwich, 1988.

3. S. Jasanoff, 'Comparative Risk Assessment - The Lessons of Cultural Variation', in 'Toxic Hazard Assessment of Chemicals', ed. M.L. Richardson, Royal Society of Chemistry, London, 1986, p.259.

4. J.M. Ratcliffe, 'Lead in Man and the Environment', Ellis Horwood, Chichester, 1981.

5. DoE Press Notice, 7 April 1983.

6. 'Lead and Health', the report of a working party, Department of Health and Social Security, HMSO, London, 1980.

7. A. Holmes, 'A Changing Climate', Financial Times Business Information Ltd., London, 1987.

8. Royal Commission on Environmental Pollution. Ninth Report, Cmnd 8852, HMSO, London, 1983.

9. UK Blood Lead Monitoring Programme - Interim Report on Results for 1985; Pollution Report No. 24, DoE, London, 1987.

10. M.J. Quinn and H.T. Delves, UK Blood Lead Monitoring Programme - Preliminary Results for 1986, Proceedings of the 6th International Conference on Heavy Metals in the Environment, CEP Consultants, Edinburgh, 1987, Vol. II, p.206, and Digest of Environmental Statistics, Chapter 6, 1988.

11. P.C. Elwood, *Sci. Total Environ.* 1986, **52**, 1.

12. UK Blood Lead Monitoring Programme 1984-87; Results for 1984, Pollution Report No.22, HMSO, London 1986.

13. P.C. Elwood, J.E.J. Gallacher, K.M. Phillips, B.E. Davies, and C. Toothill, *Nature (London)*, 1984, **310**, 138.

14. ENDS, Report 155, p.17, Environmental Data Services Ltd., London, December 1987.

15. Interim Guidance on the Implications of Recent Revisions of Risk Estimates, NRPB-GS9, HMSO, London, 1987.

16. L.A. Sagan, *Health Phys.*, 1987, **52**, 521.

17. J.S. Hughes and G.C. Roberts, NRPB-R173, HMSO, London, 1984.

18. 'Cost-Benefit Analysis in the Optimisation of Radiological Protection', ASP9; NRPB, Chilton, 1986.

19. Reactor Risk Reference Document, NUREG-1150, Vol.1, US Nuclear Regulatory Commission, Washington DC, 1987.

20. P.W. Mummery and A.R. Anderson, in 'ALARA, Principles, Practice and Consequences', Adam Hilger, Bristol, 1987, p.104.

21. Radioactive Waste Management Advisory Committee, 7th Annual Report, HMSO, London 1986, p.10.
22. J. Simaeve, G. Clemente, and M. O'Riordan, *Radiation Protection Dosimetry*, 1984, 7, 15.
23. L.E.J. Roberts, *Nature (London)*, 1988 (in press).
24. 'Exposure to Radon Daughters in Dwellings', NRPB-GS6, HMSO, London, 1987.
25. ICRP Publication 37, *Ann. ICRP*, 1983, **10**, No. 2/3, Pergamon Press, Oxford.
26. Royal Commission on Environmental Pollution, 6th Report, HMSO, London, 1976.
27. House of Commons. First Report from the Environment Committee, HMSO, London, 1987.
28. 'Assessment of Best Practicable Environmental Options for Management of Low and Intermediate Level Disposal of Radioactive Wastes', Department of the Environment, London, 1987.
29. Radioactive Waste Management Advisory Committee, 8th Annual Report, HMSO, London, 1987.
30. R.V. Kemp, M. Purdue, and T.O. O'Riordan (in press).
31. 'The Way Forward', UK Nirex Ltd., Harwell, Oxfordshire, 1988.
32. L.E.J. Roberts, *Proc. Royal Inst.*, 1988, **59**, 259.
33. A. Irwin and K. Green, *Policy and Politics,* 1983, **11**, 439.
34. Further Review of the Safety for use in the UK of the Herbicide 2,4,5-T, MAFF, London, 1980.
35. F.H. Tschirley, *Sci. Am.*, 1986, **254**, 21.
36. E. Ashby, The Lord, Reconciling Man with the Environment, Oxford University Press, 1977.
37. 'The Public Understanding of Science', The Royal Society, London, 1985.
38. D. Collingridge and C. Reeve, 'Science Speaks to Power', Frances Pinter, London, 1986.
39. I. Smart, *Nature (London)*, 1987, **328**, 209.
40. R. Peto, in 'Assessment of Risk from Low Level Exposure to Radiation and Chemicals', Plenum Press, London, 1985, p.1.

3
Risk Assessment — Prediction and Reality

Sir Frederick Warner

UNIVERSITY OF ESSEX, WIVENHOE PARK, COLCHESTER, ESSEX CO4 3SQ, UK

1 Introduction

The techniques for probabilistic risk assessment rely on databases such as those operated by the United Kingdom Atomic Energy Authority (UKAEA) to carry out studies on projects in the design state. Uncertainties are not important for the principal task of eliminating major hazards by following fault trees. The more general field of risk assessment gives more problems, especially for low risks. Predictions of death cannot be relied on where dispersion models, dose-response relations, and criteria such as LD_{50} give wide bands. However, these uncertainties are not reflected in the explanation of risk to the public nor adequately qualified in relation to the benefits to be expected in comparison with the risk and its position in relation to others which are greater and easier to reduce.

2 Risk Assessment

The subject of risk assessment was the topic of a report from a study group of the Royal Society in 1983.[1] This followed a two-day Discussion Meeting on the Assessment and Perception of Risk, which had this to report:[2]

> 'The techniques of risk assessment go back to the need for reliability in military equipment and air transport. They can now use failure rates held in databanks containing rapidly increasing amounts of information. The use of fault trees in risk analysis gives a logical basis for reducing risk during the conceptual state of projects. A number of papers deal with the problems of assessment shown up in conventional, unconventional, and nuclear power plants. The problems have existed much longer in the risks of failure in bridges, dams, buildings, chemical and petroleum installations, and transport. Risks associated

with drugs and medical procedures are complicated by the benefits weighed against them; the risks also show up only over long periods as a result of epidemiological studies and finally in mortality tables. The papers discuss not only the risk that is final, that of death, but also of injury up to fates worse than death, with special references to risks at work'.

The report still represents concisely the methodology needed for the study of risk and the examination of the reliability of the estimates which can be made. It made many observations of the variability of the perception of risk and has led to many studies since by applied psychologists into public attitudes (see also Chapters 2 and 6).

The Royal Society report begins with a number of definitions and these are quoted in the 'Glossary of Terms'. Areas of prediction are those in engineering risks, biological risks, and observation of the risks which affect man.

2.1 Detriment (see Glossary). Although detriment may represent the only numerical way of comparing different events associated with the same hazard, or the combined effects of events from different hazards, the fact that any such comparison is an arbitrarily weighted total of incommensurables must never be forgotten. Total detriment (per individual or per population) may be an aid to decision, especially when reasonable alternative systems of weighting lead to the same conclusion, but should not be regarded as a substitute for reasoned judgement.

2.2 Risk Assessment. The general term used to describe the study of decisions subject to uncertain consequences is **Risk assessment** (see Glossary). It is conveniently sub-divided into **Risk estimation** and **Risk evaluation**. The former includes:

(a) The identification of the outcomes;
(b) The estimation of the magnitude of the associated consequences of these outcomes;
(c) The estimation of the probabilities of these outcomes.

2.3 Risk Estimation. In approaching risk estimation, three classes may be identified, as suggested by Cohen and Pritchard:[3]

(a) Risks for which statistics of identified casualties are available;
(b) Risks for which there may be some evidence, but where the connection between suspected cause and injury to any one individual cannot be traced (*e.g.*

cancer long after exposure to ionizing radiation or a chemical);
(c) Experts' best estimation of probabilities of events that have not yet happened.

Additionally, there are risks which were not foreseen, for which causal conditions are sought after new effects on casualties appear.

In engineering, there has been a move away from a deterministic to a probabilistic basis for assessing the risk of failure of an engineering design. A simple example of the deterministic approach is to be found in the criteria used for the design of pressure vessels, such as boilers or reactors. Fifty years ago the material in common use was mild steel with an ultimate tensile strength of 28-32 tons per square inch. To allow for uncertainties in use, the design of the pressure vessel was based on limiting the stress to 7-8 tons per square inch - a safety factor of 4. This was later changed to a basis of 50% over the limit of proportionality - the stress at which the material ceases to behave elastically and follow Hooke's law (stress is proportional to strain).

This deterministic approach incorporates the concept of variability of stress and strength but implies that there is a level of probability of failure that is negligible for design purposes, without quantifying that level. The probabilistic approach recognizes that there is a distribution of stresses and strengths having mean values with variations around the means. The search is then made for the upper end of the stress distribution and the lower end of the strength distribution. This still involves some judgement at the cut-off point to be taken in each case so that a value judgement is in fact implicit in both approaches on the acceptable level of risk. What a probabilistic approach achieves is a lower level of risk through a sophisticated examination of the different mechanisms which may result in failure.

The advantage of the probabilistic approach in risk assessment is not just in the closer approximation of reality to prediction but in the process involved in the assessment which systematically identifies the levels of risk in the sub-sets of the complete process and weighs their importance. The definition of **Risk** (see Glossary) in the Royal Society report is followed by the statement that in statistical theory it obeys all the formal laws of combining probabilities. So far, therefore, the setting of prediction against reality in risk assessment has dealt with the parameters which are capable of fairly exact measurement and prediction which do not have frequent fluctuations in time.

Some of the principles discussed at the Conference on 'Risk Assessment of Chemicals in the Environment' can be compared with risks in engineering. The emphasis on what has become known as major hazards can as a classification be misleading. Problems in assessment of risk are really related to psychological and sociological aspects (see also Chapter 7 and reference 4) of the perception of risks with the political pressures. These pressures then require to be resolved in consideration of the availability and allocation of resources rather than taking full note of the proposals resulting from a risk-benefit analysis.

The Flixborough disaster with 28 sudden deaths is a case in point. The drama surrounding this event and the exhaustive investigation which followed contrasts with the lack of attention given following important reduction of real risks. The introduction of legislation in the UK to make compulsory the wearing of seat-belts in the front of motor cars resulted in the reduction of deaths in motor accidents from 7000 to 5000 per annum - in risk terms from a probability of 1.4×10^{-4} to 1.0×10^{-4} per annum.

The study of risk and predictions for the effects of chemicals in the environment does not have the actuarial base which comes from years of statistics. These statistics have the advantage of recording death, which is certain, final, and arguable to a small extent as to cause. The comparison of actuarial data with predictions has been dealt with in Chapter 2. Succeeding chapters will deal with the much more difficult problem of predicting the effects of chemicals in the environment. It is hard enough to put figures on the effect of exposure to chemicals in a working environment. The gross effects which were identified early on as occupational diseases led to the immediate elimination or control of exposure to materials such as mercury, nickel, β-naphthylamine, cyclic and polycyclic hydrocarbons, amines, nitro-compounds, and vinyl chloride monomer to take a few classic cases.

Outside the working environment, risk assessments depend on dispersion mechanisms, dose-response relations, and assumptions made on mortality and morbidity. Dispersion mechanisms come down to models in fluid and particle mechanics.

Succeeding chapters deal with the problems in estimating the levels of chemicals in the environment, the mechanisms by which concentration can occur as in the food-chain, and gross criteria such as LD_{50} against the complex problems of small doses, their elimination or transformation. This chapter focuses on the transport of materials in air and water and the amounts reaching the target and neglects the mechanisms through which they are concentrated or attenuated in the soil or biota.

The prediction of the dispersion of pollutants in the atmosphere and the comparison with reality as shown by measured amounts has a sharp tool in radionuclides. Their detection in very small amounts is made possible by the sensitivity of the instruments which can measure disintegrations per second although the characterization of individual species is more difficult. Nevertheless, the success of the modelling in relation to observed amounts has been dramatically shown by the European effort devoted to the Chernobyl releases. There have now been seven reports on Atmospheric Dispersion based on the work begun in 1977 under the chairmanship of D.A. Jones at the National Radiological Protection Board.[5-11] The latest report (1986), on the uncertainty in dispersion estimates obtained from the working group models, examines the Gaussian plume model which is acceptable where the distribution of material can be described by the parameters σ_y and σ_z, which represent the spread in the crosswind and vertical dimensions respectively. The model cannot be used below wind speeds of about one m s^{-1} nor where the vertical wind velocity cannot be neglected. The latter assumption is inapplicable for terrains which are not smooth and level. The former means that calm conditions, particularly under inversions, are not capable of prediction. There are many such examples (*e.g.* North India) where long periods persist in winter with clear daytime skies and rapid cooling of the earth at night. In the UK there are typical episodes in November of about ten days when these conditions occur and ground level concentrations of pollutants like sulphur dioxide can increase tenfold.

The uncertainties in the models are listed in Figure 1 of ref. 11 and detailed in Tables 1-6 of that report. One of the tables shows the variation between predicted and observed concentrations for annual averages at a particular site to be 0.5-2.0, and for another site between 0.33 and 3.0 for 89% of the samples and 0.1-10 for 100% of the samples. The predictions which are made with these models also assume that no significant settling occurs of aerosols with an activity median aerodynamic diameter (AMAD) of less than 10 μm. There are considerable problems in dealing with calculation of wet deposition because of the spatial and temporal variabilities of rainfall rate and vertical air movements within a rainstorm. No quantitative statement of the uncertainty on wet deposition calculations can be given. Their reliability is likely to increase with increasing spatial and temporal averaging and in persistent light rain.

The difficulties in this area of micrometerology are great. They extend across the whole area of prediction. This chapter has looked at some of those which are connected with assigning a stability category to any situation before a prediction can be made of the contribution of turbulence to the spread and dilution of material in the atmospheric environment. This depends on the sensible heat flux at the Earth's surface. In itself this is subject to diurnal variation but in daytime measurements alone at Cardington (UK) the mean flux was 81 W m^{-2} with an error of ±55 W m^{-2}. The complications in such modelling have been demonstrated in the most recent workshop on environmental consequences of nuclear war on the calculation of surface temperatures where smoke clouds reduce solar radiation and produce inversions. The non-expert in these fields can only conclude that the gap between prediction and reality is quite large.

An opportunity to examine these problems occurred at a workshop held recently in Moscow. The author led a party of scientists from many countries on a visit to Chernobyl. The fall-out pattern here was complicated by the discharge of solid along with gaseous radionuclides. The pattern of fall-out depended greatly on particle size distribution. Near the damaged No.4 reactor, immediate ground level activities as high as 1000 Roentgens $hour^{-1}$ were measured. Two years later, after removal of 0.5 m of topsoil the level measures was between 10^{-1} to 10^{-6} of the original value.

The small amounts of fall-out over Europe have been measured and fitted into dispersion models which reasonably account for the results of the release. The much smaller quantities reaching Japan and the USA were also reported at the workshop. As a result, a proposal is being discussed to carry out a study on biogeochemical pathways for the transfer of radionuclides in the environment from discharges at Chernobyl and from installations handling spent nuclear fuel, such as power stations and re-processing plants.

There is an area where other mechanisms for the spread of chemicals must be sought beyond the modelling of Gaussian plumes. Where failure of equipment results in the sudden release of materials from a container in which the pressure and temperature conditions are above boiling point at atmospheric pressure, a two-phase jet can be propelled from the point of failure as a result of vapour-lifting of the vessel contents. This happened at Flixborough with cyclohexane under reaction at 155 °C and 8.8 kg cm^{-2} (0.86 MPa) above atmospheric pressure. The momentum of the jet containing liquid issuing through a 500 mm hole carried cyclohexane in a concentration within the flammable limits to a point of ignition 130 m upwind.

A number of studies on the sudden release of materials from containers in which they are above the boiling point at atmospheric pressure assess the movement and dispersion of dense clouds. Many of them are concerned with toxic as distinct from flammable materials. The storage of chlorine and ammonia were typical cases, which were considered in the First[12] and Second[13] Reports of the Health and Safety Executive on Canvey and in the 1982 Netherlands study of the Rijnmond area.[14] A critical study of the uncertainties was made by Griffiths and Mapson.[15] The first section deals with those associated with sudden or intermittent releases, large or small, for which mathematical models of dispersion can be made.

3 Human Toxic Response

This is illustrated by the effects of ammonia and chlorine, as shown in Tables 1 and 2.

Table 1 *Health hazards of ammonia, from Chemical Industries Association Ltd. (CIA) (1974)*

p.p.m. (v/v)	mg m^{-3} (*25 °C and* 1 atm)	*General effect*	*Exposure period*
25	17.5	Odour detectable by most persons	Maximum for 8 h working period
100	70	No adverse effect for average worker	Deliberate exposure for long periods not permitted
400	280	Immediate nose and throat irritation	No serious effect after ½-1 h
700	490	Immediate eye irritation	No serious effect after ½-1 h
1700	190	Convulsive coughing, severe eye, nose, and	Could be fatal after ½ h
2000-5000	1400-3500	throat irritation	Could be fatal after ¼ h
5000-10,000	3500-7000	Respiratory spasm, rapid asphyxia	Fatal within minutes

Table 2 *Chlorine vapour concentrations causing injury, from Braker and Messman (1971)*[15]

Vapour concentration		
p.p.m. *v/v*	mg m^{-3} *(25 °C and* 1 atm*)*	*Effect*
1	2.9	Maximum allowable for 8 h exposure
3.5	10.2	Minimum detectable by odour
15	43.5	Minimum causing throat irritation
30	87	Minimum causing coughing
4	11.6	Maximum that can be breathed for one hour without damage
40-60	116-174	Dangerous after 30 min exposure

From all the references studied by the authors,[15] the conclusions drawn as to concentrations C and time of exposure t resulting in death are that the results are highly inconsistent and limited by the experimental evidence on human deaths. As an example, they quote Weiss[16] that a concentration of 5000 p.p.m. of ammonia can cause 'immediate death from spasm', whereas Sax[16] states that 'the lowest published lethal concentration' (LC_{20}) for humans is 10,000 p.p.m. for 3 h. The ideal way to express combinations of C and t would be in the form of a probit function[17] $Pr = a + b\ln(C^n t)$, where the probit value Pr is a measure related to the percentage of an exposed population that suffer a given level of damage and the coefficients a and b and the index n are determined from controlled tests. However the data needed to establish the coefficients and the index for human fatalities are not available. Examination of the various attempts made show ranges of distance and areas for LD_{50} contours with a 50 t release of chlorine for stability condition D (wind speed 5 m s^{-1}) of 1-5.3 km and 0.52-3.8 km^2 respectively. Chlorine is used in tonnage quantities at waterworks as a chemical disinfectant to prevent the risk of water-borne disease. The risks in the transportation and storage of chlorine are negligible in relation to the health benefits. Research is, however, being undertaken to produce chlorine on-site at waterworks by electrolysis.

The uncertainty surrounding the use of probit functions is reflected in published material. Marshall[19] examines historical evidence and in a personal communication criticizes the use of probits applied to the chlorine attack at Ypres on April 22 1915 made by Withers and Lees.[20]

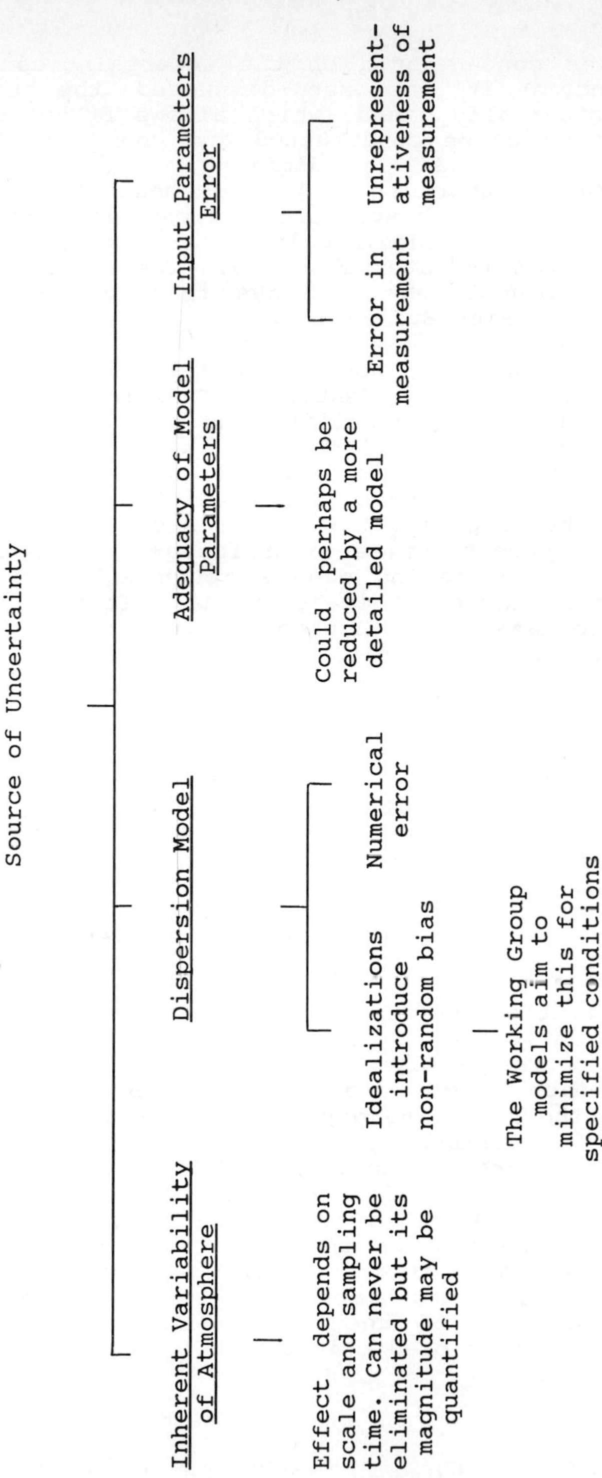

Figure 1 Sources of uncertainty in dispersion modelling

4 Conclusions

This chapter has concentrated on the uncertainties in risk assessment. It has not discussed the firm information, actuarially based, which allows frequency-consequence curves to be constructed and compared with those derived from model predictions. The Royal Society Study Group Report on Risk Assessment[1] examined the evidence on risk available from laboratory experiments and from epidemiological studies. These risks are considered in later chapters. The Report can provide a conclusion in what it says in a section of Risk Estimation and Risk Evaluation:

> 'There is an inherent uncertainty [as can also be seen in Figure 1] in identifying hazards and in making estimates of probability and consequence of events. Risk estimation should therefore indicate the limitations of this quantification. Risk estimation is a process of prediction and is unlikely to be precise. For some risks of very low probability, uncertainties by a factor of ten are commonplace. But making even a rough estimate is better than making no estimate at all. It can draw attention to hazards which have not previously been recognized, or mistakenly dismissed as trivial. This aspect may be more important than the further refinement of calculations of risks for hazards that are well recognized. A view has to be taken or judgement formed on the basis of imperfect scientific information. Safety factors have to be adopted either explicitly or implicitly, and great caution has to be applied to very new products or processes whose full potential has yet to be judged by experience. Risk evaluation should take account of the natures and qualities of different hazards, and differences in public attitudes even though these attitudes vary with time, sometimes rapidly'.

The concentration on studies of dispersion in the atmosphere is deliberate. It is a region which is universal and reasonably homogeneous, but with fluctuations which give rise to great problems in modelling with the resulting gap between prediction and reality. The distribution of materials in other systems such as rivers is complicated by the difference in scales between island and continental rivers - up to three orders of magnitude between those of the UK and of Europe. In these systems there are also additional mechanisms such as transfer in sediments and bio-accumulation in organisms. Many succeeding chapters discuss some of these and demonstrate even greater difficulties in prediction.

5 References

1. 'Risk Assessment', A Study Group Report, The Royal Society, London, 1983.

2. F. Warner, Preface, 'The Assessment and Prediction of Risk', *Proc. R. Soc. London, A*, 1981, **376**, 1-206.

3. A.V. Cohen and D.K. Pritchard, 'Comparative Risks of Electricity Production' (HSE Research Paper II), HMSO, London, 1980.

4. S. Jasanoff, in 'Toxic Hazard Assessment of Chemicals, ed. M.L. Richardson, The Royal Society of Chemistry, London, 1986, p.259.

5. R.H. Clarke, The first report of a Working Group on Atmospheric Dispersion: a model for short and medium range dispersion of radionuclides released to the atmosphere, Harwell, NRPB-R91, HMSO, London, 1979.

6. J.A. Jones, The second report: a procedure to include deposition in the model for short and medium range atmospheric dispersion of radionuclides, Chilton, NRPB-R122, HMSO, London, 1981.

7. J.A. Jones, The third report: the estimation of long range dispersion and deposition of continuous releases of radionuclides to atmosphere, Chilton, NRPB-R123, HMSO, London, 1981.

8. J.A. Jones, The fourth report: a model for long range atmospheric dispersion of radionuclides released over a short period, Chilton, NRPB-R124, HMSO, London.

9. J.A. Jones, The fifth report: models to allow for the effects of coastal sites, plume rise and buildings on dispersion of radionuclides and guidance on the value of deposition velocity and washout coefficients, Chilton, NRPB-R157, HMSO, London 1983.

10. J.A. Jones, The sixth report: modelling wet deposition from a short release, Chilton, NRPB-R198, HMSO, London, 1986.

11. J.A. Jones, The seventh report: modelling wet deposition from a short release, Chilton, NRPB-R199, HMSO, London, 1986.

12. Health and Safety Executive, 'Canvey - an Investigation of Potential Hazards from Operations in the Canvey Island, Thurrock Area', HMSO, London, 1978.

13. Health and Safety Executive, 'Canvey - a Second Report', HMSO, London, 1981.

14. COVO Steering Committee, 'Risk Analysis of Six Potentially Hazardous Industrial Objects in the Rijnmond Area, a Pilot Study', Reidel, Dordrecht, 1982.

15. R.F. Griffiths and L.C. Megson, *Atmos. Environ.*, 1984, **18**, 1195.

16. 'Hazardous Chemicals Data Book', ed. G. Weiss, Noyes Data, New Jersey, 1980.

17. N.I. Sax, 'Dangerous Properties of Industrial Materials', 5th Edn., van Nostrand Reinhold, Scarborough, CA, 1979.

18. D.J. Finney, 'Probit Analysis', 3rd Edn., Cambridge University Press, Cambridge, 1971.

19. V.C. Marshall, 'Major Chemical Hazards', Ellis Horwood, London, 1987.

20. R.M.J. Withers and F.P. Lees, *J. Haz. Mat.*, 1987, **15**, 301.

4

Current Procedures for the Evaluation of Chemical Safety

D.V. Parke, D.F.V. Lewis, and C. Ioannides

DEPARTMENT OF BIOCHEMISTRY, UNIVERSITY OF SURREY, GUILDFORD, SURREY GU2 5XH, UK

1 Introduction

The procedures currently employed to evaluate the safety of chemicals vary widely from country to country and from agency to agency, though the principles vary less, being dependent generally on the intended use of the chemical and the annual tonnage to be produced. Two major aspects generally have to be addressed, namely (i) safety for humans of those chemicals which are intended for direct or indirect human consumption, *e.g.* food additives, medicines, veterinary medicines, animal feed additives, cosmetics, and agrochemicals, together with safety for any workforce involved, and (ii) ecological and environmental safety, especially of pesticides, agrochemicals, and industrial chemicals, *e.g.* dyestuffs, polymers, plasticizers, and detergents. At present, the extent of testing of new chemicals is far greater for medicines, food additives, and pesticides than it is for industrial chemicals, but the degree of testing of the latter is steadily increasing (see Table 1).

The need to evaluate chemicals for safety is overwhelming, as instanced by the numerous deaths of babies dosed with the untested antibiotic chloramphenicol during the 1946 European epidemic of infantile diarrhoea, the many fatalities among agricultural workers using the then relatively untested pesticide parathion, and the indiscriminate use of benzene and carbon tetrachloride as solvents and cleaning agents until some 25 years ago when they were shown to be toxic and carcinogenic by human epidemiological and animal studies. However, safety evaluation is highly costly, in both time and money, and should not be restrictive of new therapeutic treatments, beneficial new foods and agricultural processes, or of international trade; hence it is generally related to the tonnage produced and to the probable extent of human exposure. For example, a prescription medicine such as a β-adrenergic antagonist that may have to be taken every day for life, or a non-steroidal, anti-inflammatory, analgesic drug that can be purchased 'over-the-counter' will generally be tested more extensively than say a local anaesthetic which may be administered only once or twice in a life-time. Similarly, a food-additive used as a flavouring adjunct, with a world-wide production of less than ten kilograms per year and a probable average annual human consumption of a milligram or less per person, may not be tested at all, whereas the widely used antioxidants butylated hydroxyanisole (BHA) and butylated hydroxytoluene (BHT) have been examined extensively and continuously for safety since they were first introduced more than twenty-five years ago. The extent of testing of industrial chemicals and intermediates similarly depends on the annual tonnage that is to be produced and, more recently, on the environmental persistence, since highly persistent non-metabolizable chemicals such as the polychlorinated biphenyls (PCBs) will progressively accumulate in the environment, and in the human body, to an extent where this becomes more critical than the annual level of production (see also Chapter 23).

Table 1 *Typical procedures used in the safety evaluation of different categories of chemicals*

Safety evaluation Procedures	*A*	*B*	*C*	*D*	*E*
Purity and specification	++	+	++	+	-
Physiochemical properties and QSAR	-	-	-	-	++
Acute animal toxicity studies					
LD_{50}	+	+	+	+	+
Skin and eye studies	+	-	+	++	-
Acute human studies					
Clinical pharmacology	++	-	-	-	-
Metabolism and pharmacokinetics	++	-	-	-	-
Chronic animal toxicology Studies					
3 months (2 species)	++	++	++	+	±
6/12 months (2 species)	++	++	++	-	-
Metabolism and toxicokinetics	++	++	++	+	±
Reproduction (2 generations)	++	++	++	-	-
Teratology	++	++	++	-	-
Mutagenicity	++	+	+	+	±

Table 1 (continued)

	A	B	C	D	E
Carcinogenicity (2 species)	++	++	++	-	-
Neurotoxicity	-	-	++	-	-
Immunotoxicity	±	-	-	-	-
Ecotoxicology	-	-	++	-	+
Chronic human studies					
Therapeutic efficacy	+	-	-	-	-
Human toxicity (post-marketing surveillance)	±	-	±	-	-
Pharmacokinetics	+	-	-	-	-

A Medicines
B Food additives
C Pesticides and agrochemicals
D Cosmetics
E Industrial chemicals.

This table is provided to give a broad indication of the importance of the various safety evaluation procedures for different categories of chemicals, which nevertheless do vary from country to country and also in the course of time. The importance of each safety evaluation procedure varies from: (++), critically important; to (+), required; (±), desirable; (-), not generally required.

<u>1.1 Safety Evaluation Procedures</u>. Chemical toxicity may be acute, as in the neurotoxicity of organophosphorus esters or the hepatotoxicity of carbon tetrachloride, or may be chronic as in the hepatotoxicity of mono-(2-ethylhexyl) phthalate (MEHP), the fetotoxicity of diethylstilboestrol, or the carcinogenicity of 2-acetylaminofluorene. Evaluation of both the acute and chronic potential toxicity of chemicals is undertaken by using animals as experimental models for man (see Table 1), based on one of the following assumptions, namely that (i) anything likely to be toxic to man will manifest some degree of toxicity in other living systems provided that the dose is high enough, (ii) if a sufficient number of different animal species are dosed with the chemical, one at least is likely to exhibit the toxicity that would be seen in man, and (iii) if an animal species exhibits a similar pattern of metabolism and toxicokinetics of the chemical to that seen in man, then this species would be likely to exhibit similar toxicity.

Acute toxicity is evaluated by a number of short-term procedures, from which the LD_{50} may be calculated; they include short-term repeated dose and subchronic toxicity studies, and skin and eye irritancy tests (see

Table 1 and also Chapter 14). Chronic toxicity includes procedures for repeated administration of the chemical to appropriate animal species for three, six, or twelve months or more, with the accompanying pathological examinations, including post-mortem examination, histopathology, clinical chemistry, and haematology, at specified intervals during the study and at termination. In addition, specific procedures are undertaken to test for teratogenicity, reproductive effects, mutagenicity, carcinogenicity, and, in certain instances, immunotoxicity and neurotoxicity (see Table 1). As the toxicity of chemicals is determined, to a major part, by the rates and extents of absorption, tissue distribution, and excretion, and also by the pathways and rates of metabolism, the study of the toxicokinetics and metabolism of the chemical is now required for all animal-based safety evaluation procedures.

The choice of animal species for safety evaluation studies is of paramount importance as the animals are serving as surrogate models for man, and although, by convention, rats and dogs are preferred for chronic toxicity studies, rats and mice for life-span carcinogenicity, and rats and rabbits for reproduction and teratogenicity studies, other species including guinea pigs, hamsters, ferrets, monkeys, and other primates may also be used in the toxicity testing of chemicals. Validation and/or selection of appropriate species can be achieved by comparative metabolism and toxicokinetic studies undertaken in man and the animals to be studied, and although it is rarely possible to find an animal which, for a given chemical, is a perfect model for man, suitable approximations can usually be found from which biological extrapolations can be made to predict effectively the probable toxicity in man.

The routes of administration of the chemical used in toxicity testing procedures depend largely on the intended use of the chemical, but the most frequent route used is via the diet or by oral intubation; industrial solvents may be administered by inhalation, and cosmetic chemicals by topical application. In toxicokinetic studies it is also necessary to administer the chemical intravenously to ensure instant and complete absorption, so that the rates and extent of absorption after other routes of administration may be quantified, and the metabolism compared with that following administration by other routes. The level of dosage in the various species is also critical, as the rates of oxidative metabolism of chemicals vary inversely with the size of the animal species. Most chronic safety evaluation studies are undertaken at three dose levels, the highest being sufficient to elicit frank toxicity and to identify the target organ(s)/tissue(s) (the maximum tolerated dose), the lowest being toxicokinetically equivalent to, or a simple multiple of, the level of human exposure, and the middle dose approximating to the highest dose likely not to result in any frank toxicity (but see Chapters 8, 9, 14, and 15).

For industrial chemicals and agrochemicals, it is customary to evaluate materials also for environmental safety (ecotoxicology) (see Chapters 24 to 26). In Japan this is accomplished by conducting toxicity studies in fish and other indicator species. In the United States, the Environmental Protection Agency (EPA) evaluates the probable safety of new industrial chemicals from considerations of their molecular structure, physicochemical properties (such as log *P* values), probable routes of metabolism, predictions of toxicity from quantitative structure-activity relationships (QSARs), and any known biological/ toxicological properties (see also Chapters 5, 8, 15, 21, 23, 25, and 26). If from any of these considerations there are good indications of possible toxicity then biological studies to investigate these predictions are requested. However, in many instances it is economically more expedient for the chemical manufacturer to redesign his product to attempt a greater probability of safety than to undertake expensive and time-consuming safety evaluation studies.

Different chemicals have their safety evaluated by different procedures and although these may also vary from one country to another, international agencies such as WHO, FAO, EEC, and OECD, and international trading, have resulted in some degree of harmonization (see Chapters 5 to 7). As may be seen from Table 1, human studies play a much more major role in the safety evaluation of medicines, particularly in the early post-marketing surveillance, when individual idiosyncracies due to pharmacogenetic differences, impaired clearance due to age or disease states, and immunotoxicological effects, are first seen; human observations are also made for agrochemicals and industrial chemicals, but these are limited to incidental exposures of industrial and agricultural workers. Chemical specification and purity may be critical to toxicity, since there are known instances where five per cent of impurities can increase toxicity ten-fold (*e.g.* malathion); chemical specification and purity are thus highly important in the safety evaluation of new medicines, where diastereoisomers are now being regarded as impurities, but have little meaning in the safety evaluation of industrial chemicals which are generally mixtures of unknown and varying composition. Other major differences in the safety evaluation procedures for different chemicals, as one might expect, are the greater emphasis placed on physicochemical properties and QSARs in the safety evaluation of industrial chemicals, the greater emphasis in acute topical toxicity for cosmetics, the ecotoxicological studies generally required for pesticides, agrochemicals, and industrial chemicals, the neurotoxicity studies for pesticides and, increasingly, the immunotoxicity potential of new medicines.

The theoretical backgrounds to all these procedures, and the practical details for carrying them out, go far beyond the scope of this chapter and interested readers are referred to recent standard works.[1-13]

<u>1.2 Failures of the Present System</u>. The present system of basing the safety evaluation of chemicals on the results of toxicological investigations in experimental animals, although generally successful, has unfortunately experienced some failures (see Chapter 7). These include both false negatives and false positives, all due, at least in part, to the imperfection of the animal models as surrogates for man, or to the toxicologist's inability to make appropriate extrapolations of human risk assessment from animal data. Most of the known failures have occurred with medicines, but this is because (i) medicines are the most intensively tested in animal toxicology studies of all new chemicals, and (ii) the results obtained are soon checked in human studies. Hence, the failures with medicines serve as a reliable guide for the probable rate of failure in all other categories of chemicals, which are mostly not tested as extensively in animal studies as are medicines and, furthermore, the findings are not ultimately checked in defined, controlled, and monitored human studies as are those for medicines.

Among the many possible reasons for the false negatives, *i.e.* failure to identify the potential toxicity of a chemical to humans from toxicological studies in experimental animals, are the following:

(1) Lack of monitoring. Thalidomide teratogenicity was not predicted because it was not looked for; similarly, the delayed neurotoxicity resulting from exposure to certain pesticides was not predicted because it was not expected, it is seen only in certain sensitive species, *e.g.* chicken, and thus was not appropriately studied.

(2) Species differences. These may occur in the metabolism of the chemical, in its toxicokinetics, or in receptor affinities, all of which may result in species differences in chemical toxicity in animals and man. These would generally be detected in the

metabolism and toxicokinetic studies, and then allowed for, but in the case of the candidate drug FPL 52757 (6,8-diethyl-5-hydroxy-4-oxo-4*H*-1-benzopyran-2-carboxylic acid) man was rather unique in all of these aspects, and the conventional procedure for safety evaluation, with pathological studies in some eight different animal species (rat, mouse, rabbit, hamster, guinea pig, ferret, baboon, and cynomolgus and squirrel monkeys) failed to detect potential toxicity or to predict toxicity in man.[14] Retrospective studies showed that determination of 'clearance' was far more reliable as a predictor of safety in humans than all the chronic toxicology studies in animals, with only the dog being a suitable model for man for this particular chemical.

In a recent book on the role of experimental animals in human risk assessment of chemical toxicity, more than twenty experienced toxicologists reviewed the problems in selecting animal species as surrogate models for man in safety evaluation studies. These problems were sufficiently numerous to exclude any significant repetition and they also offered no effective solution.[15]

(3) Genetic variations. Toxicological studies are generally conducted in animals of homogeneous genetic characteristics, *e.g.* Wistar albino rats, beagle dogs, *etc.*, whereas man is markedly heterogeneous in his genetic make-up. Consequently, chemical toxicity in human subjects due to genetic polymorphism, such as suxamethonium toxicity in subjects with impaired plasma pseudoesterase activity, isoniazid toxicity in slow acetylators, or perhexiline toxicity in individuals with slow hydroxylating ability, are unlikely to be predicted from animal toxicity studies.

Similarly, genetic variations may occur in inbred strains of experimental animals, *e.g.* the DBA/2 'non-responsive' mice are much less susceptible to the carcinogenicity of polycyclic aromatic hydrocarbons than are the responsive strains of mice.[16]

(4) Non-linear toxicokinetics. Chemical toxicity is highly dependent on dosage, often because different pathways of metabolism, particularly activation, come into play at high doses when the normal pathways of detoxication have become saturated, as when endogenous substrates for conjugation have become depleted. A similar situation may occur from repeated low exposure to chemicals which are poorly metabolized and excreted, and hence accumulate in the body tissues, *e.g.* polychlorinated biphenyls. Many lipophilic compounds are excreted in the bile and undergo enterohepatic circulation, so that if the elimination of the chemical depends on a pathway of metabolism that is readily

saturated, such as glucuronide conjugation, repeated exposure will lead to a build-up of the body burden of the chemical, with high tissue concentrations and consequent toxicity. This pattern of non-linear kinetics was responsible for the toxicity of a number of drugs such as benoxaprofen, which is entero-hepatically circulated and has a long half-life, especially in the elderly.[17]

Non-linear kinetics are also seen when chemicals inhibit or induce drug metabolism. The hepatocarcinogen 2-acetylaminofluorene is not toxic to rats when only a single dose is administered, but on repeated dosing the activating enzyme is induced and the amide is *N*-hydroxylated to yield the proximate carcinogen which initiates carcinogenesis.

(5) Oxygen radical toxicity. It is now well accepted that increased generation of oxygen radicals (hydroxyl radical and singlet oxygen) may result in tissue necrosis, mutations, and malignancy, and an increase in those natural disease states, such as cardiovascular disease, diabetes, renal failure, cataract, and male infertility, which are known to be associated with oxygen radical toxicity. Oxygen radicals produced by redox cycling of quinones, or futile cycling of the microsomal mixed-function oxidases, may be the consequence of exposure to a wide variety of xenobiotic chemicals. However, as the effects may largely be an augmentation of the natural processes of ageing and of age-associated disease, such toxicity has been difficult to assess and evaluate.

Similarly, among the possible reasons for false positives, *i.e.* manifestations of toxicity in experimental animals that are probably of little or no significance to humans, are the following:

(1) Species differences. It is well known that for a large number of chemicals small rodents are poor surrogates for man in toxicity testing. For some chemicals the toxicity in small animals is 10- to 50-fold greater than that seen in man, although for other chemicals the level of toxicity is much less than that seen in man. The reason for this is that small mammals have high rates of basal metabolism, high levels of tissue oxygen, and hence high rates of oxidative metabolism of chemicals; where oxidative metabolism results in detoxication, experiments in small rodents, with doses based only on body weight, may greatly underestimate the probable level of toxicity in human subjects, and conversely, where oxidative metabolism leads to activation, the level of predicted toxicity may be greatly exaggerated. An excellent example of this pitfall is the carcinogenicity of chloroform in small rodents, which

has been attributed to its oxidative metabolism to carbonyl chloride; as the extent of oxidative metabolism of chloroform in small rodents is 10 to 50 times greater than that which is likely to occur in man, the weak carcinogenicity of the halocarbon in rodents is probably of no consequence to man[18] (see also Chapter 9).

(2) High potential for carcinogenicity in small rodents. In the absence of human epidemiological data it is considered that life-span carcinogenicity assays in two rodent species, namely rats and mice, are the most definitive method for assessing the carcinogenic potential of chemicals in humans. However, small rodents have high rates of xenobiotic oxygenation and as most carcinogens require oxidative metabolism for activation, the rationale of employing small rodents to assess the carcinogenic potential of chemicals in man is theoretically unsound. Mice have an even more notorious reputation than rats for yielding spurious positive results in carcinogenicity testing, and the facile induction of tumours in mice associated with the feeding of chlorinated pesticides, such as DDT, and numerous other chemicals is well known and is widely considered to indicate a hyper-responsiveness of this species. Among the probable reasons for this are that mice have (i) high tissue O_2 uptake and hence a high ability for the oxidative activation of carcinogens and the production of oxygen radicals, which are known mutagens and carcinogens and may also act as promoters, (ii) high levels of cytochrome P-448, the enzyme which specifically activates many mutagens and carcinogens (man and rat have lower levels), and (iii) a high ratio of glutathione S-transferase/epoxide hydrolase activities, so that protective tissue glutathione is more readily depleted in the detoxication of mutagenic/carcinogenic epoxides and oxygen radicals (man has a low ratio and conserves his glutathione).

Indeed, in an evaluation of 266 long-term and carcinogenicity studies the overall concordance between the two rodent species selected to evaluate the carcinogenic potential of chemicals in man was only 74% (198/266); 131 chemicals were negative in both species, 32 were positive in rats only, 36 were positive in mice only, and 67 were carcinogenic in both rats and mice.[19] However, of the 135 chemicals which were carcinogenic in at least one sex of either rats or mice, only 38 chemicals were positive in both sexes of both species, 71 chemicals were carcinogenic in both sexes of mice, and 65 chemicals were carcinogenic in both sexes of rats. If positives and negatives in both sexes of both species (4/4 positive or negative) are required as criteria, the overall concordance between rats and mice exposed to the same chemical becomes only 64% (169/226). This reappraisal of the study of Haseman and Huff[19] supports the views expressed by Di Carlo that 'carcinogenesis in mouse cannot be predicted from bioassay data from rat, and *vice versa*' and 'it is premature to attempt extrapolation of carcinogenesis bioassay results from rodents to humans until satisfactory extrapolations can at least be made between rodent species'.[20]

(3) Receptor differences. Toxicity may be considered to be mediated through 'receptors', whether these are enzymes, steroid receptors, or other hormonal receptors such as the cytosolic Ah receptor. The carcinogenicity of steroid hormones is known to be associated with specific hormone receptors and these may differ both in abundance and in their affinity for the hormone, in different species. Similarly, there are species differences in the levels of the enzyme neurotoxic esterase, which may be regarded as the receptor responsible for the delayed neurotoxicity of certain pesticides. This importance of species differences in receptor activities may be illustrated by the species difference in the carcinogenicity of a number of progestogens, such as lynoestrenol, which gives rise to breast tumours in the beagle dog but exhibits no increased tumorigenicity in rodents and is considered to present no risk to human subjects.[21] Similarly, selection of an appropriate animal species, with similar levels of neurotoxic esterase to man, is important in the testing of pesticides and other chemicals for delayed neurotoxicity.

1.3 Objections to the Present System. In addition to the lack of reliability of the present system of toxicity testing, due to the false negatives and false positives encountered in the use of experimental animals, objections have also been raised over (i) the high cost in time and money, (ii) the largely subjective nature of the assessments and their vulnerability to human error, and (iii) animal rights.

The cost of a full programme of safety evaluation for a single chemical is of the order of £5 million and takes a minimum of 5 years to complete if properly organized. The investment in time and money is therefore very considerable and requires a large potential market for the new chemical in order for the venture to be viable. Even the most modest programme of toxicity testing, say a 30 day study in two species and full mutagenicity, will cost around £100,000 and take a year. Where chemicals are patented, the considerable length of time required to complete the animal studies means a loss of patent life and hence of income.

The lack of scientific credibility of experimental animal studies is further exacerbated by the high degree of inherent errors in these procedures. The maintenance of large numbers of animals, on different doses, for long periods of time, gives ample opportunity for errors to occur, as they do only too frequently. Furthermore, not only error, but frank bad practice has occurred in the past, so that an expensive and rigorous monitoring system of Good Laboratory Practice is now mandated by most regulatory agencies. Stringent control of the total environment of the experimental animals is essential, since it has been shown that diet, the atmosphere, and the physical environment can profoundly affect chemical toxicity; indeed, dietary deficiencies in folate or riboflavin and contamination of the air intake with chlorinated solvent vapour have all been known to reverse the true findings in life-span carcinogenicity studies. Finally, the evaluation of chemical toxicity by histopathology is essentially subjective and, despite recent standardization of procedures and terminology is frequently more a matter of opinion of the pathologist than an objective scientific measurement. In the ultimate process of risk assessment, it is not unknown for statistical analyses to be modified by the inclusion of additional control groups and by other devices to modulate further the highly subjective original evaluation.

The past decade or so has seen the emergence of various animal welfare and animal rights groups, some of them militant, and most of them highly motivated and politically active. The public worldwide are therefore being persuaded against the continuance of animal experimentation, new legislation giving further protection to animals against non-essential experimentation has been enacted, and governments are making available funds for developing alternative approaches and methods for the safety evaluation of chemicals. Although it is possible to support the need for animal experimentation in fundamental medical research, in view of the empiricism of the present safety evaluation procedures, the numerous errors and the many failures, it is less feasible to support the continued use of large numbers of experimental animals in the safety assessment of chemicals. Alternatives are being developed and may, eventually, partly replace the present system of testing chemicals in live animal models.

<u>1.4 Alternative Procedures</u>. In the wish to reduce the cost in time and money of the present traditional methods of chemical safety evaluation, with additional endeavours directed to placation of animal welfare groups, a number of alternatives to experimentation with living animals have been proposed, and these may broadly be classified into three major categories, namely (i) human studies, (ii) *in vitro* studies, and

(iii) quantitative structure-activity relationship (QSAR) studies.

1.4.1. Human studies. The first of these alternatives, namely human studies, would obviously be scientifically more desirable and more acceptable than animal studies, but is limited by ethical considerations. Possibilities for experimentation in humans have been debated at great length but for obvious reasons would be limited to new medicines, where human investigations are already used quite extensively in pre-clinical volunteer studies, clinical trials, and post-marketing studies. The expansion of the post-marketing studies of new drugs indicates a possible way forward for the more successful safety evaluation of chemicals, in that this greatly facilitates the identification of toxicity associated with human genetic polymorphisms and hypersensitivity, two of the major problems of drug safety.

1.4.2. *In vitro* studies. *In vitro* studies are undoubtedly the most favoured of the various alternatives but unfortunately still retain many of the disadvantages of live animal studies, such as species differences in the pathways of metabolism, and in the toxicokinetics, and receptor activities, as well as the problems of impaired conjugation and excretion mechanisms, impaired defence against oxygen toxicity, and limited survival. Perfused isolated livers, kidneys, and other organs, tissue and cell cultures, fresh hepatocyte suspensions, and tissue homogenates and fractions, are all highly useful in elucidation of the mechanisms of toxicity, but all suffer from being essentially non-physiological and consequently more vulnerable to potential chemical toxicity as they are largely deprived of the benefit of the normal physiological detoxication systems (see Table 2).

Table 2 *Limitations of toxicity studies in isolated cell systems*

1. Mixed-function oxidase activities are lost, and the detoxicating cytochromes P-450 are replaced by the activating cytochromes P-448.

2. The responses of hepatocytes to cytochrome P-450 inducers are markedly different from those observed *in vivo*.

3. Conjugation pathways are impaired due to non-replacement of glucuronide (UDPGA), sulphate (PAPS), *etc.*

4. Glutathione is depleted and not adequately replaced, impairing the antioxidant and radical-scavenging defence system, and leading to autoxidative damage and cytotoxicity.

5. Plasma membrane permeability is markedly increased, resulting in loss of nucleotides, nutrients, *etc.*, leading to loss of viability.

6. Clearance is impaired due to absence of excretion mechanisms and loss of detoxication, resulting in enhanced exposure.

7. Cell metabolism is not subject to hormonal and homeostatic regulation, decreasing the natural defensive response to xenobiotics.

Hence bromobenzene is markedly hepatoxic to isolated rat hepatocytes at concentrations which would be without any effect to the well-nourished intact rat, and this chemical and many others such as carbon tetrachloride and paracetamol have been studied in various isolated organ, cellular, and other *in vitro* systems for the explicit purpose of magnifying the toxicity in order that the mechanisms may be better understood.

1.4.3 QSARs. Since toxicity is determined by the chemical nature of the compound, and by its interactions with a variety of biological macromolecules, from drug-metabolizing enzymes to hormonal receptors, the ultimate potential for toxicity must be determined by the intrinsic properties of the chemical and the characteristics of the actual biological species exposed, these latter being dependent on the genetics of the individual animal and on its past and present environments. Hence, in using animal models to assess chemical toxicity in humans we are actually compounding the problem by introducing new variables, namely the genetic and environmental differences between experimental animals and man, which are then largely ignored in the subsequent risk assessment process. It is therefore not incongruous to approach the problem of safety evaluation of chemicals from the standpoint of molecular structure, including QSARs with related chemicals of known similar toxicity, together with their physicochemical properties including lipophilicity, and probable pathways and rates of metabolism, procedures which are currently employed by the US EPA in the safety evaluation of new industrial chemicals. As lipophilicity (estimated by log P, where P is the n-octanol-water partition coefficient) may be calculated from the molecular structure, as also may any predictions of metabolism and toxicokinetics, a knowledge of the molecular structure of the chemical is all that is necessary to embark on the QSAR approach to safety evaluation. This is of particular value in the assessment of industrial chemicals, which are often mixtures of variable composition within known structural limits, and is also invaluable as an adjunct in the safety evaluation of new medicines, food additives, and pesticides.

A more esoteric approach to the QSAR method for the safety evaluation of chemicals has been developed recently from the concept that carcinogenicity, mutagenicity, and many forms of chronic toxicity are due to the endogenous activation of chemicals and the interaction of the parent compound and/or its 'reactive intermediates' with specific intracellular receptors.[22] It has now been recognized that oxygenation of a chemical, or its conversion into a free radical, is a general prerequisite to its biological activation and hence to its chronic toxicity. Oxygenation at **unhindered** carbon centres in the molecule leads to the formation of metabolites which are suitable substrates for the enzymes, epoxide hydrase(s), and glutathione S-transferase(s), and hence are readily conjugated and excreted. In contrast, oxygenation at **hindered** carbon and at nitrogen centres results in formation of oxygenated products that are not readily conjugated, and hence, as electrophilic reactive intermediates, interact with vital intracellular nucleophiles, such as sulphydryl enzymes, proteins, and DNA, resulting in cytotoxicity, mutations, and malignancy. Recent studies have shown that these metabolic oxygenations of xenobiotic chemicals are mostly carried out by one or more of a superfamily of hepatic microsomal enzymes, the cytochromes P-450,[23] of which the two major families are the polycyclic aromatic hydrocarbon (PAH)-inducible cytochromes P-448 (rat liver cytochromes P-450 c and d) and the phenobarbital-inducible cytochromes P-450 (rat liver cytochromes P-450 b and e).[24] These two families have many distinct differences,[25] and whereas the phenobarbital-inducible cytochromes P-450 result mainly in unhindered oxygenations and the detoxication of xenobiotic chemicals and carcinogens, the cytochromes P-448 result in hindered oxygenations and the activation of toxic chemicals and carcinogens.[26]

By the use of specific substrates, inhibitors, and inducers of the cytochromes P-450 and P-448 it has been possible to characterize these families of enzymes, and from the molecular conformations of the substrates, determined by computer graphics, to assign a particular chemical substrate to the cytochrome most likely to be involved in its oxidative metabolism.[27–29] In a theoretical study based on molecular orbital calculations and computer graphics it has been established that most cytochrome P-448 substrates contain fused aromatic or heteroaromatic rings, giving rise to overall molecular planarity with relatively small molecular depth.[30] In contrast, substrates of the cytochromes P-450 are globular molecules with greater conformational freedom and an ability to bind to the enzyme at more than one point of attachment.[30] As the cytochromes P-448 activate many potentially toxic chemicals and carcinogens, and the other families of cytochromes P-450 generally lead to detoxication, the corollary of these findings is the possibility of prediction of the toxicity of chemicals on the basis of their molecular dimensions.[28-30] Furthermore, the substrates of the cytochromes P-448 have been shown to have an identical, or very similar, molecular conformation to those of the inducers of the cytochromes P-448, which manifest their enzyme regulatory effect by interaction with the Ah receptor.[31] Chemicals with the characteristic molecular conformation to fit the active site of the cytochromes P-448 may therefore experience the following: (i) activation by cytochromes P-448 to a reactive intermediate (chemical toxicity) or a proximate carcinogen (mutagenesis), (ii) interaction with the Ah receptor to effect DNA transcription and induction of cytochrome P-448 (co-carcinogenesis), and (iii) interaction with the Ah and possibly other growth factor receptors to effect the transcription of regulatory RNA, replication of DNA, and mitosis (promotion). Hence, determination of the molecular conformation of a chemical by computer graphics, together with determination of its possible induction and inhibition of the cytochromes P-448 by enzymological and immunological methods, yields valuable predictive information concerning the potential toxicity of the chemical, its mutagenicity, and its carcinogenicity.

2 Methods

Molecular geometries were obtained from either crystallographic data or, when these were not available, from energy minimization by procedures incorporated in the Modified Intermediate Neglect of Differential Overlap, 3rd version (MINDO/3) method for electronic structure calculation.[32,33] Starting geometries were obtained from tables of standard bond lengths and angles.[34] Molecular graphical plots to obtain molecular dimensions were produced using the PLUTO crystallographic package with standard van der Waals atomic radii. Views parallel and perpendicular to the main molecular plane were plotted on a CALCOMP plotter on a scale of 1 cm = 1 Ångström unit. The molecular orbital calculations by the MINDO/3 method were performed on a CDC 7600 mainframe computer at the University of Manchester (UMRCC) via a remote site link from the University of Surrey. Graphical plots were produced using the Prime 750 computer at the University of Surrey.

Electronic structure calculations were performed by the Complete Neglect of Differential Overlap (CNDO) method, which form part of the COSMIC package generously made available by Drs. J.G. Vinter, A. Davis, and M.R. Saunders of Smith, Kline and French Ltd. COSMIC was also used to obtain minimized molecular geometries by methods essentially similar to those of MINDO/3. The COSMIC system runs on a Sigmex 6130 molecular graphics workstation coupled to a MicroVax II computer.

3 Results

3.1 Prediction of Toxicity from Chemical Structure. The spatial parameter (area/depth^2 (a/d^2) electronic structures (E(HOMO), E(LEMO) and ΔE), and the specificities for the cytochromes P-448 of one hundred different medicines, food additives, pesticides, and industrial and environmental chemicals are given in Table 3, and a plot of the values of a/d^2 against ΔE is shown in Figure 1.

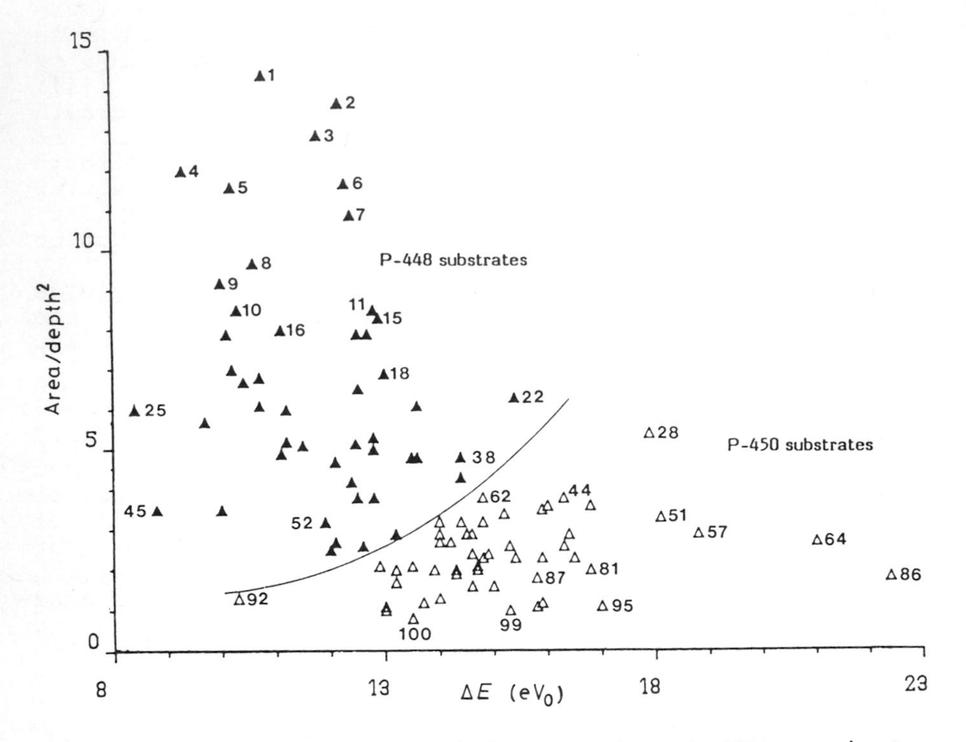

Figure 1 Plot of the spatial parameter (a/d^2) against the electronic parameter (ΔE) for 100 chemicals. Chemicals are identified by numbers as designated in Table 1. Numerical data for a/d^2 and ΔE for all 100 chemicals are given in Table 1; the numbers of some chemicals have been omitted for clarity.

Table 3 *The spatial parameters, electronic structures and cytochrome P-448 specificities of a variety of xenobiotic chemicals*

Compound	*A*	*B*	*C*	*D*	*E*
1 Dibenz(*a,h*)anthracene	14.4	-9.7	1.1	10.8	+
2 α-Naphthoflavone	13.7	-10.6	1.6	12.2	+
3 β-Naphthoflavone	12.9	-10.4	1.4	11.8	+
4 Benzo(*a*)pyrene	12.0	-9.0	0.3	9.3	+
5 Dibenzobenzimidazole	11.7	-9.9	2.4	12.3	+
6 1-Nitropyrene	11.6	-10.2	0.0	10.2	+
7 Anthraflavic acid	10.9	-11.4	1.0	12.4	+
8 *N,N*-Dimethylaminoazobenzene	9.7	-9.7	0.9	10.6	+
9 2-Aminoanthracene	9.2	-9.2	0.8	10.0	+
10 Ellipticine	8.5	-9.2	1.1	10.3	+
11 Trp-P-1*	8.5	-9.9	2.9	12.8	+
12 4-Aminobiphenyl	8.3	-10.3	2.6	12.9	+
13 5,12-Dimethylchrysene	8.0	-9.8	1.3	11.1	+
14 β-Naphthylamine	7.9	-10.4	2.1	12.5	+
15 TCCD	7.9	-10.9	1.8	12.7	+
16 3-Methylcholanthrene	7.9	-9.2	0.9	10.1	+
17 9-Hydroxyellipticine	7.0	-9.1	1.1	10.2	+
18 Coumarin	6.9	-11.4	1.6	13.0	+
19 Cytembena*	6.8	-8.0	2.6	10.6	-
20 *o*-Aminoazotoluene	6.8	-9.8	0.9	10.7	+
21 Dimethylbenzidine	6.4	-9.5	2.9	12.4	+
22 Styrene	6.3	-12.4	3.0	15.4	+
23 Zoxazolamine	6.1	-10.4	3.2	13.6	+
24 Glu-P-1*	6.1	-9.2	1.5	10.7	+
25 7-Ethoxyresorufin	6.0	-9.1	-0.7	8.4	+
26 MeIQx*	6.0	-9.2	2.0	11.2	+
27 Quinacrine	5.7	-9.4	0.3	9.7	+
28 Benzene	5.4	-13.9	4.0	17.9	+
29 α-Carboline	5.3	-10.0	2.8	12.8	+
30 MeIQ*	5.2	-9.2	2.0	11.2	+
31 IQ*	5.1	-9.2	2.3	11.5	+
32 Diacetylbenzidine	5.1	-10.0	2.4	12.4	+
33 2-Acetamidofluorene	5.0	-10.4	2.4	12.8	+
34 5,6-Dimethylchrysene	4.9	-9.8	1.3	11.1	+
35 Caffeine	4.8	-10.7	2.8	13.5	+
36 Theophylline	4.8	-10.7	2.9	13.6	+
37 Paracetamol	4.8	-10.7	3.7	14.4	+
38 Chlorpromazine	4.7	-10.0	2.1	12.1	+
39 Phenacetin	4.3	-10.6	3.8	14.4	+
40 Propranolol	4.2	-10.2	2.2	12.4	+
41 Phenothiazine	4.1	-10.0	2.6	12.6	+
42 7-Ethoxycoumarin	3.8	-10.9	1.9	12.8	+
43 Oestradiol	3.8	-11.0	3.8	14.8	-
44 Aniline	3.8	-12.4	3.9	16.3	-
45 Retinol	3.6	-8.4	0.3	8.7	-
46 Cholesterol	3.6	-11.6	4.4	16.0	-
47 Imidazole	3.6	-11.9	4.9	16.8	-
48 *p*-Xylene	3.5	-12.2	3.7	15.9	-
49 9-t-Butylanthracene	3.5	-9.3	0.7	10.0	-

Table 3 (continued)

50	Pregnenolone	3.4	-11.3	3.9	15.2	-
51	Lauric acid	3.3	-13.3	4.8	18.1	-
52	Aflatoxin B1	3.2	-10.8	1.1	11.9	+
53	Isoniazid	3.2	-12.1	2.3	14.4	-
54	Lignocaine	3.2	-11.2	3.6	14.8	-
55	Feprazone	3.2	-10.7	3.3	14.0	-
56	Lindane	2.9	-13.3	0.7	14.0	-
57	Perhydrofluorene	2.9	-12.4	6.4	18.8	-
58	Azoprocarbazine	2.9	-10.6	2.6	13.2	+
59	Safrole	2.9	-10.8	3.8	14.6	+
60	Pregnenolone 16α-carbonitrile	2.9	-11.6	2.9	14.5	-
61	Pyrazole	2.9	-11.5	4.9	16.4	-
62	Dimethylnitrosamine	2.9	-12.8	1.7	14.5	-
63	Butylated hydroxyanisole (BHA)	2.7	-10.5	3.7	14.2	-
64	Pentane	2.7	-14.1	6.9	21.0	-
65	Phenylbutazone	2.7	-10.7	3.3	14.0	-
66	2,5-Diphenyloxazole	2.7	-10.0	2.1	12.1	+
67	Arachidonic acid	2.6	-12.0	4.3	16.3	-
68	Lanosterol	2.6	-11.1	4.2	15.3	-
69	Warfarin	2.6	-10.9	1.7	12.6	-
70	Benoxaprofen	2.5	-10.8	1.2	12.0	-
71	Nitrosopiperidine	2.4	-12.7	1.9	14.6	-
72	SKF-525A*	2.4	-11.4	3.5	14.9	-
73	Butylated hydroxytoluene (BHT)	2.3	-11.0	3.8	14.8	-
74	Oleandomycin	2.3	-12.0	3.9	15.9	-
75	Phenylhexane	2.3	-12.7	3.8	16.5	-
76	Benzphetamine	2.3	-11.6	3.8	15.4	+
77	Dexamethasone	2.1	-11.2	1.7	12.9	-
78	Isosafrole	2.1	-10.4	3.1	13.5	+
79	*t*-Stilbene oxide	2.1	-11.2	3.5	14.7	-
80	Testosterone	2.0	-11.6	2.3	13.9	-
81	Acetone	2.0	-13.1	3.7	16.8	-
82	Ethylmorphine	2.0	-10.5	3.8	14.3	-
83	Ketoconazole	2.0	-11.2	2.0	13.2	-
84	Clofibrate	2.0	-11.7	3.0	14.7	-
85	Metyrapone	1.9	-11.9	2.4	14.3	-
86	Ethanol	1.8	-15.0	7.4	22.4	-
87	Hexobarbital	1.8	-12.2	3.6	15.8	-
88	Chlordane	1.7	-12.2	1.0	13.2	-
89	Phenylimidazole	1.6	-11.8	3.2	15.0	-
90	Monoethylhexyl phthalate	1.6	-12.8	1.8	14.6	-
91	Miconazole	1.3	-12.1	1.9	14.0	-
92	Rifampicin	1.3	-9.2	1.1	10.3	-
93	Ascorbic acid	1.2	-11.0	2.7	13.7	-
94	Camphor	1.2	-12.1	3.8	15.9	-
95	Allylisopropylacetamide	1.1	-12.1	4.9	17.0	-
96	Aldrin	1.1	-11.9	1.1	13.0	-
97	Phenobarbital	1.1	-12.5	3.3	15.8	-
98	Clotrimazole	1.0	-9.6	3.4	13.0	-
99	Phenytoin	1.0	-12.0	3.3	15.3	-
100	DDT*	0.8	-12.7	0.8	13.5	-

A Spatial parameter (area/depth^2)
B Electronic structure *E*(HOMO)
C Electronic structure *E*(LEMO)
D ΔE
E Cytochrome P-448 specificity

+ Substrates indicated as (+) interact with cytochromes P-448 but may also interact with other cytochrome P-450 proteins. Substrates indicated as (-) have either been shown not to interact with cytochromes P-448 or have not been studied.

* Abbreviations are as follows: 11. Trp-P-1, 3-amino-1,4-dimethyl-5*H*-pyrrido[4,3-*b*]indole; 15. TCDD, 2,3,7,8-tetrachlorodibenzo-*p*-dioxin; 19. cytembena, 3-*p*-anisoyl-3-bromoacrylic acid; 24. Glu-P-1, 2-amino-6-methyldipyrido[1,2*a*-3',2'-*d*]imidazole; 26. MeIQx, 2-amino-3,8-dimethylimidazo[4,5-*f*]quinoxaline; 30. MeIQ, 2-amino-3,5-dimethylimidazo[4,5-*f*]quinoline; 31. IQ, 2-amino-3-methylimidazo[4,5-*f*]quinoline; 72. SKF 525-A, proadifen, β-diethylaminoethyldiphenylpropyl acetate; 100. DDT, dichlorodiphenyltrichloroethane.

The most significant determinant for interaction with the cytochromes P-448, or with the Ah receptor, is the spatial parameter, a/d^2. Cytochrome P-448 substrates, inhibitors and inducers are planar, rigid molecules with values of a/d^2 of >4; substrates of other isozymes of the superfamily of cytochromes P-450 are generally globular, flexible molecules with a/d^2 values of about 1.0. Chemicals with a/d^2 values between 1 and 4 appear to be substrates for both the cytochromes P-448 and P-450.

As the cytochromes P-448 result in the metabolic activation of toxic chemicals, mutagens, and carcinogens, many chemicals with high a/d^2 values which are substrates of the cytochromes P-448 are predictably potent carcinogens and mutagens, *e.g.* dibenzanthracene (a/d^2 = 14.7), benzo[*a*]pyrene (12.0), dimethylaminoazobenzene (9.7), β-naphthylamine (7.9), and 3-methylcholanthrene (7.9). In contrast, typical substrates of the phenobarbital-inducible cytochromes P-450, which have no significant mutagenic or carcinogenic activity, have low values of a/d^2 *e.g.* hexobarbital (1.8), phenobarbital (1.1), phenytoin (1.0), and DDT (0.8).The food pyrolysis products, recently shown to be potent carcinogens, are all substrates of the cytochromes P-448 with high values of a/d^2, *e.g.* Trp-P-1 (8.5), Glu-P-1 (6.1), and IQ (5.1) (see also Table 3). Similarly, several drugs of known toxicity are seen to be substrates of the cytochromes P-448 with high values of a/d^2, *e.g.* ellipticine, a cytotoxic anticancer drug (8.5), paracetamol, hepatoxic at high dosage (4.8), and phenacetin, nephrotoxic at high dosage (4.3).

α-Naphthaflavone (a/d^2 = 13.7), β-naphthoflavone (12.9), and 9-hydroxyellipticine (7.0) are all potent inhibitors of the cytochromes P-448 but are not confirmed carcinogens, although β-naphthoflavone and 9-hydroxyellipticine are also potent inducing agents of the cytochromes P-448 and hence may be regarded as co-carcinogens. TCDD is the most potent inducer known of the cytochromes P-448 and, although a co-carcinogen, does not appear to be a carcinogen, probably because the presence of the four chlorine atoms makes it resistant to metabolic oxygenation.

Safrole (a/d^2 = 2.9) and isosafrole (2.1) are substrates of both the cytochromes P-450 and P-448 and, as might have been predicted, are weak carcinogens. Benzene (5.4) is a radiomimetic toxin and carcinogen, and is probably a substrate of both cytochromes P-450 and P-448. Similarly, aflatoxin B_1 (3.2) is a substrate of both major families of the cytochrome.

The electronic structure of a chemical also determines, to a limited extent, the type of cytochrome P-450 with which the chemical interacts. Low values of E(LEMO) and ΔE characterize the cytochrome P-448 substrates, whereas higher values of ΔE are characteristic of cytochrome P-450 substrates. For example, lauric acid (a/d^2 = 3.3, ΔE = 18.1) is a cytochrome P-450 substrate while aflatoxin B_1 (a/d^2 = 3.2, ΔE = 11.9) is a substrate of both cytochromes P-450 and P-448. Hence, both spatial and electronic parameters should be considered, as shown in Figure 1, in the prediction of the interaction of chemicals with the cytochromes P-448, and therefore their potential toxicity, mutagenicity, and carcinogenicity.

<u>3.2 QSARs of Chemical Structure with Receptor Affinity and Cytochrome P-448 Induction</u>. Correlations of the molecular dimensions of a series of polychlorinated biphenyls with their binding to rat TCDD cytosolic receptor (Ah receptor) and their induction of cytochrome P-448 (7-ethoxyresorufin *O*-deethylase activities) are shown in Figure 2. 3,3'4,4',5-Penta- (1) and 3,3',4,4'-tetra-chloro-biphenyl (2) are planar molecules with no chlorine atoms in the 2-positions, and have high a/d^2 values. 2,3,4,4'5-Penta- (3), 2,3,3',4,4'-penta- (4), 2,3,3',4,4',5'-hexa- (5), 2,3,3',4,4',5-hexa- (6), 2,3',4,4',5-penta- (7), 2',3,4,4',5-penta- (8), 2,3',4,4',5,5'-hexa- (9), 2,3,4,4'-tetra- (10), and 2,3,4,5-tetra-chlorobiphenyl (14), all with one chlorine atom in the 2-positions, are less planar and have lower a/d^2 values. Lastly, 2,2',4,4',5,5'-hexa- (11), 2,3',4,4',5',6-hexa- (12), and 2,2',4,4'-tetra-chlorobiphenyl (13) all have two chlorine atoms in the *ortho*-positions, and also are less planar.

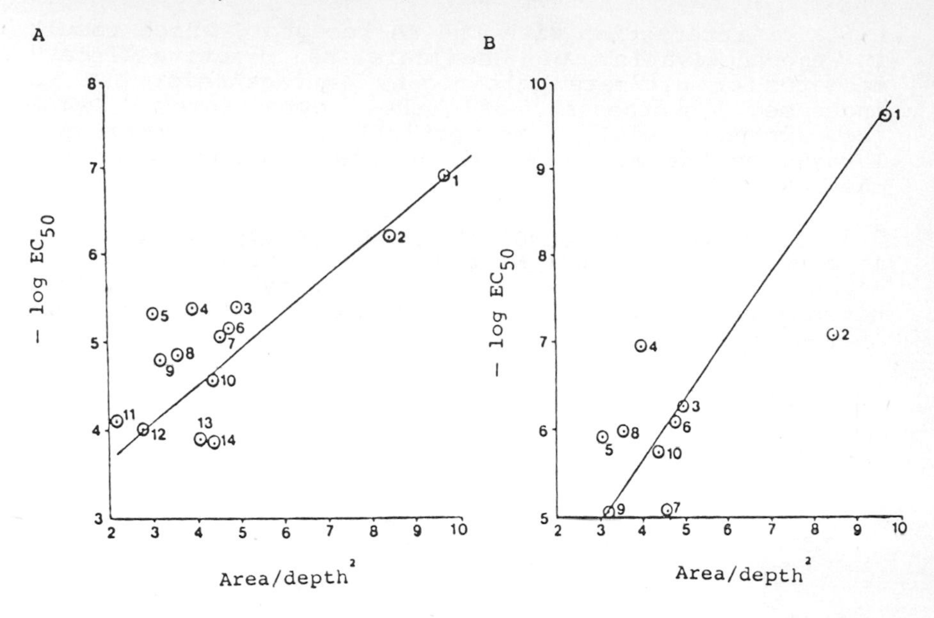

Figure 2 Correlation plot of molecular dimensions (a/d^2) of a series of polychlorinated biphenyls with (A) rat TCDD cytosolic receptor binding and (B) cytochrome P-448 induction as measured by increased 7-ethoxyresorufin deethylase (EROD) activity. Data for rat TCDD cytosolic receptor binding (-log EC_{50}) and for cytochrome P-448 induction (EROD) activity as -log EC_{50}) were taken from Safe et al.[36] Molecular dimensions were determined by computer graphics as described in Section 2. Identities of the PCBs are as follows: (1) 3,3′,4,4′,5-penta; (2) 3,3′,4,4′-tetra; (3) 2,3,4,4′,5-penta; (4) 2,3,3′,4,4′-penta; (5) 2,3,3′4,4′,5′-hexa; (6) 2,3,3′,4,4′,5-hexa; (7) 2,3′,4,4′,5-penta; (8) 2′,3,4,4′,5-penta; (9) 2,3′,4,4′,5,5′-hexa; (10) 2,3,4,4′-tetra; (11) 2,2′,4,4′,5,5′-hexa; (12) 2,3′,4,4′,5′,6-hexa; (13) 2,2′,4,4′-tetra; (14) 2,3,4,5-tetra.

Excellent correlations are shown between the molecular dimensions (a/d^2) of the 14 polychlorobiphenyls and their binding to the TCDD cytosolic receptor for the induction of the cytochromes P-448 (correlation coefficient = 0.80), and also with the induction of cytochrome P-448 as measured by increased 7-ethoxyresorufin *O*-deethylase activity (correlation coefficient = 0.83). This presents further evidence that acceptance at the active sites of the cytochromes

P-448 or interaction with the Ah receptor, which result in the activation of chemicals to reactive intermediates or ultimate carcinogens (mutagenesis) and to increased synthesis of the cytochromes P-448 (co-carcinogenesis), respectively, is determined largely by the molecular dimensions (planarity) of the chemical.

3.3 QSARs of Carcinogenicity with Electronic Structure in a Series of Dialkyl Nitrosamines. Correlation of the electronic structure of a series of dialkyl nitrosamines with their carcinogenic potential is shown in Figure 3.

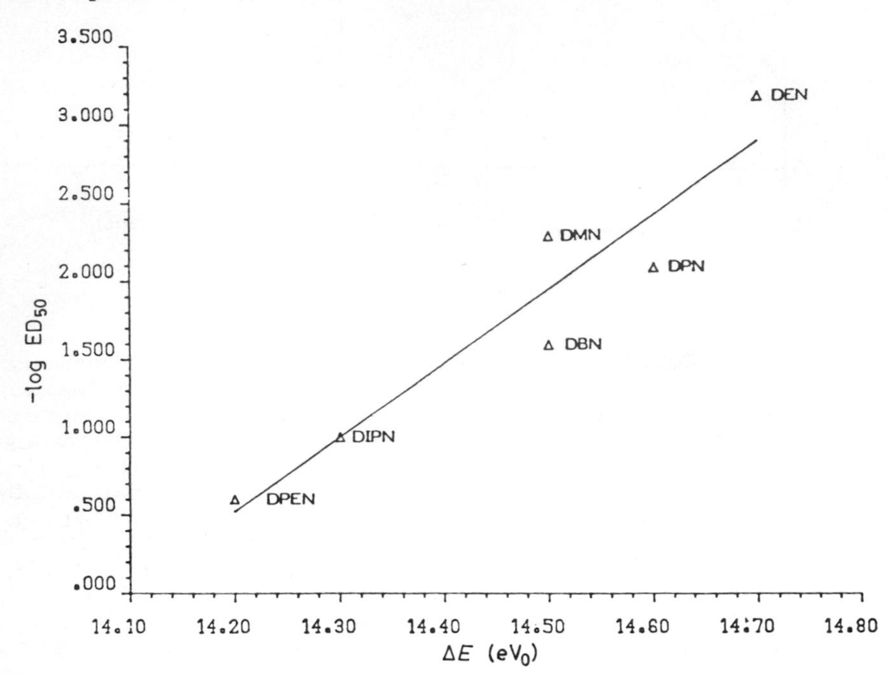

Figure 3 Correlation plot of electronic structures of a series of dialkylnitrosamines with their carcinogenesis in rat. The ED_{50} data are the mean total doses required to produce tumours in 50% of treated rats, expressed as mol kg^{-1} body weight, and are taken from Wishnok et al.[35] Data for ΔE values (E(HOMO)) - E(LEMO)) were determined by computer graphics as described in Section 2. The nitrosamines are identified as follows:
DMN, dimethylnitrosamine;
DEN, diethylnitrosamine;
DPN, di(n-propyl)nitrosamine;
DIPN, di(iso-propyl)nitrosamine;
DBN, di(n-butyl)nitrosamine;
DPEN, di(n-pentyl)nitrosamine

The mean total dose of nitrosamine required to produce tumours in 50% of exposed rats shows a linear relation (r = 0.95) with the ΔE (E(HOMO) - E(LEMO)) values for dimethyl-, diethyl-, di-n-propyl-, di-iso-propyl-, and di-n-pentyl-nitrosamines. Data for the carcinogenic potential of the nitrosamines is taken from Wishnok et al.[35] No correlation of carcinogenic potential with a/d^2 was observed.

This confirms the previous observations from the data in Table 3 that the electronic structure of a chemical, in addition to its spatial conformation, determines its carcinogenic/toxic potential.

4 Discussion

Conventional methods for the evaluation of chemical safety from studies in experimental animals are costly, time-consuming, fallible due to species differences in chemical toxicity, and increasingly unacceptable to society. Their supplementation or partial replacement by *in vitro* methods, including tissue and cell culture, has met with very limited success, again due to species differences in the enzymes involved, and also to the loss of physiological integrity and detoxication ability, which together generally result in a gross overestimate of potential toxicity (see Table 2). As most aspects of chemical carcinogenicity and many forms of chronic toxicity are the results of biological activation, with oxygenations by the cytochromes P-448 being among the most frequent mechanisms involved, an alternative QSAR approach to determe the potential toxicity/carcinogenicity of chemicals has been made based on their ability to interact with the active site(s) of the cytochromes P-448 and with the cytosolic Ah receptor which regulates this 'toxicating' enzyme. A similar, though much less esoteric, approach has been used for the past ten years in the assessment of the potential toxicity of new industrial chemicals, by the US EPA, although their procedure has been physicochemical and empirical instead of being scientifically based on known molecular mechanisms of chemical toxicity as in the method that has now been devised.

As the cytochrome P-448 enzymes and the Ah receptor, on which the QSAR computer graphic procedure is based, are present in humans, the approach is valid for determining the potential carcinogenicity/toxicity of chemicals in man, although no account is taken of genetic variations of these enzymes/receptor, so no allowance is made for the human variations in susceptibility to chemical toxicity/carcinogenicity. Human cytochromes P-448 are homologous to those present in rat, mouse and rabbit, and have been shown to be immunologically indistinguishable. Furthermore, the cytochromes P-448 in human liver and several other tissues have been shown to be inducible by the same agents which induce this family of enzymes in experimental animals. The advantages of the QSAR computer graphic approach are that it avoids the use of experimental animals which are less responsive than man (*e.g.* DBA mouse) or are hyper-responsive (*e.g.* mouse and hamster).

The present QSAR computer graphic method is based on the following observations:

(i) Most known instances of the carcinogenicity and chronic toxicity of chemicals are the results of the metabolic activation of these chemicals in the exposed individual, to 'ultimate carcinogens' and 'reactive intermediates'.[22]

(ii) Most cases of the metabolic activation of chemicals to reactive intermediates involve oxygenations at hindered positions, which are effected primarily by the cytochromes P-448, or by oxygen radicals.[26]

(iii) Substrates of the cytochromes P-448 are rigid planar molecules with molecular dimensions of $a/d^2 > 4$, and electronic structures with relatively small differences (ΔE) between E(HOMO) and E(LEMO) values.[28,29]

(iv) Structural requirements of the Ah receptor, which directly regulates the biosynthesis of the cytochromes P-448, and indirectly regulates nuclear chromatin, DNA replication, and mitosis, are the same or very similar to those of the active site of the cytochromes P-448.[29]

(v) Hence there are excellent correlations of chemical carcinogenicity/toxicity with metabolism by the cytochromes P-448, induction of the cytochrome P-448, and with the molecular dimensions ($a/d^2 > 4$) and electronic structure (low values of ΔE) of the chemical concerned.

In contrast, chemicals which are metabolized by the other major families of the cytochromes P-450 are oxygenated in unhindered positions of the molecule, are mostly conjugated and detoxified, and generally do not interact with the Ah receptor, do not affect DNA replication or mitosis, and are characterized by molecular dimensions in which a/d^2 approximates to unity and by electronic structures in which ΔE is high. However, the cytochrome P-450$_{ALC}$ (P450 II E) sub-family, which has a specific affinity for ethanol, benzene, aniline, *etc.*, activates the nitrosamines, which are not substrates of the cytochromes P-448.

As the cytochrome P-450 superfamily contains many different P-450 families,[23] there are a number of exceptions to these broad generalizations. For example, benzene is highly toxic to the haematopoietic system and gives rise to malignancy in human subjects in a greater number of different tissue sites than any other chemical carcinogen; it is metabolized by both a cytochrome P-450 (P450 II E) and by cytochrome P-448, and although 'hindered' oxygenation of benzene is not possible little is known of the pathways of the metabolic activation of benzene. Similarly, the dialkylnitrosamines, which are highly carcinogenic in rodents, are metabolized selectively by a cytochrome P-450 (P450 II E) and do not have the molecular dimensions or electronic structures (dimethylnitrosamine, $a/d^2 = 2.9$ $\Delta E = 14.5$ eV) characteristic of cytochrome P-448 substrates or chemical toxicity, nor do they induce the synthesis of cytochrome P-448. Ethanol is another similar case; it has been mooted as a human carcinogen, is highly cytotoxic at high concentrations, but is metabolized by a cytochrome P-450 (P450 II E) and does not have molecular dimensions or electronic structures characteristic of cytochromes P-448 substrates (ethanol), $a/d^2 = 1.8$, $\Delta E = 22.4$ eV).

The linear correlation of the induction of cytochrome P-448 with molecular conformation (a/d^2) in a series of polychlorinated biphenyls (see Figure 2) shows that molecular conformation is a major determinant in the interaction of chemicals with the Ah receptor, and similarly in their interaction with the active site of cytochrome P-448. The polychlorinated biphenyls may interact with cytochromes P-448 or with cytochromes P-450, according to their planarity. Polychlorinated biphenyls which are planar, that is have chlorine substituents only in the *para* and *meta* positions, induce the cytochromes P-448, whereas those which are non-planar, that is have chlorine substituents in the *ortho* positions, induce the cytochromes P-450.

In the case of the dialkylnitrosamines, which are probably all activated by the same cytochrome P-450 (P450 II E), no correlation of carcinogenic potential with molecular conformation (a/d^2) was observed. Instead a linear correlation was observed with the electronic structures (ΔE) of the nitrosamines (see Figure 3). The carcinogenicity of the nitrosamines could possibly be a function of their oxygenation, and as ΔE is related to the energy of activation of the chemicals it will be related to their ease of interaction with the active site of the cytochrome P-450, and thus to their ease of oxygenation. Hence, with this series of toxic chemicals which are all activated by the same cytochrome, carcinogenicity is a function of electronic structure.

In conclusion, this novel QSAR computer graphic approach, developed from fundamental concepts of conformation biology and the known mechanisms of chemical carcinogenicity/ toxicity, offers an effective alternative, or at least a valid adjunct, to animal experimentation in the safety assessment of chemicals.

5 Acknowledgements

The authors gratefully acknowledge the financial support of their computing facilities by the Humane Research Trust, and the support of Dr. D.F.V. Lewis, a Senior Research Fellow, by Food and Veterinary Laboratories Ltd., Guildford, Surrey, UK. Patents are pending on the computer graphic procedure (UK Patent No. 8727572).

6 References

1. World Health Organization, 'Principles and Methods for Evaluating the Toxicity of Chemicals, Part 1'. Environmental Health Criteria 6, WHO, Geneva, 1978.

2. World Health Organization, 'Principles for the Safety Assessment of Food Additives and Contaminants in Food', Environmental Health Criteria 70, WHO, Geneva, 1987.

3. W.H.W. Inman, 'Monitoring for Drug Safety', MTP Press, Lancaster, UK, 1980.

4. Organization for Economic Co-operation and Development, 'Guidelines for Testing of Chemicals', (1981 and continuing series), OECD, Paris, 1981.

5. K. Snell, 'Developmental Toxicology', Academic Press, London, 1982.

6. International Agency for Research on Cancer, 'Evaluation of the Carcinogenic Risk of Chemicals to Humans', IARC Monograph, Suppl. 4, International Agency for Research on Cancer, WHO, Lyons, 1982.

7. M. Balls, R.J. Riddell, and A.M. Worden, 'Animals and Alternatives in Toxicity Testing', Academic Press, London, 1983.

8. European Chemical Industry Ecology and Toxicology Centre, 'Risk Assessment of Occupational Chemical Carcinogens', ECETOC Monograph No.3, European Chemical Industry Ecology and Toxicology Centre, Brussels, 1982.

9. W.B. Clayson, D. Krewski, and I. Munro, 'Toxicological Risk Assessment', Vol. 2, General Criteria and Case Studies, CRC Press, Boca Raton, Fl, 1985.

10. D.R. Laurence, A.E.M. McLean, and M. Weatherall, 'Safety Testing of New Drugs: Laboratory Predictions and Clinical Performance', Academic Press, London, 1984.

11. J. Ashby, F.J. de Serres, M. Draper, M. Ishidate Jnr., B. Margolin, B. Matter, and M.D. Shelby, 'Evaluation of Short-Term Tests for Carcinogens: Report of the International Progress on Chemical Safety Collaborative Study on *In Vitro* Assays', Elsevier, Amsterdam, 1985.

12. J.H. Dean, A.E. Munson, M.I. Luster, and H.E. Amos, 'Toxicology of the Immune System, Target Organ Toxicity Series', Raven Press, New York, 1985.

13. A. Worden, D. Parke, and J. Marks, 'The Future of Predictive Safety Evaluation', Vols. 1 and 2, MTP Press, Lancaster, UK, 1986.

14. A.J. Clarke, B. Clark, C.T. Eason, and D.V. Parke, *Reg. Toxicol. Pharmacol.*, 1985, **5**, 109.

15. M.V. Roloff, 'Human Risk Assessment. The Role of Animal Selection and Extrapolation', Taylor and Francis, London, 1987.

16. R.E. Kouri, H. Ratry, and C.E. Whitmire, *Int. J. Cancer*, 1974, **13**, 714.

17. D.N. Bateman and S. Chaplin, *Br. Med. J.*, 1988, **296**, 761.

18. D.M. Brown, P.F. Langley, D. Smith, and D.C. Taylor, *Xenobiotica*, 1974, **4**, 151.

19. J.K. Haseman and J.E. Huff, *Cancer Lett.*, 1987, **37**, 125.

20. F.J. Di Carlo, *Drug Metab. Rev.*, 1984, **15**, 409.

21. Committee of Safety of Medicines, 'Medicines Act Information Letter No. 25', Department of Health and Social Security, London, 1979, p.2.

22. D.V. Parke, *Arch. Toxicol.*, 1987, **60**, 5.

23. D.W. Nebert and F.J. Gonzalez, *Ann. Rev. Biochem.*, 1987, **56**, 945.

24. D.R. Nelson and H.W. Strobel, *Mol. Biol. Evol.*, 1987, **4**, 572.

25. C. Ioannides and D.V. Parke, *Biochem. Pharmacol.*, 1987, **36**, 4197.

26. C. Ioannides, P.Y. Lum, and D.V. Parke, *Xeniobiotica*, 1984, **14**, 119.

27. D.F.V. Lewis, *Drug. Metab. Rev.,* 1986, **17**, 1.

28. D.V. Parke, C. Ioannides, and D.F.V. Lewis, *Fed. Eur. Soc. Toxicol.*, 1986, 14.

29. D.F.V. Lewis, C. Ioannides, and D.V. Parke, *Biochem. Pharmacol.*, 1986, **35**, 2179.

30. D.F.V. Lewis, C. Ioannides, and D.V. Parke, *Chem. Biol. Interact.*, 1987, **64**, 39.

31. D.W. Nebert and F.J. Gonzalez, *Trends Pharmacol. Sci.*, 1985, **6**, 160.

32. R.C. Bingham, M.J.S. Dewar, and D.H. Lo, *J. Am. Chem. Soc.*, 1975, **97**, 1285.

33. D.F.V. Lewis, *Chem. Rev.*, 1986, **86**, 1111.

34. L.E. Sutton, 'Tables of Interatomic Distances and Configurations in Molecules and Ions', Special Publication No. 18, The Chemical Society, London, 1965.

35. J.S. Wishnok, M.C. Archer, A.S. Edelman, and W.M. Rand, *Chem. Biol. Interact.*, 1978, **20**, 43.

36. S. Safe, S. Bandiera, T. Sawyer, B. Zmudzka, G. Mason, M. Romkes, M.A. Denomme, J. Sparling, A.B. Okey, and T. Fujita, *Environ. Health Persp.*, 1985, **61**, 21.

5
Risk Assessment of Chemicals: A Global Approach

M. Mercier

INTERNATIONAL PROGRAMME FOR CHEMICAL SAFETY, WORLD HEALTH ORGANIZATION, PALAIS DES NATIONS, CH-1211 GENEVA 10, SWITZERLAND

1 Introduction

One may think of life on this planet as a constant flow of waves, *i.e.* the propagation of the species - after one starts the previous one stops. Each generation brings forth its own peculiarities; *e.g.* one that is of current concern in the present generation is the 'risk culture'. This culture has produced and continues to produce a profusion of literature dealing with an array of concepts which weighs the benefits of an unprecedented explosion of scientific and technological advances against accompanying disadvantages.

There are two elements in the risk culture which may be difficult to grasp. The first is the terminology in current use in the literature (see below), and the second is the drive. In this context is the impelling acquired concern, which appears to derive its motivation from the desire to enjoy the benefits without upsetting the fundamentals of the preservation of life. Not all the intentions are clear. In effect, one has the impression that in some instances the drive for preservation relates more to the perpetuation of the 'selfish gene' than to the protection of the overall wave formation through which life flows. The zero-risk concept, together with the practical measures which may be derived from it, has the appearance of supporting a generation's unreasonable wish to live forever.

This chapter describes the process of risk assessment in the field of chemical safety as it has been and is carried out at the World Health Organization, and specifically within the International Programme on Chemical Safety (IPCS), a co-operative programme of the United Nations Environmental Programme (UNEP), the International Labour Office (ILO), and the World Health Organization (WHO). It also attempts to elaborate a rational basis for a global approach as a contribution to international harmonization within the field of chemical testing and also for deriving support from scientific safety evaluations or 'risk assessment'. The WHO expert committees, Joint Expert Committee on Food Additives (JECFA) and Joint Meeting on Pesticide Residues (JMPR), have probably contributed more to the elaboration of sound national food regulations than any other international bodies aimed at harmonizing often divergent national approaches to the problem of risk from ingestion of food chemicals, and related problems of food safety, food technology, and food control. JECFA and JMPR achieved this by providing recommendations based on scientific evidence and by establishing a rational model of risk and safety assessment that is widely acknowledged and accepted.

Hundreds of highly skilled international specialists have given, and continue to give, their time and talents freely to foster advances in toxicological methodologies and analytical procedures, to consolidate accessible presentation of data, and to keep abreast with scientific developments, which often require revision of previous conclusions. Via reports, toxicological monographs, and profiles of chemical specifications for identity and purity prepared by JECFA and JMPR, national food regulatory authorities and the Codex Alimentarius Commission are provided with all the necessary elements for making the best decisions on the rational use of chemicals in food and agriculture.

In 1980, all JECFA and JMPR activities related to the safety and risk assessment of food additives and pesticide residues were administratively and technically incorporated into the IPCS. Since then IPCS has developed a keen interest in all aspects pertaining to the toxicological evaluation of food additives, contaminants, and pesticide residues in food, including methodological aspects for testing and safety assessment of chemicals used in food. These activities are part of the responsibilities of IPCS insofar that its objectives include the formulation of guiding principles for exposure limits, such as Acceptable Daily Intakes (ADIs) for food additives and pesticide residues and acceptable levels for toxic substances in food, air, water, soil, and the working environment.

1.1 The JECFA/JMPR Model of Safety Evaluation. This model is illustrated in Figure 1. It has been derived from a critical analysis of the *modus operandi* of JECFA and JMPR expert committees and it represents the result of many years of experience in the field of safety evaluation of a variety of food chemicals. The model has been tested against the usefulness and practical applicability of an end-point of safety assessment known as the Acceptable Daily Intake (ADI) for man, or the amount of a chemical, expressed in milligrams per kilogram body weight (mg kg^{-1} b.w.) that can be taken daily in the diet, even over a lifetime, without risk.

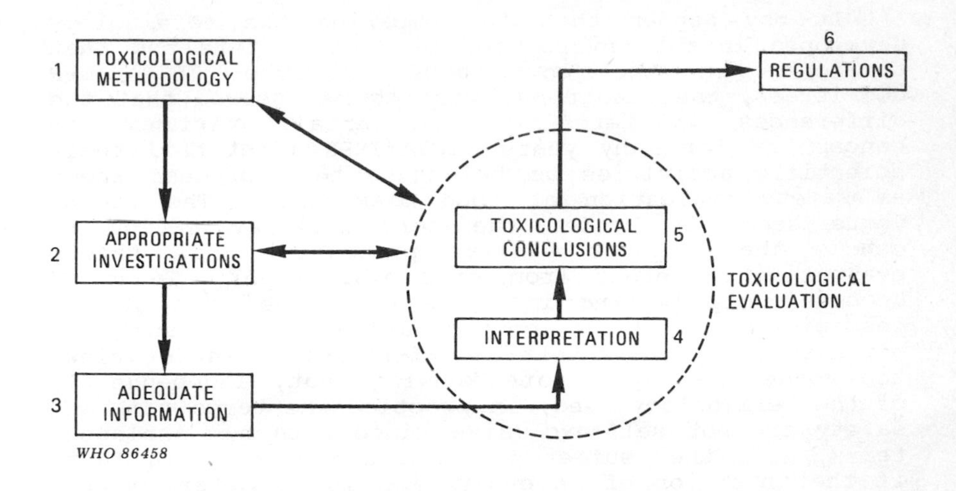

Figure 1 Flow diagram identifying the critical points and objectives of toxicological assessment of xenobiotics

This model of safety evaluation is regarded as a complex process having a dynamic character since new scientific evidence may at any time challenge the results of a previous evaluation. This is particularly true in regard to the safety assessment of those chemicals which, because of their nature of their expansion of use, need to be kept under periodic review. Figure 1 is interpreted as follows: the toxicological methodology (1) leads to the design of appropriate investigations (2) which should supply adequate information (3) which, after proper interpretation (4) will assist in the formulation of toxicological conclusions (5), thus providing a rational basis for regulations (6) on the safe exposure to the chemical in question. Steps 4 and 5 may be or may not be partially combined, since the interpretation of the results obtained by the investigator(s) who conducted the scientific work would have to be taken into account by the group or individuals charged with responsibility of formulating the toxicological conclusions. Steps 4 and 5 thus represent the phase of the process denoted as safety evaluation of a chemical.

The complexity of the overall evaluative process may often require the adoption of administrative measures designed to ensure the continuous awareness of the parties interested in the production or use of a particular chemical. The adoption of the concept of the temporary ADI and the related provisions for submission of additional information are two examples of such administrative overtones.

It may happen that in comparing the terminology developed in the environment of risk culture and that developed in the environment of the JECFA/JMPR Committees, the reader may wish to be assured that the differences in terms do not entail variance in concepts. For many years JECFA/JMPR identified their scientific activities as belonging to a process known as safety evaluation of food chemicals. The recent vogue seems to prefer the term risk evaluation to denote the same or similar processes. Is safety evaluation different from risk evaluation? To avoid unconsciously falling into the Byzantine trap of the pessimist's (risk = bottle-half-empty) versus the optimist's (safety = bottle-half-filled) point of view, the reader may appreciate knowing that, independently of the terminology used, in practice the terms risk and safety are not self-exclusive since both are analogous terms, *i.e.* they suffer pluses and minuses. In fact, if the intention of an evaluation is to determine the dose of a chemical which does elicit toxic effects it will suffice to determine this dose (risk evaluation). If, in turn, the scope of an evaluation is that of establishing the safety of a compound then, after having identified the dose causing concern, it will be necessary to ascertain if a dose does exist, at which the chemical does not exert any toxic effect; if this is the case, and this dose is found to be compatible with the level of exposure and proves to be effective for the intended use at this level, then the chemical may be accepted as safe at the level of use *i.e.* (safety evaluation).*

However, there may be cases where the available data do not permit a clear differentiation between a toxic and a safe dose, *e.g.* when there is an ambiguity between negative and positive findings, then considerations other than toxicological ones may be brought into play (risk assessment and management).

It is by keeping in mind the latter situation that few persons would disagree that absolute risk is less of a myth than absolute safety; this is confirmed by the natural tendency (evident in most safety evaluation processes) to rely more readily on positive than negative results.

*Note the terms hazard, risk (and associated terms), and safety are included in the Glossary of Terms at the end of this book.

<u>1.2 The JECFA/Codex Committee on Food Additives (CCFA) and JMPR/Codex Committee on Food Residues (CCPR)</u>. The results of safety or risk evaluations may be academic exercises if they are not illustrated in societal frames. As it has been rightly stated, to make a decision on any particular safety issue three different factors must be considered. *(i)* There is a need to know the potential effects and results of the hazardous situation or event under consideration, *i.e.* the consequences must be evaluated. *(ii)* There is a need for information on the likelihood that this event will occur *i.e.* estimate the risk. *(iii)* it is necessary to know how to judge the results of the deliberations, *i.e.* criteria of acceptability. Unless all three factors are appreciated no decision can ever be justified or defended (see Hogh, 1987).

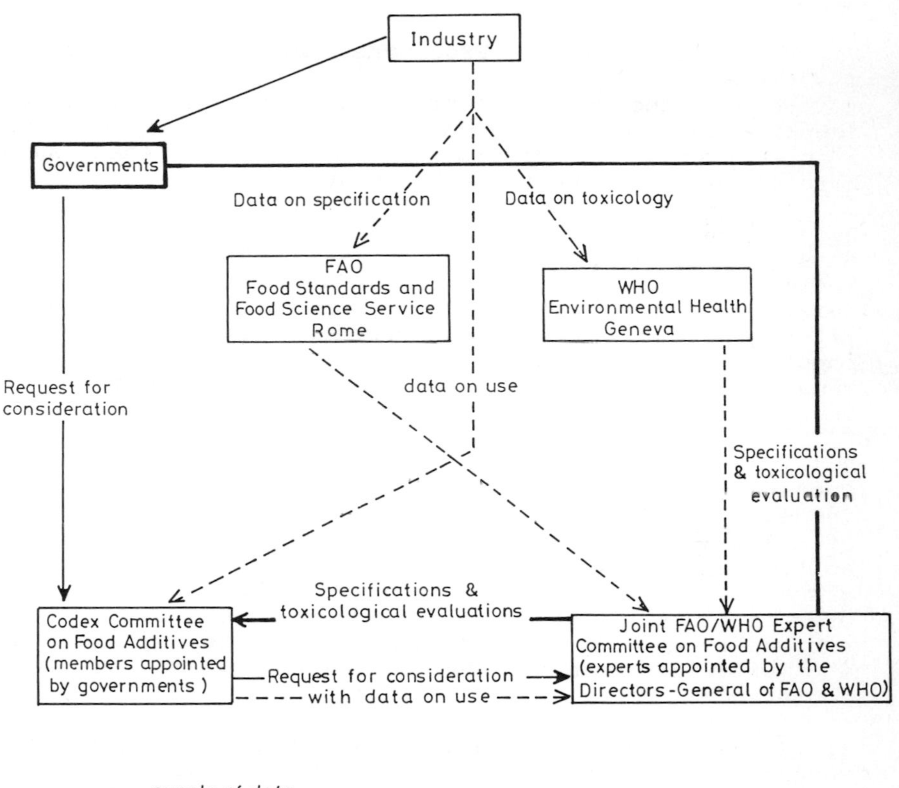

Figure 2 Operational model developed by JECFA and CCFA

In the global environment of food chemicals these philosophical principles are applied through the Joint Food and Agriculture Organization (FAO)/WHO Expert Committee on Food Additives/Codex Committee on Food Additives (JECFA/CCFA and the Joint FAO/WHO Meeting on Pesticides Residues/Codex Committee on Pesticide Residues (JMPR/CCPR) systems. The JECFA/CCFA and the JMPR/CCPR systems have been functioning since 1963 and have proved to be of value in providing the right framework for credibility and international acceptance: JECFA/JMPR international independent scientific-technical committees serving as the advisory bodies to CCFA/CCPR which, in turn, are international political-administrative organizations endeavouring to harmonize the legislation of Member States on food chemicals. Figure 2 illustrates the international operative model which combines the two concepts of risk assessment and risk management.

2 The International Programme on Chemical Safety (IPCS)

It is the purpose of a hazard assessment to identify the harmful effects that could arise, whereas a risk assessment determines the chances that such harmful effects will arise in practice.

2.1 Genesis. From 1972 onwards significant events outside the JEFCA/CCFA and JMPR/CCPR systems have influenced the international outlook in the field of safety and risk evaluation of chemicals. These were: *(i)* the advent of the environmental impact, *(ii)* the popularization of toxicological sciences, and *(iii)* the managerial shifts or adjustments triggered by innovative concepts.

2.1.1 The environmental impact. The United Nations Conference on the Human Environment held in Stockholm in 1972 called for a combined approach to the effects of chemicals and environmental factors in air, water, soil, and the working environment. This unified and global approach was reflected quickly in the organizational changes which then took place both nationally and internationally in order to implement on a worldwide basis the recommendations of the conference.

2.1.2 Popularization of the toxicological sciences. During the past two decades the toxicological sciences have developed into many specialized branches reflecting the complexity of the unified approaches of the health assessment of chemicals and environmental factors in air, water, and food, especially when these factors are considered in relation to the human target. The unified concept of 'total body burden' became the underlying objective of any safety or risk assessment of chemicals. At the operational level, the multi-disciplinary approach required to attain the objectives created the need to integrate programme areas into fully comprehensive programmes. The trend towards integration inevitably demanded organizational changes at both national and international levels.

2.1.3 Innovative concepts. These concepts take into account the fast changes occurring in the modern world environment requiring the need for a balanced contribution of social, political, and technical roles between peoples and their governments, between individual countries, and between countries and groups of countries, in order to adjust structure-function relationships of existing managerial frameworks.

These novel concepts draw their innovative character from their practical execution rather than from their theoretical formulation, since there was not enough experience for their implementation. However, they accounted both for the integration of on-going environmental programmes into more comprehensive programmes, and also for the formulation of close co-operation among the countries themselves, under a global organization's directing, co-ordinating, and technical co-operation functions. The end result was the development of comprehensive co-operation environmental programmes.

In the field of international chemical safety and risk a comprehensive co-operative programme was instituted and named the International Programme on Chemical Safety (IPCS).

2.2 Historical Sketch and Objectives. The considerable international concern expressed by many national representatives at the Stockholm Conference on the dangers of chemicals to humanity and to the natural environment encouraged the Conference to recommend that programmes, guided by WHO, should be undertaken for the early warning and prevention of harmful effects of the various environmental agents, acting both singly and in combination, to which humans were being exposed increasingly, directly or indirectly, and for the assessment of the potential risk to human health.

As a specialized agency for health in the United Nations system, WHO has a mandate from its Member States to address all the factors which have an impact on human health, and this includes chemicals. At that time, WHO was promoting health, based on balanced social and economic development.

In 1977, the World Health Assembly, the governing body of the World Health Organization, requested the Director-General to study the problem of long-term strategies to control and limit the impact of chemicals on human health and the environment. On this basis IPCS was developed and structured by WHO. The interest of other international organizations in chemical safety was clearly demonstrated by the ILO and the United Nations Environment Programme (UNEP), joining with WHO in the IPCS, which was formally launched in 1980 when a Memorandum of Understanding was signed between the three organizations.

The IPCS is now a co-operative programme of the United Nations Environmental Programme (UNEP), the International Labour Office (ILO), and the World Health Organization (WHO). WHO is the executing agency for the programme and the Central Unit of IPCS is located in the WHO Division of Environmental Health in Geneva, Switzerland.

The overall objectives of IPCS are to catalyse and co-ordinate activities in relation to chemical safety.

In order to ensure efficient use of resources and integration of the results, IPCS works closely with other international and WHO programmes which are also involved in the area of safe use of chemicals. Examples are collaboration with the Council for Mutual Economic Assistance (CMEA), the Organization for Economic Research and Development (OECD), the Commission of the European Communities (CEC), and the WHO programmes on environmental pollution, occupational health, safe use of pesticides, and food safety. There is close collaboration with the Food and Agriculture Organization of the United Nations (FAO) for the joint safety evaluation of food additives and contaminants, pesticide residues in food, and veterinary drug residues in food of animal origin.

2.3 Global Approach in Risk Assessment. Specific chemicals and groups of chemicals are currently reviewed for toxicity, risk, and safety by IPCS. Seventy-five Environmental Health Criteria Documents have been published, ten are in press, and thirty-nine are in various stages of preparation (see Appendix I). Of those published there are fifty-five monographs on chemicals, ten on toxicological and safety methodology, and eight on physical hazards. Among the chemicals reviewed or under review are economically important chemicals as well as natural environmental contaminants.

Safety considerations related to human health and the environment (including risk assessment) may affect in varying degrees the production, availability, and international trade of many chemicals. Therefore it is paramount that the conclusion reached in this field be based on sound data, good science, and credible risk and safety evaluations.

The proliferation of groups undertaking toxicological evaluations and safety assessments of chemicals is today a fact that is looked upon with concern by national regulatory authorities, as well as industry, since the end-points of these evaluations often diverge from one another. Such discrepancies could be due to different interpretation of data, but generally they arise because of a difference in the data available for the risk of safety evaluation. This is also poor use of a scarce expertise. Internationally reputed expert groups recently issued recommendations asking for the establishment of mechanisms to effect a better liaison between the various evaluative groups in order to ensure a greater degree of uniformity of their conclusions (see WHO, 1978).

The IPCS, by promoting a global approach in safety and risk assessment of chemicals represents a mechanism whereby a higher degree of uniformity and harmonization may be achieved among the various national and international assessment groups. The role of IPCS is illustrated in the following sections.

<u>2.3.1 Toxicological methodology</u>. Methodology is the subject of considerable controversy among different experts. In experimental toxicology, the controversy considers topics such as the most suitable animal species to be utilized in any particular investigation, the number of animals in the experiment, dose range, and the duration of a toxicological test. These related aspects are discussed and recommendations are issued in the IPCS Environment Health Criteria (EHC) documents (methodology series) (see Appendix II).

By presenting an updated set of considerations and recommendations developed by reputed international experts, the IPCS is promoting global harmonization of methodology of testing chemicals for safety.

<u>2.3.2 Appropriate investigations</u>. While there is sufficient consensus on the principle that experimental tests should simulate the route of human exposure, there is still no universal agreement on the type and numbers of tests necessary to establish conclusively the risk and safety of a chemical. The international expert committees of JECFA and JMPR suggests the following series of tests as useful for risk and safety evaluation of food chemicals: biochemical studies, metabolism, effects on enzymes, acute toxicity, reproduction studies, carcinogenicity, mutagenicity, and teratogenicity studies, sub-chronic and chronic toxicity tests, and observations in human beings. Additional recommendations for other tests for chemicals in general are contained in the set of

methodological documents listed in Appendix II, particularly in EHC Nos 6, 57, 60, and 70.

2.3.3 Adequate information. The qualitative and quantitative adequacy of data is essential for safety and risk evaluation. In spite of the fact that no general agreement exists on the value attributed to the unpublished literature, IPCS agrees to review unpublished documents in the course of risk and safety evaluations of chemicals whenever the information contained in these documents is relevant for the evaluation and when assurance is given on its validity. In toxicology this literature is represented by reports of original studies developed in the majority of cases by commercial testing laboratories and commissioned by the industry seeking registration of its products. The unpublished literature used in the assessment of chemicals is peer reviewed by the experts participating in task groups who are responsible for the final evaluation. By reviewing and summarizing valid unpublished literature on risk and safety of chemicals IPCS contributes to international harmonization in the acceptance by one country of studies conducted in another country.

2.3.4 Interpretation. This is probably the most controversial criterion in experimental toxicology. What is the value of statistical versus biological significance in determining the seriousness of an observed effect? What interpretation should be given to an effect observed in laboratory animal species or strains when it has been demonstrated that the animal species or strain in question metabolizes the chemical in a different fashion from man? These and many other similar questions which are debated in scientific meetings, symposia, congresses, and round-tables find their place in several of the EHCs in the methodological series previously mentioned.

Considerable controversy exists on the value to be attributed in the course of a risk and safety evaluation to *in vitro* mutagenicity tests and on their assumed role in establishing the carcinogenic potential of a chemical. Recommendations and advice to help in this scientific and technical predicament can be found particularly in EHC Nos. 46, 47, and 51.

2.3.5 Toxicological conclusions. In this area the most sensitive issue revolves on the practical significance of extrapolating from animal data to man. The long-standing experience, for example, of utilizing the concept of the Acceptable Daily Intake (ADI) and its worldwide acceptance clearly indicate that this index of safety has played and still plays a significant role in the harmonization of toxicological conclusions with regard to food additives, food contaminants, and pesticide residues in food as well as veterinary drugs residues in food. IPCS is sponsoring and actively supporting technically and scientifically the international committees of JECFA and JMPR, which are charged with the responsibility of evaluating chemicals in food. Similarly, IPCS agreed to follow the same model in elaborating drinking water quality guidelines.

2.3.6 Regulations. Toxicological conclusions resulting from the evaluation model indicated in Figure 1 are solely concerned with the levels of a chemical acceptable or tolerated by the human body. These are not directly relevant to the establishment of acceptable or tolerable levels in the environment (water, food, air, or soil). The transcription of levels acceptable by the body into levels acceptable in the environment is one of the most difficult tasks to undertake. Harmonization in this sector goes beyond the exclusive domain of toxicology since it lies at the interface between the science and the law. Science and, consequently, toxicology and the law operate in territories which are often diametrically opposed since science thrives in uncertainties while the law thrives in certainty. Currently, the harmonization in this area may only advance by developing formulae for a *modus vivendi*.

2.4 Rational Basis for Harmonization. This chapter acknowledges that current understanding of harmonization is difficult. The difficulty arises in part from the complexity of interests affected by regulatory decisions when acceptable or tolerable levels of chemicals are proposed or established.

Regardless of possible divergent views on the significance, need, and type of harmonization in the fields of safety and risk evaluations of chemicals, it may be of some use to attempt to identify certain parameters which may contribute to clarification of the issue. This may help lead the way to the building of a rational basis for harmonization. The following parameters are proposed: *(i)* a potential harmonizing group may wish to acquire a working knowledge of what other assessment groups are doing or propose to do. *(ii)* A potential harmonizing group may wish to achieve an in-depth perception of all the reasons underlying the conclusions reached by other groups. *(iii)* A potential harmonizing group may wish to exercise a critical judgement on whether that which has been decided by another group represents the best decision.

Finally, harmonizing groups seeking a global approach in solving problems related to risk and safety assessments of chemicals may find it convenient and profitable to join forces with the International Programme on Chemical Safety.

3 Reading List

F. Bro-Rasmussen 'Hazard and Risk Assessment of Chemicals', IPCS Working paper, International Programme on Chemical Safety, WHO Geneva, Switzerland, 1987.

'Risk Assessment and Risk Management of Toxic Substances', A Report to the Secretary, Department of Health and Human Services (DHHS) from the Executive Committee of the DHHS Committee to Co-ordinate Environmental and Related Programs (CCERP), DHHS, Washington, DC, 1985.

FAO/WHO, Rome/Geneva, 'General Principles Governing the Use of Food Additives', 1st Report of JECFA, *WHO Technical Report Series,* 1957, No. 129.

FAO/WHO, Rome/Geneva, 'Procedures for the Testing of Intentional Food Additives to Establish their Safety for Use', 2nd Report of JECFA, *WHO Technical Report Series,* 1958, No. 144.

H.W. Gottinger, 'Hazard: An Expert System for Risk Assessment of Environmental Chemicals', *Meth. Inform. Med.* 1987, **26**, 13-23.

M.S. Hogh, 'Cost-effective Safety', *BP Quarterly Bulletin on Health, Safety and the Environment*, 1987, **3**, 1-3.

IPCS, 'Glossary of Terms to be Used in IPCS Publications', International Programme on Chemical Safety, WHO, Geneva, Switzerland, 1986, (in press).

IPCS, 'Report of the IPCS Technical Review Meeting to Compare CMEA and OECD Approaches to Toxicity Testing and Risk Assessment', International Programme on Chemical Safety, WHO, Geneva, Switzerland, May 11-16, 1987.

D. Krewski and E. Somers, 'Risk Assessment and the Control of Toxic Chemicals in the Environment', Paper presented at the Joint CMEA/IRPTC/IPCS Seminar on Methodology of Optimal Use of Internationally-Prepared Health Risk Evaluations, Moscow, USSR, November 19-30, 1984.

L.B. Lave, 'Health and Safety Risk Analysis: Information for Better Decisions', *Science*, 1987, **136**, 291-295.

T.R. Lee, 'The Public's Perception of Risk and the Question of Irrationality', Paper presented at a meeting on the Assessment and Perception of Risk at the Royal Society, London, 12-13 November, 1980.

I.C. Munro, and D.R. Krewski, 'Risk Assessment and Regulatory Decision Making', Food Cosmet. Toxicol., 1981, **19**, 549-560.

D. Okrent, 'Comments on Societal Risk', *Science*, 1980, **208**, 372-375.

'Toxic Hazard Assessment of Chemicals', ed. M.L. Richardson, The Royal Society of Chemistry, London, 1986.

M. Russell and M. Gruber, 'Risk Assessment in Environmental Policy-making', *Science*, 1987, **236**, 286-290.

P. Slovic, 'Perception of Risk', *Science*, 1987, **236**, 280-285.

E. Smith, 'The International Programme on Chemical Safety: An Overview', IPCS, WHO, Geneva, Switzerland, 1987.

E. Somers, 'Making Decisions from Numbers', *Regul. Toxicol. Pharmacol.*, 1987, **7**, 35-42.

R.A. Squire, 'Human Risk Assessment from Animal Data', in "The Pesticide Chemist and Modern Toxicology", American Chemical Society, Washington, 1981, pp.493-501.

C. Starr, and C. Whipple, 'Risks of Risk Decisions', *Science*, 1980, **208**, 1114-1119.

G. Vettorazzi, 'Xenobiotic Substances: Harmonization of Toxicological Conclusions', *Regul. Toxicol. Pharmacol.*, 1988, (in press).

G. Vettorazzi, 'Advances in the Safety Evaluation of Food Additives', *Food Add. Contam.*, 1987, **4**, 331-356.

WHO, Copenhagen, 'Risk Assessment', Interim Document No.6. World Health Organization, Regional Office for Europe, Copenhagen, 1982.

WHO, Geneva, 'Evaluation of Certain Food Additives and Contaminants', 22nd Report of JECFA, *WHO Technical Report Series*, 1978, No.631.

R. Wilson and E.A.C. Crouch, 'Risk Assessment and Comparison: An Introduction', *Science*, 1987, **236**, 267-270.

Appendix 1

International Programme on Chemical Safety

Health and Safety Guide Series

Guides have been published as follows:

Acrylonitrile	(1986)
Kelevan	(1987)
Methylene Chloride	(1987)
Tetrachloroethylene	(1987)
1-Butanol	(1987)
2-Butanol	(1987)
2,4-D	(1987)
Epichlorohydrin	(1987)
t-Butanol	(1987)
Isobutanol	(1987)
Tetradifon	(1987)

Further Guides are in preparation on the following: Chlordane, Heptachlor, Quintozene, Tecnazene, Endosulfan, Ethylene Oxide, Propylene Oxide, Dimethoate, Aldrin and Dieldrin, Phosphorus Trichloride, Phosphorus Oxychloride, Phosphine, Ammonia, Dimethyl Sulphate, Dichlorvos, Cypermethrin, Hydrazine, 1,2-Dichloroethane, Magnetic Fields, Diaminotoluene, Toluene Diisocyanates, Toluene, Camphechlor, Mirex, Chlordecone, Hexachlorobenzene, PCBs/PCTs, Resmethrins, Allethrins, Tetramethrin, *d*-Phenothrin, Deltamethrin, Permethrin, Fenvalerate, 1-Propanol, 2-Propanol, Phenol, Dimethylformamide, Propachlor, Isobenzan, Endrin, Xylenes, Ethylbenzene, Acetaldehyde, Acroleine, Hexachlorobutadiene, Hexachlorocyclopentadiene, Aldicarb, Malathrin, Fenitrothrin, Trichlorfon, Chlorophenols, Chlorobenzenes, Selected Glycol Ethers.

Appendix 2

International Programme on Chemical Safety

Environmental Health Criteria Series

Published issues:

1	Mercury	(1976)
2	Polychlorinated Biphenyls and Terphenyls	(1976)
3	Lead	(1977)
4	Oxides of Nitrogen	(1977)
5	Nitrates, Nitrites, and *N*-Nitroso compounds	(1977)
6	*Principles and Methods for Evaluating the Toxicity of Chemicals, Part 1	(1978)

7	Photochemical Oxidants	(1978)
8	Sulfur Oxides and Suspended Particulate Matter	(1979)
9	DDT and its Derivatives	(1979)
10	Carbon Disulfide	(1979)
11	Mycotoxins	(1979)
12	Noise	(1980)
13	Carbon Monoxide	(1979)
14	Ultraviolet Radiation	(1979)
15	Tin and Organotin Compounds	(1980)
16	Radiofrequency and Microwaves	(1981)
17	Manganese	(1981)
18	Arsenic	(1981)
19	Hydrogen Sulfide	(1981)
20	Selected Petroleum Products	(1982)
21	Chlorine and Hydrogen Chloride	(1982)
22	Ultrasound	(1982)
23	Lasers and Optical Radiation	(1982)
24	Titanium	(1982)
25	Selected Radionuclides	(1983)
26	Styrene	(1983)
27	*Guidelines on Studies in Environmental Epidemiology	(1983)
28	Acrylonitrile	(1983)
29	2,4-Dichlorophenoxyacetic Acid (2,4-D)	(1984)
30	*Principles for Evaluating Health Risks to Progeny Associated with Exposure to Chemicals during Pregnancy	(1984)
31	Tetrachloroethylene	(1984)
32	Methylene Chloride	(1984)
33	Epichlorohydrin	(1984)
34	Chlordane	(1984)
35	Extremely Low Frequency (ELF) Fields	(1984)
36	Fluorine and Fluorides	(1984)
37	Aquatic (Marine and Freshwater) Biotoxins	(1984)
38	Heptachlor	(1984)
39	Paraquat and Diquat	(1984)
40	Endosulfan	(1984)
41	Quintozene	(1984)
42	Tecnazene	(1984)
43	Chlordecone	(1984)
44	Mirex	(1984)
45	Camphechlor	(1984)
46	*Guidelines for the Study of Genetic Effects in Human Populations	(1985)
47	*Summary Report on the Evaluation of short-Term Tests for Carcinogens (Collaborative Study on *In Vitro* Tests)	(1985)
48	Dimethyl Sulfate	(1985)
49	Acrylamide	(1985)
50	Trichloroethylene	(1985)
51	*Guide to Short-Term Tests for Detecting Mutagenic and Carcinogenic Chemicals	(1985)
52	Toluene	(1986)

* Methodology document (yellow cover)

53 Asbestos and Other Natural Mineral Fibres (1986)
54 Ammonia (1986)
55 Ethylene Oxide (1985)
56 Propylene Oxide (1985)
57 *Principles of Toxicokinetic Studies (1986)
58 Selenium (1986)
59 *Principles for Evaluating Health Risks from Chemicals During Infancy and Early Childhood: The Need for a Special Approach (1986)
60 *Principles and Methods for the Assessment of Neurotoxicity Associated With Exposure to Chemicals (1986)
61 Chromium (in press)
62 1,2-Dichloroethane (1987)
63 Organophosphorus Insecticides - A General Introduction (1986)
64 Carbamate Pesticides - A General Introduction (1986)
65 Butanols - Four Isomers (1987)
66 Kelevan (1986)
67 Tetradifon (1986)
68 Hydrazine (1987)
69 Magnetic Fields (1987)
70 *Principles for the Safety Assessment of Food Additives and Contaminants in Food (1987)
71 Pentachlorophenol (1987)
72 *Principles of Studies on Diseases of Suspected Chemical Etiology and Their Prevention (1987)
73 Phosphine and Metal Phosphides (in press)
74 Diaminotoluenes (1987)
75 Toluene Diisocyanates (1987)

* Methodology document (yellow cover)

Further issues are in preparation on the following: Thiocarbamate Pesticides - A General Introduction, Dithiocarbamate Pesticides - A General Introduction, Dichlorvos, Cypermethrin, Pyrrolizidine Alkaloids, TCDD, Vanadium, Dimethoate, Aldrin and Dieldrin, Man-Made Mineral Fibres, Mycotoxins, Allethrins, Resmethrins, Tetramethrin, *d*-Phenothrin, Deltamethrin, Permethrin, Fenvalerate, 1-Propanol, 2-Propanol, Vinylidene Chloride, Formaldehyde, Dimethylformamide, Cadmium, Isobenzan, Propachlor, Chlorofluorocarbons, PCBs/PCTs, Phenol, Chlorobenzenes, Hexachlorobenzene, Chlorophenols, Aldicarb, Hexachlorocyclopentadiene, Photochemical Oxidants (Update), Sulfur Oxides and Suspended Particulate, Matter (2nd edition), Selected Glycol Ethers, Phthalates, LAS.

Appendix 3

WHO New Water Quality Guidelines

ANNEX 1

Organics

i) *Chlorinated organics*

(a) Chlorinated alkanes: carbon tetrachloride[1]; 1,2-dichloroethane[1]; 1,1,1-trichloroethane[3]; dichloromethane[3].

(b) Chlorinated ethenes: 1,1-dichloroethene[1]; 1,2-dichloroethene[3]; trichloroethene[1]; tetra-chloroethene[1]; vinyl chloride[2].

ii) *Non-chlorinated solvents*

Benzene, lower alkyl-benzenes, and vinylbenzene: styrene[2]; toluene[2]; xylene[2]; ethylbenzene[2].

iii) *Polynuclear aromatic hydrocarbons*[1]

iv) *Chlorination by-products including trihalomethanes*

(a) Trihalomethanes: chloroform[1]; bromoform[2]; dichlorobromomethane[2]; dibromochloromethane[2].

(b) Chlorophenols: 2,4,6-trichlorophenol[1].

(c) Other chlorination reaction products[2].

v) *Pesticides*

Aldrin/dieldrin[1]; chlordane[1]; 2,4-D[1]; DDT[1]; γ-HCH (lindane)[1]; heptachlor and heptachlor epoxide[1]; methoxychlor[1]; atrazine[2]; simazine[2]; alachlor[5]; bentazon[5]; MCPA[5]; metalachlor[5]; molinate[5]; pendimethalin[5]; propanil[5]; pyridate[5]; trifluralin[5]; pyrethroids[4]; ethylene dibromide[4]; DBCP[4]; 1,2-dibromo-3-chloropropane[4]; 1,3-dichloropropane[4]; 1,2-dichloropropane[4]; 1,3-dichloropropene[4]; aldicarb[4]; carbofuran[4].

vi) *Miscellaneous*

Acrylamide[2]; plasticizers, diethylhexylphthalate and diethylhexyladipate[3]; hexachlorobutadiene[4]; epichlorohydrin[4]; EDTA[4]; NTA[4]; amines[4].

1. Compound with a guideline value; recommended for re-evaluation at Water Research centre (WRc) (Medmenham) consultation.

2. Compound with no guideline value; recommended for consideration with **high priority** at WRc (Medmenham) consultation.

3. Compound with no guideline value; recommended for consideration with **medium priority** at WRc (Medmenham) consultation.

4. Compound with no guideline value; recommended for consideration with **low priority** at WRC (Medmenham) consultation; only to be reviewed if documentation readily available or can be produced at low cost.

5. Evaluated at one of two consultations sponsored by EURO in Rome in 1987; not given a priority rating at WRc (Medmenham) consultation.

Inorganics

i) *Heavy metals, fluorides, sodium, etc.*: aluminium, ammonia; arsenic; asbestos; barium; beryllium; cadmium; chromium; cyanide; fluoride; hardness; iodine; lead; mercury; nickel; selenium; silver; sodium.

ii) *Nitrate and nitrite*

Aesthetics (all re-evaluations)

Aluminium; chloride; colour; copper; hardness; hydrogen sulphide; iron; manganese; oxygen, dissolved; pH level; sodium; sulfate; taste and odour; temperature; total dissolved solids; turbidity; zinc.

ANNEX II

Possible structure of summaries of the available information on organic constituents, which will be published in revised Volume 2 of WHO Water Quality Guidelines. Summaries of inorganic constituents could also use many of these headings. It is not expected that all the headings would be used in each case.

1. General aspects, including:
    - 1.1 Source
    - 1.2 Occurrence in water

2. Routes of exposure
    - 2.1 Water
    - 2.2 Food
    - 2.3 Air
    - 2.4 Skin
    - 2.5 Relative significance of different routes of exposure

3. Health effects
    - 3.1 Biochemical aspects including:
        - 3.1.1 Absorption
        - 3.1.2 Distribution
        - 3.1.3 Biotransformation
        - 3.1.4 Elimination

3.2 Acute, subchronic and chronic toxicity
3.3 Carcinogenicity
3.4 Mutagenicity
3.5 Teratogenicity
3.6 Human studies

4. Proposed Guideline Value, including justification

5. References

Remarks: i) the summary will be 2-5 pages in length;
ii) support documentation should be provided to serve as a basis for evaluation by the review group;
iii) re-evaluation of compounds already included in Volume 2 will only include data reported since the last review.

6
Reasoning About Risk

S. Jasanoff

CORNELL UNIVERSITY, PROGRAMM ON SCIENCE, TECHNOLOGY AND SOCIETY, ITHACA. NY 14853, USA

1 Introduction

Current approaches to assessing the risks of chemicals in the environment in both industrialized and industrializing countries reflect nearly two decades of scientific and policy development. A technological determinist might reasonably expect international risk assessment policies to have converged during this period as a result of advances in such fields as analytical chemistry, toxicology, epidemiology, and biostatistics.[1] This chapter suggests, however, that although some convergence has indeed occurred, substantial differences still remain in the methods and procedures by which different countries assess risk. The absence of an internationally accepted terminology for environmental risk management highlights the lack of uniformity.[2] Cross-national divergences can be attributed to differences in the legal and institutional arrangements for formulating regulatory policy, as well as the goals that different national authorities seek to promote through risk assessment.

The primary focus of this chapter is on significant evolutions in the strategy of risk assessment adopted by the United States. During the 1980s, there has been a marked retreat, at least on the part of US federal agencies*, from the no-risk philosophy that previously dominated certain areas of risk management, most notably programmes for controlling environmental carcinogens. The quest for greater flexibility in coping with such hazards is reflected in a variety of federal risk assessment policies, such as the revised guidelines for carcinogenic risk assessment issued by the Environmental Protection Agency (EPA) in 1986.

*In the US context, agencies of the national government are called 'federal agencies'. Agencies belonging to the fifty state governments are called 'state agencies'.

EPA has also taken the lead in developing a number of techniques that enable regulators to focus on those risks that are genuinely worth regulating and to set aside those that society can either ignore or treat as matters for individual decision-making. Concepts that have loomed as increasingly significant in the US policy environment include comparative risk assessment, *de minimis* risk, and risk communication.

This chapter argues that US risk assessment policies have moved closer to those of other comparably industrialized countries in two respects: a lowered emphasis on generic policies and an increased willingness to exclude small risks from the regulatory agenda. Nevertheless, a number of important differences persist. On the whole, the US risk assessment process remains, as before, more highly formalized, quantitative, and inclined to use conservative assumptions than that of most other countries. These points are illustrated by means of contrasts with risk assessment practices in Britain and India. Moreover, a more flexible outlook on risk decisions at the level of US federal policy-making appears to have been offset in some instances by a heightened perception of risk among state agency officials and the general public. Because of these residual differences, US decisions with regard to specific environmental pollutants can still be expected to diverge in many instances from those adopted by other countries.

2 Moderating Influences on US Policy

In the aftermath of the oil crisis, double-digit inflation, and President Reagan's landslide victory, US regulators began the 1980s with a discernibly more sceptical attitude towards the need for strict environmental standards than was evident in the previous decade. Deregulation became the prevailing philosophy of government. The Office of Management and Budget (OMB) emerged as an increasingly powerful advocate of the new administration's determination to limit the economic impacts of regulation. Cutbacks in staff and funding placed pressure on the major federal agencies to narrow their regulatory agendas.

By the time the Reagan administration took office, federal policies for assessing the risks of chemicals had already developed into a focus of scientific and political controversy. Particularly contentious were the attempts by some of the most important federal agencies, *e.g.* the Environmental Protection Agency (EPA), the Occupational Safety and Health Administration (OSHA), the Consumer Product Safety Commission (CPSC), and the Interagency Regulatory Liaison group (IRLG), to develop guidelines for the regulation of carcinogens. The US chemical industry perceived these efforts as scientifically misguided and economically unsupportable, a view that was wholeheartedly endorsed by most European chemical companies as well. In early 1981, a sympathetic new administration responded by suspending OSHA's generic cancer policy, potentially the most draconian chemical control initiative of the Carter years. The IRLG was disbanded, and the Office of Science and Technology Policy (OSTP) took the lead in an effort to bring greater scientific credibility and uniformity to the assessment of carcinogenic risks.

Scientific developments also stimulated a reconsideration of the risk assessment strategies adopted by federal regulators in the 1970s. Analytical techniques capable of detecting trace amounts of chemicals, sometimes in the parts per trillion range, dispelled the notion that any environmental medium, such as drinking water, could be shown to be absolutely risk-free. Concurrently, burgeoning testing programmes in industry and government generated information implicating a growing roster of substances as potentially hazardous to health. It became clear that American society could not afford simply to eliminate all human exposure to every suspected environmental health hazard. Federal decision-makers turned to more sophisticated forms of risk analysis in order to separate trivial risks from those that merited serious regulatory concern.

Advances in cancer research also proved extremely significant for the developing field of risk assessment. The 'one-hit' model of carcinogenesis was displaced by an emerging scientific consensus that cancer was the result of a multi-stage process in which chemicals could be implicated in a variety of different ways (*e.g.* as initiators or as promoters). These findings, coupled with a growing awareness of inter-species differences in the response to environmental carcinogens, suggested that existing agency guidelines for assessing carcinogenic risk might be in need of considerable refinement.

3 New Approaches

The major modifications in federal risk assessment policy during the 1980s can be seen largely as a response to the complex of scientific, economic, and political pressures outlined above. Within EPA, in particular, there was a notable increase in high-level administrative support for risk assessment, and a corresponding increase in the resources committed to the agency's varied risk assessment programmes. These policies signalled a significant maturation in the agency's thinking with respect to the assessment and management of environmental hazards.

3.1 Revised Guidelines. In September 1986, EPA issued five guidelines for assessing the health risks of environmental pollutants. They covered, respectively, carcinogen risk assessment, exposure assessment, mutagenicity risk assessment, assessment of developmental toxicants, and assessment of chemical mixtures. The carcinogenic risk assessment guidelines[3] represented a particularly notable achievement. They were the first revision of the 'interim' guidelines in use at EPA since 1976 and reflected nearly three years of intensive staff work and scientific review within the agency.

As a principled attempt to bridge uncertainty and to combine science with policy, EPA's guidelines for assessing carcinogenic risk have been a focus of heated controversy for more than a decade.[4] For many years, the process of assessing carcinogens was tightly controlled by the agency's Carcinogen Assessment Group (CAG), a unit of the Office of Research and Development (ORD). Between 1976 and 1983, CAG assessed the carcinogenic potential of some 150 compounds, including such major items on EPA's regulatory agenda as arsenic, benzene, coke oven emissions, and vinyl chloride. In describing CAG's monopoly over risk assessment at EPA, one agency official has remarked that 'it was not unlike American influence at the end of World War II'.[5]

Over the years, however, CAG's philosophy of risk assessment came to be perceived as unduly rigid and resistant to change. Pressure mounted both inside and outside the agency to weaken CAG's hold on the risk assessment process and, at the same time, to make what some viewed as long overdue revisions in the agency's approach to regulating carcinogens. Soon after the Reagan administration took office, for example, high-level EPA officials attempted to rewrite the agency's carcinogen assessment policy by drawing a controversial distinction between 'genotoxic' and 'epigenetic' compounds.[6] But this effort which was widely seen as a relaxation of existing regulatory standards, had to be abandoned in the face of charges that the agency was changing an established policy without adequate scientific or public review.[7,8]

In 1984 an agency-wide task force began revising the carcinogen risk assessment guidelines with much greater attention to public consultation and scientific peer review. But, although the new guidelines were carefully drafted and deliberated, they had to overcome a significant political hurdle before they could be published in final form. Like any major regulatory initiative, the guidelines were subject to review by OMB's cost-conscious and generally adversarial Office of Information and Regulatory Affairs (OIRA). After months of delay, OIRA raised a number of objections to the newly reformulated guidelines and asked EPA to make some further revisions to the text. The agency ultimately succeeded in dislodging the guidelines from the OIRA bottleneck and issued them without substantial modification, but not until Congress, increasingly impatient with OIRA's regulatory review process, threatened to cut off funding for the unit.[9]

The 1986 carcinogenic risk assessment guidelines can thus be seen as the product of a lengthy scientific and political learning process. A comparison of some key provisions in the 1986 document with earlier versions of the guidelines shows how the agency's views of risk assessment changed during this period.

The relative lengths of the 1976 interim guidelines and the revised 1986 version provide an instructive, though crude, measure of change. Whereas the earlier document occupied fewer than four pages in the *Federal Register*, including explanatory remarks by the EPA administrator and an appendix on economic impact assessments, the latter runs to three times that length. Most of this difference can be attributed to changes in the scientific basis for risk assessment during the decade intervening between the two policy statements. The 1986 guidelines display on the whole much more sensitivity to the possibility that new data will necessitate case-by-case deviations from general principles, and they acknowledge the need to incorporate all available information into risk assessments.

In a number of instances, the 1986 guidelines treat possibilities that were either unknown or not explicitly addressed ten years earlier. Thus, the 1976 cancer policy made no mention of the special problems involved in interpreting animal data obtained at very high exposure doses. In contrast, the 1986 guidelines state that positive studies at levels above the maximum tolerated dose (MTD) 'should be carefully reviewed to ensure that the responses are not due to factors which do not operate at exposure levels below the MTD'; further, evidence that tumour responses at high exposures may have been triggered by indirect mechanisms 'should be dealt with on an individual basis.'[10]

The classification system for carcinogens set forth in the 1986 guidelines is also substantially different from that outlined in the 1976 interim policy. The earlier guidelines presented at best a skeletal approach to making 'weight of evidence' determinations for potential human carcinogens. EPA implied that the primary factors involved in making such judgements were 'considerations of the quality and adequacy of the data and the kinds of responses induced by the suspect carcinogen.'[11] Thus, epidemiological studies, in conjunction with confirmatory animal tests, were viewed as the 'best' evidence of possible human carcinogenicity. Positive evidence from animal tests alone was regarded as either 'substantial' or 'suggestive', depending on the nature of the tumours observed in the test animals.

By 1986, EPA's weight of evidence guidelines reflected far greater awareness both of possible complexities in the available information and of developments outside the United States. The classification system currently used by EPA is an overt adaptation of that established by the International Agency for Research on Cancer (IARC) in evaluating human and animal evidence of carcinogenicity.[12] In describing this system, moreover, the US agency indicated that it was not to be applied 'rigidly or mechanically':

> 'At various points in the above discussion, EPA has emphasized the need for an overall, balanced judgement of the totality of the available evidence. Particularly for well-studied substances, the scientific database will have a complexity that cannot be captured by any classification scheme'.[12]

Again, this language emphasizes the need for individualized judgement based on all the evidence, an approach that contrasts sharply with earlier attempts by US agencies to exclude evidence on particular chemicals rather than exempt them from the application of general risk assessment guidelines.[13]

<u>3.2 Quantitative Risk Assessment</u>. EPA's early experiences with regulating environmental carcinogens, especially pesticides, led to a correspondingly early interest in the use of quantitative risk estimation methods.[14] The 1976 guidelines, endorsed the use of varied extrapolation models in estimating cancer risks to humans. They were, however, largely silent about the considerations that should drive the selection of such models in particular cases.

The caution and respect for data that characterize the 1986 guidelines as a whole are particularly apparent in the discussion of mathematical models for high- to low-dose extrapolations. The agency still subscribes to CAG's policy of using a linearized multistage model 'in the absence of evidence to the contrary'.[15] But the guidelines also provide that relevant biological and statistical evidence will be reviewed in selecting a suitable model and that EPA will state a justification for the model actually selected.

Most noteworthy, perhaps, are the explicit indications that in individual cases the data may necessitate a departure from the linearized multistage model. For example, the agency acknowledges that this model may be inappropriate in cases where the observed data 'are non-monotonic or flatten out at high doses' or when pharmacokinetic, metabolic, or mechanistic data support the use of a different model.

The current EPA guidelines admit that the risk estimates produced by the linearized multistage model represent merely plausible upper limits and that the 'true' risk could be much lower, maybe even as low as zero. This observation appears to confirm the chemical industry's frequent complaint that EPA uses overly conservative assumptions in its risk assessments, *i.e.* assumptions that systematically overstate the degree of risk. Supporters of the guidelines in and out of the agency, however, see such conservatism as a defensible response to scientific uncertainty and one that is warranted by EPA's mission to protect public health. As one analyst has noted, the guidelines encourage the use of conservative assumptions only by default when the data support no other approach to risk assessment: 'The use of such assumptions has a higher hurdle to jump than previously'.[16] From this perspective, the caveats expressed about the linearized multistage model are significant, because they indicate the agency's willingness, at least in theory, to set this conservative procedure aside in favour of one that is based on available data and produces a more accurate estimate of risk. This point will be considered below in connection with EPA's recent assessment of the health risks of formaldehyde.

3.3 Preliminary Consequences. To what extent are the foregoing changes in EPA's philosophy of risk assessment reflected in risk estimates associated with particular toxic substances? A systematic answer to this question obviously lies beyond the scope of this chapter. Two recent health risk assessments carried out by EPA (dioxin and formaldehyde) suggest, however, that refinements in the theory and practice of risk assessment may indeed have major repercussions on the agency's appraisal of the risks associated with some pervasive environmental contaminants.

3.3.1 Dioxin (2,3,7,8-tetrachlorodibenzo-*p*-dioxin). Few man-made chemicals have generated as much public concern in the United States in recent years as dioxin. In New York State, reports that a new waste incinerator would emit minute quantities of dioxin into the air triggered vehement public protest despite expert reassurances that the facility would pose no significant risk to health. In Missouri, the discovery of dioxin contamination forced the federal government to buy out property-owners in the entire town of Times Beach. Across the country, hundreds of lawsuits were brought to recover damages for health injuries allegedly caused by dioxin, including the much publicized action by Vietnam veterans against the manufacturers of Agent Orange.

Scientific opinion about the risks presented by dioxin, however, has been far from uniform, even within the federal regulatory establishment. EPA concluded in its 1985 Health Assessment Document that dioxin was a probable human carcinogen, but EPA's risk estimates for the compound differed by one or two orders of magnitude from those of other federal agencies and foreign governments.[17] Some experts were convinced that these differences resulted from a failure to make full use of the available scientific information. Michael Gough, a biologist formerly employed by the congressional Office of Technology Assessment (OTA), summed up these views in his widely cited study of dioxin:

> 'The absence of detectable harm in exposed people has reduced the level of concern about dioxin. On the other hand, high estimates of cancer risk derived from the EPA's and other agencies' extrapolations of risk from animal tests still maintain the level of concern about environmental exposures higher than is justified. The extrapolations are based on what many scientists widely perceive to be misinterpretations of the mechanism by which dioxin causes cancer'.[18]

Well aware of such scepticism within the independent scientific community, EPA in 1986 appointed an expert panel to report on current findings concerning the health impacts of dioxin, including its mechanism of action and other factors bearing on risk assessment.[19] The panel's recommendations led, in turn, to a formal review of the agency's dioxin risk assessment and the preparation of a revised study by the end of 1987. Preliminary reports indicate that EPA has sharply reduced its estimate of dioxin's cancer risk in this draft report, although the compound remains the most toxic substance regulated by the agency.[20] The revised estimate reflects, at least in part, a new sensitivity to mechanistic considerations. In particular, the new risk assessment reportedly incorporates EPA's current belief that dioxin is non-genotoxic and hence probably acts only as a promoter of cancer. At the time of preparing this chapter, however, the draft risk assessment still had not undergone external peer review.

3.3.2 Formaldehyde. From the standpoint of risk assessment, formaldehyde, like dioxin, has proved to be a frustrating compound for US regulatory agencies. Although formaldehyde unquestionably causes nasal cancer in rats, it does not produce a statistically significant increase in tumours in mice, and epidemiological studies have failed to establish a definitive link between formaldehyde and human cancer. Moreover, the sharp non-linearity of the dose-response curve in rats has led observers to speculate that exposure at high doses triggers a secondary mechanism of cancer induction in this species that may not be implicated at lower doses or in other species. Lack of knowledge about this mechanism has been a primary source of uncertainty in agency efforts to select a mathematical model for low-dose extrapolation.

EPA's efforts to develop a credible risk assessment for formaldehyde followed upon several years of scientific, political, and legal controversy about the appropriate way to manage this ubiquitous compound.[21–24] During this period, one attempt to regulate formaldehyde on the basis of a quantitative risk assessment was overturned by a federal court.[25] In 1983, an international workshop convened to review the state of scientific knowledge concerning formaldehyde failed to produce agreement about a model for use in quantitative risk assessment. EPA, it was clear, could not reasonably hope for an emerging scientific consensus to ratify its approach to assessing formaldehyde's carcinogenic potential.

While EPA and other regulatory agencies pondered what to do about formaldehyde, the scientific picture with respect to the substance continued to change. By 1986, a number of new epidemiological studies, including a US survey of 20,000 workers exposed to formaldehyde, had been completed. At the same time, researchers at the Chemical Industry Institute of Toxicology (CIIT) had carried out a line of research with potentially more significant consequences for risk assessment: an investigation of the pharmacokinetic properties of formaldehyde.[26] Focusing on formaldehyde-DNA interactions in the nasal lining of rats, the CIIT studies sought to measure the dose delivered to the target cells. CIIT researchers contended that this delivered dose, rather than the dose actually administered to the test animals, should form the basis for high- to low-dose extrapolation in EPA's risk assessment methodology.

In the face of continuing technical and political uncertainty, EPA sought to legitimize its risk assessment strategy through partly scientific and partly procedural means. In its draft risk assessment completed in 1985, EPA elected not to incorporate

CIIT's pharmacokinetic data on the ground that these results had not yet been sufficiently validated. The draft, however, was reviewed by a sub-committee of the agency's Science Advisory Board (SAB), which concluded that EPA should have given fuller consideration to the published findings of the CIIT researchers. In response, EPA convened a special expert panel to review whether the available pharmacokinetic information on formaldehyde could reasonably have been factored into a risk assessment. This panel agreed with EPA that the CIIT studies, including the methodology for measuring the delivered dose, were not yet scientifically ripe for incorporation into risk assessments. Armed with this support, EPA completed its assessment of formaldehyde on the basis of the linearized multistage model, as prescribed by its risk assessment guidelines.

EPA's critics see these events as illustrative of a general problem: that the conservatism of the agency's risk assessment guidelines impedes scientific progress by failing to respond to significant new research results. From a public health perspective, however, EPA's decision not to use the emerging pharmacokinetic data on formaldehyde can be seen as prudent policy. Since even the scientific community could not agree that this information was well validated, there were good grounds for continuing to use the conservative assumptions of the multistage model by default. Consultation with the SAB and the pharmacokinetics panel also helped guard the agency against charges of scientific arbitrariness. The example demonstrates, however, that the revised risk assessment guidelines still leave room for considerable policy debate, especially around the question of when conservative risk assessment assumptions should be set aside in response to new scientific information. This issue is sure to be reopened as EPA proceeds with its planned revision of the risk assessment guidelines in 1988.

4 Risk Management Considerations

As US agencies, with EPA in the forefront, have begun to modify their approaches to risk assessment, opinion has also begun to change concerning the ways in which risk estimates should be incorporated into public policy. This section describes two new developments in the field of environmental risk management, one at the federal level and one at the state level, in order to draw some preliminary conclusions about the probable impacts of risk assessment on regulatory policy and the public's perception of risk.

4.1 *De minimis* Risks. One of the most noteworthy developments in the field of environmental risk management in the past decade is the growing support for defining a *de minimis* level below which risks should not be regulated. At common law, the *de minimis* concept was invoked to bar claims that the courts regarded as too trivial to merit a remedy (*de minimis non curat lex* or 'the law does not concern itself with trifles'). US courts have applied the *de minimis* principle more broadly to prevent the regulation of insignificant risks by government agencies. The Supreme Court, for example, ruled in 1980 that the Occupational Safety and Health Administration (OSHA) could not tighten the workplace standard for benzene unless the agency could show that workers were exposed to a 'significant risk' at existing exposure levels.[27] In a similar vein, the US Court of Appeals for the District of Columbia Circuit held that the Food and Drug Administration (FDA) could not regulate plastic beverage containers on the basis of a purely theoretical argument that trace quantities of the packaging material might migrate into the liquid.[28]

Beginning in the 1980s, both FDA and EPA have attempted to incorporate the *de minimis* principle into their regulatory policies, particularly those relating to suspected carcinogens. Both agencies have accepted as a general policy that risks lower than one in one million are insignificant for regulatory purposes.[29,30] To date, some of the most visible decisions not to regulate risks below this *de minimis* threshold have emanated from FDA. In 1986, for example, FDA sought to establish an explicit *de minimis* exception to the Delaney clause by permitting the use of several colour additives that were shown to induce cancer in animal tests (a strict interpretation of the Delaney clause would prohibit adding even one molecule of a suspected carcinogen to food). Quantitative analysis suggested that the risks posed by the additives in question were negligibly small.[31] But a reviewing court overturned the FDA initiative on the ground that Congress intended the Delaney clause to serve as an absolute prohibition against carcinogenic dyes in food.[32]

While EPA applies no uniform *de minimis* risk policy across its varied regulatory programmes, the concept is relevant to the agency's standard-setting practices in several areas. Thus, in establishing 'recommended maximum contaminant levels' (RMCLs) for carcinogens under the Safe Drinking Water Act, EPA relies, where necessary, on quantitative risk assessments based on animal bioassay data. Risk levels for RMCLs established in this way have ranged between one in a million and one in one hundred thousand. EPA presumably regards these risks as insignificant *(de minimis)* from the standpoint of public health.

Another area where a *de minimis* policy may come into play is the regulation of hazardous air pollutants under the Clean Air Act. Here, EPA must comply with a recent court decision that requires the agency to establish a 'safe' level of exposure to such substances, including such non-threshold pollutants as carcinogens. Presumably, the agency will still be free to ignore risks that are shown to be insignificant. EPA has indicated that it will use a new risk assessment method in making the necessary determinations of safety and significance, although David Doniger, a noted US environmentalist, has attacked the agency for proposing to 'pour more money down the rathole of risk assessment'.[33]

There is considerable administrative, judicial, and industrial support for a *de minimis* policy, based on quantitative risk assessment, in regulating hazardous products and pollutants. This consensus reveals, perhaps more clearly than any other aspect of current US risk management policy, that the ideology of 'zero risk' no longer pervades the regulatory debate, even on the perennially sensitive issue of environmental carcinogens. Doubts do persist about the capacity of the agencies to develop scientifically valid and socially equitable policies based on a risk management strategy that still presents many unanswered questions.[34] But agencies are likely to respond to these ethical and methodological quandaries by sharpening and formalizing their *de minimis* strategies rather than by abandoning them.

<u>4.2 Proposition 65</u>. In the US scheme of risk assessment and management, significant policy initiatives are ordinarily mounted by the federal regulatory agencies. Occasionally, however, a development at the state level departs sufficiently from federal practices to be worthy of special attention. This is the case with Proposition 65, a controversial environmental statute overwhelmingly approved by California voters in the election of November, 1986. Briefly, the law provides that chemicals certified by the state as carcinogens or reproductive toxins will be presumed hazardous to health and that their discharge into potential sources of drinking water will be banned. In addition, persons who might be exposed to any of the designated chemicals have a right to be warned of their possible exposure. Businesses using the listed chemicals can escape the discharge prohibition and the duty to warn only by establishing that their products or discharges are 'safe'.[35] In effect, the law shifts the traditional burden of proof with respect to the health hazards of environmental chemicals; substances listed in accordance with Proposition 65 are presumed hazardous until proved safe.

Besides the impact it is guaranteed to have on the chemical industry in California, Proposition 65 may have nationwide repercussions if a substantial number of states follow California's example. Similar measures are currently under consideration in at least twenty states. Their adoption would in effect create a two-tier system for chemical risk management, with the states in some instances imposing stricter standards on environmental discharges than the federal government.

The most interesting feature of Proposition 65 for purposes of this chapter is the potential it creates for divergent risk assessments and control strategies between the federal government and the states. The California law, for example, appears to require listing of all substances deemed carcinogenic by IARC, OSHA, and the US National Toxicology Program. Its regulatory reach may accordingly prove more comprehensive in some respects than EPA's.

The provision permitting businesses to prove that their discharges are safe opens up another area of possible federal-state conflict. According to the new law, a discharge may be considered safe if it does not cause detectable amounts of the chemical to enter into a drinking water source or if the amount, while detectable, does not pose a 'significant risk' of cancer over a lifetime of exposure. (There is a parallel provision, expressed in terms of no observable effects, governing reproductive toxins.) The statute thus contains a built-in *de minimis* exemption for chemicals presenting a less than significant risk to human health.

California, however, has no settled policy on what levels of risk will be regarded as 'significant' or whether such determinations will be made pursuant to the state's own policy for carcinogenic risk assessment. Depending on the way these issues are resolved, California's determinations of 'significant risk' may or may not conflict with those underlying EPA's regulatory policies, such as RMCLs for drinking water.

5 An International Perspective

As the preceding discussion indicates, US policy-makers, both state and federal, remain firmly committed to using formal risk assessment as a basis for developing regulatory policy. US principles for assessing risk have changed over time to take account of new scientific knowledge and new techniques for reducing uncertainty. Some of these modifications have brought US decision-making more in line with that of other industrial countries and international agencies. Examples of outright convergence include EPA's modelling of its weight of evidence determinations on IARC's and the incorporation of IARC's determinations of carcinogenicity into California's list of chemicals subject to Proposition 65.

Nevertheless, the US stands apart from other industrial and industrializing nations in the extent to which its risk management practices are built on quantitative risk assessment. This raises a number of interesting questions for policy analysts in other countries. Should the US approach to risk assessment be seen in some sense as the wave of the future? Will the techniques developed in the United States, especially quantitative policy analysis, assume increasing importance for decision-makers in other countries? Or will US risk assessment practices remain, as they did through the 1970s, an exceptional strategy produced in response to a unique constellation of legal, political, and cultural influences?[36-39] A brief comparison of the US situation with those of Britain and India suggests some preliminary answers to these questions.

<u>5.1 Risk in Britain</u>. The differences between chemical risk assessment in Britain and the United States have attracted attention in the past primarily because decision-makers in the two countries have often reached different conclusions about the health impacts of the same substances. Comparisons between the two countries indicate that British experts draw more conservative inferences from empirical data than their American counterparts. Put differently, British decision-makers have generally required a higher quantum of proof before designating chemical substances as hazardous to human health.[2] The issue addressed here is somewhat different. It focuses on a divergence in the process rather than in the outcome of risk assessment. Specifically, why is there so much less emphasis on the quantitative assessment of chemical hazards in Britain than in the United States?

<u>5.1.1 Case 1: Formaldehyde</u>. Recent assessments of formaldehyde in the two countries suggest at least a partial answer to this question. In Britain as in the United States, the health risks of formaldehyde, including especially its carcinogenic potential, were reassessed in light of the CIIT studies showing that the substance causes cancer in rats.[40] A toxicity review carried out for Britain's Advisory Committee on Toxic Substances (ACTS) concluded, however, that 'there was no evidence suggesting that formaldehyde has produced cancer in humans'.[41] Subsequently, the completion of an epidemiological study of exposed workers seemed to ratify the official British position that formaldehyde did not present a cognizable risk of cancer to humans. Although the British government tightened its controls on formaldehyde, this result appeared to be motivated by questions of technological

feasibility rather than the perceived risk of cancer.[42]

In the United States, the outcome was substantially different. EPA completed a controversial reassessment of formaldehyde in 1987 and classified the substance, in accordance with the carcinogen risk assessment guidelines, as a 'probable human carcinogen'. The classification was based on EPA'S determination that the data on formaldehyde presented limited evidence of carcinogenicity to humans, sufficient evidence of carcinogenicity in animals, and additional supporting evidence (*i.e.* positive mutagenicity studies).[41] EPA also carried out a quantitative risk assessment of formaldehyde based on the CIIT rat bioassay.

5.1.2 Case 2: Lead. Another example that illustrates interesting variations between British and American risk assessment methodologies is that of lead in the environment. In Britain, environmental lead was the subject of the Ninth Report of the Royal Commission on Environmental Pollution,[44] while in the United States airborne lead has maintained a conspicuous position on EPA's risk management agenda since the early 1970s. One issue that experts in both countries have tackled with varying success is the impact of low-level exposure to lead on children's behaviour and intelligence.

The Royal Commission took up this question after the available evidence had already been reviewed by governmental and non-governmental experts. The Commission elected not to re-analyse the data in detail, but concluded that no definite causal relationship could be established on the basis of what was known. Even future research, the Commission felt, was unlikely to shed much more light on the issue, especially in view of the many confounding social factors affecting children's IQ, and the fact that the impact on children's health appeared 'at the most small at the concentrations found in the general population'.[44]

Given the multiple unknowns about the effects of lead on health, the Commission decided to formulate its policy recommendations on the basis of what was knowable. Sir Richard Southwood, then chairman of the Royal Commission, has provided an interesting account of the risk management approach that his colleagues adopted.[45] In a step-by-step analysis, the Commission first considered the physiology of lead and its accumulation in the environment. This analysis indicated grounds for concern, since lead is both highly persistent and genotoxic with no apparent threshold. Further, comparisons of blood-lead levels in the general population with levels at which health effects are known to occur suggested that the margin of safety between the two levels was already too small, especially for urban populations. Finally, the Commission carried out an economic impact analysis to show that a gradual phase-out of lead in petrol would not prove overly burdensome to the automobile industry (see also Chapter 2). On the basis of these findings, the Commission recommended generally that the British government should seek to reduce anthropogenic lead in the environment and, more specifically, that lead additives in petrol should eventually be eliminated.

EPA's approach to dealing with the scientific uncertainties concerning the health impact of lead has been notably different, especially in the context of the air quality standards programme. The US agency found the data on lead and young children sufficiently incomplete, inconsistent, and indirect to rule out ordinary statistical analysis. But EPA's response has been to sample and quantify the divergences in expert opinion on this issue by means of a new analytical technique known as 'judgemental probability encoding'.[46] In other words, despite the paucity of data, EPA remains committed to the methodology of quantitative risk analysis and is prepared to modify this method as necessary.

<u>5.1.3 Implications</u>. The most striking feature of the British approach in both the formaldehyde and lead cases is that the risk to human health was deemed insufficient to merit further analysis. This result was arrived at somewhat differently in the two cases, but the ultimate effect in both was to rule out any need for formal risk assessment. As far as formaldehyde was concerned, the epidemiological data evidently did not meet the relatively high threshold requirements applied to potential carcinogens in Britain, requirements that appear to correspond to IARC and EPA guidelines for 'known human carcinogens'.[47] Since formaldehyde was not characterized as hazardous to human health, the question of quantitative risk assessment was understandably moot. By contrast, formaldehyde did satisfy EPA's criteria for designation as a 'probable human carcinogen', thus paving the way to further assessment of its risks.

In the case of lead, EPA and the Royal Commission seem implicitly to have agreed that definite conclusions about the degree of risk to children could not be drawn on the basis of existing (or even projected) studies. But whereas the Commission sought an alternate basis for decision-making, one founded on risks that were better understood (persistence, genotoxicity), EPA tried to gain a better measure of the risk to children by subjecting the range of expert disagreement to quantitative analysis. EPA's need for numbers may be attributed in part to that agency's obligation to set numerical air quality standards for pollutants like lead. Britain has not enacted a comparable standard-setting programme. But the case also illustrates a lower tolerance in Britain than in the United States for using statistical techniques on data that are viewed as qualitatively too weak or ambiguous to yield meaningful measures of risk.

5.2 Risk in India (see also Chapter 7). The catastrophic accident in Bhopal in late 1984 painfully dramatized the inadequacies of chemical risk assessment and management policies in much of the Third World. Within India, chemical risks emerged for the first time as an issue for public debate and the government was pressured to rethink its role in managing hazardous technologies. Given its high level of technical and industrial sophistication, and its experiences as both a producer and importer of chemicals, India is likely to set an example of some importance to other Third World nations concerned with problems of risk management. As yet, however, India's own policies for regulating hazardous chemicals remain in a state of flux, and it is too early to tell whether India will be able to formulate a theory and practice of risk management that genuinely respond to the needs of other developing countries.

The Bhopal disaster revealed many gaps and weaknesses in the Indian government's power to regulate hazardous enterprises effectively.[48] The passage of a new law, the Environment (Protection) Act 1986, promised to remedy some of these deficiencies. But although the law has, on its face, increased the central government's authority to establish and enforce environmental standards, it remains essentially inoperative until implementing regulations are adopted by the Indian Parliament.

The only concrete indications as to how the 1986 Act might be implemented to date are contained in a report prepared under the auspices of the Indian Law Institute (ILI) in 1987.[49] While this document in no way represents official Indian policy, it provides a basis for identifying areas where Indian risk assessment and management policy is likely to follow available Western models and where it is likely to diverge substantially. For example, the ILI report recommends that:

> 'The import or use of any substance that is banned in the country of its origin or the country where the substance was first manufactured for its hazardous impact on the environment ... should also be banned in India.'[50]

If the Indian government adopts this recommendation, India will in effect be accepting certain risk management decisions of other countries and, by implication, the risk assessments on which such decisions are based.

With regard to environmental standard-setting, however, the ILI report suggests that India may follow a more autonomous policy. The report envisages that both new and existing chemicals will be subject to testing requirements specified by the central authority charged with implementing the Act. If the ILI recommendations are followed, such testing will include epidemiological and laboratory studies of health effects such as carcinogenicity, mutagenicity, teratogenicity, behavioural disorders, and psychic impacts. The report recommends that testing methods officially adopted in India be approved in advance by the Indian Council for Medical Research. If a significant programme of chemical testing and assessment is undertaken in India, it is not unlikely that the results will differ in some cases from those reached in other countries. Indeed, such divergences would be entirely consistent with what is already known about the impact of legal, political, and cultural factors on the perception and assessment of risk.

6 Prospects for Harmonization

In an increasingly interdependent world economy, there are strong reasons for advocating internationally harmonized approaches to the assessment, and even management, of chemical risks. The global impact of such widely dispersed chemical products as DDT or chlorofluorocarbons contributes a further rationale for seeking such communal solutions. The directions taken by US risk policy in the 1980s, however, suggest that cross-national *rapprochements* in the area of risk assessment are likely to occur slowly at best. Complete harmonization of risk assessment policies is in any event an unrealistic goal, given the considerable disparities that still remain in the risk management programmes of very similar countries, such as Britain and the United States.

Scientists whose work impinges on chemical risk assessment can take satisfaction from the fact that public policy does indeed change in response to new knowledge. Recent modifications in EPA's principles for assessing carcinogenic risk provide a striking illustration of such movement. Although EPA has not abandoned its commitment to the use of general principles, the agency clearly recognizes that the conservative assumptions embedded in these principles represent default positions to be adopted only in the absence of adequate information. Of course, disagreements may still arise between EPA and other authorities (whether in the United States or elsewhere) over the point at which the evidence justifies a departure from the principles of risk assessment. The formaldehyde case, briefly discussed above, exemplifies such a conflict. It is worth noting that the position eventually adopted by EPA with respect to formaldehyde diverged not only from that advocated by the US chemical industry, but also from that espoused by regulatory authorities in Britain.

Finally, the contrasts between American, British, and Indian policy again emphasize the point that scientific progress can play only a limited, albeit important, role in determining how different countries evaluate and control the risks of toxic and hazardous substances. Many of the factors that shape even the largely technical process of risk assessment remain, at bottom, culturally and politically conditioned. Science may facilitate, but will not force, convergence on such fundamental issues as the degree of conservatism with which one should assess a hazard about which much still remains unknown. Under the circumstances, **risk analysts in different countries should perhaps devote more attention to communication than to harmonization, particularly by clarifying to what extent their assessments are driven by scientific considerations and to what extent by factors lying beyond science.**

7 References

1. US National Research Council. 'Risk Assessment in the Federal Government: Managing the Process', National Academy Press, Washington, DC, 1983.

2. S. Jasanoff, 'Comparative Risk Assessment: The Lessons of Cultural Variation,', in 'Toxic Hazard Assessment of Chemicals', Royal Society of Chemistry, ed. ML Richardson, London, 1986, p.262.

3. *Federal Register,* September 24, 1986, **51**, pp. 33992-34003.

4. E. Rushefsky, *Politics Life Sci.*, 1985, **4**, 31.

5. T. Yosie, 'The Culture of Risk Assessment at the Environmental Protection Agency', Annual Meeting of the American Chemical Society, New York, April 15, 1986, p.8.

6. E. Marshall, *Science*, **218**, 1982, 975.

7. E. Marshall, *Science,* **220**, 1983, 36.

8. M. Wines, *National J.*, June 18, 1983, 1264.

9. National Academy of Public Administration, 'Presidential Management of Rulemaking in Regulatory Agencies', Washingdon, DC, 1987, pp.6-7.

10. *Federal Register*, 1986, **51**, 33995.

11. *Federal Register,* 1976, **41**, 21404.

12. *Federal Register,* 1986, **51**, 33996.

13. R. Brickman, S. Jasanoff, and T. Ilgen, 'Controlling Chemicals: The Politics of Regulation in Europe and the United States', 1985, Ithaca, NY, Cornell University Press, p.205.

14. E.L. Anderson et al., and the Carcinogen assessment Group of the US Environmental Protection Agency, *Risk Analysis,* 1983, **3**, 278.

15. *Federal Register,* 1986, **51**, 33997.

16. T. Yosie, 'Science and Sociology: The Transition to a Post-Conservative Risk Assessment Era', paper presented at Annual Meeting of the Society for Risk Analysis, Houston, November 2, 1987, p.3.

17. T. Yosie, ref.16, p.12.

18. M. Gough, 'Dioxin, Agent Orange', Plenum Press, New York, 1986, p.219.

19. T. Yosie, ref.16, p.11.

20. P. Shabecoff, *New York Times,* December 9, 1987, p.1

21. F. Perera and C. Petito, *Science*, 1982, **216**, 1285.

22. N. Ashford, C. Ryan, and C. Caldart, *Harvard Environ. Law Rev.*, 1983, **7**, 297.

23. S. Jasanoff, 'Risk Management and Political Culture', Russell Sage Foundation, New York, 1986, p.41.

24. J.D. Graham, L.Green, and J. Roberts, 'Seeking Safety', Harvard University Press, Cambridge, MA, forthcoming.

25. *Gulf South Insulation v. Consumer Product Safety Commission,* 701 F. 2d 1137 (5th Cir. 1983).

26. M. Casanova-Schmitz, T.B. Starr and M. D'A Heck, *Toxicol Appl. Pharmacol.*, 1984, **76**, 26-44.

27. *Industrial Union Department, AFL-CIO v. American Petroleum Institute,* 448 US, 607, (1980).

28. *Monsanto v. Kennedy*, 613 F.2d 947 (D.C. Cir. 1979).

29. A. Kessler, *Science,* 1984, **223**, 1037.

30. E. Marshall, *Science*, 1984, **224**, 851.

31. R.W. Hart et al., *Risk Anal.*, 1986, **6**, 117.

32. *Public Citizen v. Young,* No.86-1548 (D.C. Cir. 1987)

33. 'Benzene Rules to Heed Vinyl Chloride Decision, Though Controls May be Same, EPA Analyst Says', Bureau of National Affairs, *Environ. Reporter*, Current Developments, January 15, 1988, p.2012.

34. J. Mumpower, *Risk Anal.*, 1986, **6**, 437.

35. Nossaman, Guthner, Knox & Elliott and Roger Lane Garrick, 'Surviving Proposition 65', Los Angeles, CA, 1987.

36. R. Brickman, S. Jasanoff, and T. Ilgen 'Controlling Chemicals: The Politics of Regulation in Europe and the United States', Cornell University Press, Ithaca, NY, 1985. (See also ref.13).

37. S. Jasanoff, 'Risk Management and Political Culture', Russell Sage Foundation, New York, 1986.

38. D. Vogel, 'National Styles of Regulation', Cornell University Press, Ithaca, NY, 1986.

39. G. Wilson, 'The Politics of Safety and Health', Oxford University Press, Oxford, 1985.

40. W.D. Kerns, K.L. Parkov, D.J. Donofrio, E.J. Gralla, and J.A. Swenberg, 'Carcinogenicity of Formaldehyde in Rats and Mice after Long-Term Inhalation Exposure', *Cancer Res.*, 1983, **43**, 4382-92.

41. Health and Safety Executive, 'Formaldehyde', 'Toxicity Review', 1981. pp. 2, 11.

42. S. Jasanoff, 'Cultural Aspects of Risk Assessment in Britain and the United States', in B.B. Johnson and V.T. Covello, 'The Social and Cultural Construction of Risk', Reidel, New York, 1987, p. 372.

43. US EPA, Office of Pesticides and Toxic Substances, 'Assessment of Health Risks to Garment Workers and Certain Home Residents from Exposure to Formaldehyde', April 1987, pp. 106.

44. Royal Commission on Environmental Pollution, Ninth Report, 'Lead in the Environment', HMSO, London, 1983.

45. T.R.E. Southwood, *Marine Pollut. Bull.*, 1985, **16**, 346.

46. T.S. Wallsten and R.G. Whitfield, 'Assessing the Risks to Young Children of Three Effects Associated with Elevated Blood-Lead Levels', Argonne National Laboratory, Argonne, IL, December, 1986.

47. R. Brickman et al., ref. 35. pp.187-217.

48. S. Jasanoff, *Environment*, 1986, **28**, 12-16, 31-38.

49. Indian Law Institute, 'Environment Protection Act: An Agenda for Implementation', ed. Upendra Baxi, Tripathi, Bombay, 1987.

50. Ref. 47, p.17.

7
A Systems Approach to the Control and Prevention of Chemical Disasters

C.R. Krishna Murti

SCIENTIFIC COMMISSION FOR CONTINUING STUDIES OF THE EFFECTS OF BHOPAL GAS LEAKAGE ON LIFE SYSTEMS, CABINET SECRETARIAT, GOVERNMENT OF INDIA CANCER INSTITUTE, MADRAS, INDIA

1 Introduction

The insatiable demand for chemicals by industry, agriculture, defence, and consumer needs has led to a rapid expansion of the chemical and petrochemical industries. Thus, worldwide production or organic chemicals registered a nine-fold increase from 7 to 63 x 10^9 kg during the twenty year period 1950 to 1970 and a further four-fold increase to 250 x 10^9 kg from 1970 to 1985.[1] When seen in the light of benefits accruing to mankind, this progress is indeed spectacular but has not been without its dark side manifesting as a series of disastrous events. The regularity and frequency of these episodes have prompted the author of one recent study to refer to them as 'normal accidents'.[2] Indeed, accidents and hazards appear to be 'norms' rather than exceptions of the contemporary High-Tech Society.[3]

In particular, the decade 1974-84 witnessed an unusually large number of accidents leading to massive releases of hazardous chemicals. The Flixborough and Beek explosions, the Seveso disaster that spewed dioxins over a large area, the Missisauga collision of rail cars containing chlorine and propane, the Houston spill of anhydrous ammonia, the Sommerville spill of phosphorus trichloride, the Mexico City liquefied gas explosion and, the worst in history, the Bhopal disaster involving the deadly methyl isocyanate, all happened in a recent span of ten years.[4]

A disaster is defined as 'an act of nature or an act of man which is or threatens to be of sufficient severity and magnitude to warrant emergency assistance'[5] or 'a disruption of the human ecology which the affected community cannot absorb with its own resources'.[6] Many of the accidents referred to above fulfil the requisites of this definition of a disaster.

Admittedly, in the wake of these catastrophic events, chemical industry cannot afford the luxury of complacency on a supposedly good track-record of safe performance.[7,8] A recent review of the public health risks of the rapidly accelerating mobility of chemical hazards raises many issues relevant to the social accountability of the chemical industry, particularly in controlling the release of toxic chemicals.[9] In the context of the 1988 3rd FECS Conference on Chemistry and the Environment, it is noted that the chemical industry is now expending $ million per year in the recognition of perceived societal risks, but in view of the severe nature of the disasters alluded to above could be doing more.

2 Natural versus Man-made Disasters

Even as nature continues to rebel and bring infinite suffering to Earth's children in the form of natural disasters, man has made significant advances in understanding earthquakes or cyclones. Though he cannot prevent their occurrence, thanks to many scientific developments, he can at least be better prepared to face their consequences. He has today, thus, means within his reach by which earthquake-susceptible regions can be kept under constant surveillance. Again, he can predict the possibility of a devastating cyclone crossing a coastal region well in advance to enable the people to be evacuated to safer zones. These advances and the resultant tools facilitate the implementation of preparedness programmes so that the suffering caused by the disaster can be substantially mitigated. Above all, global experience accumulated over the years can be objectively brought into service while facing emergencies such as the outbreak of epidemics or the psychiatric trauma induced by the loss of shelter and security.

In contrast, the consequences of man-made disasters, *e.g.* fire, explosion, chemical pollution, or radiation, on public health and the environment are not readily predictable. Even if an alarm sounds it is often too late and hence there is very little time to prepare the exposed community to face the outcome. More often than not, even if the cause of the accident can be traced, its relationship to human health effect is difficult to establish. The result is that measures of emergency medical response have to be initiated on arbitrary considerations. This, in turn, adversely influences the post-emergency initiatives for the rehabilitation of victims.

The man-made chemical disasters mentioned above presented scenarios which differed from episode to episode. Their tragic sequelae and the magnitude of

Table 1 *Major accidental releases of hazardous materials*

Location	*Year*	*Material*	*Source*	*Sequelae*
Hamburg, Germany	1928	Phosgene	Storage	10 Dead, 7200 ill
Chicago, USA	1978	Hydrogen sulphide	Tanker	8 Dead, 29 ill
Memphis, USA	1979	Parathion	Storage	150-200 evacuated
London, UK	1980	Cyanide	Factory	4000 evacuated
Fitchburg, USA	1982	Monovinyl chloride	Factory	3000 evacuated
Bhopal, India	1984	Toxic cloud	Factory	More than 2500 dead More than 3000 diseased

Adapted from Bowonder *et al.*[7] and Vilain.[10]

their impact on public health and the environment exhibited, as illustrated in Table 1, a wide spectrum of diversity of the chemicals involved and the intensity of effects elicited. While recognizing its uniqueness, Vilain[10] calls a chemical accident 'a bizarre event, albeit with repetitive patterns but great variability in terms of cause/consequence ratio'.

The uncertainties inherent in a disaster involving 'runaway' chemical reactions pose a number of unanswered questions related to the mechanism that triggered them, the nature of products formed, their environmental pathways, and effects on living systems. National, regional, and international conferences have addressed many of these issues and have attempted to focus greater attention on chemical hazards.[11-19]

The ecological disaster of the Rhine in November 1986 caused by an accidental discharge of toxic chemicals from a Sandoz manufacturing plant at Basel, Switzerland, is an unwelcome warning that such episodes can occur even in a highly industrialized country with a long reputation for efficient and safe management of chemical industries. The unpleasant prospect of having to live with more such hazards makes it all the more necessary to intensify scientific efforts to study chemical accidents in their totality. An attempt is made in this chapter to project models for resolving the complexities of chemical disasters with a view to develop methods for their prevention and control.

3 Conceptual Approach

The elements of a chemical disaster can be represented as a cause-effect sequence:

Event ⟶ Impact on Target ⟶ Sequelae

Using a highly simplified systems approach as shown in Figure 1, one can, in principle, characterize the outcome of a disaster caused by the accidental release of large quantities of toxic chemicals. This simple model helps us to describe accident scenarios at the place of manufacture, processing, or formulation, or at storage sites, during transportation, or disposal of wastes. Superimposing the model on the actual site of a given accident, the targets at the site or outside the site can be identified. While explosion, fire ball, and loss of human life represent sequelae at the site, diffusion of smoke into air or discharge of effluents into the aquatic environment illustrate the trans-frontier dimensions of chemical disasters.[20-22] The sequelae of chemical disasters are brought out diagramatically in Figure 2. One can also visualize the potential for a recurrence if the release is not contained immediately.[23]

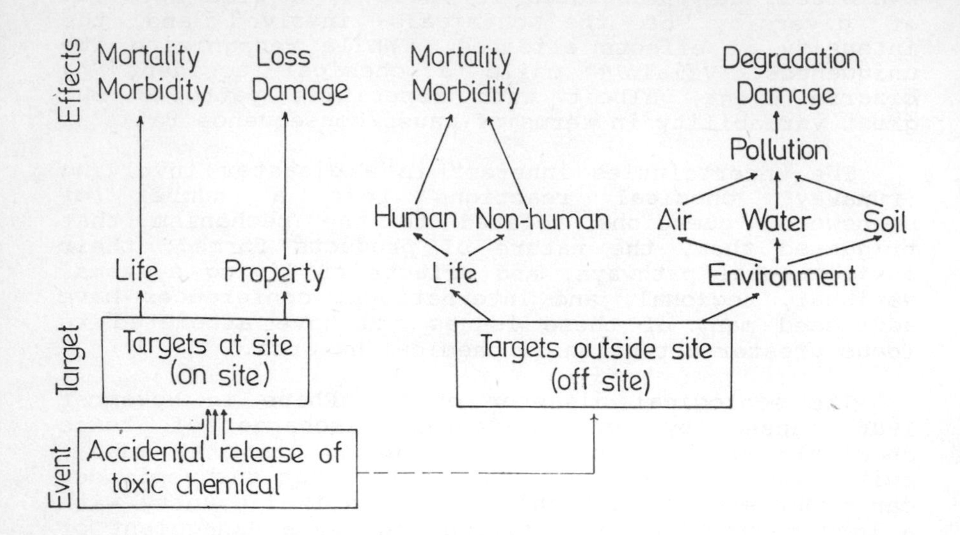

Figure 1 Elements of a disaster caused by release of a toxic chemical

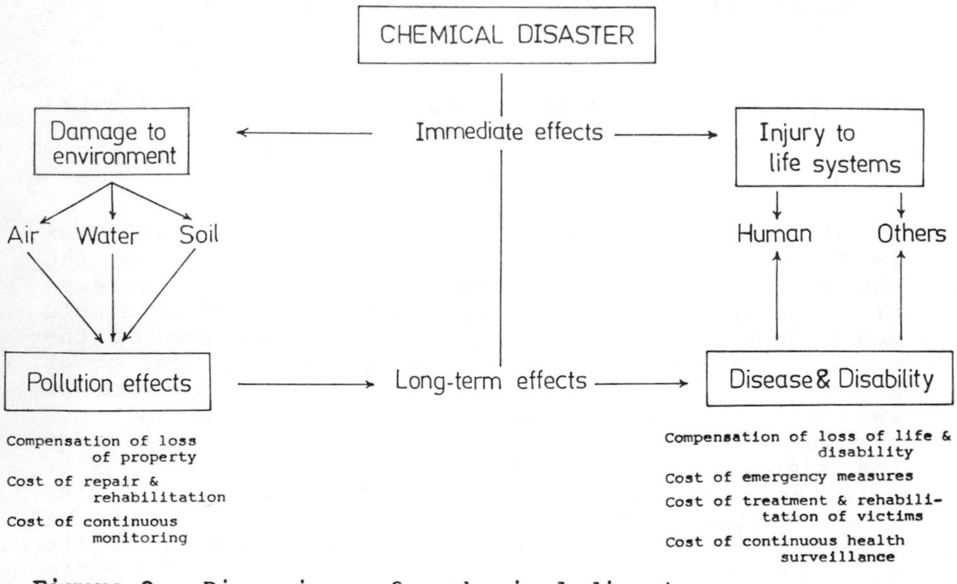

Figure 2 Dimensions of a chemical disaster

4 The Scenario

Presumably, once the extensive information available within the industry or regulatory agencies is made accessible for computer analysis, scenarios can be reconstructed on a mini-scale. Retrospective study can then be made of the dynamics of diffusion of the toxic chemical in the environment and hopefully a better understanding of the nature of human or systems failure acquired which previously has led to accidental releases.[24] From records of accidents entered in log books, check lists can be prepared and a 'fault-tree' diagram drawn to trace the cause to its very roots.[25–28] Alternatively, computer simulation enables the diagnosis of the failure, *i.e.* whether due to defects in design, construction, operation, or maintenance of process equipments or failure of control mechanisms to respond to alarm signals.[29–33]

Notwithstanding all this progress, there are still many uncertainties, and convincing mechanisms of the disaster cannot be enunciated. The unresolved complexities of 'runaway' reactions involving highly reactive intermediate chemicals complicate the issue further.[34] As became tragically apparent in Bhopal, immediate emergency response was given without any idea of the nature and intensity of toxic exposure and the community was totally unprepared to face all aspects of a tragedy of such magnitude.[35,36]

5 The Nature of Disaster and its Targets

The following situations are typical of chemical disasters:

i) Leak of corrosive material or toxic chemical from a reaction vessel, storage tank or carrier affecting land or water bodies

ii) Explosion and fire ball formation involving explosive and inflammable chemicals

iii) Release of a highly toxic cloud of gaseous/particulate material spreading by diffusion to neighbouring human settlements.

Installations, buildings, and private property are the main targets of explosion and fire ball formation or leak of corrosive fluids. Life systems in the vicinity of the site are ready victims in all the situations whether they are 'on site' or 'off-site', or when the released toxic chemicals contaminate the three compartments of the environment, air, water, and soil.

6 Health Effects of Accidental Releases of Toxic Chemicals

The effects produced by a chemical disaster will depend primarily upon the nature of the chemical, its toxicity, its pathway in the environment after release, and its fate in the human body after exposure and absorption. The effects can vary widely from target to target. They would also depend upon the medium in which the resultant chemical contamination takes place. If air and water are contaminated even for a short duration, the resultant acute effects can include high mortality and morbidity of human, animal, avian, and aquatic life. An ecokinetic model for gaining insight into the interplay of several elements and the consequent effects on living beings of exposure to accidentally released toxic chemicals is proposed in Figure 3. It is evident that the fate of the released chemicals is determined by both biotic and abiotic interactions.

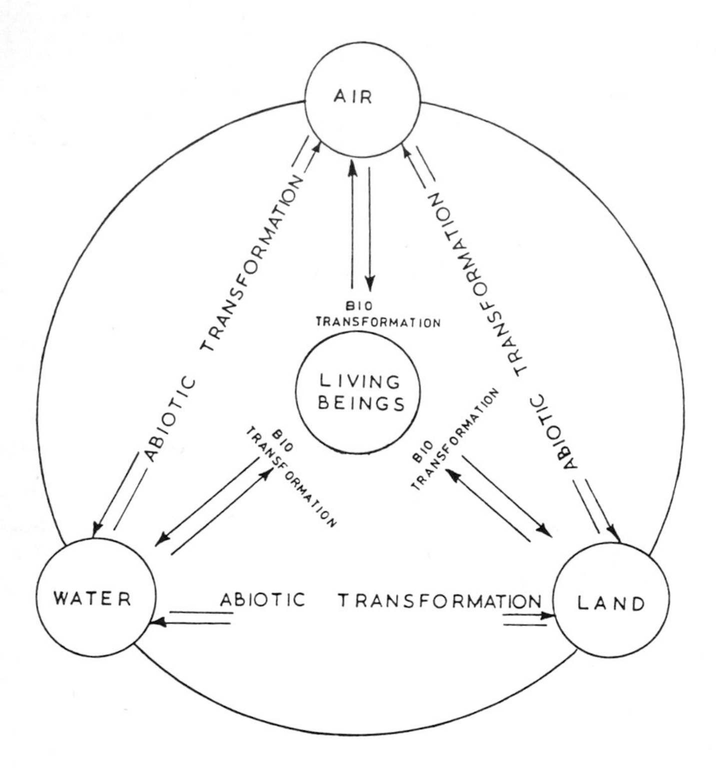

Figure 3 Ecokinetic model for transport and transformation of an accidentally released toxic chemical

Untoward biological responses elicited by the absorbed chemical will depend upon the degree and duration of exposure and the property of the chemical to reach the target organs in the human body: lungs, brain, skin, gastro-intestinal tract, liver, kidneys, reproductive organs, and others. After combining with the target organ, the chemical can cause serious disturbances in their physiological functions. Significant changes have been noticed to occur after toxic chemical exposure in the following functions: respiration, nerve conduction, metabolism including digestion, excretion and assimilation, endocrine control, blood cells and blood protein regeneration, and reproduction and transmission of genetic information. The effects on human health of rapid and slow transfer of impact of environmental chemicals are emphasized in Figure 4.

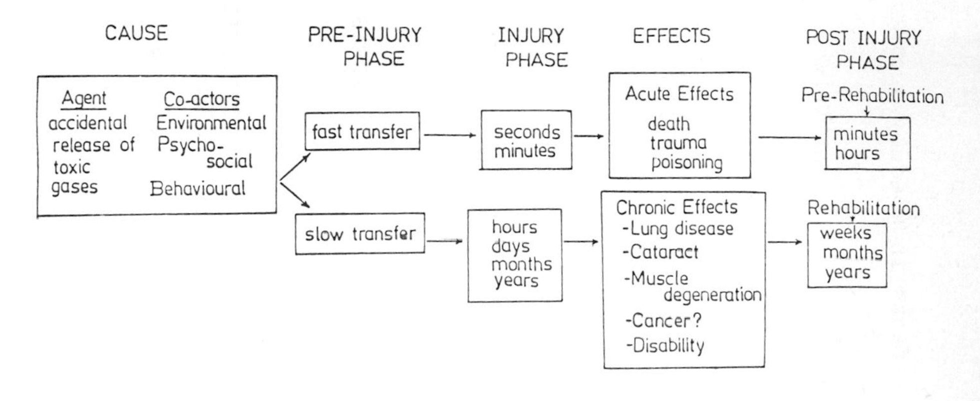

Figure 4 Effect on human body of fast and slow transfer of the impact of accidental release of toxic gases

Waller's[37] adapted model indicates the close relationship between the chemical and environmental factors on one side and the traumatic acute phase and the long-term chronic phase on the other. A list of target organs, functions affected and the disease clusters generated thus is given in Table 2. Routes of entry of chemicals into living systems are summarized in Table 3.

Table 2 *Target organs and health effects induced by accidentally released toxic chemicals*

Target organ	Health effects	Target organ	Health effects
SKIN	Altered appearance Irritation Sensitization Corrosion	NERVOUS SYSTEM	Behavioural changes Peripheral neuropathy Depression Locomotor ataxia Respiratory paralysis
EYE	Irritation Corneal opacity Retinal damage Corrosion	LIVER	MFO induction Choleostasis Neoplasia Adenoma, Carcinoma Cirrhosis Necrosis
MUCOUS MEMBRANE	Irritation Corrosion		
LUNG	Irritation Sensitization Pneumoconiosis Fibrosis	KIDNEY	Aminoacidurea Uraemia Lithiasis
	Adenoma, Carcinoma Neoplasia Asphyxiation	HAEMAPOEISIS	Bone marrow depression Leukaemia Aplastic anaemia Methamoglobinaemia
FETUS	Abortion Malformation Neonatal death	MUSCULOSKELETAL SYSTEM	Osteoporosis Cossosion
REPRODUCTIVE SYSTEM	Embryotoxicity Teratogenesis Infertility Germ cell mutation	IMMUNE SYSTEM	Suppression Alteration

Table 3 *Exposure routes of entry of accidentally released toxic chemicals*

Target	*Route*
Humans and animals in terrestrial environment	Exposure of skin and outer surface Exposure of eyes, mucous membrane of nose, mouth, windpipe Exposure of pulmonary tissues Entry through skin Entry through lungs Entry through gastro-intestinal tract Entry through ingestion of food and water Entry through ingestion of contaminated material
Plants in terrestrial environment	Exposure of above ground parts to air and to precipitation Entry through above ground parts Entry through roots
Organisms in aquatic environment	Exposure of outer surface Entry through outer surface Entry through gills Entry through gastro-intestinal tract Entry through ingestion of other organisms Entry through ingestion of suspended matter Entry through ingestion of benthic detritus

Instantaneous death and protracted illness of varying severity constitute acute poisoning effects. It is at this stage that intervention by antidotal agents is most effective. However, knowledge of the chemical nature of the toxin and its antidote is very critical. Individual susceptibility, degree and duration of exposure, and failure to reduce the residence period of the toxin in the body profoundly modify the prognosis of the exposed individual. Cancer, gene mutation, impairment of immune function, behaviour, and growth are all to be considered as possibilities in the long-term phase. The exposed population has, therefore, to be watched for any appearance of unusual symptoms of disease for many years after the accident. In disasters involving contamination of ambient air, the toxic chemical gains entry by inhalation or absorption through exposed parts of the body. The lungs and related respiratory organs will be the targets, the main impact being on the respiratory function. When water or food is contaminated by chemicals, the target will be the digestive, excretory, and assimilative systems including the gastro-intestinal tract, liver, kidneys, and pancreas. The methods available today for making estimates of exposure of man to chemicals are summarized in Figure 5. The profile of health effects to be anticipated and screened is shown in Figure 6.

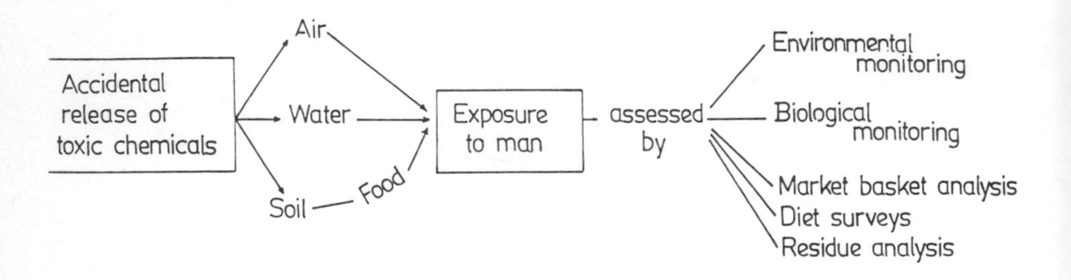

Figure 5 Human exposure to accidental release of toxic chemicals

Whereas accidental releases of chemicals in the atmosphere usually have a greater impact on human and animal life and vegetation, discharges into water bodies generally affect the non-human species to a greater extent. Similarly, accidental releases during transport are more likely to exert ecological effects than in-plant releases. It is particularly important in the assessment of ecological consequences of an accidental release to understand the pathways and transformation of the released chemical in environmental compartments.

Whole ecosystems, especially lakes and rivers, may be affected for an extended period by accidental releases of toxic chemicals. Effects on living targets, other than man, involve animals including mortality, sterility, reduced milk yield, or egg production, or lower meat quality. Effects on plants are burning, loss of foliage, growth impairment, and adverse effect on viability of seeds. Plants and animals affected by accidental releases of chemicals may also serve as indicators of both the levels of

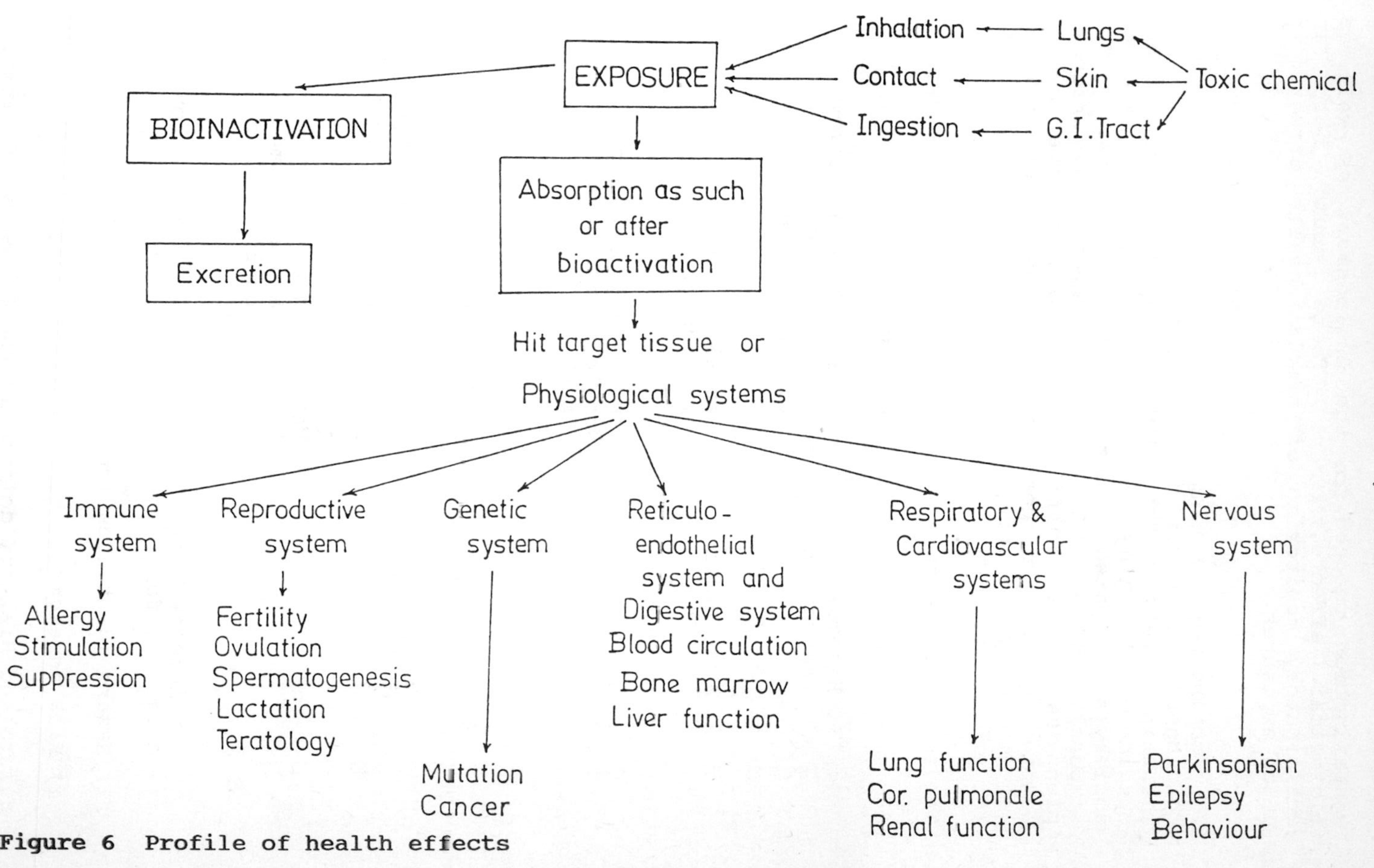

Figure 6 Profile of health effects

contamination reached in the environment as well as biological effects.[38,39] These could help in anticipatory monitoring of similar health effects in human exposures.

7 Contingency Planning

Contingency planning is indispensable not only to maintain safety within but also to ensure safety outside. Planning is also essential as part of society's preparedness to face emergencies. The components of such a chemical disaster preparedness plan designed to fulfil the Seveso Directive and as adopted by some European countries are shown in Table 4. Obviously, the pre-emergency phase is most critical for prevention.[40,41]

Table 4 *Components of chemical disaster preparedness*

	Phase	*Activity*
Before the disaster	1. Hazards Evaluation	Identification of hazards Indentification of vulnerabilities Assessment of risk
	2. Prevention	Removal of the hazard Selection of alternatives Hazard control
	3. Planning mitigation	Contingency planning Knowledge of rehabilitation methods Instituting organizational framework
After the disaster	4. Emergency	Accurate response Speed of action
	5. Follow-up	Knowledge of chemical(s) Fencing off the accident Diagnosis of needs Implementation Monitoring Feedback and adjustment Information transfer and storage

Contingency planning for accident prevention envisages the following two important activities:

i) Hazard identification and

ii) Risk assessment

Hazard and Risk are defined in the Glossary of Terms.

Many techniques are available for systematic identification of hazards and calculation of frequency of accidents although not all of them are of universal applicability. Hazard identification starts with the description of the hazardous properties of materials, special hazards of processes, and external hazards if any. This is accomplished by compiling a check list and by using more exhaustive methods such as the Hazard and Operability Study (HAZOP). Hazardous properties of certain intermediate chemicals used in chemical industry and the inherently hazardous processes used in their preparation, use or transformation are given in Table 5.

Table 5 *Properties of hazardous chemicals and hazards in processing chemicals*

Hazardous properties	*Hazards in processing*
Runaway reactions leading to intense heat generation and explosion	Alkylation
Explosivity	Amination
Inflammability	Carbonylation
Polymerization resulting in heat generation	Dehydrogenation
Oxygenation	Catalytic hydrogenation
Light-catalysed degradation	Oxidation
Release of free radicals	Polymerization
Corrosivity	Nitration
Toxicity	Sulphonation
	Manufacture of phosphorus-containing materials
	Formulation of pesticides, pharmaceutics
	Salvation
	Cryogenolysis
	Distillation
	Processing of fuel gas
	Wet metallurgy
	Electrolysis
	Electrothermal reductions
	Dry heating of carbonaceous materials

In order to define critical elements of safety, construction of fault, event trees or failure modes is also attempted. Hazards to be identified cover a wide spectrum from relatively frequent small events to large and relatively rare events. From the hazard events identified in this manner, mathematical models can be set up. Such models help in calculating liquid, two-phase, or gas release rates, liquid spill and spread and vaporization, vapour cloud formation and dispersion, explosions, or release of toxic materials. In the design stage of the plant the results of such analysis help in reducing hazards within the plant. The calculations made thus can also be used for separation distances, exclusion zones, or distances to be used for emergency planning.

Hazard analysis (HAZAL) is used to determine the frequency of individual events and the significance of each event from the point of view of safety. Within an installation, identification and predicting an inventory of all existing and potential hazards is the first stage. Remote control sensing and process control devices are used, optimizing automated procedures. When a hazard is sensed the aim should be for a fail-safe progressively managed shut-down to the necessary degree rather than an immediate total shut-down. If release of toxic materials occurs a neutralization system comes into operation immediately. To summarize, hazard potential is the criterion on which processes of production are finally allowed to be operated within equipment designed, constructed, operated, and maintained on the basis of in-depth analysis of safety.

Along with HAZOP and HAZAL, action error analysis (AEA) is also used in identifying hazards in chemical installations. HAZOP has been found to be very effective in identifying the technical factors contributing to the safety of the installation and is also a check mechanism when applied at the explorative and process design stages.

There is no single universally accepted method for risk assessment and management. Perception of risk starts from identification of hazards, their inventorization and prioritization. Together with these activities, use of failure mode and effect analysis (FMEA) and construction of a management oversight and risk tree (MORT) is necessary. Both these procedures are useful in uncovering past failures and in introducing corrective measures. Hazard identification and error or fault analysis help in evolving principles of inherent safety using the following guiding principles: avoid toxic or flammable materials where possible, reduce the inventory in process and stock, simplify the process where possible, and separate people from chemicals.[25,28,42,43] Accident ratios can be calculated and mechanisms elucidated. Their sequelae can be described with greater precision. Safety audit procedures can also be used for safety assurance.

With major hazards installations the off-site consequences of accidents can be very profound. The risk for the inhabitants and the ecosystems off-site has to be managed by local authorities. For this an objective analysis is made of the risk incorporating probability of events and their consequences in terms of injury. This then enables one to arrive at judgements of the significance of that level of risk in relation to other risks in life. Both qualitative and quantitative methods are used for risk analysis. A list of possible accidents is compiled in increasing order of severity and subsequently the likely effects at a given distance from each event are estimated. Size of failures of various components, corresponding rates of release of substances, and probabilities are arrived at by quantification of the risk. For each release the probability of injury is estimated as a function of distance, weather type, and the likely behaviour of the people of the location. These effects are then summed up by thousands of calculations involving mathematical models. With the assistance of computers the uncertainty in estimates is reduced and further refinement of procedures is attempted.[45]

The unsafe design of existing processes, shortage of land for resiting and zoning, operational problems in siting industries at remote places, clustering of industries in some highly populated urban areas, and a general shortage of trained personnel are compelling reasons for contingency planning in Third World Countries.[46] The adaptation of the procedures used for risk assessment in industrialized countries or their innovation are challenges for scientists of the Developing Countries where major chemical industries are being established or related proposals are under consideration. Four important activities to be undertaken in assessment and control of risk in chemical industries are summarized in Table 6.

Table 6 *Assessment and control of accidental release of toxic chemicals*

Hazard Analysis and Hazard Operation and accounting and safety audit should be undertaken regularly not only by industry but also by the concerned regulatory agency.

Modifications of process on plants in existing industries must conform to the same stringent requirements of safety as when they are planned and installed.

Reporting to a regulatory agency and the public of hazardous occurrences such as leaks, release of gaseous fumes, consequences, and containment measures adopted should be mandatory. All official reports on accidental releases should be published and made readily available.

Emergency plans for facing on-site and off-site emergencies caused by accidental release of toxic chemicals should be set up, revised, and updated with the acquisition of more knowledge of the mechanisms of such releases.

Environmental health is one of the main targets of major chemical accidents. The WHO Regional Office for Europe has initiated some welcome steps in setting up a study group to improve methods for Environment Health Impact Assessment of Chemical Industries.[16,17] By incorporating health impact assessment procedures into a conventional chemical risk matrix the study group has suggested a multi-step EHIA model shown in Figure 7.

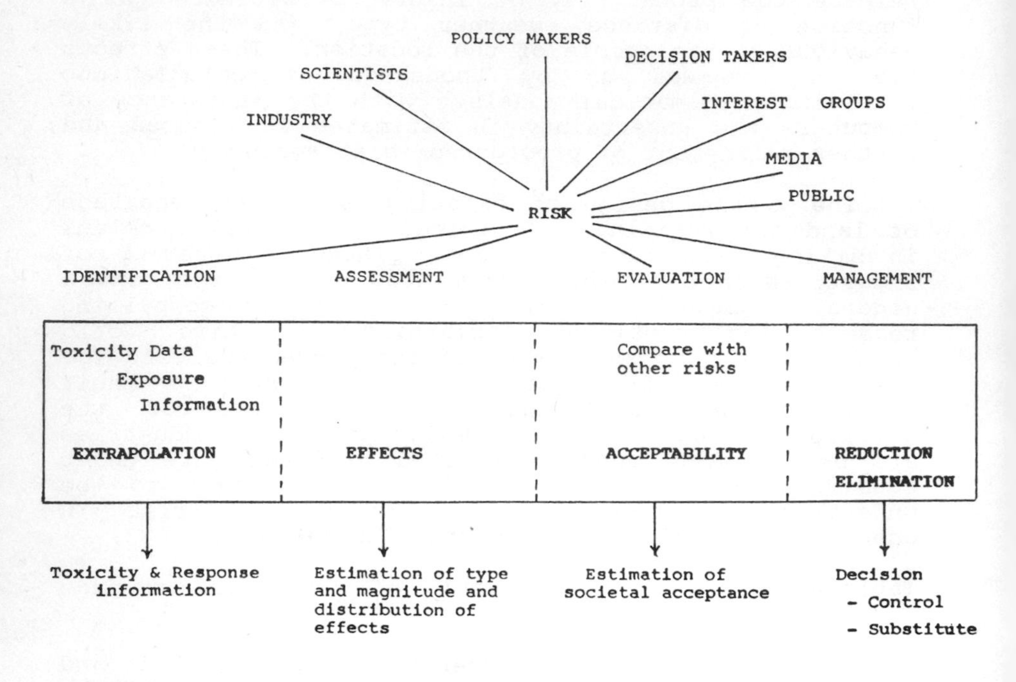

Figure 7 Risk assessment and management of chemical hazards

8 Emergency Response

Emergency response plans have to be devised at all levels: industry, locality, region, country. The very nature, complexity, and multi-dimensional impact of chemical disasters requires a multi-sectoral and multi-disciplinary approach in planning emergency response action. The response will be dictated by the specific situation encountered. Success of any response action is itself a measure of the effectiveness of preparedness planning undertaken by the country.

Due to the specific character of accidental releases and the magnitude of their off-site impact on public health and the environment, emergency response is a joint venture of civil authorities, rescue organizations, public utility and health services, regulatory agencies, community leaders, voluntary agencies or squads, the public, and the media.[47] Communication is the most vital element in all phases and hence the information support at the site must be prompt, precise, and reliable. Communication is essential not only within the network of the emergency response system but also with the public and the media.

It is recognized that the steps taken in the first few minutes/hours of an accident such as containment, 'fencing off', neutralization of effects, *etc.* determine the nature and gravity of the sequelae. This implies that response action has to be mediated by a well co-ordinated lead agency to oversee various inter-departmental efforts at every step. In addition, the need for a Central Agency has also been emphasized which, with the support of an appropriate technical infrastructure, gives directions and guidelines to the local lead agencies. Models for the local lead agency and the Central Agency are embodied in Figures 8 and 9 respectively.

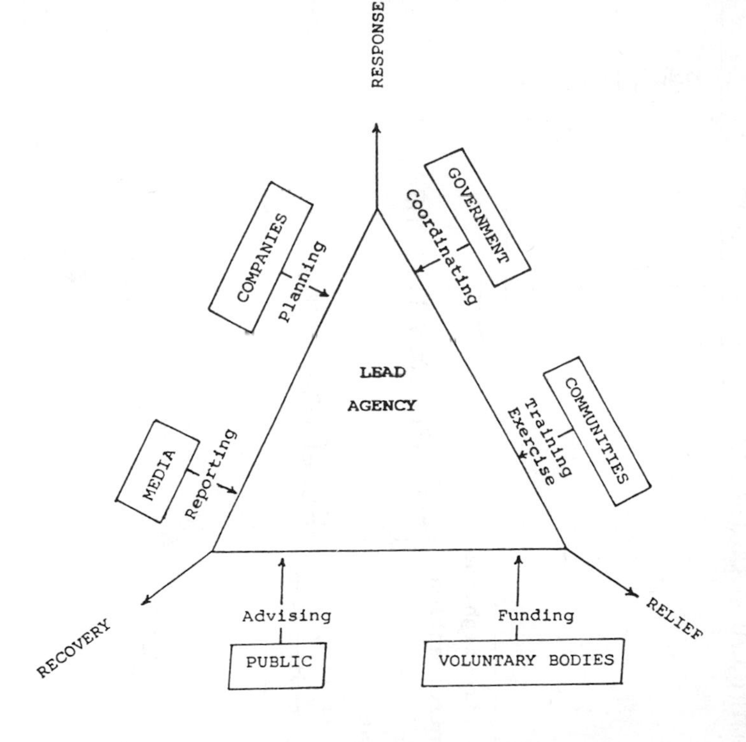

Figure 8 Lead agency for operating emergency response programme

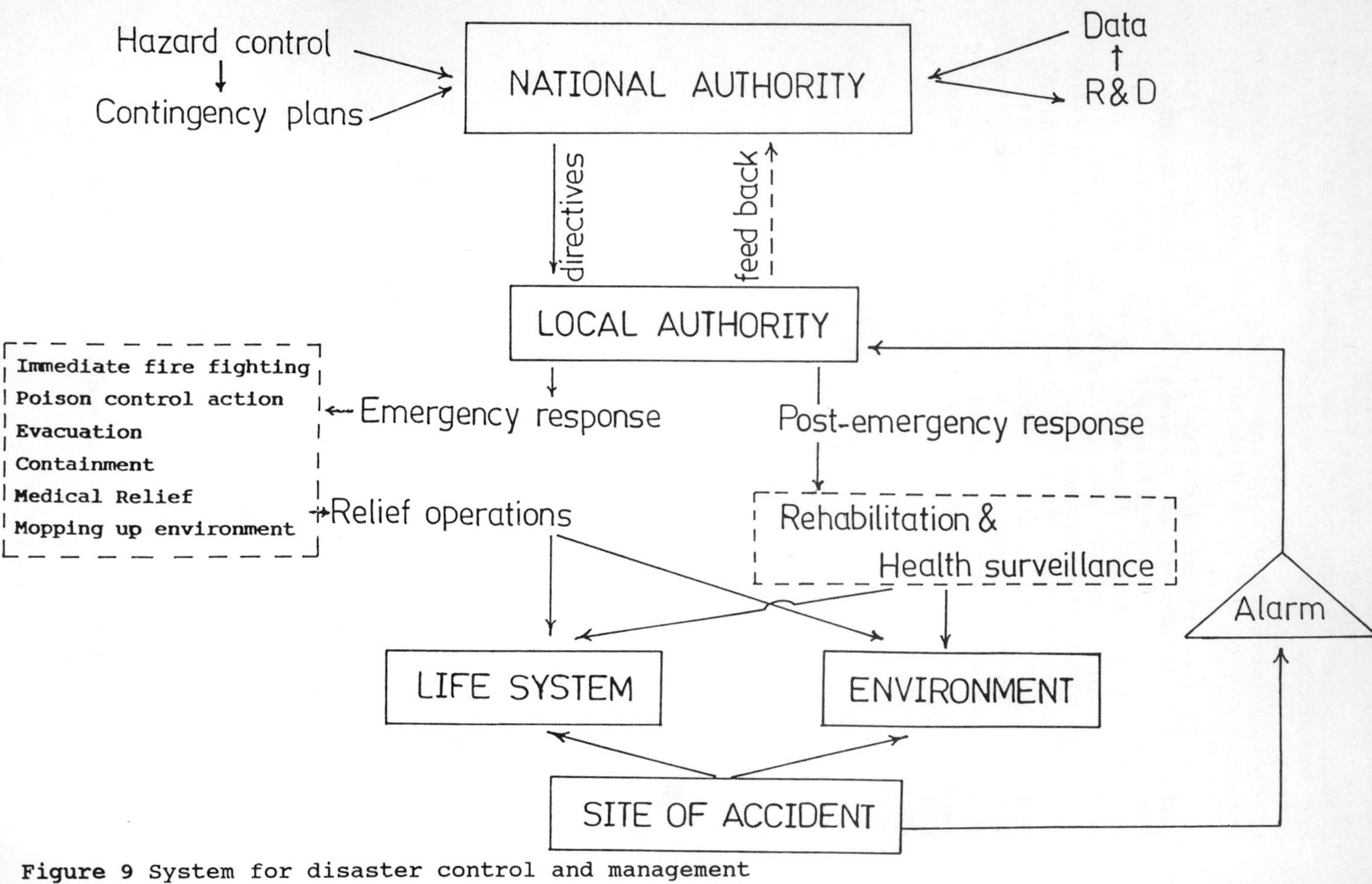

Figure 9 System for disaster control and management

A general scheme of action sequence to be triggered by a lead agency in a chemical emergency is outlined in Figure 10. The specific action sequence to be undertaken to deal with an accidental spill or release of a toxic chemical is shown in Figure 11. The construction of a chronology of the accident and action needed during emergency and post-emergency phases, as shown in Figure 12, is very helpful in designing a decision tree for the emergency command shown in Figure 13. The co-ordination of activities of a chemical response action plan requires delineation of different phases of action, allotment of the personnel to be deployed, and assignment of tasks to be completed. These are indicated in Table 7.

Table 7 *Co-ordination of activities related to chemical disaster response*

	Personnel	*Activities*
Emergency phase	Police	Law and order
	Fire Brigade	Rescue, fencing
	Transport	Evacuation
	Conservancy	Disposal of dead bodies
	Emergency squad	Decontamination
	Poison control centres	Detoxication
	Doctors, nurses	
	Social Workers	First aid
	Morgue/autopsy	Medico legal - tissue collection
	Environmental squad	Decontamination Disposal of affected vegetation, food, and soil
Post-emergency phase	Civil services	Water and food supply; civic amenities
	Hospital services	Intensive care and treatment
	Epidemiologists	Long-term health surveillance
	Environmental hygienists	Long-range environmental surveillance
	Occupational	Medical and occupational rehabilitation
	Planners, administrators	Social rehabilitation
	Community leaders	Counselling
	Psychological and legal counsellors	Claims, compensation

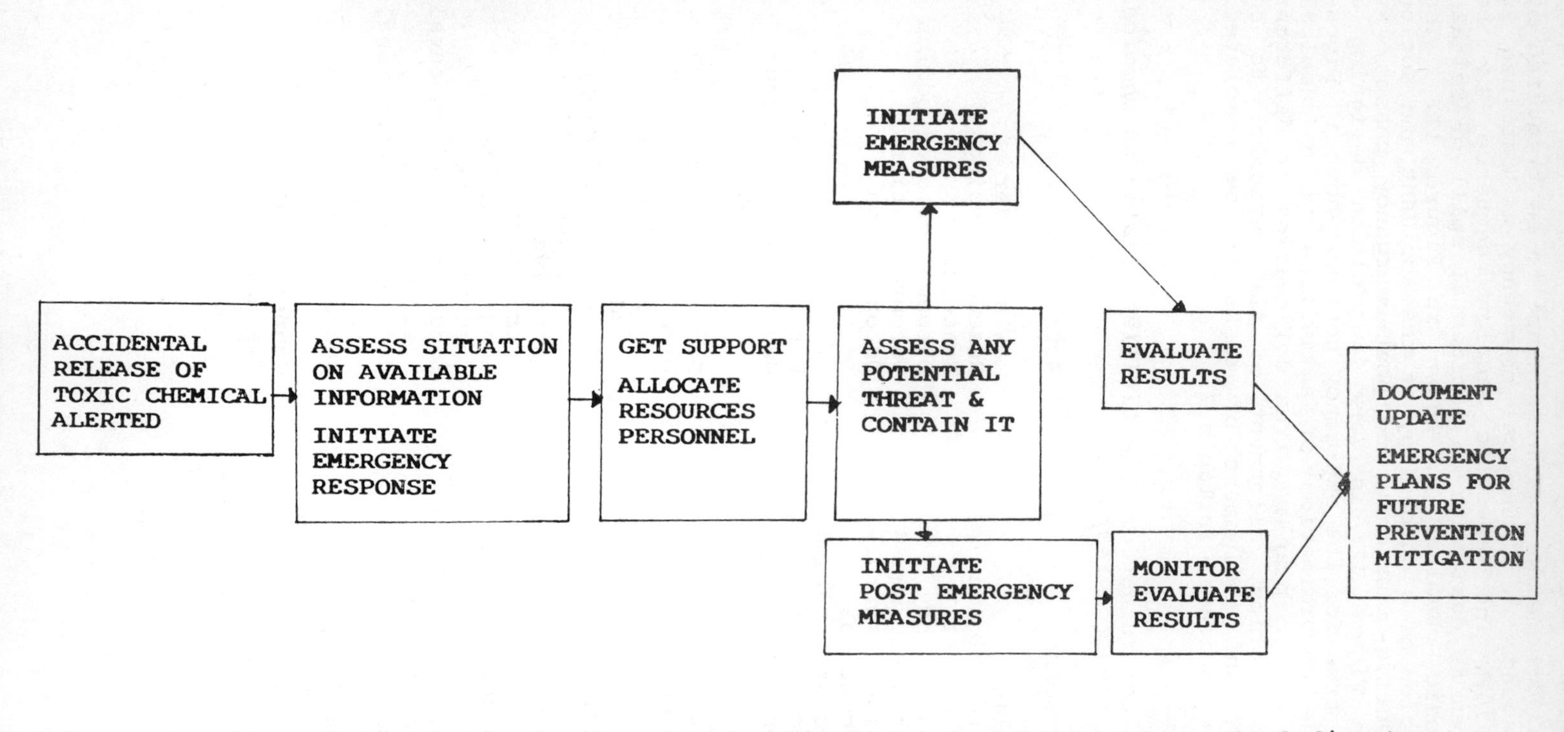

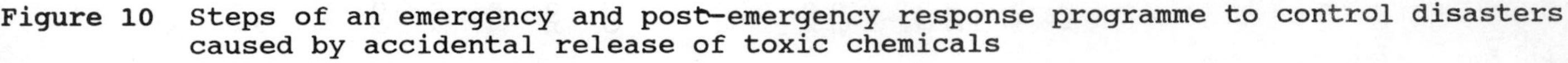

Figure 10 Steps of an emergency and post-emergency response programme to control disasters caused by accidental release of toxic chemicals

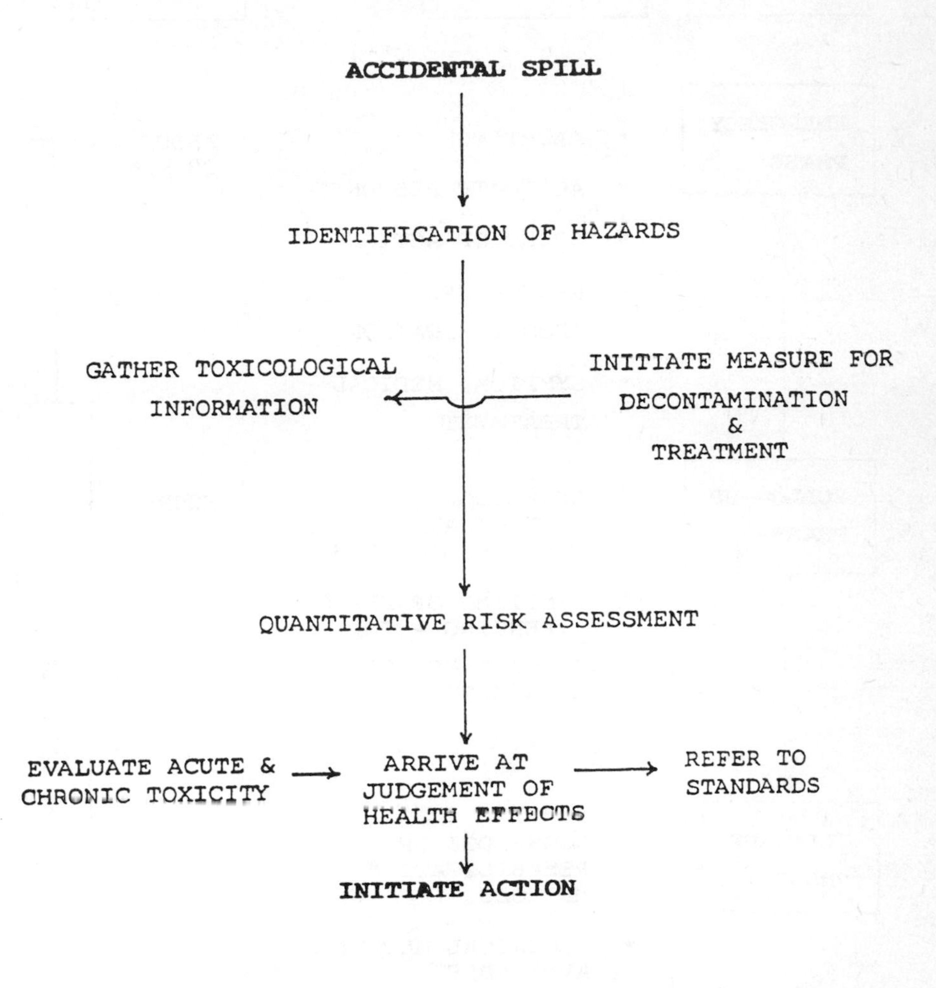

Figure 11 Sequence of tasks to be undertaken in emergency response

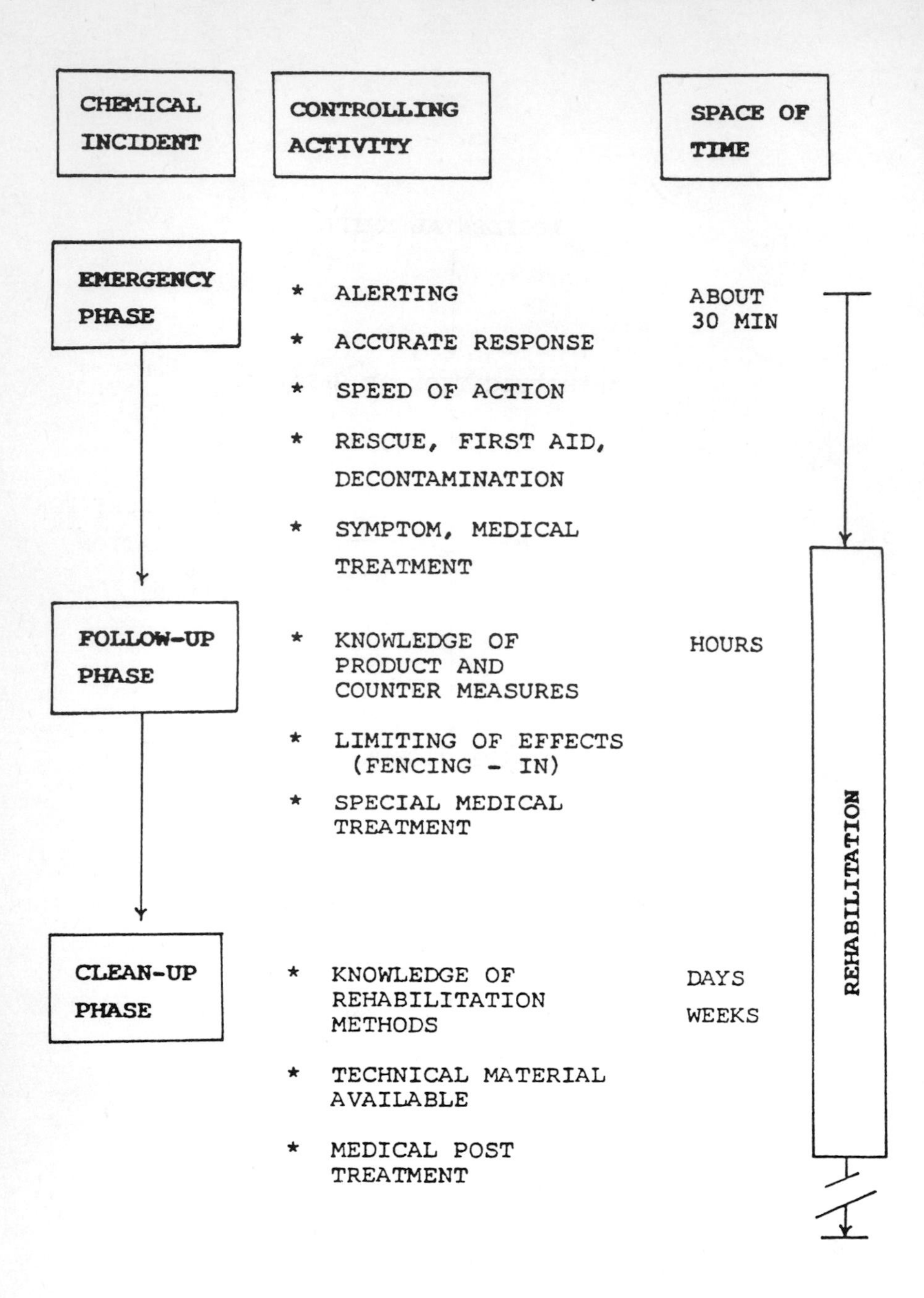

Figure 12 Chronology of a chemical accident

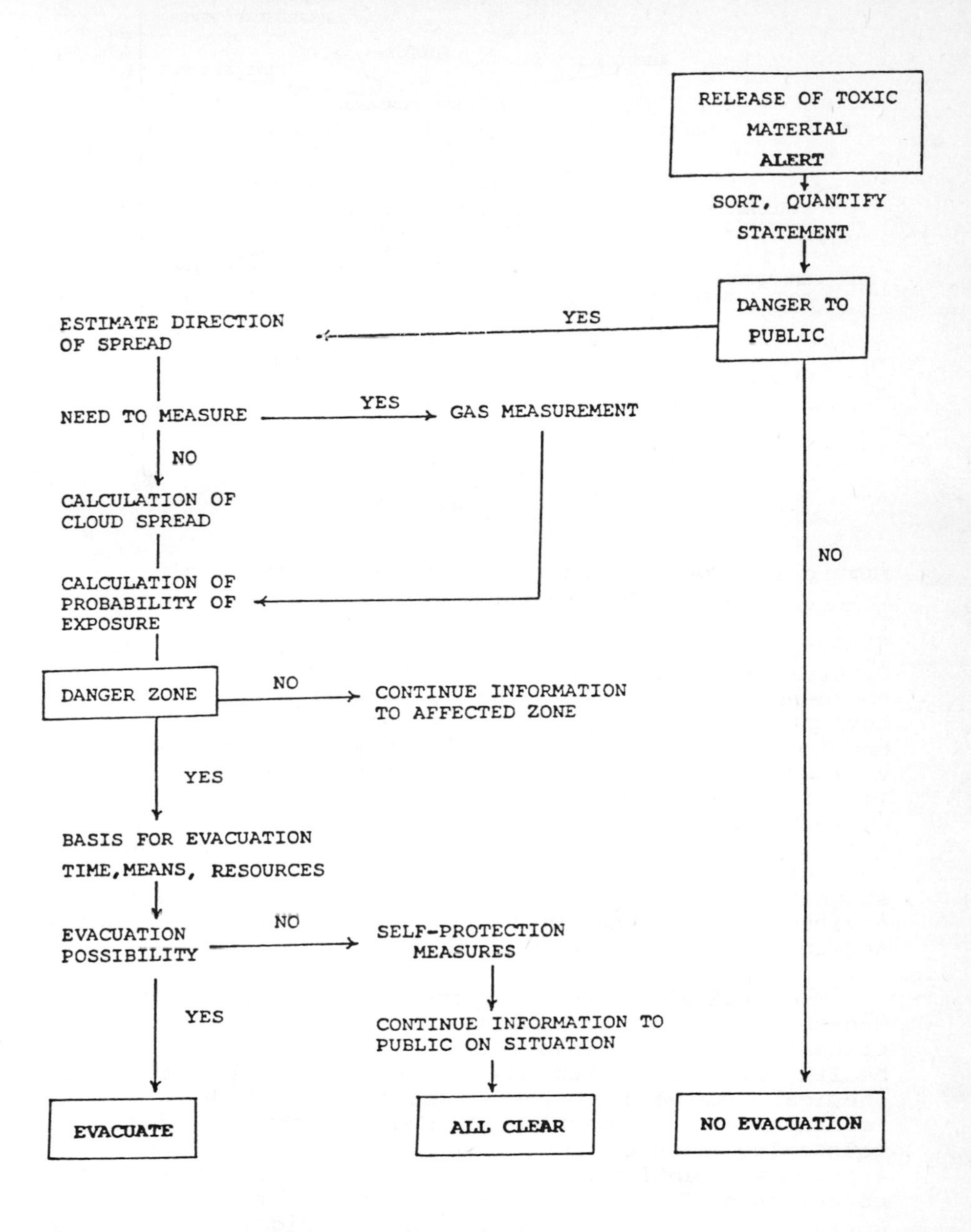

Figure 13 Decision tree for emergency command

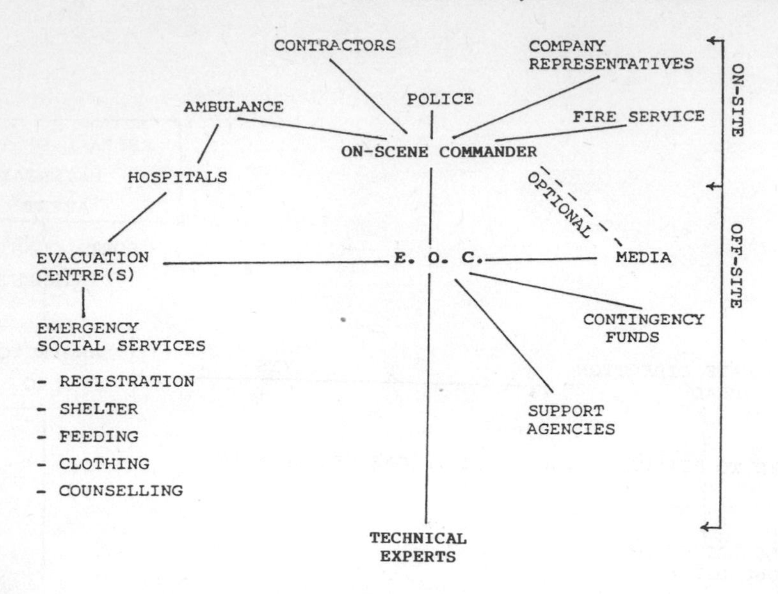

Figure 14 Command, control and co-ordination of emergency response

Toxicologists and Poison Information and Poison Control Centres have important roles to play in emergency response. The preventive role of the toxicologist is to suggest the safe level to which the chemical released in the accident should be reduced by various detoxication and mopping-up operations. The investigative role he has to play is to find out whether the population in the accident zone has been exposed to a dangerous dose of the chemical and whether serious clinical effects are likely to follow. He also suggests the procedures for decontamination of environmental compartments and detoxication of human and animal exposees of the chemical.

The medical service component of a Natural Disaster Management system is usually capable of taking care of traumatic physical injury and first aid and the public health component with the outbreak of epidemics. In chemical emergencies these services will have to be reinforced with expertise of clinical toxicologists for countering acute poisoning by administration of appropriate antidotal remedies. Some hospitals have established facilities for intensive care of people acutely poisoned by suicidal or homicidal attempts or by involuntary intakes. These require upgrading and expansion in order to deal with the clinical management of victims of acute exposure to accidentally released toxic chemicals.

In the post-emergency phase the emergency response team has to meet the challenges of medical, occupational, and social rehabilitation of human exposees. Obviously the institution of a long-range epidemiological programme becomes inevitable. The psychological impact of the calamitous event will have to be dealt with by both psychiatric treatment and counselling. Public confidence has to be restored. The legal aspects of the release and suffering will have to be attended to by appropriate procedures.

The immediate impact of the accidental release on water bodies and soil will have to be assessed and action taken to decontaminate the affected ecosystems. Structures damaged by blast or corrosion will have to be repaired. A continuous monitoring of environmental integrity and ecological balance will have to be instituted. Dairy cattle, food animals, fish, and poultry will have to be kept under vigilant observation for residual effects of intoxication. The safety of plant foods raised on soil exposed to the accidentally released chemicals and the safety of meat and milk of animals fed forage raised on contaminated soils will have to be evaluated.

9 Research Needs

Having recognized prevention as the best strategy of contingency planning, the research priorities have to be identified to devise more effective ways of prevention. Our current knowledge of the dynamics of runaway chemical reactions and mechanisms of accidents is far from complete. Similarly no adequate data on the construction materials used for fabrication of process, storage, and transport equipment in chemical industry are available. There is much more to be learned on the health sequelae of accidental exposures to chemicals and the environmental fate of the implicated chemicals. There is clearly wide scope for use of computer sciences for simulation of chemical disaster scenarios and the analysis of different aspects of technological and human failures in the safe running of hazardous installations.

Health effects of toxic chemicals might still appear years after the initial exposure. They are also modified by various environmental factors and the style of living. Longitudinal studies have to be conducted on exposees and ecosystems for several years. Environmental epidemiology is a relatively new discipline and is still in the developing phase. Inputs are urgently required from frontier areas of biomedical sciences, environmental hygiene, and statistics. In this context the initiative taken by WHO to develop environmental epidemiology is commendable.[48]

Information on the toxicity of many industrial chemicals is far from adequate. According to a National Research Council study, no reliable information is available for nearly 80% of the 48,500 chemicals listed in the US Environmental Protection Agency, inventory of toxic chemicals. Fewer than a fifth have been tested for acute effects and fewer than a tenth for chronic, reproductive, or mutagenic effects. The US National Academy of Sciences in its own review of a selected universe of 65,000 chemicals out of a *Chemical Abstracts* list of about 10^8 chemicals found toxicity information on 90% of the industrial chemicals to be far from adequate; see Table 8 for a summary of the findings published by the US National Academy of Sciences.

Table 8 *Health hazard assessment based on known toxicity information*

Category	*Size*	*Percentage*	
		Complete health hazard assessment	*No toxicity information available*
Pesticides and inert ingredients of pesticide formulation	3350	10	38
Cosmetic ingredients	3420	2	56
Drugs and excipients used in drug formulation	1815	18	25
Food additives	8627	5	46
Chemicals in commerce One million lb/year	12860	-	78
Chemicals in commerce <1 million/year	13911	-	76
Chemicals in commerce, production unknown/ inaccessible	21752	-	82

NAS Study 1984

As emphasized by Waller,[37] there have been conceptual shifts in our perception of health injury side by side with advances in biochemical toxicology and molecular mechanisms of xenobiotic interaction and transformation with living molecules. Some problem areas which could stimulate further research in related areas of molecular biology, biochemistry, and pharmacology are outlined in Figure 15. The usefulness of advanced Quantitative Structure Activity Relationship (QSARS) modules has also to be explored (see also Chapters 4 and 21).

The interlinked components of exposure assessment and toxicity evaluation require many refinements and more intensive multi-disciplinary efforts by many scientific groups at national and international levels. Above all, Chemical Disaster Management Plans at national levels will have to have the back-up support of R&D activities in the broad area of chemical safety and computerized databanks. Information stored in the databanks will not only include all accidents, hazards, and toxic (chemical) health effects from existing sources but will also be updated with more data emerging from research in the frontier areas indicated above.

10 Perspectives

Considering the frequency of occurrence of chemical accidents in many countries irrespective of their level of socio-economic development, and the anticipated phenomenal growth of the chemical industry in the Third World, the only effective solution for meeting the problems generated by chemical hazards is prevention. Analysis of past in-plant accidents in industrialized countries reveals that many of them could have been avoided, had some simple precautions been taken or basic rules of operational safety been strictly observed. 'Mini Bhopals' when allowed to build up over the years create the conditions optimal for major disasters. Since company interests normally receive the highest priority, there is an unfortunate tendency of management to downplay hazards or make them less visible. Thus it is an accepted practice with management in times of financial stringency to apply the axe first to mechanisms for maintenance of plant and worker safety or control of pollution.

Chemicals pose a real threat and even if the most extensive precautions are taken, accidents are likely to happen; hence the need for contingency planning. Chemical accidents exhibit features that call for approaches different from those used for controlling mechanical disasters or mitigating the injury caused by natural disasters. Several elements of society are affected by chemical disasters: the industrial sector,

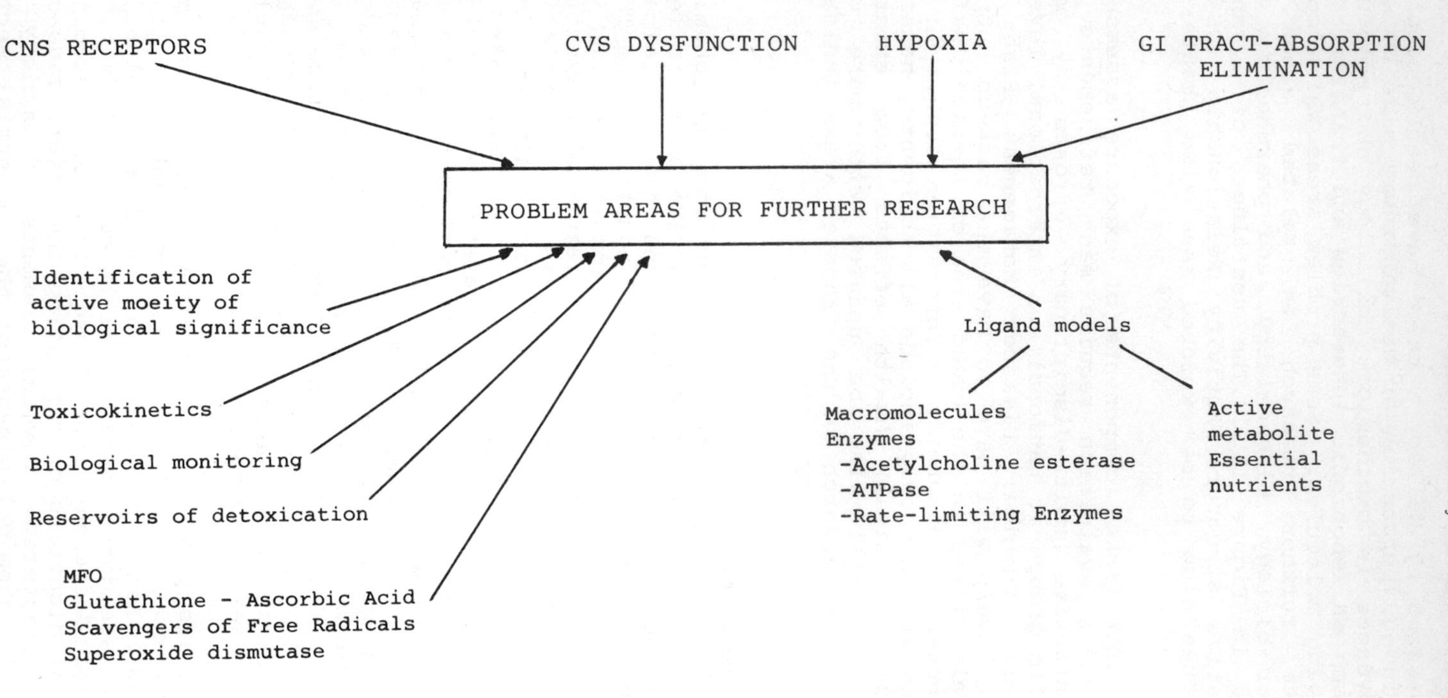

Figure 15 Unresolved problems in mechanisms of toxicity

the public health sector, environment, utilities, transport, and community welfare services; hence a multi-sectoral approach is needed.

The Commission of the European Communities has introduced a policy for prevention embodied in the Seveso Directive which is implemented jointly by the Commission and the Standing Committees of the National Competent Authorities. All industrial activities involving 178 dangerous chemicals are reported in a detailed Safety Report by the manufacturer/operator to the Competent Authorities in the form of notification. The Safety Report contains information on the dangerous substances on site, the technological processes used, and installations, together with the risk analysis, safety measures, and all information for dealing with emergencies. The Competent Authorities will include a dialogue on receipt of the Safety Report and follow-up any points as necessary. By periodic inspection they ensure that all appropriate measures for prevention of hazard have been taken. The operator has to train workers in safe handling and institute measures for emergency and arrange for the public in the vicinity to be informed in an appropriate manner of the safety measures and of the correct behaviour to adopt in the event of an accident (Seveso Directive and reference[49]).

In addition to stricter control within the industry, three major-hazard residual risk reduction strategies have been adopted in the United Kingdom by the Health & Safety Executive. These are: informing the public in the vicinity of hazardous installations of the nature of hazard and measures to be taken in event of an emergency, preparing and maintaining adequate off-site emergency plans to mitigate injury, and minimizing the number of people at risk by controlling the siting of hazardous industries. These have been fulfilled by the enactment of the Control of Industrial Major Accident Hazard (CIMAH) Regulations 1984 to implement the Seveso Directive in regard to hazardous installations and by legal and administrative arrangements with local councils.[45]

Recommendations put forward by a task force which conducted a Bhopal Aftermath Review in Canada to prevent accidents and mitigate injury caused by them include hazard analysis, siting and land-use control, safety management systems, community awareness, emergency preparedness, and emergency medical treatment.[50]

The United States of America has developed legislation requiring better emergency planning by state and local agencies. The Emergency Planning and Community Right to Know Act 1986 enjoins state commissions and local emergency planning committees to develop emergency preparedness plans for all facilities which use specific hazardous chemicals above a specific threshold quantity. Furthermore, facilities covered have to seek the appointment of a facility co-ordinator and to notify the local committee of all releases which exceed threshold levels.[51]

Effective community participation is essential for the successful implementation of any contingency plan for chemical safety. This would require creation of a better awareness among all concerned of the hazards. Policy makers in Third World Countries and bilateral and multilateral developmental agencies treat industrialization as virtually synonymous with development. It is not appreciated readily that technology transplantation and adaptation to the local *milieu* has its own inherent risks. The decision on choice of technology and risk are made on the 'usually unstated' premise that societal development necessitates the taking of acceptable risks' by those who have presently very little say in the matter.[52] This will no longer be accepted by the public of these countries who are rapidly becoming sensitive to the issues of safety and risk of developmental goals. This aspect has, however, been fully recognized by the European Confederation of Chemical Industries (CEFIC),[53] which has prepared and published 'Principles and Guidelines for the Safe Transfer of Technology' (CEFIC, April 1987).

The transboundary nature of chemical disasters calls for international efforts in prevention and mitigation of injury. A globally applicable hazard identification system has to be achieved at an early date. Analysis reveals that disasters arise out of a 'conventional not uncommon situation' and hence can be forestalled. Disasters reveal that a 'tragedy must occur before corrective measures are taken'. Disasters stress the 'severe deficiencies in national and international legal arrangements for dealing with the consequences'. Differing national standards exacerbate the problem of prevention as well as mitigation of injury. Disasters 'underscore the urgency of developing meaningful international standards and creating or strengthening international institutions' to deal with them.[54]

Although runaway reactions leading to chemical disasters are rare, when they occur their magnitude and impact unfortunately overwhelm us. Systems and human failures with serious consequences are characteristics which we have not understood thoroughly. Behavioural sciences have to answer many questions that arise on the complexities of man-machine relationship and the interface of high technology and ultimate human welfare. There is a compelling necessity to comprehend the inherent complexities and failures and overcome them. In the meantime the unending demands for certain chemicals essential for upholding the quality of our life must be met by cleaner and safer technologies.

11 References

1. UNEP 'The state of the Environment, 1986', Environment and Health, United Nations Environment Programme, Nairobi, 1986.

2. C. Perrow, 'Normal Accidents. Living with High-Risk Technologies', Basic Books, New York 1984.

3. Friedrick Naumann Foundation, International Center for Law in Development, 'Industrial Hazards in a Transnational World: Risk, Equity and Empowerment', First Report of a Joint Programme on International Standards Governing Hazardous Technologies, Council of International and Public Affairs, New York, 1987.

4. M. Abraham, 'The Lessons of Bhopal: A Community Action Resource Manual on Hazardous Technologies', International Organization of Consumers Unions, Penang, Malaysia, 1985.

5. M.F. Lechat, 'Disaster Epidemiology', Proceedings of an International Colloqium. 15-17 Dec. Prince Leopold Institute of Tropical Medicine. Antwerp, 1975.

6. J.T. Jones, 'Disaster Preparedness and Relief Development of Programmes'.

7. B. Bowonder, J.X. Kasperson, and R.E. Kasperson, *Environment*, 1985, **27**, 6-13, 31-37.

8. A.M. Natkin, 'Once is too often - Corporate Responsibility in the Aftermath of Bhopal', Essays Journal '85, World Resources Institute, Washington, 1985, pp.62-67.

9. B.I. Castleman, and V. Novarro, *Ann. Rev. Public Health,* 1987, **8**, 1-19.

10. J. Vilain, 'The Nature of Chemical Hazards. Their Accident Potential and Consequences'. Workshop on Methods to Reduce Injury Due to Chemical Accidents, New Delhi, Jan.27-Feb.2, 1987, Scientific Group on Methodologies for Safety Evaluation of Chemicals (IPCS/SCOPE), to be published.

11. ILO, Working Paper on Control of Major Hazards in Industry and Prevention of Major Accidents. International Labour Office, Geneva, 51 pp.

12. IUPAC, Report of the IUPAC Study Group on Safety in Production and Application of Chemical Products, May 2-3, New Delhi, 1986.

13. World Bank, 'Manual of Industrial Hazards Assessment Techniques', Office of Environment & Scientific Affairs, The World Bank, Washington, DC, 170 pp., 1985.

14. SGOMSEC VI Workshop. Methods to Reduce Injury due to Chemical Accidents. New Delhi Jan.27-Feb.2, 1987, Scientific Group on Methodologies for Safety Evaluation of Chemicals (co-sponsored by International Programme on Chemical and Scientific Committee on Problems of Environment). To be published as a volume in SCOPE series, John Wiley Sons.

15. UNIDO, Workshop on Hazardous Materials/Waste Management, Industrial Safety in Chemical Industry and Emergency Planning: Guidelines for Government and Industry - A Plan of Action for UNIDO, 22-26 June 1987, Vienna. United Nations Industrial Development Organization, Vienna, 1987.

16. WHO, 'Health Safety Component of Environmental Impact Assessment', Environmental Health 15, Report of a WHO Meeting, World Health Organization Regional Office for Europe, Copenhagen, 1987.

17. World Conference on Chemical Accidents, Rome, July 1987. World Health Organization Regional Office for Europe, International Programme on Chemical Safety and Instituto Superiore di Sanita, Published by CEP Consultants Edinburgh, 1987.

18. UNEP Panel of Experts on Industrial Accidents, Nairobi, 1-3 June 1987, United Nations Environment Programme. Environment Industry Office, 10897.

19. USEPA International Symposium on Preventing Major Chemical Accidents, Washington, Feb.3-7, 1987. Center for Chemical Process Safety of the American Institute of Chemical Engineers, United States Environment Protection Agency, World Bank, Washington DC, Preprint of Papers.

20. C.R. Krishna Murti, 'Impact of Chemical Disasters on Environmental Health', Nagpur, Oct.14-18, 1986. Disaster Management. A Workshop organized by Directorate General and Health Service, Ministry of Health and Family Welfare, Government of India. Sponsored by World Health Organization - Regional Office for South East Asia, New Delhi, 1986.

21. C.R. Krishna Murti, *Curr. Sci.*, 1986, 55, 1064-1066.

22. C.R. Krishna Murti, 'Preparedness Planning for Mitigating Toxic Injury Due to Chemical Disaster', in ref. 17, pp. 213-222.

23. S. Varadarajan, 'Initial Management of Event. Methods to Contain the Chemical Agents in Order to Limit the Potential Exposure Following Accidental Release of Chemicals', Scientific Group on Methods for Safety Evaluation of Chemicals Workshop 6. New Delhi, Jan.27-Feb.2, 1987.

24. J. Mcquard, 'Dispersal of Chemicals', Paper SGOMSEC VI, Workshop on Methods to Reduce Injury Due to Chemical Accidents, New Delhi, Jan.27-Feb.2, 1987. Scientific Group on Methodologies for Safety Evaluation of Chemicals, co-sponsored by International Programme on Chemical Safety & Scientific Committee on Problems of Environment, 1987.

25. H.G. Lawley, *Loss Prevention*, 1974, 8, 105.

26. 'A Guide to Hazard Operability Studies', Chemical Industries Association, London, 1977.

27. S.T. Parry, 'A Review of Hazard Identification Techniques and their Application to Major Accident Hazards', United Kingdom Atomic Energy Authority, Wigshaw Lane, Cluchtech, Warrington, 1986.

28. R.D. Turney, 'Identification and Control Accidental Sources on Chemical Plant', in ref. 17, pp.109-112.

29. Health & Safety Commission. Advisory Committee on Major Hazards, Report, 1983, Her Majesty's Stationery Office, London, 1983.

30. T.A. Kletz, 'What Went Wrong? Case Histories of Process Plant Disasters', Gulf Publishing Company Houston, Texas, 1985, pp.204.

31. K. Cassidy, *Chem. and Ind. (London)*, 1987, 80.

32. S. Belrado, A. Howell, R. Ryan, and W.A. Wallace, *Disasters*, 1983, 7, 215-220.

33. C.K. Johnson and S.R. Jordan, 'Emergency Management on Inland Oil and Hazardous Chemical Spills. A Case Study in Knowledge Engineering', 'Building Expert Systems', Addison-Wesley Publishing Co. Massachusetts, 1983.

34. J.A. Barton and P.F. Nolan, 'Runaway Reactions in Batch Processes: Institution of Chemical Engineers Symposium, Series No.85, 1984, pp.13-21.

35. The Trade Union Report on Bhopal, July 1985. International Federation of Free Trade Unions and the International Federation of Chemical, Energy, and General Workers Unions.

36. W. Morehouse and M.A. Subramaniam, 'The Bhopal Tragedy. What Really Happened and What it Means for American Workers and Communities at Risk', Council of International and Public Affairs, New York, 1986.

37. J.A. Waller, *Ann. Rev. Public Health*, 1987, 8, 21-49.

38. F. Pocchiary, V. Silano, and G.A. Zapponi, 'Rehabilitation and Reclamation after Chemical Accidents', in ref. 17, p.19.

39. A. Macri and A. Mantovani, 'Animal Health as a Component of Public and Environmental Health in Chemical Emergencies', in ref. 17, pp.215-218.

40. V. Silano, 'Evaluation of Public Health Hazards Associated with Chemical Accidents, ECO/PAHO/WHO, Pan American Centre for Human Ecology and Health, Mexico, 1985, 95 pp.

41. F. Pocchiari, V. Silano and G.A. Zapponi, 'The Seveso Accident and its Aftermath', in 'Insuring and Managing Hazardous Risks, From Seveso to Bhopal and Beyond', ed. P. Kleindorfer and H. Kunveuther, Springer Verlag, Berlin, 1985.

45. A.F. Ellis, 'Land-Use Planning in Major Hazard Control', in ref.17, pp.56-59.

46. A. Gilad and V. Silano, 'Contingency Planning and Emergencies Involving the Release of Potentially Toxic Chemicals, *J. Tox. Med.*, 1983, 3, 67-90.

45. Health and Safety Executive, London, 'Monitoring Safety', Occasional Paper Series: OP9, Her Majesty's Stationery Office, London, 1985.

46. G. Thyagarajan, 'New Approaches to Chemical Industry Planning in Developing Countries, in ref. 17, pp.37-40.

47. Guide to Emergency Planning Society of Industrial Emergency Services Officers, Paramount Publishing, 1987.

48. WHO, 'Guidelines on Studies on Environmental Epidemiology'. Environmental Health Criteria 27, published under the Joint Sponsorship of United Nations Environment Programme,International Labour Organization, and World Health Organization, World Health Organization, Geneva, 1985.

49. P. Testori-Coggi, 'The Community Policy on Major Accidents Hazards: Important Achievements in an On-going Process', in ref. 17, pp.26-28.

50. D.W. Bissett and R. Morcos, 'Major Industrial Accidents, An Assessment of the Canadian Situation', in ref. 17, pp.93-96.

51. T.C. Voltaggio, and M.J. Zickler, 'The UFPA's Post-Bhopal Review of Toxic Substances Emergency Response and Preparedness Plan in Kanawha Valley, in ref. 17, pp.203-206.

52. C.J. Dias, 'Nothing to Lose but their Lives. Distribution of Risks or Imposition of Risks', in ref.3.

53. Counseil Européen des Fédérations de l'Industrie Chimique (CEFIC), 'Principles and Guidelines for the Safe Transfer of Technology', Bruxelles, Belgium, 15 April 1987.

54. H. Gleckman, 'International Standards and the Role of the United Nations System', First Workshop of the Joint Programme on International Standards Governing Hazardous Technologies, June 25, 1986 and on United Nations Economic and Social Council, Commission on Transnational Corporations, In 'Transnational Corporations and Issues Relating to the Environment', Report of the Secretary General, New York, The Commission, 13th Session April 7-16, 1987, E/C 10/1987/12 February 4, 1987.

Section 2: Contribution of Toxicology to Risk Assessment

8
Risk Assessment Techniques for Carcinogenic Chemicals

S.C. Batt and P.J. Peterson

MONITORING AND ASSESSMENT RESEARCH CENTRE, THE OCTAGON BUILDING, 475a FULHAM ROAD, LONDON SW10 0QX, UK

1 Introduction

One of the major environmental issues of concern to both scientists and administrators is the control of risks due to the manufacture, use, and disposal of chemicals. This concern arises from the increasing number of chemicals which are being shown to have toxic properties in addition to increasing levels of production and hence potential for exposure. This situation has led to the realization that increased legislative control is required to ensure adequate protection of human health. Control measures can limit the presence of hazardous chemicals in environmental media or regulate their mode of use, thus reducing human exposure and potential health risks. Absolute health risk assessment, *i.e.* the assessment of whether a chemical may cause an adverse health effect and at what level of exposure and with what frequency or probability, provides the fundamental basis for appropriate regulatory/control measures.

Since the historic observation, more than two centuries ago, by Percival Pott that chimney sweeps exhibited a high incidence of scrotal cancer (see also Chapter 11), with the resulting identification of soot as the causative agent, it has been established that a high proportion of human cancers are associated with chemicals. Information is still lacking on what proportion of human cancer is due to occupational, environmental, and consumer exposure. Estimates for the USA have attributed 25-30% of all cancer deaths to tobacco, 2 - 4% to alcohol, 10 - 70% to diet, and 2 - 10% to occupational exposures.[1] The steady, almost exponential, increase in suspected or proven occupational carcinogens over the past 10 to 15 years (Figure 1) serves to illustrate the importance of assessing potential carcinogenic risks associated with both existing and new chemicals present in the environment.

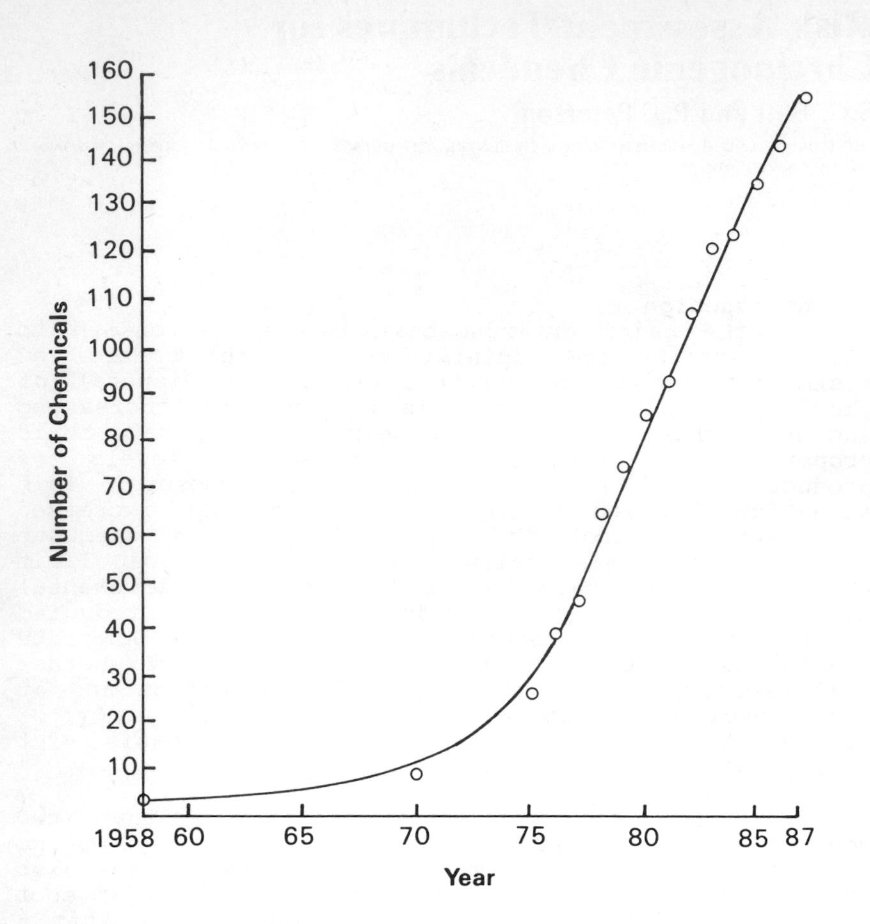

Figure 1 Increase of suspected or proven occupational carcinogens in the German list of MAK-values (data adapted from Ref. 2)

In general terms the process of health risk assessment consists of three components - hazard prediction/identification, risk estimation, and risk evaluation (for definitions refer to Glossary of Terms). Figure 2 shows the interrelationship of these components.

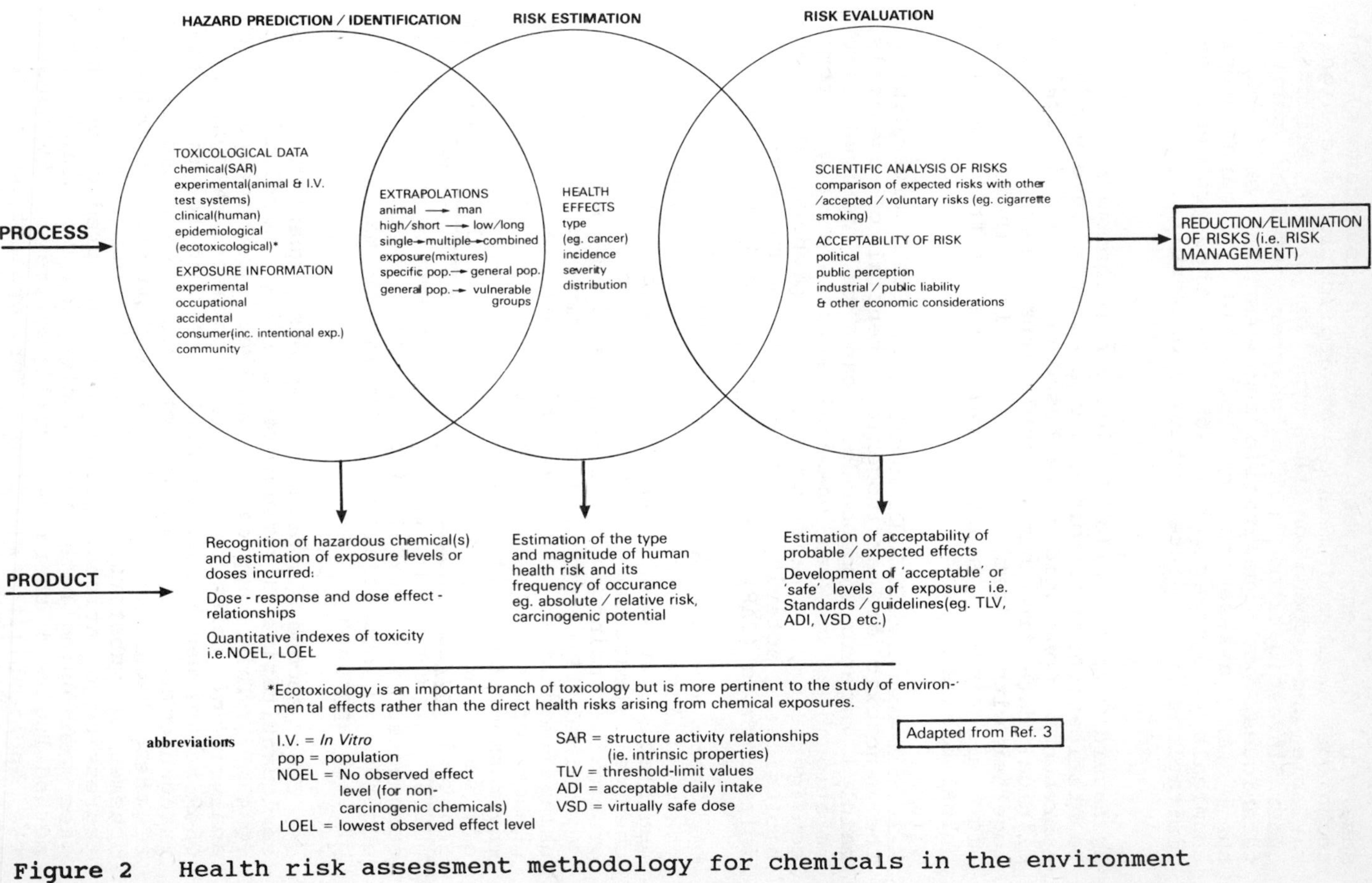

Figure 2 Health risk assessment methodology for chemicals in the environment

Although the overall methodology is basically the same for different types of chemical substances, the techniques used and methods of application are highly variable and depend on the type of health effect(s) under investigation. This chapter outlines current techniques, both qualitative and quantitative, used in the risk assessment of potentially carcinogenic chemicals, together with some of the problems and limitations in the interpretation of results.

2 Hazard Prediction/Identification Techniques

Identification of carcinogenic chemicals involves many approaches. Conclusions are based on epidemiological or clinical data from human populations, where available, together with information obtained from *in vivo* and *in vitro* experimental studies, particularly from long-term animal bioassays.

2.1 Chemical Structure Analysis. Predictive data on the genotoxic or carcinogenic potential of chemicals, mainly organic compounds, can be derived from relationships between chemical structure (*e.g.* atomic configuration, physicochemical properties) and biological activity [so-called structure-activity relationships (SAR)] (see also Chapters 4 and 21).

There are two main approaches to hazard prediction that utilize SARs. The first is essentially a qualitative approach and involves a comparison of the chemical structure with that of other compounds whose genotoxic/carcinogenic potentials have already been determined. Procedures for predicting carcinogenicity using qualitative SARs have been reviewed for a large number of chemicals, many of which are potential environmental pollutants.[4,5] The scope and usefulness of such analyses depend on a knowledge of the genotoxicity/carcinogenicity for a sufficiently large number of chemicals which have common structures or subgroups. The qualitative nature of the resulting estimate of potential toxicity is the main limitation of this approach.

The second approach employs quantitative or semiquantitative structure-activity relationships (QSARs). This approach involves the computerized analysis of relatively large databases. A number of QSAR techniques have been well developed for predicting mutagenic and carcinogenic potential. A basic feature of these techniques is the use of pattern recognition schemes or substituent weighting factors, coupled with regression equations. Although useful predictions have been made using both quantitative and semiquantitative methodologies,[4] there are still severe limitations to the applicability of QSARs for predicting toxicity profiles.

At present the most common problem with QSAR techniques is that of 'outliers', *i.e.* certain chemicals in a series of 'closely-related structures' which do not fit the derived toxicity relationship. Although reasons have been proposed to explain why known outliers do not fit certain QSAR's these are mainly speculative and until this problem has been resolved the potential of these techniques will remain limited.[6]

An interesting development in structure analysis is the attempt to predict metabolic pathways, and hence potential active metabolites of chemical compounds. In view of the known importance of metabolic activation in chemical carcinogenesis, such an approach promises considerable potential in carcinogenic prediction.

2.2 Short-term Tests. Much effort has been devoted to the development of rapid screening tests for the prediction of chemical carcinogens. Based on current understanding of carcinogenic mechanisms, most established tests are designed to evaluate the genetic activity of a substance and employ as end-points, well-defined genetic markers in prokaryotes, lower eukaryotes, and in mammalian cell lines. Tests can be grouped according to the end-point detected, *i.e.* DNA damage, gene mutation, chromosome effects, cell transformation effects, *etc.*

In a recent review of the literature McCann et al.[7] identified over 100 different short-term testing methods although less than 20 are in regular use. For a detailed appraisal of short-term testing techniques the reader is referred to IARC 1980[8] and IPCS 1985.[9]

Carcinogenic chemicals may induce one or a number of genetic changes and therefore the assessment of genotoxic hazard should include a combination of tests with different end-points. However, not all carcinogens are genotoxic.

Before short-term assays can be used to assess potential human carcinogenic activity with any degree of confidence, their validity should be thoroughly evaluated. The most ambitious validation exercise to date was the International Program for the Evaluation of Short-term Tests for Carcinogenicity,[10] in which some 30 *in vitro* and *in vivo* assays were evaluated for their ability to discriminate between carcinogenic and non-carcinogenic compounds. Bacterial mutation assays gave the best overall performance with a predictive value for carcinogenic potential of around 90%. Other assays that discriminated well between carcinogens and non-carcinogens included *in vivo* and *in vitro* tests for chromosome damage, and for unscheduled DNA synthesis, and assays using yeasts and results from *Drosophila* tests and *in vitro* gene mutation assays. Other more recent publications have indicated a lower predictive potential for mutagenicity assays.[11-13]

The potential for short-term tests to provide quantitative data is much less certain than their potential to produce qualitative information. There is, however, suggestive evidence that some quantitative correlation in potency between short-term tests and animal carcinogenicity bioassays may exist.[14] It is currently recognized that more data are needed on dose-response relationships from short-term tests in order to develop techniques for estimating 'genotoxic potential'. Nevertheless, efforts have been made to construct a database of 'carcinogenic potency' based on a quantitative analysis of mutagenicity data for a variety of chemicals.[7]

In general, short-term tests are a useful tool in carcinogenic hazard prediction. However, these tests have a number of limitations that should be fully realized in the interpretation of results. As an example, specific potentials and limitations of three different assays used to assess mutagenic activity are shown in Table 1.

Table 1 *Specific potentials and limitations of selected short-term tests for mutagenic activity*[4]

Salmonella typhimurium (Ames Test)

Potentials

1. Reliable in testing known carcinogens and non-carcinogens: false positives, <10%; false negatives, <10%
2. Has extensive database
3. Tester strains are extensively engineered to enhance sensitivity
4. Results in different tester strains provide information about the mechanism of mutation (*e.g.* frameshift *vs.* base pair substitution.
5. Known human chemical carcinogens are positive. These include 2-naphthylamine, benzidine, bis(chloromethyl) ether, aflatoxin B_1, vinyl chloride, 4-aminobiphenyl
6. Useful as a tool in rapidly obtaining information about the potential mutagenic/carcinogenic activity of uncharacterized compounds in complex mixtures; *e.g.* has been applied to cigarette smoke condensate and fractions, hair dyes, soot from city air, *etc.*
7. Valuable as a bioassay to direct the chemical fractionation and identification of mutagenic components in complex mixtures

Limitations
1. Considered qualitative in nature
2. Not responsive to metals, chlorinated hydrocarbons (long chain), asbestos and like particles
3. Sample must be sterile (problems in sample sterilization)
4. Highly toxic components of complex samples may mask mutagenic properties of other chemicals
5. Organic solvents used in chemical fractionation (*e.g.* methylene chloride) may be mutagenic and therefore may require removal (*e.g.* by solvent exchange) prior to bioassay

Sex-linked recessive lethal test in *Drosophila melanogaster*

Potentials
1. A large number of the test organisms can be raised easily and economically
2. Metabolic activation is endogenous
3. Genotoxic effects are evaluated in the germ cells
4. Many loci in the genome are monitored at the same time
5. Has large database on chemical mutagens

Limitations
1. Responds poorly to mutagenic polycyclic aromatic hydrocarbons
2. Insecticides cannot be evaluated

Mammalian cell mutagenesis *in vitro*

Potentials
1. Allows observation of the full range of mutation responses of somatic cells, not just point mutations
2. Faster and Less expensive than *in vivo assays*
3. Can be coupled with *in vitro* metabolic activation systems from various sources
4. Mutations measured in cultured mammalian cells are more relevant to the intact mammal than are mutations measured in bacterial systems
5. Responds to particulates (clastogenic effect)
6. May offer better quantitative correlation with carcinogenicity than the Ames test (preliminary results)

Limitations
1. The cell lines used for experimentation are transformed lines and therefore do not represent 'normal' cells
2. These cells do not undergo meiosis
3. Only one or two loci from the entire genome are monitored
4. Cells have limited endogenous metabolic activation capability

5. There are some doubts about the genetic nature of the events observed at certain loci

2.3 'Limited *in vivo* Bioassays'. This level of assessment employs *in vivo* tests that provide further evidence in a relatively short period (*i.e.* 40 weeks or less) of potential carcinogenicity. Some tests also provide relative potency ratings when the design includes known carcinogens. Unlike short-term tests, these assays are not applied as a 'battery', but are selected according to information available on the chemical. Such tests include: the induction of skin and pulmonary neoplasms in mice, breast cancer induction in female rats, and altered foci induction in rodent liver. An evaluation of these assessment techniques has been undertaken by Wiesburger and Williams.[15] It was concluded that negative results in any of these tests cannot be taken as strong evidence of non-carcinogenicity, since all are limited in their end-point.

2.4 Interpretation of Data from Predictive Studies. Although positive data from structure analysis, short-term tests, and limited bioassays are suggestive of a potential carcinogenic hazard, negative data cannot be taken as strong evidence of non-carcinogenicity since all studies are limited in their end-points.

2.5 Long-term (Chronic) Animal Bioassays. It is generally considered that in the absence of adequate epidemiological data, long-term, animal bioassays provide the most important information on the potential carcinogenic risks from exposure to chemicals. Experience has shown that most known human carcinogens are also carcinogenic in one or several animal species, even though the tumour type may be different from those elicited in man. The basic premise of carcinogenic bioassays is therefore that chemical substances producing tumours in rodents and other experimental animals will affect human cells similarly. It is beyond the scope of this chapter to review the principles and methods involved in chronic animal carcinogenicity testing and the reader is referred to references 16 and 17 for further details. Generalized issues with respect to the use of experimental animal studies in risk assessment are discussed in Chapter 9. In addition to indicating the presence or absence of tumours, long-term animal bioassays provide important quantitative information, the most fundamental of which is the excess incidence of tumours in exposed animals compared to control groups and the proportion of malignant to benign. More sensitive information is obtained if the parameters for assessment include the latency period or 'time to tumour'. The latter is often expressed as the time for 50 or 100% animals to manifest tumours. Such a classification provides an indication of potency.[18]

Other relevant parameters of effect include the number of species and strains affected, and the type and numbers of tumours which, when considered in relation to dosage, yield a more refined estimate of dose-response (see section 3.1). In view of the well-known role of biotransformation in carcinogen activation, supplementary metabolic and pharmacokinetic data are also extremely important in assessing the relevance of results from chronic bioassays to man (see Chapter 9 for examples)

2.6 Clinical and Epidemiological Studies. The strongest evidence of a chemical's carcinogenicity to humans is derived from clinical and epidemiological studies. Epidemiological approaches to the assessment of carcinogenic risks are dealt with in Chapters 10 to 12 and will not be discussed here.

The main weakness of both clinical and epidemiological studies is that they are relatively ineffective in proving that certain cancers are directly caused by exposure to a particular chemical. Unless the carcinogenic effect of the chemical is very unusual, as in the case of angiosarcoma of the liver caused by exposure to vinyl chloride, it may pass unnoticed in a normal survey. Another limitation to studies of this kind is the lack of reliable exposure data and consequently dose-response relationships can only rarely be established. Factors such as smoking and concomitant exposure to other chemicals are often important confounding variables in human studies and present a particular problem with respect to interpretation of carcinogenicity data.

Potentially carcinogenic chemicals may be studied by monitoring genetic effects in exposed populations.[19] This is most conventionally carried out by taking blood samples and examining peripheral lymphocytes for genetic damage.[20] The International Agency for Research on Cancer (IARC) reviews epidemiological evidence for the carcinogenicity of chemicals on a continuous basis. Chemicals that have been positively identified as human carcinogens are listed in Table 2.

Table 2 *Chemical substances established as human carcinogens*[21]

Agent	*Exposure*[a] *A*	*B*	*C*	*Site of cancer*
Aflatoxin			+	Liver
Alcoholic drinks			+	Mouth, oesophagus

Table 2 (continued)

	A	*B*	*C*	
Aromatic amines				
4-Aminobiphenyl	+			Bladder
Benzidine	+			Bladder
2-Naphthylamine	+			Bladder
Arsenic[b]	+	+		Skin, lung
Asbestos	+			Lung pleura, peritoneum
Azothrioprine				Reticulo-endothelial
Benzene	+			Marrow
Bis(chloromethyl)ether	+			Lung
Busulphan (Myleran)[c]		+		Marrow
Cadmium[b]	+			Prostate
Chlorambucil				Marrow
Chlornaphazine		+		Bladder
Chromium[b]	+			Lung
Cyclophosphamide		+		Bladder
Erionite			+	Lung pleura
Melphalan		+		Marrow
Methyl-CCNU				Marrow
Mustard gas (sulphur mustard)	+			Larynx, lung
Nickel[b]	+			Nasal sinuses, lung
Oestrogens				
Unopposed		+		Endometrium
Transplacental (DES)[d]		+		Vagina
Phenacetin		+		Kidney (pelvis)
Polycyclic hydrocarbons	+	+		Skin, scrotum, lung
Steroids				
Anabolic (oximetholone)		+		Liver
Contraceptives		+		Liver (hamartoma)
Tobacco (smoke and smokeless products)			+	Mouth, pharynx, larynx, oesophagus, lung, bladder
Treosulphan		+		Marrow
Vinyl chloride	+			Liver (angiosarcoma)

A Occupational
B Medicinal
C Environmental/consumer

[a] A plus sign indicates where the main evidence of carcinogenicity was obtained.
[b] Certain compounds or oxidation states only.
[c] Butane-1,4-diol dimethanesulphonate.
[d] Diethylstilboestrol.

2.7 Interpretation of Data from Chronic Animal Bioassays and Human Studies. A number of factors may constrain the analysis and interpretation of data from animal and human carcinogenicity studies. A variety of statistical techniques have been developed to adjust for confounding factors, and these, together with methods for estimating absolute cancer rates, confidence intervals and significance tests, have been reviewed.[22-24] Significance tests are used to assess neoplastic response in relation to control or background cancer incidence. Background or so-called spontaneous levels of cancers in both animals and humans due to other factors (both exogenous and endogenous) must be controlled for when estimating absolute cancer risk. Similarly extrapolations of risk from animal studies to humans and from one population group to another should take into account differences in background levels. In general, the higher the background cancer incidence, the larger the size of the study group required to detect increases due to the chemical exposure under investigation (see also Chapter 28).

Confidence in positive results from chronic animal and epidemiological studies is increased if a dose-response relationship is demonstrated and if the same end-point has been reported from a number of independent studies. Demonstration of a decline in risk after cessation of exposure in individuals or in whole populations also supports a causal interpretation.

2.8 Exposure Assessment. Assessment of exposure, in individuals as well as populations, is a prerequisite for quantification of risk (see Section 3). Estimation of human exposure to a given environmental pollutant involves an initial consideration of the possible sources of the chemical, *i.e.* industrial, domestic, consumer products, *etc.* A good inventory of sources will provide important information on possible environmental levels, critical exposure pathways, levels of exposure, and populations at particular risk.

The following general methods can be used to estimate either levels of exposure or values that are proportional to levels of exposure.

(a) Estimates of duration of exposure as well as ambient levels of the chemical may be the best or only estimate of dose following chronic exposures; it is recognized that such estimates are crude, but, in some situations (*i.e.* occupational exposure) they may be all that is available. It is important to consider all likely absorption routes, *e.g.* dermal as well as inhalation exposure (such as risks from solvents; see Chapter 13).

(b) Direct measurement of the chemical in environmental media (*e.g.* air, water, or food sampling) may be used.

(c) Direct measurement of the chemical or its metabolites in body fluid and tissues, (*e.g.* blood, breast-milk, and hair) and in excretion products (*e.g.* urine and expired air) to estimate body burden.

(d) Observation of pathological evidence of organ or tissue damage may be helpful in estimating doses, *e.g.* cytogenetic damage in blood lymphocytes.

(e) Assessment of a variety of biochemical indicators, *e.g.* enzyme activities, DNA, and protein adducts.

(f) The use of questionnaires, interviews, or records related to work histories and lifestyles may assist in determining the exposure of concern or other exposures.

Data obtained from these methods together with information on lifestyle factors can be used in models to estimate the integrated exposure of humans.

3 Risk Estimation Techniques

3.1 Dose-Response Assessment and Risk Extrapolation Techniques. Knowledge of the dependence of biological effects on exposure is essential for the quantitative health risk estimation of environmental chemicals. The aim of environmental health risk estimation is to apply dose-effect data, available from specific study situations (mainly animal bioassays and heavily exposed population groups), to the general population in order to calculate the possible risk to the latter. Therefore cancer risk assessment generally involves extrapolating risk from the relatively high exposure levels employed in animal studies or from occupational studies, where cancer responses can be measured, to risks at relatively low exposure levels that are of environmental concern.

A major problem in dose-response and risk extrapolation is the determination of an appropriate mathematical model to predict effects at hypothetical low levels of exposure. Several models have been developed, and those in common use in carcinogenic risk assessment are categorized in Table 3.

Table 3 *Models used in dose-response and risk extrapolations*[18]

Distribution models
Log-probit
Mantel-Bryan
Logit
Weibull

Mechanistic models
One-hit (linear)
Gamma multihit
Multistage (Armitage-Doll)
Linearized multistage

Other models
Statistico-pharmacokinetic
Time-to-tumour

3.1.1 Distribution models. These are based on mathematical functions of presumed population characteristics *i.e.* on the assumption that every member of a population has a critical dosage (threshold) below which the individual will not respond to the exposure in question. The log-probit model assumes that log dose-responses have a normal distribution. This model serves as the basis for the Mantel-Bryan risk extrapolation procedure.[25] Other distribution models on which carcinogenicity dose-response models have been based include the logit and Weibull models.

3.1.2 Mechanistic models. Such models are based on the presumed mechanism of carcinogenesis. Each model reflects the assumption that a tumour originates from a single cell. The concept underlying the one-hit model is that a tumour can be induced by exposure of DNA to a single molecule of a carcinogen. This model is essentially equivalent to assuming that the dose-response is linear in the low-dose region. The gamma multihit model[26] is an extension of the one-hit model, which assumes that more than one hit is required at the cellular level to initiate carcinogenesis.

The biological justification for the multistage (Armitage-Doll) model[27] is that cancer is assumed to be a multistage process that can be approximated by a series of multiplicative linear functions. The dose-response predicted by this model is approximately linear at low doses resulting in estimates of potential risk that are similar to those of the one-hit model.

3.1.3 Other models. As has already been mentioned it is known that many chemicals are carcinogenic by virtue of biotransformation to reactive metabolites. The statistico-pharmacokinetic model arises from a consideration of competing metabolic activation and deactivation (*e.g.* detoxification and DNA repair systems) reactions and estimates the 'effective dose', *i.e.* the level of reactive metabolites formed, rather than the administered dose.[28,29]

An alternative derivation of the probit model relates it to the time at which a tumour is detected. In assessing risk this so-called 'time-to-tumour' model uses the time-to-observance (latency) in addition to the proportion of animals possessing tumours at each dose.[30]

3.2 Limitations to Mathematical Modelling Techniques. Because the mechanisms of carcinogenesis are not sufficiently understood, none of these models has a fully adequate biological rationale and therefore the limitations in risk extrapolation should be appreciated. All require extrapolation of risk-level responses in the observable range to that area of the dose-response curve where the responses are not observable. Matters are further complicated by the fact that the risk-level relations assumed by the various procedures are practically indistinguishable in the observable range (5-90% incidence) but diverge substantially in their projection of risks to the non-observable range. In the final analysis, the criteria used to select the most appropriate model should take into account biological characteristics of the chemical (*e.g.* metabolic profile). Mathematical models and methods used for low-dose risk extrapolations have been reviewed in detail.[18,30,31] Risk estimation methods have also been described for mixtures of chemicals.[30,32]

3.3 Interspecies Dose Extrapolations. The basis of extrapolations from doses administered to test animals to corresponding levels of human exposure is an important consideration in the risk estimation process. The most frequently used dose extrapolation procedures are outlined in Chapter 19 and have been evaluated by the Food Safety Council.[30] For carcinogens, an accepted model is that based on surface area extrapolation. If the exposure is expressed in mg kg^{-1} body weight, this model can be written as follows:

$$d_H = [(d_A \times W/70)/70]^{2/3} = d_A(W/70)^{1/3}$$

where d_H and d_A are the human and animal daily exposure levels (mg kg^{-1}) respectively, 70 is the assumed average human weight (kg), and W is the animal weight (kg). It is assumed that surface area is proportional to body weight raised to the $^2/_3$ power.

<u>3.4 Statistical Presentation of Risk Estimates</u>. Estimates of carcinogenic risk can be presented in a number of ways. First, an estimate of incremental (excess over background) unit risk can be made. Under an assumption of low-dose linearity, this estimate states the excess lifetime cancer risk in terms of continuous exposure over an average lifetime corresponding to a particular carcinogen dose. Secondly, estimates can be made of the dose corresponding to a given level of risk. Thirdly, risk can be expressed in terms of excess individual lifetime risk (*i.e.* risk at different ages), and fourthly, risk can be linked to the excess incidence of cancer per year in an exposed population. Relative risk, frequently used in environmental health research, is the ratio between the risk or rate of cancer incidence or mortality among those exposed to the hazard and the risk among the control population. Calculations expressed in unit risk estimates have the added advantage of providing a means of comparing potencies of different carcinogens.

<u>3.5 Cancer Risk Estimation Models in Current Use</u>. Incremental unit risk estimates for carcinogenic chemicals are commonly based on absolute or relative risk calculations using the linear (multistage) risk model. The model currently used by the United States Environmental Protection Agency (EPA)[33] is based on the establishment of upper-bound risks, expressed in terms of excess individual risk in exposed populations or as nationwide impact in terms of annual increase in cancer incidence.

The upper bound is calculated using conservative exposure estimates and a linear non-threshold mode at low doses. It is assumed that the lower bound may approach zero. Human data are used where possible and extrapolated to low doses using the linear non-threshold model. Negative human data are used to place an upper bound on risk, defined as that maximum risk which would have gone undetected in the particular study considered.[33] Where human data are available, incremental risk estimations have been calculated using a method known as the 'average relative risk model' where the unit lifetime risk (UR), *i.e.* the risk associated with a lifetime exposure to 1 μg m^{-3} of an air pollutant, can be calculated using the equation

$$UR = P_0(R-I)/X$$

Where: P_0 = background lifetime risk; this is obtained from age/cause-specific death or incidence rates found in national vital statistics tables using the life table methodology, or from a matched control population.

R = relative risk, being the ratio between the observed (O) and expected (E) number of cancer cases in the exposed population; the relative risk is sometimes expressed as the standardized mortality ratio SMR = $(O/E) \times 100$.

X = lifetime average exposure (standardized lifetime exposure for the study population on a lifetime continuous exposure basis); in the case of occupational studies, X represents a conversion from the occupational 8 hour, 240 day exposure over a specific number of working years and can be calculated as X = 8 hour TWA x 8/(24 x 240)/365 x (average exposure duration (in years)/ life expectancy (70 years)), where TWA is the time-weighted average ($\mu g\ m^{-3}$).

Because this method has several advantages over many other more sophisticated extrapolation models, the average relative risk method was recently adopted by WHO to establish carcinogenic risk estimates for a number of organic and inorganic air pollutants[34] (see Table 4). A similar calculation may be used for exposure by other routes.

Table 4 *Carcinogenic risk estimates[a] based on human studies[34]*

Substance	*IARC Group[b] classification*	*Unit risk[c]*	*Site of tumour*
Acrylonitrile	2A	1×10^{-5}	Lung
Arsenic	1	4×10^{-3}	Lung
Benzene	1	4×10^{-6}	Blood (leukaemia)
Chromium(VI)	1	4×10^{-2}	Lung
Nickel	1	4×10^{-4}	Lung
Polynuclear aromatic hydrocarbons (carcinogenic fraction)[d]		9×10^{-2}	Lung
Vinyl chloride	1	1×10^{-6}	Liver and other sites

[a] Calculated with average relative risk model.
[b] For more details refer to Ref. 21.
[c] Cancer risk estimates for lifetime exposure to a concentration of 1 $\mu g\ m^{-3}$ (in air).
[d] Expressed as benzo[*a*]pyrene (based on benzo[*a*]-pyrene concentration of 1 $\mu g\ m^{-3}$ (in air) as a component of benzene-soluble coke-oven emissions).

3.6 Interpretation of Risk Estimates. Quantitative risk estimates should not be regarded as being equivalent to the 'true' cancer risk, but as an approximate estimate of the magnitude of the risk which can serve as a basis for establishing control measures, setting priorities, balancing risk and benefit, and estimating the extent of the public health problem posed by a given carcinogen.

4 Risk Evaluation Techniques

The primary aim of health risk assessment is to provide a basis for protecting public health from adverse effects from exposure to chemicals. This section considers current approaches, which utilize the data obtained from the risk identification and estimation processes to establish so-called 'safe' or 'acceptable' levels of exposure to carcinogens.

4.1 Threshold Considerations. A controversial issue with respect to risk evaluation of chemical carcinogens is the existence of threshold levels *i.e.* whether or not a level exists at which no response will take place. The argument against a threshold assumes that a single molecule of a genotoxic chemical is capable of causing a mutation of DNA and that a single point mutation can lead to cancer. This assumption is the basis for the 'conservative' approach to mathematical modelling techniques. The arguments for threshold levels are based on the existence of gene repair mechanisms, immune defences, and epigenetic mechanisms (see Chapter 9). There may also be metabolic thresholds below which rapid detoxification occurs so that there is no response, *e.g.* for 1,4-dioxane, or functional thresholds for organ damage to which tumour formation is secondary, *e.g.* 3-amino-1,2,4-triazole. Proof has not been provided for any carcinogen that no threshold exists and, in fact, no-observed-effect levels (NOEL) have been documented in both experimental animals and humans, although these have generally been disputed on the grounds of insufficiently large study groups. A considerable amount of research has gone into developing information on the 'actual' shape of the dose-response curve at low levels of exposure. Such information is of paramount importance in the selection of the most appropriate evaluation approach.

4.2 Approaches Based on Safety Factors. Exposure data are usually lacking or inadequate for human studies, mainly due to the latency of cancer induction of up to 20-30 years for some compounds. In the absence of adequate human data the traditional approach for establishing 'safe' levels or tolerances for chemical exposures is to reduce the NOEL obtained from chronic animal bioassays by a safety factor, usually up to 100, that takes into consideration both intra- and interspecies variation as well as the seriousness of the response. This is illustrated in Figure 3 with respect to estimating the acceptable daily intake (ADI) for a chemical.[35] The solid line to point A is the dose-effect curve determined by a multiple dosing experiment. In setting an ADI concentration (point C) a selected safety factor is applied at point A. If the curve T is the 'true' dose-response curve (*i.e.* with a threshold for effect) then a safety factor approach would be appropriate. If however curve NT applies (*i.e.* no threshold for effect) then some individuals in the population would be at risk.[35]

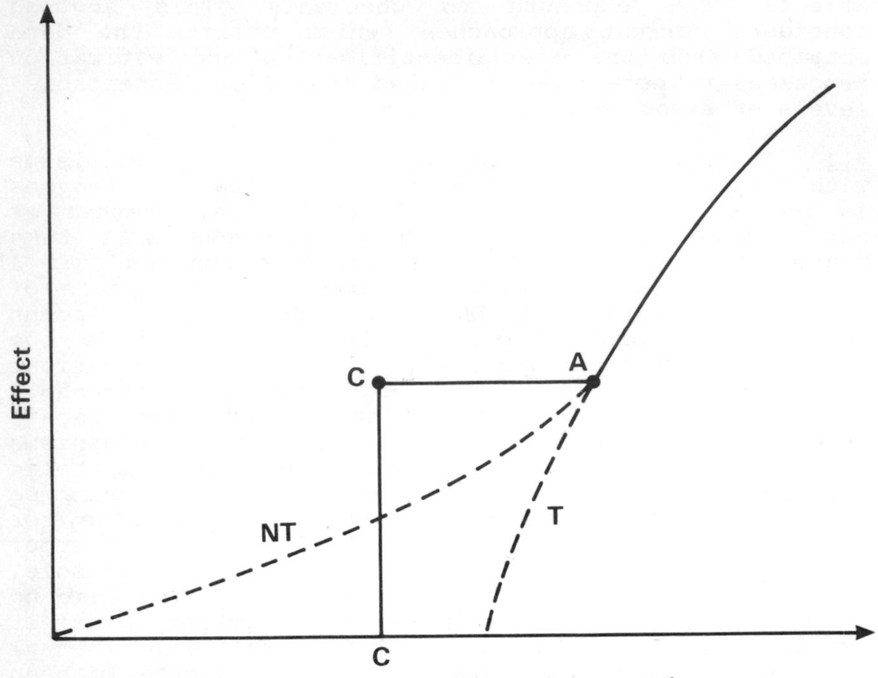

Figure 3 Exposure effect curve showing the different possible estimates at lower dose levels

Point A = No-observed-effect-level (NOEL)
Curve T = Possible dose-response curve (threshold) at lower doses.
Curve NT = Possible dose-response curve (no-threshold at lower doses.

The application of safety factors to the NOEL has been criticized on the grounds that not only is the NOEL highly dependent upon the number of animals in the study group but also that no account is taken of the slope of the dose-response curve, *i.e.* a safety factor that provides an acceptable margin of safety for a chemical substance with a steep dose-response slope will provide a smaller safety margin if the curve is relatively shallow. In addition, because the NOEL is considered a sub-threshold dosage level, it is evident from Figure 3 that the safety factor approach may not be suitable for non-threshold carcinogens.

The safety factor approach has also been used to control limits or 'threshold limit values' for industrial exposures.

4.3 Approaches Based on Risk Extrapolation. This is the major approach currently used for regulating carcinogens in the environment. It is based on the concept of 'virtual safety', introduced by Mantel and Bryan[25] and utilizes the 'conservative' mathematical models for low-dose risk extrapolation described previously.

Such an approach requires that a level of 'acceptable' lifetime risk be arbitrarily pre-selected as a criterion of response. Presently one fatality in 100,000 or 1,000,000 is most often considered an acceptable risk for exposure to a potential carcinogenic chemical. The exposure level or dosage that is predicted to produce this low frequency of excess response is referred to as the virtually safe dose (VSD).

This method has been utilized by WHO[35,36] to establish water quality guidelines for a number of potentially carcinogenic chemicals (Table 5).

Table 5 *Water quality guidelines for substances considered to be potential carcinogens*[35,36]

Constituent	*Guideline value* $\mu g\ l^{-1}$
Alachlor	0.3
Benzene	10
Benzo[*a*]pyrene	0.01
Carbon tetrachloride	3
Chloroform	30
1,1-Dichloroethane	0.3
1,2-Dichloroethane	10
Hexachlorobenzene	0.01
Tetrachloroethene	10
Trichloroethene	30
2,4,6-Trichlorophenol	10

Guideline values were calculated from animal and human data using the linear multistage extrapolation model (see Chapter 5 for details), assuming an excess lifetime cancer risk of 1 in 100,000 and an average consumption of two litres of drinking water per day. It should be emphasized that the uncertainties involved in such calculations are considerable and variations of up to two orders of magnitude could exist.

4.4 Establishing Acceptable Levels of Risk. As already stated there is no firm biological basis for current mathematical techniques used in low-dose cancer risk extrapolations. The main alternative is the setting of 'absolute' safety levels (zero tolerance) for potentially carcinogenic chemicals. To achieve a 'zero risk', chemical exposure must be reduced to below the risk threshold. However, where a threshold cannot be demonstrated it is assumed that a finite risk may occur at any exposure level and consequently absolute control is only possible if the source of exposure is eliminated altogether. The Delaney Clause of the Federal Food and Drug Act (see Chapter 6) is an example of a 'zero risk' approach for food additives.[37] This clearly is not a reasonable alternative for many environmental pollutants, such as, for example, benzene (present in the environment as a result of its emission from motor vehicles). Risk assessment is therefore unavoidable and must implicitly or explicitly involve a balance of risk and benefit. Some of the main factors considered in establishing acceptable risk levels for chemical exposures are outlined in Table 6.

Table 6 *Some factors considered in establishing acceptable risk levels*[18]

Beneficial aspects of the chemical

Economic growth
Employment
Increased standard of living
Increased quality of life
Taxes generated

Detrimental aspects of the chemical

Decreased quality of life
Emotional difficulties
Health effects
Lawsuits
Loss of environmental resources
Loss of work
Medical payments

5 Conclusions

The successful completion of and the nature of the health risk assessment process are dependent on the quality and availability of toxicological data from epidemiological studies, animal bioassays, and *in vitro* studies.

Quantitative assessment of human cancer risk is a difficult process, complicated by many factors. Among these are: the long induction/latent period of tumourogenesis; the different possible mechanisms and therefore the unknown suitability of dose-response models; the paucity of epidemiological data on most chemical substances; (detailed quantitative data on exposure in particular are lacking) the mixed nature of most chemical exposures and the modification of carcinogenic effects by a variety of 'life-style' factors.

In the final evaluation of human cancer risks both the application of safety factors and extrapolations from mathematical models suffer from their dependence on unverifiable assumptions. Since the correct form of the dose-response relationship cannot presently be determined with any degree of certainty, linear extrapolation methods have been recommended. These so-called 'conservative' methods may greatly overestimate the low-dose risk and over simplify the toxicological database. Extrapolation from animals to man will often be aided by detailed studies of metabolic profiles together with mechanistic observations as metabolically-based thresholds of tumour induction have been demonstrated (see Chapters 9, 14, and 15). Appreciation of such qualitative information in combination with wise use of mathematical models can lead to a more satisfactory approach to cancer safety evaluation.

6 References

1. Technologies for determining cancer risk from the environment, USEPA, Washington DC, US Printing Office, Publication No. OTA-H-138, 1981.

2. D. Henschler, 'Carcinogenicity testing - existing protocols are insufficient', *Trends Pharmacol. Sci.*, (TIPS), Nov. 1985, Federation of European Societies of Toxicologists, p.26-28.

3. *Science of the Total Environment* 1986, **51**, vii-x.

4. Structure Activity Correlation as a Predictive Tool in Toxicology, 'Fundamentals, Methods and Applications', ed. L. Golberg, Chemical Industry Institute of Toxicology Series, Hemisphere Pub. Corp., 1983.

5. J.C. Arcos, Y.T. Woo, M.F. Argus, and D.Y. Lai, 'Chemical Induction of Cancer', Vols I, IIA, IIB, IIIA, IIIB. Academic Press, New York, 1968-1985.

6. 'Structure-Activity Relationships in Toxicology and Ecotoxicology: An Assessment', ECETOC Monograph No. 8, Feb. 1986, pp.88.

7. J. McCann, L. Horn, G. Litton, J. Kaldor, R. Macgaw, L. Bernstein, and M. Pike, 'Short-term Tests for Carcinogens and Mutagens: Chapter 15, A Data Base designed for Comparative Quantitative Analysis'. In ref. 4.

8. International Agency for Research on Cancer, Supplement 2, Long-term and Short-term Screening Assays for Carcinogens: A Critical Appraisal', IARC Monographs on the Evaluation of Carcinogenic Risk of Chemicals to Humans, Lyon, 1980, pp.426.

9. International Programme on Chemical Safety, (IPCS), 'Guide to Short-Term Tests for Detecting Mutagenic and Carcinogenic Chemicals', WHO Environmental Health Criteria Report No. 51, WHO, Geneva, 1985, pp.208.

10. 'Evaluation cf Short-term Tests for Carcinogens, Progress in Mutation Research Vol. 1., ed. F.J. De Serres and J. Ashby, Amsterdam, Oxford, New York, 1981.

11. B.L. Pool and D. Schmäl, *Path. Res. Pract.*, 1987, **182**, 704-712.

12. E.J. de Serres, in 'Pesticides - Minimizing the Risks', ed. N.N. Ragsdale and R.J. Kuhr, American Chemical Society, 1987, Ch. 4.

13. R.W. Tennant et al., *Science*, 1987, **236**, 933-941.

14. M. Meselson and K. Russell, 'Comparisons of Carcinogenic and Mutagenic Potency', in 'Origins of Human Cancer', ed. H.H. Hiatt, J.D. Watson, and J.A. Winston, Cold Spring Harbor Laboratory', New York, 1977, pp.1473-1487.

15. J.H. Weisburger and G.M. Williams, 'Carcinogen testing. Current problems and new approaches', *Science*, 1971, **214**, 401-407.

16. J.F. Robens, J.J. Joiner, and R.L. Schueler, 'Methods in Testing for Carcinogenicity', in 'Principles and Methods of Toxicology', ed. A.W. Hayes, Raven Press, 1986, p.79-105.

17. International Programme on Chemical Safety (IPCS), 'Principles and Methods for Evaluating the Toxicity of Chemicals', Part 1, Environmental Health Criteria Report No.6, WHO, Geneva, 1978, pp. 272.

18. 'Casarett and Doull's Toxicology - the Basic Science of Poisons', 3rd Edn., ed. Klaassen et al., Collier Macmillan, 1986, pp. 974.

19. International Programme on Chemical Safety (IPCS), 'Guidelines for the Study of Genetic Effects in Human Populations, Environmental Health Criterial Report No. 46, WHO, Geneva, 1985, pp. 126.

20. V.S. Zhurkov, K.N. Yakovenko, and M.A. Pilinskaya, 'Analysis of Cytogenetic Damage in Human Lymphocytes as a Biological Indicator of Mutagenic Effect'. In ref. 32.

21. International Agency for Research on Cancer, Supplement No.7, 'Overall evaluations of carcinogenicity: An updating of IARC Monographs Volumes 1-42'. IARC Monographs on the Evaluation of Carcinogenic Risks to Humans, Lyon, 1987, pp. 440.

22. R. Peto, M.C. Pike, N.E. Day, R.G. Gray, P.N. Lee, S. Parish, J. Peto, S. Richards, and J. Wahrendord, in Ref. 8, pp. 311-426.

23. N.E. Breslow and N.E. Day, 'Statistical Methods in Cancer Research, Vol. 1, The Analysis of Case-control Studies', IARC Scientific Publications No. 32, International Agency for Research Cancer, Lyon, 1980.

24. N.E. Breslow and N.E. Day, 'Statistical Methods in Cancer Research Vol. 2, The Design and Analysis of Cohort Studies', IARC Scientific Publications No. 82, International Agency for Research on Cancer, Lyon, 1987.

25. N. Mantel and W.R. Bryan, *J. Cancer Inst.*, 1961, **27**, 455.

26. J. Cornfield, F.W. Carlborg, and J. Van Ryzin, 'Setting Tolerances on the Basis of Mathematical Treatment of Dose-Response Data Extrapolated to Low Doses', in Proceedings of the First International Congress on Toxicology, ed. G.L. Plaa and W.A.M. Duncan, Academic Press, New York, 1978, pp.143-164.

27. P. Armitage and R. Doll, 'Stochastic Models for Carcinogenesis', in Proceedings of the Fourth Berkeley Symposium on Mathematical Statistics and Probability, Vol. 4, ed. W. Lecam and J. Neyman, California Press, Berkeley, California, 1961, pp. 19-38.

28. J. Cornfield, *Science*, 1977, **198**, 693.

29. B. Altschuler, *Environ. Health Perspec.*, 1981, **42**, 23-27; *Fund. Appl. Toxicol.*, 1981, 1, 124-6.

30. Food Safety Council, 'Quantitative Risk Assessment', *Food Cosmet. Toxicol.*, 1980, **18**, 711.

31. E.L. Anderson, 'Quantitative Approaches in Use in the United States to Assess Cancer Risk', in Methods for Estimating Risk of Chemical Injury: Human and Non-human Biota and Ecosystems', ed. V.B. Vouk, et al., John Wiley and Sons, New York, SCOPE No. 26, 1985, pp. 712.

32. 'Methods for Assessing the Effects of Mixtures of Chemicals', joint symposia No. 6, ed. V.B. Vouk, et al., John Wiley and Sons, New York, SCOPE No.30 (SGOMSEC 3), 1987, pp. 894.

33. US EPA *Federal Register*, 1980, 79318-79379.

34. 'Air Quality Guidelines for Europe', WHO Regional Publication, European Series No. 23, WHO, Regional Office for Europe, Copenhagen, 1987, pp. 7-19.

35. 'Setting Environmental Standards', 'Guidelines for Decision-Making', ed. H.W. de Koning, WHO, Geneva, 1987, pp. 98.

36. WHO Environmental Health Series No. 27, Drinking Water Quality: Guidelines for Selected Herbicides, WHO, Geneva, 1987, pp. 23.

37. F. Coulson, Regulatory aspects of carcinogens and food additives: the Delaney Clause, Academic Press, New York, San Francisco and London, 1979.

9

Carcinogenic Risk Assessment Using Animal Data: The Importance of Mechanisms of Toxic Action

I.F.H. Purchase

IMPERIAL CHEMICAL INDUSTRIES PLC, CENTRAL TOXICOLOGY LABORATORY, ALDERLEY PARK, NR. MACCLESFIELD, CHESHIRE SK10 4TJ, UK

1 Introduction

The ultimate purpose of risk assessment of environmental carcinogens is to provide a rational basis on which preventative measures may be introduced into society. In particular, quantitative risk assessments allow a comparison of the relative risks of different events so that the most risky can be controlled first. Quantitative risk assessment will also contribute to the judgement of the cost to human health of a particular activity in relation to the benefit which society, or individuals, obtains as a consequence of that activity.

In the field of environmental carcinogenesis, quantitative risk assessment is particularly important. An expert committee of the World Health Organization reported in 1964 on how existing knowledge could be applied to the prevention of cancer.[1] The committee noted that 'the categories of cancer that are thus influenced, directly or indirectly, by extrinsic factors include many tumours of the skin and mouth, the respiratory, gastrointestinal and urinary tracts, hormone dependent organs, haemopoetic and lymphopoetic systems, which collectively account for more than three quarters of human cancers'. It would seem, therefore, that the majority of human cancer is potentially preventable.[1] Further work on this subject by Doll and Peto[2] confirms that a large proportion of human cancer is preventable. The important factors identified by them include diet (35% of cancer deaths), tobacco (30%), occupation (4%), alcohol (3%), pollution (2%), medicines and medicinal procedures (1%), and industrial products (<1%). Reproductive and sexual behaviour, geophysical factors and infection together contribute about 20% of cancer deaths, suggesting that some 80% of cancer deaths are attributable to factors which are related to chemical carcinogens.

In this chapter the risk assessment process is described and the method currently used in toxicological risk assessment discussed. The mathematical models used in carcinogen risk assessment are described and improvements to the current methods by the incorporation of mechanistic data are illustrated.

2 The Risk Assessment Process

Data for carcinogenic risk assessment are derived primarily from the results of long-term animal experiments and from human epidemiology studies. By far the majority of data used are derived from animal carcinogenicity studies and this chapter will concentrate primarily on risk assessment from animal data.

The assessment of risk is derived from a definition of the carcinogenic potency and an estimate of exposure. All risk assessments using animal data start with the experimentally determined incidence of cancer in various groups of animals treated with the chemical of interest. On the basis of these data an assessment of the nature and magnitude of the carcinogenic effect can be made, and with the knowledge of this and other data, an assessment of the likelihood that man would be susceptible to the carcinogenic effect. Finally, the magnitude of the response of different doses from the experimental data may allow the assessment of the response at doses below those used in the experiment.

The assessment of exposure is a complex and difficult task and will be referred to only in passing in this chapter.

3 Generalized Issues in Risk Assessment

Whenever using experimental animal data for human risk assessment a number of fundamental assumptions have to be made, not all of which are testable experimentally. Nevertheless, these assumptions are widely used because risk assessment would be difficult, if not impossible, without them.

3.1 Extrapolation Between Species. In the absence of any other mechanistic information, it is often assumed that the data derived from an animal experiment are useful in assessing human carcinogenic risk. There are well recognized differences in the physiology and anatomy of laboratory animals and man. For example, the lifespan of the laboratory rodent is approximately two years whereas that of man is approximately seventy years. Cancer appears to develop in these species over a timescale that is proportional to the lifespan and it is generally assumed that this will apply to all chemicals being assessed. The assumption that man will respond in a similar fashion to laboratory animals is frequently shown to be inappropriate. Mechanistic studies demonstrate that qualitative and quantitative aspects of metabolism, tissue susceptibility, and immune and other defence mechanisms can explain observed differences in the response of laboratory species and man to the effect of carcinogens. Expert judgement may be required to assess the nature of the end-point or the mechanisms of carcinogenic action of the chemical in the experimental animal and to decide whether they are relevant to the human situation.

3.2 Extrapolation from High to Low Dose. Most laboratory experiments are carried out in relatively small groups of animals (between 50 and 100) and at high doses. In many long-term animal carcinogenicity experiments it is the aim to select the top dose so that it is the 'maximum tolerated dose' in order to ensure that the results from the animal studies do not overlook a carcinogenic response simply because the dose is too low. Most environmental exposure occurs at doses which are several orders of magnitude lower than those used in the experiment. Clearly, the shape of the dose-response relationship is vitally important in establishing the likely effects at doses substantially below those in the observable range.

However, as the majority of carcinogenicity experiments use only two or three doses, it is impossible to assess the shape of the dose response with any degree of precision. Risk assessment must rely on some arbitrary assumption about the shape of the dose-response relationship at low doses.

3.3 Extrapolation from Controlled Experimental Conditions to Variable Human Situations. Most animal experiments have all critical factors carefully controlled by the experimenter in order to enhance the reproducibility of the experimental observations. This is in contrast to differences, sometimes very large differences, in the environmental exposure and other circumstances of the human population. The complexity of these circumstances precludes an assessment of each one individually and hence risk assessment will normally extrapolate directly from the animal data unless there is good evidence to suggest that an important confounding factor has been introduced.

3.4 Individual Variability. The human populations exposed to chemicals may have differences in age, sex, and ethnic background and will certainly have a more heterogenous genetic make-up than will experimental animals. For the majority of carcinogenic risk assessments individual variability can only be taken into account by the application of a safety factor.

3.5 Dosimetry. The problems of estimating environmental exposure have already been mentioned. There are equally difficult problems in extrapolating the doses used in animal experiments to those which might be experienced in the human situation. Doses may be extrapolated from animal experiments to man by:

(a) Expression of dose as a function of bodyweight (usually mg kg^{-1})

(b) Expression of dose as a concentration in food or water (usually parts per million)

(c) Expression of dose as a concentration in inhaled air (usually as parts per million)

(d) Correction of the dose for surface area. This is achieved by raising the bodyweight (b.w.) to the power of $^2/_3$. The correction requires that doses expressed as concentrations in air or food are converted to mg kg^{-1} b.w. in the laboratory species. The dose is then corrected for surface area and converted back to concentrations in air or food using appropriate conversions for man. Corrections using surface area are based on the observation that metabolic rate is proportional to body surface area. For acute toxicity this method may have some merit, but the rate of metabolism will affect the potency of a carcinogen in different ways depending on its metabolic fate. Using this method provides figures which characterize man as 5-12 times more susceptible to carcinogens than laboratory animals when using the extrapolation on the basis of bodyweight.

(e) Extrapolation assuming that the tissue dose is the primary determinant of carcinogenic response. This approach requires a study of the metabolic pathways and the kinetics of the pathway which generates the proximate carcinogen. *In vitro* studies may be necessary to obtain the appropriate human data. Examples of this approach are given later in this chapter.

There are further problems if the experiment is carried out to a protocol which is dissimilar from the human experience, for example if the animal experiment is for a lifetime and the human exposure is for a shorter period or if the animal experiment uses an exposure route which is not experienced in man. As a general rule for chemical carcinogens, where the tissue dose is presumably the most important in determining the carcinogenic or other responses the total body burden may be computed from the various routes of exposure in order to assess the overall dose.

3.6 Mixtures. When animals or man are exposed to two or more chemicals further complications are introduced. In general the response of the mammal will depend on whether the chemicals' activity is interactive or not and whether the effects of the chemicals are similar or not. If the chemicals are not interactive and the response is similar, it may be assumed that an additive effect will occur. If they are not interactive and the toxicity is dissimilar the overall response may be less than additive. If there is interaction between the chemicals this may intensify or weaken the response observed. This subject is so complex that it warrants a separate discussion and it will not be attempted here.

3.7 Biological Factors in Risk Assessment. There are a variety of biological factors which impact on the assessment of carcinogenic hazard. Amongst these are the quality of the experiment, including the quality of the pathology, environmental control, estimation and standardization of chemical administration, and many other factors (which together make up compliance with good laboratory practice). A further problem may be encountered if there are differences in outcome from different experiments. In the majority of cases where this occurs there is some aspect of the exceptional experiment which may explain the reason why the response was different (*e.g.* the use of a different strain of animal). There is a tendency to utilize the positive data and ignore the negative data in carcinogenic risk assessment on the basis that this will provide a conservative estimate of risk.

The results of experimental animal studies also provide a variety of data in addition to the simple indication of the presence or absence of cancer. This information includes the number of neoplasms per animal, the number of different types of neoplasms observed, and the number of species affected. The organ or target tissue in which the carcinogenic response occurs is also important as some rodents have extremely high and variable incidence of certain types of tumour in control animals. Where a chemical induces neoplasms which normally occur in high and variable incidence, the response carries less weight than the appearance of tumours in other organs. The time to development of tumour will also give an indication of potency.

Information about the way in which the chemical is metabolized may be critical in assessing the relevance of the results to man. All of these factors have to be taken into account in the overall process of risk assessment.[3]

4 Quantitative Methods in Human Risk Assessment

When safety evaluation relied primarily on human observation, there were no attempts to assess the magnitude of risk in any statistical sense.[4] Once the safety assessment relied for its primary data on animal testing, a different treatment of the data was required in order to assess human risk. In the 1940s the idea of a safety factor, which was to be applied to the results from animal studies, was developed.

4.1 Safety Factors. The approach that has been used since that time for the majority of assessments of toxicological risk has been based on the establishment of a no observed effect level (NOEL) and the application of a safety factor. The NOEL is the highest dose in animal experiment which produced no effect in the most sensitive animal study conducted with that chemical. Once a NOEL had been determined, a safety factor (SF), usually of 100-fold, was applied to determine the acceptable daily intake (ADI). This safety factor was justified on the basis of a 10-fold difference to reflect an inter-species difference in susceptibility and a 10-fold difference to reflect inter-individual variation in susceptibility.

The NOEL-SF approach assumes that as the dose reduces below the no-effect level the response continues to reduce. It is obviously a powerful technique to use for phenomena which exhibit threshold effects as there is a genuine safety factor of 100-fold below the NOEL. The technique also assumes, as indicated above, that the toxicological data are relevant to the human situation and that the extrapolation of dose is reliable. It has been used now for many years in Western Europe and the USA and indeed has worked sufficiently well to be adopted internationally as the standard procedure for assessing the ADI. It does, however, require expert judgement and usually relies on a consensus committee decision in order to avoid too much subjectivity.

The NOEL-SF approach can be criticized on several grounds. The experimental group size affects the determination of the NOEL; the smaller the group size the less likely it is that an effect will be observed. This phenomenon has the effect of rewarding poor experimental design as small group sizes in experiments will tend to produce higher NOELs. Much of the information from an experiment can be wasted as there is a great tendency to use the information from the NOEL as the sole basis for setting the ADI. The type of toxic effect and the shape of the dose-response curve should be taken into account in setting the size of the safety factor. It is also obvious that the NOEL must be one of the doses used in the experiment; if this is not the case the experiment may well have to be repeated.

In spite of the obvious limitations of this technique, it has proved to be easy to understand and its usefulness is demonstrated by its widespread application in the safety assessment of chemicals.

<u>4.2 'Bench-Mark' Dose</u>. In 1984, Crump[5] reported on a method which would enable the toxicologist to compute the dose which would give a defined effect (the so called bench-mark dose). The technique is applicable for both quantal (proportion of animals affected) and continuous (*e.g.* organ weights) data.

The same basic experimental data that are used in the NOEL-SF procedure can be used for determining the bench-mark dose. A variety of mathematical models can be used (for example, polynomial regression) to calculate the dose likely to produce a pre-determined effect - say a 10% or a 1% reduction in bodyweight. A safety factor can then be applied to the bench-mark dose in the traditional way, or a bench-mark dose for a suitably small effect (for example, a reduction in bodyweight of 0.1%) can be used to determine the ADI. The mathematical models can be expressed in a form which allows calculation of a threshold dose.

The great advantage of this technique is that it does not assume a fixed relationship between the no-effect level and the safe dose. The precision of estimation of the bench-mark dose will increase with increased group size and therefore the experimenter is rewarded for good experimental design. It is possible to calculate a bench-mark dose in experiments where there is no NOEL. The calculation of the bench-mark dose also uses all of the data points in the experiment under consideration, and thus (unlike the NOEL-SF approach) takes account of the pattern of dose-response relationship in the risk assessment.

It is still too early to assess whether the bench-mark dose will prove to be as useful as the NOEL-SF method of determining ADIs. The technique certainly provides a much greater flexibility and is intellectually more rigorous.

<u>4.3 Mathematical Models Used in Carcinogenic Risk Assessment</u>. Carcinogenicity is often considered a special case in toxicology largely because of the long latency between the beginning of exposure and the development of disease, and the irreversible nature of the disease. Cancer is not unique in having these characteristics (*e.g.* some chronic neurotoxins share these properties) and it must be assumed that other characteristics of carcinogenicity have resulted in its special treatment. One of these must be the emotional

impact of cancer on the general public because 20-25% of all deaths are due to cancer. It is further supported by the hypothesis that, because carcinogens can produce an effect by inducing a single mutation, no threshold dose exists. For these and possibly other reasons the use of mathematical models for assessing low-dose risk for carcinogens has developed as a separate component of risk assessment in toxicology.

The development of mathematical models for quantifying human risk from carcinogenic substances can be traced back to the 1940s and 1950s.[4] These models were based on theories of one-hit or multiple-stage mechanisms in carcinogenesis. In 1961 Mantel and Bryan[6] proposed a probit model for assessing low-dose carcinogenic risks by extrapolation and further suggested that a risk of one in 100 million should be considered as 'a virtually safe dose'. In 1973, the Food and Drug Administration proposed an improved Mantel-Bryan Method,[4] including the use of the virtually safe dose, for computing the level of carcinogens allowable in food. Before that time, carcinogenic substances were controlled in food using the concept of no detectable levels. This placed the emphasis on analytical methods and with advances in analytical methodology the acceptable levels were continually changing. The Mantel-Bryan procedure allowed a much more satisfactory method of determining the analytical sensitivity needed. The consequence of the introduction of the Mantel-Bryan technique by the FDA was to stimulate the development of a variety of different mathematical models for carcinogenic risk assessment.

The basis of all the methods is to apply a mathematical model to the tumour incidence derived experimentally from long-term animal carcinogenicity studies, usually conducted in rats or mice.

The methods now in use include:

(a) The Probit Model. This model assumes individual thresholds with a log-normal distribution in the population at risk and produces an S-shaped dose-response curve.

(b) Logit Model. This model again produces an S-shaped dose-response curve and the dose for a defined risk is lower than that computed from the probit model.

(c) One-Hit Model. This model assumes that the dose-response relationship is linear at low dose and as a consequence tends to produce a very low calculated virtually safe dose. The one-hit model often fails to fit the experimental data well.

(d) Multi-Hit Model. This is a generalized version of the one-hit model.

(e) Multi-Stage Model. This is based on the Armitage-Doll multi-stage hypothesis. It assumes that the effect of a carcinogenic agent occurs in multiple steps and that the effect at each step is additive.

(f) Weibull. This is a generalization of the one-hit model.

The application of these statistical models to the assessment of low-dose risk produces estimates of risk from the same experimental data which may vary by factors as great as 100,000.[7,8]

4.4 Conservatism. Because of the uncertainty in carcinogenic risk assessment, it has become the policy of some USA agencies (particularly the EPA) to introduce a series of conservative assumptions in order to provide a greater protection for those exposed to the carcinogenic chemical. The conservative approach assumes that a single molecule of a carcinogen can interact with DNA producing cancer and therefore that there can be no threshold for chemical carcinogenesis. The conservative assumptions include the insistence on using a maximum tolerated dose, the use of positive data only (ignoring all negative results), the use of the upper 95% confidence limit of risk at a given dose rather than the maximum likelihood estimate of the carcinogenic risk as a basis for decision making, the assumption that the dose response is linear at low dose, and the selection of the most sensitive response irrespective of evidence of mechanisms of action.

The data from animal carcinogenicity studies are used for low-dose risk assessment and consist typically of an experiment in which there are two or three dose groups and one control. This type of experiment is inadequate to assess the suitability of the mathematical model on the basis of goodness of fit. There has, therefore, been a tendency to select the most conservative model for risk assessment.

When conservative assumptions are made at every stage in the risk assessment process, a systematic bias is introduced which overstates the risk; because the degree of conservatism is unknown at each step and each step amplifies the previous bias in the assumption, the magnitude of the overstatement of risk is unknown but may run to several orders of magnitude.

4.5 Genotoxic and Non-Genotoxic Carcinogens. In the late 1970s, when the majority of these mathematical models were being developed, it was reasonable to assume that the majority of chemical carcinogens were genotoxic.[9] The idea that chemical carcinogens interacted with DNA and therefore that there could be no threshold was widely accepted.

The perception today is substantially different. The most reliable database on which to assess the genotoxicity of chemical carcinogens is that developed by the National Toxicology Program in the USA. Ashby and Tennant[10] have analysed the results of 230 chemicals tested for carcinogenicity by the NTP. The results of *in vitro* mutagenicity studies (principally the Ames test) have demonstrated that a much higher proportion than was formerly believed of chemical carcinogens are negative in genotoxicity assays. In general, where a chemical carcinogen affects multiple organs or both sexes of both rat and mouse, it has a high probability of being genotoxic. If however, a single organ in the single sex of a single species is affected it is more likely that the carcinogen will be non-genotoxic. This observation challenges the perception that all carcinogens are genotoxic and therefore have no threshold.

There are of course numerous examples of chemicals which produce tumours in animal experiments by mechanisms which are unlikely to involve direct interaction of the chemical with the DNA. These examples include the induction of subcutaneous sarcomas on repeated injection of hypertonic glucose or saline, the induction of bladder cancer after the implantation within the bladder of any solid material,[11] or the induction of skin cancer after repeated damage to the skin.[12] Recently a number of chemicals, including butylated hydroxyanisole and propionic acid,[13] have been shown to produce fore-stomach cancer in rats after repeated gavage administration of high concentrations. A further series of chemicals, exemplified by di-(2-ethylhexyl) phthalate, is known to produce liver cancer in rodents probably as a consequence of the induction of high levels of intracellular organelles known as peroxisomes. Finally, there are a number of chemicals, some of which are naturally occurring hormones, which induce cancer in endocrine organs. In all of these examples, the dose of carcinogen was sufficiently high to produce repeated hyperplasia in the organ affected.[15] The assumption is that a repeated hyperplastic response results in derangement of normal cell control with the result that cancer develops.

Any assessment of carcinogenic risk of these chemicals must take into account the mechanism by which they produce their carcinogenic effect. Hence, for example, if a naturally occurring hormone given at a high dose produces cancer after prolonged hyperplasia but does not produce cancer where there is no hyperplasia, it is important to establish as a component of the risk assessment the dose at which hyperplasia occurs. It can be assumed that at doses below this level there will be a much smaller likelihood that cancer will develop.

4.6 Examples of Non-Genotoxic Carcinogens.

4.6.1 Trichloroethylene. Trichloroethylene is non-genotoxic in the majority of *in vitro* and *in vivo* mutagenicity assays. It is also not carcinogenic to the rat at doses of up to 2000 mg kg^{-1}/day.[16] Recently the reasons for the difference in susceptibility of the rat and mouse to trichloroethylene carcinogenesis have been described.[16-18]

Trichloroethylene is metabolized to trichloroacetic acid by both the mouse and the rat after gavage administration. In the case of the rat there is a threshold dose above which no further conversion of trichloroethylene into trichloroacetic acid takes place. However, in the mouse there is a linear conversion of trichloroethylene into trichloroacetic acid. In the mouse there is a substantial increase in liver weight after administration of trichloroethylene and this is accompanied by a dramatic proliferation of hepatic peroxisomes. It was found that trichloroacetic acid was the chemical responsible for inducing peroxisome proliferation and that trichloroacetic acid would induce peroxisome proliferation in both mice and rats with approximately equal potency.

In the mouse, therefore, trichloroethylene administration results in metabolism to trichloroacetic acids which induces peroxisome proliferation. In the rat, the metabolism of trichloroethylene to trichloroacetic acid is saturable at doses of trichloroethylene of 500 mg kg^{-1} and above; at this dose there is insufficient trichloroacetic acid produced to induce the proliferation of peroxisomes.

The results can be reproduced *in vitro*. Hepatocytes from mice produce a 30-fold greater amount of trichloroacetic acid from trichloroethylene than do hepatocytes from rats. Human hepatocytes produce a third of the trichloroacetic acid that rat hepatocytes produce. In addition it is possible to demonstrate that the mouse and rat hepatocytes *in vitro* respond equally to trichloroacetic acid by the production of peroxisomes. However human hepatocytes are unresponsive to peroxisome proliferation.

The response of human liver to trichloroethylene differs from that of mouse liver in two important respects; firstly, it is a hundred times less capable of metabolizing trichloroethylene to trichloroacetic acid and secondly, peroxisome proliferation does not occur after administration of trichloroethylene or trichloroacetic acid.

It is uncertain by precisely what mechanism peroxisome proliferation leads to liver cancer in trichloroethylene-treated mice. It has been suggested that the massive proliferation of peroxisomes, which results in stimulation of certain oxidation pathways but has no effect on peroxisomal catalase, produces an increase in active oxygen species in the liver cell.[20] The increase in reactive oxygen species is thought to lead to the induction of liver cancer.

Any risk assessment of the carcinogenic effects of trichloroethylene should take into account this difference in susceptibility between the mouse, rat, and humans, and would probably conclude that at doses of trichloroethylene which do not induce peroxisome proliferation there would be no additional risk of cancer. In man, that implies that, as no peroxisomal proliferation occurs, there is also no susceptibility to the development of liver cancer.

4.6.2 Methylene dichloride. Methylene dichloride is mutagenic in the Ames test and other *in vitro* mutagenicity assays. However, there is no evidence that methylene dichloride is genotoxic in *in vivo* systems, including assessment of its effects on chromosome aberrations, unscheduled DNA synthesis, dominant lethal effects, and binding to DNA.[21] For practical purposes it can be considered as non-genotoxic *in vivo*.

Methylene dichloride produces high incidences of lung and liver cancer in the B6C3F$_1$ mouse but does not increase the incidence of these cancers in rats.[22] Studies of the mechanism of action of methylene chloride have demonstrated that it is metabolized by two pathways.[21] The first of these is the cytochrome P-450 oxidation which is a high-affinity, low-capacity pathway. It is saturable at doses above about 500 p.p.m. by inhalation exposure. The second pathway involves conjugation with glutathione and is a low-affinity, high-capacity pathway which is non-saturable up to the maximum doses used in the animal carcinogenicity studies (4000 p.p.m.). The cytochrome P-450 pathway is saturated at doses which produce cancer and is as active in those species which are not susceptible to the carcinogenic effects of methylene dichloride as it is in the susceptible species. The glutathione pathway, however, is most active in the mouse and is much less active in all the other non-susceptible species. It is concluded, therefore, that the glutathione pathway is the one responsible for producing the proximate carcinogen.

In vitro studies have confirmed the results observed *in vivo*. The capacity of the cytochrome P-450 pathway to metabolize methylene dichloride is similar in mouse, rat, hamster, and man. The glutathione pathway is far more active than the cytochrome P-450 pathway in the mouse at doses above 1000 p.p.m.; the glutathione pathway is also more active in the mouse than it is in the rat, hamster, or man. In assessing the carcinogenic risk of methylene dichloride to man, the absence of genotoxicity *in vivo* and the fact that human liver cells metabolize methylene dichloride through the glutathione pathway at a lower rate than the rat and hamster (neither of which develops liver cancer) and at a much lower rate than that observed in the mouse, which does develop liver cancer, must be taken into account.

These results indicate that extrapolation from high dose to low dose even with the experimental range is confounded by the different behaviour of the two metabolic pathways. However, the expression of dose as tissue dose of the active carcinogen, *i.e.* in terms of the metabolites produced by the glutathione pathway, is a more appropriate technique for extrapolation from high to low dose and from one species to another. It seems unlikely that man is susceptible to methylene dichloride carcinogenicity at doses normally encountered, which are substantially below those used in these experiments, because of the negligible activity of the glutathione pathway in man.

5 Conclusions

The current methods used by toxicologists for risk assessment are best described as pragmatic and none of them are seen to be without flaws. The NOEL-SF approach, although used very widely for the assessment of acceptable daily intakes of chemicals, has many disadvantages for assessing the risk of carcinogenic chemicals. The 'bench-mark' dose concept has been introduced for non-carcinogenic end-points, but is not used for carcinogen risk assessment.

The introduction of mathematical models to the assessment of human carcinogenic risk is based on a whole variety of assumptions, some of which are questionable. The different mathematical models may provide a very wide range of estimates of risk, but it is possible to give a single figure estimate of risk which may be grossly misleading.

The idea that risk assessments should be conservative has distorted the proper evaluation of the use of mathematical models in carcinogenic risk assessment. The use of several assumptions or rules, each of which over-estimates the risk assessment, is a scientific exercise. Accuracy should be the aim and under- or over-assessment of risk will not be in the best interests of society.

The use of numerical mathematical models in the assessment of carcinogenic risk has tended to overshadow the importance of consideration of biological factors in risk assessment. The biological response of the experimental animal in terms of the number, site, frequency, and latency of occurrence of cancers can be as important as the numerical assessment in providing a judgement of acceptable risks. More recent information demonstrating that many carcinogens are non-genotoxic further emphasizes the importance of considering biological factors in carcinogenic risk assessment. If the non-genotoxic mechanism of action is understood, doses below those which trigger the mechanism are unlikely to produce a carcinogenic effect.

When mathematical models are used in carcinogenic risk assessment they should be as accurate as possible, should include estimates of error, and should provide estimates of carcinogenic risk based on more than one mathematical model.

6 References

1. WHO, Geneva, 'Prevention of Cancer', WHO Technical Report Series, No. 276, 1984.

2. R. Doll and R. Peto, *J. Natl. Cancer Inst.*, 1981, **66**, 1191.

3. European Chemical Industry Ecology and Toxicology Centre, 'Risk Assessment of Occupational Chemical Carcinogens', Monograph No.3, ECETOC, Brussels, 1982.

4. P.B. Hutt, 'Use of Quantitative Risk Assessment in Regulatory Decision Making under Federal Health and Safety Statistics' in 'Risk Quantitation and Regulatory Policy', D.E. Hoel, R.A. Merrill, and F.P. Pevera, Banbury Report 19, Cold Spring Harbor Laboratory, pp.15-20.

5. K.S. Crump, *Fund. Appl. Toxicol.*, 1984, **4**, 854.

6. N. Mantel and W.R. Bryan, *J. Natl. Cancer Inst.*, 1961, **27**, 455.

7. Food Safety Council Scientific Committee, Quantitative Risk Assessment, *Food Cosmet. Toxicol.*, 1980, **18**, 711-734.

8. I.F.H. Purchase, J. Stafford, and G.M. Paddle, 'Toxicological Risk Assessment, ed. D.B. Clayson, CRC Press, Boca Raton, Florida.

9. B.N. Ames, W.E. Durston, E.Yamasaki, and F.D. Lee, *Proc. Natl. Acad. Sci.,* USA, 1973, **70**, 2281.

10 J. Ashby and R.W. Tennant, *Mutation Res.*, 1988, **204**, 17.

11. C.S. Weil, C.P. Carpenter, and H.F. Smyth, *Arch. Environ. Hlth.*, 1965, **11**, 569.

12. T.S. Argyris, *CRC Crit. Rev. Toxicol.*, 1985, **14**, 211.

13. W. Greim, *Bundesgesundheitsplat*, 1985, **28**, 322.

14. See J.W. Bridges, 'Toxic Hazard Assessment of Chemicals', ed. M.L. Richardson, The Royal Society of Chemistry, London, 1986, p.233.

15. P. Grasso, personal communication, 1987.

16. National Toxicology Programme, NTP Draft Report Abstracts on Nine Chemical Carcinogenesis Bioassays, *Chem. Reg. Reporter,* 1983, **6**, 767.

17. C.R. Elcombe, *Arch. Toxicol.*, 1985, **8**, 6.

18. T. Green and M.S. Prout, *Toxicol. Appl. Pharmacol.*, 1985, **79**, 401.

19. M.S. Prout, W.M. Provan, and T. Green, *Toxicol. Appl. Pharmacol.,* 1985, **79**, 389.

20. J.K. Reddy, D.L. Azarnoff, and C.E. Hignite, *Nature (London),* 1980, **283**, 397.

21. European Chemical Industry Ecology and Toxicology Centre, 'The Assessment of Carcinogenic Hazard for Human Beings Exposed to Methylene Chloride', Technical Report NO.26, ECETOC, Brussels, 1987.

22. National Toxicology Program, NTP Technical Report on the Toxicology and Carcinogenesis Studies of Dichloromethane in F344/N Rats and $B6C3F_1$ mice, NTP TR306, 1986.

10
The Contribution of Epidemiology to Risk Assessment

Sir Richard Doll

IMPERIAL CANCER RESEARCH FUND, CANCER EPIDEMIOLOGY & CLINICAL TRIALS UNIT, THE RADCLIFFE INFIRMARY, UNIVERSITY OF OXFORD, OXFORD OX2 6HE, UK

1 Introduction

Hazards of disease due to man-made agents or to agents distributed by Man ought, in principle, to be avoided without putting any humans at risk. On these grounds, the epidemiology of non-infective disease was, until fairly recently, considered in the USSR to be an unacceptable branch of medicine, as it implied that humans had been unnecessarily put at risk. In practice, many risks have been produced that could not have been anticipated or which, on balance, were thought to be justified because they were side effects of procedures that were socially beneficial. These risks have caused a variety of diseases or disabilities (such as the Spanish toxic oil syndrome, some outbreaks of asthma, and mental impairment due to lead); see also Chapters 11 and 12. Epidemiologists need to be permanently on the look out for the occurrence of new epidemics of new or old diseases. In most instances, however, small amounts of chemicals that may occur in the environment can be dealt with by the body without producing disease or disability, so long as the amounts to which individuals are exposed are below the relevant threshold value. The principal problem today is the assessment of the risk of producing cancer or hereditary disease which, it is thought, is linearly proportional to the degree of exposure down to infinitessimally low levels.

2 The Role of Epidemiology

The role of epidemiology in this situation is limited. It provides powerful tools for determining the major causes of cancers, all of which vary greatly in incidence from one community to another (always five times, often 50 times, and occasionally 500 times) but its methods are generally weak and unsatisfactory for determining acceptable limits of exposure to the thousands of chemicals that may be found in the environment in small amounts.

Epidemiology may sometimes be able to help when physical or chemical causes of cancer have been discovered, to which some individuals have been sufficiently exposed to cause a high risk of the disease, by demonstrating a dose-response relationship that will enable risks to be extrapolated down to very low levels (see also Chapter 8). Even in these circumstances, however, extrapolation is often unreliable as human exposure is commonly complex and the biological availability of carcinogenic chemicals may vary in the different circumstances in which Man is exposed to large and small doses, as, for example, with exposure to tobacco smoke and the polycyclic hydrocarbons produced by the combustion of fossil fuels.

3 Cancer Risks

Direct assessment of a small risk of a common cancer is normally impossible, unless investigation can be focused on cases in people who have not been exposed to the principal causes. One can investigate small risks of lung cancer by focusing on cases that occur in life-long non-smokers, so long as the exposures of interest are frequent and we can obtain some quantitative or qualitative measures of the degrees to which different individuals have been exposed (as can be done with environmental tobacco smoke and radon in buildings; see also Chapter 16). In the same way, one can assess risks of uncommon cancers under similar conditions and can set upper limits to any likely risk, as in the cases of formaldehyde and nasal sinus cancer and phenoxyacetic herbicides and soft tissue sarcomas. Epidemiology will not, however, produce any worthwhile evidence of small environmental risks of such common cancers as cancer of the stomach and large bowel, as their principal causes are still unknown. Direct assessment of a small risk is easiest in the rare circumstances in which the agent is the only, or almost only, cause of the disease, as was the case when some angiosarcomas of the liver occurred in the vicinity of a plant manufacturing polyvinyl chloride.

4 Conclusions

In other circumstances, epidemiology should be able to set an upper limit of risk, so long as it is possible to define exposure to an agent without it being confounded with exposure to other agents that may be causes of the disease. The upper limit of such risk may, however, be relatively high (such as lifetime risks of between 1×10^{-3} and 1×10^{-2}) as in the case of bladder cancer associated with the consumption of saccharin. Epidemiology cannot be expected to prove that no risk exists, and unless society is prepared to eliminate all agents that are carcinogenic in laboratory animals or mutagenic *in vitro* or that may secondarily create situations in which carcinogenic agents are formed in the environment or *in vivo*, it will need to accept the concept of a negligible risk and give it a numerical value.

5 Reference

R. Doll, *Chem. Br.*, 1987, **23**, 847.

11
Epidemiological Methods

J.A.H. Waterhouse

UNIVERSITY OF BIRMINGHAM, INSTITUTE OF OCCUPATIONAL HEALTH, UNIVERSITY ROAD WEST, PO BOX 363, BIRMINGHAM BA5 2TT, UK

1 Introduction

Everyone knows the story of the first established occupational cancer, written up by Percival Pott, surgeon, of Bart's Hospital.[1] Only a dozen years ago, the bicentenary of his publication linking cancer of the scrotum to chimney sweeps was celebrated. Was this an example of epidemiology? Yes, because epidemiology can range from the very simple to the highly complex in its methodology. A single case of scrotal cancer in a chimney sweep, or two, or even three, might not strike a chord, but beyond this number the strength of the relationship increases. It is in fact the repetition - the numerical aspect - that impresses. Then of course one looks at other aspects, medical, biological, aetiological. The soot which lodges in the rugae of the scrotum, the peculiar sensitivity of that skin in comparison with other, the long and continued intensive exposure to soot in their early apprentice days, these are among the other aspects which are supportive of a relationship. But the first clue was the numerical one.

2 Epidemiology

Epidemiology is a form of quantitative medicine. It deals with the group possessing some characteristics in common rather than with the individual patient which is the subject of most medical consultations. For this reason, it must be concerned with numbers and thus with statistics. It measures the impact of disease on a population, and thus was very much involved with the evolution, growth, and pattern of epidemics. Much of the early application of mathematics to medicine was in this field, and many notable mathematicians have contributed to it. But there is also a reciprocal impact which the population can make on the disease and its manifestation. Epidemiology therefore is concerned to describe and to analyse the principal quantitative aspects of this interaction. It is closely linked with demography, and would claim much of the latter's subject matter and techniques for its own. Births and deaths form a common ground, together with much of the Census data, but the detailed analysis of mortality and even more of morbidity would be within epidemiology's domain. Nevertheless there are many similarities, especially in that branch of epidemiology known as descriptive, in contrast to the other main branch which is analytical.

3 Scrotal Cancer

About a hundred years after Pott's paper the same tumour was shown to occur as an occupational disease among those distilling oil from coal, and also among Scottish shale oil workers - this latter by the well-known Dr. Bell,[2] the original of Conan Doyle's Sherlock Holmes. Later it became known in this country, mainly in the Lancashire area, as mule-spinners' cancer. In the earlier years of this century the Manchester Committee on Cancer was set up to investigate it more closely. After the last war the process was fully mechanized and the occupation ceased to exist. About twenty years ago Dr. Miles Kipling had come to be senior medical inspector of factories in the Birmingham area and came to enquire of the author, as Director of the Birmingham and West Midlands Cancer Registry, whether there might now be appearing in this area an excess of scrotal cancers.[3] His idea was that former mulespinners from Lancashire might have gravitated to the Birmingham area as one of full employment (as it was then), and that they might develop scrotal cancers, due actually to their former employment, while in another job in Birmingham. An analysis of the records showed that there was an excess of scrotal cancers, but not for that reason. This excess, together with the means of demonstrating it, were required almost immediately, via a court case,[4] brought by a Trade Union on behalf of a widow whose husband had recently died from a scrotal tumour. The data that were used are shown in Table 1 and they demonstrate another essentially simple yet cogent use of epidemiology for this purpose. The two southern health regions of London, known then as the South West and South East Metropolitan regions, shared a single cancer registry which was modelled very much on the same lines as in Birmingham. The successive comparisons of figures for the two registries clearly demonstrated that in round figures the South Metropolitan Cancer Registry (SMCR) was close to twice the size of the Birmingham Regional Cancer Registry (BRCR). This relationship applied to overall population, to the males only, to the total numbers of cancers and again for the males only, and for certain specific cancer sites of the male genitalia - except for scrotum, where the ratio was reversed. Thus, it was demonstrated that there were approximately four times the number of cases which would be expected to occur if the experience had been the same as in the SMCR. Note that no specialized data or techniques were needed for this demonstration, nor very specialized knowledge to appreciate its importance.

Table 1

	BRCR	*SMCR*
Population (millions)	4.76	8.22
Male only	2.35	3.88
Total cases of malignant disease (3 years)	40,069	77,393
Male only	20,625	38,697
Total cases of:		
Testis (3 years)	152	292
Penis (3 years)	77	139
Scrotum (8 years)	113	55

BRCR = Birmingham Regional Cancer Registry
SMCR = South Metropolitan Cancer Registry

Why then was there this excess in the Midlands, if it was not attributable to displaced mulespinners? It was shown that the prevalent occupational group was the toolsetters. Not only did this description of their occupation figure frequently in the cancer registry notes (taken from the original hospital case notes), but it became apparent from seeing the nature of their work how their clothing, especially in the groin area, was almost soaked in oil.

4 Hazard and Disease

In its simplest form the detection of a relationship between a potential toxic hazard and a consequential disease amounts to demonstrating an irregularity in the ratios of numbers shown in Figure 1. For the horizontal division of the table, the upper layer (+) includes those who have been exposed to the hazard and the lower layer (-) those unexposed: of the vertical division, the left-hand column (+) includes those who have developed the disease and the right-hand column (-) those without it. If the ratio of those with the disease to those without it is the same for both rows, the exposed and the unexposed, then no relationship exists between exposure to the hazard and development of the disease. If they differ, and furthermore, if they differ to a statistically significant extent, then a relationship is said to exist, presumptively with the disease more commonly found in those exposed to the hazard, though it is possible to find the reverse. Usually, those with the disease, even in the exposed group, are in a small minority of the total, so that *a* and *c* are both small in comparison to *b* and *d*.

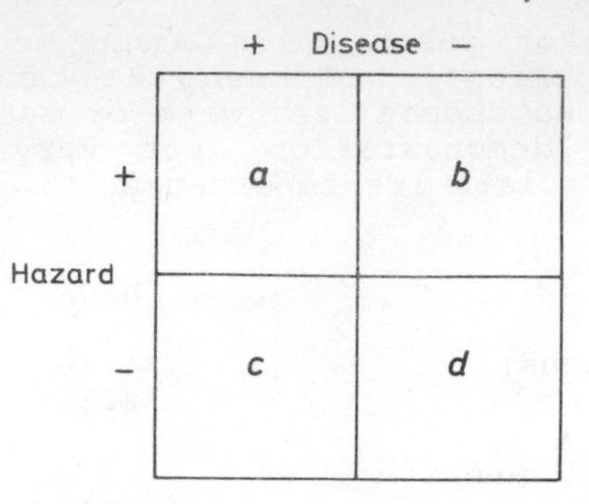

Figure 1 Distribution of factory population according to whether or not exposed to hazard and whether or not diagnosed with disease under consideration: disease-hazard contingency table

The proportion of those with the disease in each group can thus be approximated as *a/b* and *c/d*, and these can be regarded as the odds of getting the disease respectively for each group. The ratio of these two, *i.e.* *a/b:c/d = ad/bc* is then known as the 'Odds Ratio' and provides a measure of the difference in risk between the two groups. It is a measure also of what is often referred to as the 'Relative Risk', again between the two groups, exposed and not exposed.

Let us take as an example the data in Table 1, in order to compare the relative risk of scrotal cancer in the Birmingham region with that in the South Metropolitan region. Table 2 sets out the figures, taking Birmingham as 'exposed to risk' and South Metropolitan as 'not exposed', together with cancer of the scrotum as 'the disease' and all other cancers as 'without the disease'. Then the odds ratio is (113 x 38642)/(55 x 20512) = 3.87, which is very close to the rough ratio of four times, which was obtained by a simpler approach to the same figures. Since there is only concern with the ratios of corresponding quantities, the fact that the scrotal cases were accumulated over an eight-year period for each registry and the total male cancers over three years is immaterial to the result. The statistical significance of the result can be evaluated by the X^2 test in the normal manner, yielding a X^2 (1 degree of freedom) of 78.5, which is so remote as a chance event that a real difference can clearly be said to be well established.

Table 2

	Cancer of scrotum	*All other cancers*	*All male cancers*
BRCR	113	20,512	20,625
SMCR	55	38,642	38,697

<u>4.1 A Pilot Trial</u>. Any test of this kind, however, can only be regarded as in the nature of a pilot trial. If the result is statistically significant and seems worth pursuing, then it represents only the beginning of a fuller investigation, as it did for the Birmingham study. Moreover, it is essentially a crude test, which could be refined for instance by examining the possibly differential effects of age, or of other factors which can 'confound' the relationship - that is, can serve to confuse or distort an otherwise directly discernible relationship. Age is probably the most important 'confounding factor' in epidemiology, since virtually all diseases are sensitive to it. Nearly all cancers show an incidence rate which rises exponentially with age. Figure 2 shows the age-incidence graph for squamous cell skin cancer in males and females, first on an arithmetical vertical scale and then on a logarithmic scale, since on the latter an exponentially increasing curve becomes a straight line.

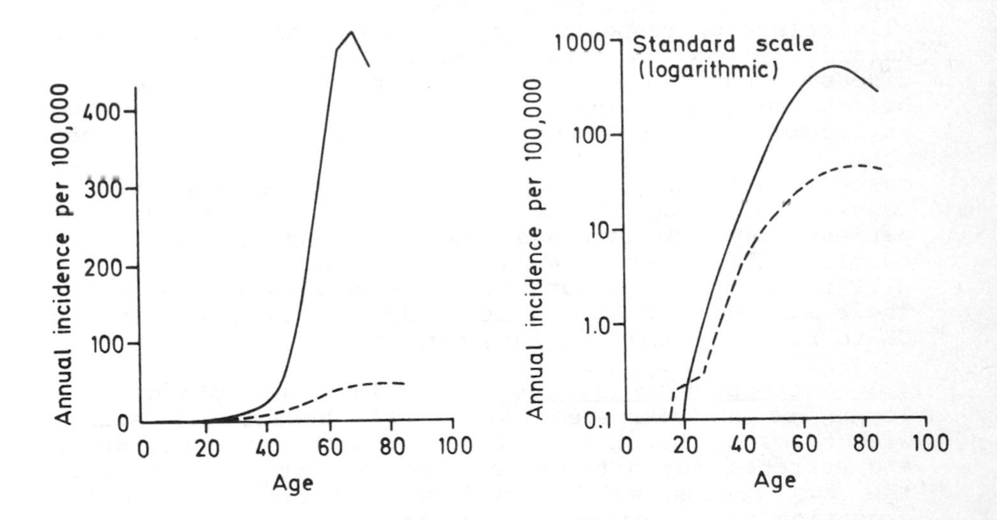

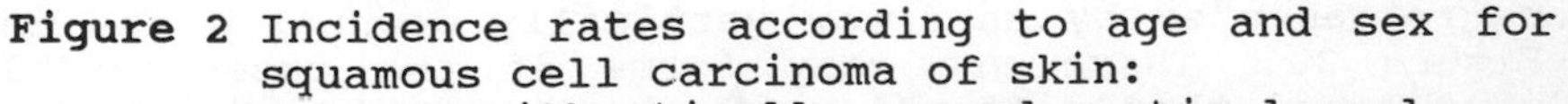

Figure 2 Incidence rates according to age and sex for squamous cell carcinoma of skin:
(a) on arithmetically spaced vertical scale
(b) on logarithmically spaced vertical scale

A big difference in average age between two places being compared could thus very easily distort any other difference of a lower size, so that the place with the older population would exhibit a much higher apparent risk. There are several methods of compensating for such differences, usually by means of some form of stratification or standardization. When the confounding factor's distortion has been effectively removed, it becomes possible to evaluate separately any other effect.

4.2 Case-control Studies. When there is a group of cases of a disease which is considered possibly to be related to an industrial or environmental hazard, another group is selected, unexposed to the putative hazard, to act as a 'control' and thereby to provide information about the contrasting potential hazard. In effect this method is an extension into the human domain of one of the powerful techniques used so successfully in the scientific renaissance to demonstrate and evaluate the separate contributions of single factors in the production of a resultant effect. When all factors but one are held constant, any resultant variation must be attributable to variation of that one excluded factor. Clearly it is not in all circumstances possible to hold all other influences constant, nor even always to know which others are relevant - and these considerations are of especial importance in the investigation of human disease. A strict application of the method would require that the control group be chosen to parallel the 'case' group in all relevant respects except that of the suspect hazard. Among the first parallels in nearly all human disease studies are age and sex; next after these will be those determined by the disease and the environmental situation. Doll and Hill[5] for instance, when first investigating the rapid increase in lung cancer deaths in the UK, chose to parallel lung cancer cases in a group of London hospitals each by another patient matched for age and sex and in the same hospital, but with a non-cancer diagnosis. They were all interrogated about their occupations and about their smoking history. This study was the first in the UK to relate smoking to lung cancer.

4.3 Retrospective Studies. This is an example of what is called a 'retrospective' study, because it starts with cases of the disease together with a control group and searches for differences in the past experience of the two groups which could be contributing to the causation of the disease. In the simplified diagram of Figure 1, it approaches this table from above and could be represented by an arrow vertically downwards, as in Figure 3. There are many forms of the case-control study, and they differ largely in the type and methods of selection of the control group.

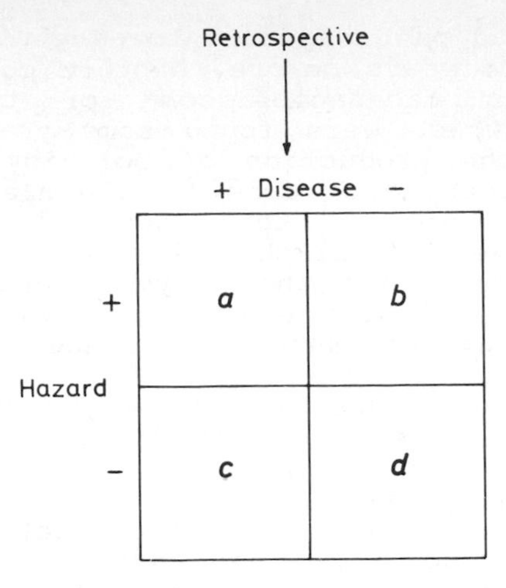

Figure 3 Disease-hazard contingency table approached first through disease: retrospective studies

The more closely they can be made to parallel the case group the more efficient the study - except that the parallelism may, inadvertently, include one or more of the aetiological factors, when of course the efficiency is reduced. It is especially useful as a method when the number of cases is small, when, to attempt to compensate for the inevitable irregularities of small numbers, it is usual to select two or more controls for each case: beyond four controls per case there is virtually no further gain in efficiency. Whatever the number of cases, however, it is often the choice of control which will be found to be circumscribed by numerous difficulties, and in consequence it is not the optimum which is ultimately available for analysis. It will be seen later that there is another technique which is, in its basic approach, closely akin to the case-control method, but utilizes a rather more subtle and sophisticated multivariate analysis to obtain a higher degree of efficiency.

4.4 Rare Conditions. There are of course some conditions which stand out by their very rarity. Prolonged exposure to asbestos dust can result in asbestosis, which is a specific form of pneumoconiosis, a fibrotic progressive disease reducing the efficiency of the lungs, or in lung cancer, or in an otherwise very rare tumour of the pleura, mesothelioma. This last condition was exceedingly rare in its occurrence until it was found in association with the rapid increase of use of asbestos, though following after a long time lapse. Nowadays, almost any case of mesothelioma will be found to have had past exposure to asbestos, perhaps up to forty or more years before. Another condition of extreme rarity is haemangiosarcoma of the liver, several cases of which were found some years ago in association with the production of polyvinyl chloride (PVC). It was the cause of considerable alarm because of the widespread use of this material. Careful investigation however established that the association was with the monomer, not the polymer, and that the slight contamination of some PVC products with residual monomer was inadequate to cause the disease.

4.5 Latent Period. One of the major differences between cancers and other types of disease in relation to a potential industrial hazard is the long latent period required for the development of a cancer. Twenty years is often cited as a representative figure, but it can be very much more than that or occasionally less. It is one of the reasons for the lateness of their recognition as industrial diseases, since the long time lag or latency leads to their confusion with 'naturally occurring' cancers which are more common with increasing age. In the case of chimney sweeps' cancer, they had of course been exposed at very early ages when climbing the chimneys to dislodge the soot. There are analytical techniques available now both to allow for latency and to heighten the sensitivity of the method. It is in fact from certain industrial carcinogenic hazards that it has been possible to obtain valuable information about latency itself. It is an important aspect of the whole process of carcinogenicity, but one particularly difficult to investigate because of the dearth of reliable information. Well-kept industrial records of first exposure to any potential hazards can therefore be of considerable value.

4.6 Prospective Studies. It might appear desirable to maintain full records of each new substance or process introduced into industry, especially the chemical industry, and of all those who might be exposed to them, forwards for an indefinite period. Some such systematic approach has been under consideration at various times, both in the UK and elsewhere, but its cost and its ramifications quickly become prohibitive. It would of course be thus possible to conduct a prospective study, beginning with the potential hazard and determining the incidence of any consequential disease after a lapse of time. This method in fact amounts to approaching the basic diagram of Figure 1 from the left-hand side, in contrast to the retrospective approach, from above, as in Figure 4.

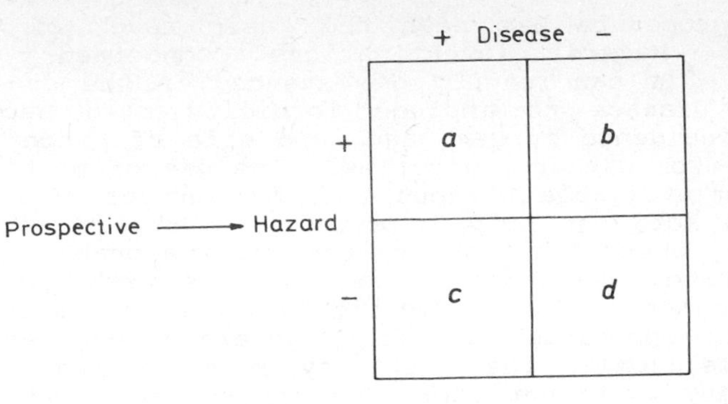

Figure 4 Disease hazard contingency table approached first through hazard: prospective studies

While prospective studies in the true sense of 'wait and see' are uncommon, an adapted form, known sometimes as a 'retrospective prospective' or a 'historical prospective' study is more frequently found. The indication for such a study may be the suspicion of a specific disease as being process-related, or merely to investigate the health record of an industry. A prerequisite condition is of course the evidence of good past records, going back twenty years or more. Given the situation at the time good records began, it should be possible to travel forward from that point to the present time, knowing also the personnel changes in that period. It thus becomes the equivalent of a prospective study, effectively begun many years ago.

4.7 Mortality Rates. The Registrar General publishes annually tables of mortality rates by sex, age groups, and disease. If the mortality rate for men in the age group 50-54 from rectal cancer is 12 per 100,000, then in a corresponding group of 100,000 men followed for a year, approximately 12 would be expected to die from that cancer. In a similar way, a pattern of 'expected' deaths can be drawn up for each disease or disease-group, taking note of the sex and age composition of a factory population over the period available for study. Naturally, allowance must be made for the changes in age over the period, for differences in mortality rates by calendar year, and for the addition of new cases and for the deaths known to have occurred. Then the observed or actual pattern of deaths by cause can be compared with the expected, and any notable differences evaluated in a statistical manner. Expressed as a ratio of the observed number of deaths to those expected from a specific cause, and multiplied by 100, it is known as the Standardized Mortality Ratio or SMR for that cause. In this form it is the analogue of the Relative Risk, discussed earlier.

This procedure has been the basic model of many industrial hazard investigations concerned with mortality. It can readily be extended in the case of malignant disease to include morbidity also because rates of incidence by sex, age, and site of cancer are available from cancer registries. The use of morbidity data, when available without infringement of the Data Protection Act, can help to extend a mortality study, especially when the cancer has a reasonably good survival rate, which is seldom the case with lung or stomach cancer. Cancer morbidity data are also of particular importance for skin cancers where, except for the melanomas, the mortality rate is negligibly small. They would not therefore appear in a mortality analysis, yet they are by no means uncommon in a number of occupations, and deserve closer attention and study.

4.8 Reservations. The principal objection that can be raised against the use of standard mortality rates to calculate expectations is that a factory population is not equivalent in a number of respects to the general population of the country. Frequently the hazard study will be chiefly concerned with production workers, who will probably be in the Registrar General's social class grades III, IV, and V, excluding I and II. There will often be marked geographical differences in mortality rates, again distinguishing the factory from the general population. Both of these types of differences would contribute 'biases', distorting the basis of the comparison. Sometimes the extent of a difference in mortality pattern is sufficiently large to override these biases. Other biases include the Healthy Worker Effect which emphasizes the fact that factory workers must in general be in good health and often physically strong, in contrast to the general population which will include all the disabled and handicapped. Another is the Survivor Population Effect which again leads to a better health record for a factory population because those whose infirmity or illness renders them incapable of working in the factory will have left after only a relatively short time.

Some of these reservations could be overcome by the use of a 'combined factory population', made up of that sub-group of the general population. Except possibly in close relation to a Census, it is not easy to construct such a population, and to establish its mortality experience. What is in fact a much more soundly based method is afforded by the use of the factory's own population as an internal control. Clearly it is not possible in this way to discover any overall mortality difference for the factory as a whole, but to compare mortality and morbidity between different sections within a factory the method can usually be held to avoid most of the biases already described. The method is largely due to Sir David Cox[6] and can also be described as that of 'regression models in life tables' (RMLT). In a manner similar to a regression equation it takes account of each distinctive item of information available for an employee: quantitative, such as age, time since joining factory, duration of exposure to a specific substance, *etc.*, and qualitative, such as work history by department or job, lifestyle factors, smoking, *etc.* Workers exposed to specific hazards are grouped, and in effect compared with all other workers not exposed to those hazards, but taking account of possible perturbations of the comparison due to any of the other factors included. For optimum use it essentially requires good and reliable data over a long period for every employee. Despite its greater efficiency therefore it tends to be rather more expensive to conduct, since it is also much more extensive.

5 An Example from the Rubber Industry

A large-scale study, mainly of the tyre section of the British Rubber Industry, was recently completed.[7] More than 36,000 workers were studied, who entered the industry between 1946 and 1960, and were followed up for deaths occurring up to 1985. The original aim of the study had been to verify that the removal of 2-naphthylamine from use as an anti-oxidant ingredient, which took place in 1949, had reduced the excess of bladder cancers which Professor Case[8] had earlier demonstrated. The workers were divided into three five-year entry groups or 'cohorts': 1946-50; 1951-55; 1956-60. Only the first cohort had been exposed, and it was shown that in the two later cohorts there was no excess over the 'natural' or sporadic rate, but that the first cohort still showed an excess, made clearer when allowance was made for latency. Such a demonstration is in fact the most satisfying result of an industrial hazard investigation: Case had implicated 2-naphthylamine as the culprit responsible for the excess, and when it was removed, workers never exposed to it showed no excess. In the course of this study, however, there were found excesses of several other cancers, and some deficits also. They were all based on SMRs obtained by comparison with the general population, the largest differences being for lung and stomach cancers. When analysed internally, by the RMLT method, only the lung and stomach excesses remained, the deficits found for cancers of the prostate and testis, and for Hodgkins disease for instance being attributable to the social class effect, since they are all found more commonly in the higher social classes. The excesses remaining were found to be related to the duration of time spent in either 'fume and/or solvent exposed' jobs (lung cancer), or in 'dust exposed' jobs (stomach cancer), both of which correspond well to known similar exposures.

6 Conclusions

The analysis of human 'experimental data' discussed in this chapter presents no insuperable ethical problems, since it does not amount to deliberate experimentation for its own sake. The costs in time and money are high, both for investigative studies and for subsequent screening and compensation. Clearly some alternative method, more objective and less time-consuming, would be highly desirable. Bacterial mutagenesis (the 'Ames test') was at one time thought to provide the answer, but although it is a valuable test in many ways it is no surrogate for human experience. Nor is animal experimentation, though it can also be of considerable value. Ultimately it is necessary to obtain human data and so to organize its collection and the efficiency of its analysis as to maximize the use of the information available. The epidemiological methods already exist, though they can no doubt be yet further refined. Still, however, the alert observer with an active and well-trained mind is the most important single factor in the initial detection of new toxic hazards.

7 References

1. P. Pott, 'Chirurgical Observations', Hawes, Clarke and Collings, London, 1775.

2. Joseph Bell, 'Paraffin Epithelioma of the Scrotum', *Edinburgh Med. J.*, 1876, **33**, 135-137.

3. J.A.H. Waterhouse, 'Cutting Oils and Cancer', *Ann. Occup. Hyg.*, 1971, **14**, 161-170.

4. *Times*, London, 5th October 1968; Stokes *v.* Guest, Keen and Nettlefold, Ltd: Birmingham Assizes.

5. R. Doll and A.B. Hill, 'Smoking and Carcinoma of the Lung', *Brit. Med. J.*, 1950, *ii*, 739.

6. D.R. Cox, 'Regression Models and Life Tables', *J. Roy. Stat. Soc., Ser. B,* 1972, **34**, 187-219.

7. T. Sorahan, et al. 'Mortality in the British Rubber Industry, 1946-85', *Brit. J. Ind. Med.*, 1988, (in press).

8. R.A.M. Case and Margery E. Hosker, 'Tumours of the Urinary Bladder as an Occupational Disease in the Rubber Industry in England and Wales', *Brit. J. Prev. Soc. Med.*, 1954, **8**, 39-50.

12

An Epidemiological Approach for the Risk Assessment of Chemicals Causing Human Cancer and other Disorders

S. Li

DEPARTMENT OF SCIENTIFIC INFORMATION, CHINESE ACADEMY OF PREVENTIVE MEDICINE, 10 TIAN TAN XI LI, BEIJING, CHINA

1 Introduction

Since the 1970s, environmental problems have become a world-wide concern. China and many other countries face the challenges of these problems. In particular, more attention is being devoted to studying the effects of environmental pollution on human health. Over the past 50 years, as China developed industrially, increases in environmental pollution and changes in social habits have changed the disease pattern. The most striking changes are the decline in the proportion of deaths due to infectious diseases and the increase in the proportion of death due to diseases which are related to potentially hazardous environmental factors including smoking, atmospheric pollutants, indoor air pollutants (sulphur dioxide, carbon monoxide, NO_x *etc.*), water hardness, trace elements such as selenium, noise *etc.*, *e.g.* cancer, cardiovascular disease, and cerebrovascular accidents. Table 1 shows the changes between 1953 and 1973 in the composition of total deaths in China. From this Table, one can see that in 1953 (based on statistics from Shanghai), pulmonary tuberculosis and measles accounted for 28.8% of all deaths, whereas in 1973 malignant tumours accounted for 28.4% of all deaths, cerebrovascular accidents 19.4%, and heart disease 16.8%. Together these three causes account for 64.6% of all deaths while pulmonary tuberculosis accounts for only 3.9% of all deaths.[1]

Current epidemiological research indicates that approximately 80% of human diseases are related to environmental factors including air pollution, water pollution, hazardous solid waste, food-borne diseases, food poisoning, noise, radiation, and endemic diseases such as fluorosis *etc.* Therefore, exploring the relationship between environmental factors and disease causation is one of the major components of preventive medicine.

As is well known, environmental pollution is due to human activity and in turn causes a decline in environmental quality, *i.e.* it causes the physical, chemical, and biological changes of atmosphere, water, and soil, and results in hazardous effects on the normal life and development of humans and animals.

Table 1 *The changes in the disease pattern in China*

	1953		*1973*	
Code No	*Cause of death*	*%*	*Cause of death*	*%*
1	Tuberculosis	14.8	Malignant tumours	28.4
2	Measles	14.0	Cerebrovascular accidents	19.4
3	Cerebrovascular accidents	13.1	Heart disease	16.8
4	Heart disease	9.9	Respiratory disease	12.0
5	Malignant tumours	9.4	Gastro-intestinal disease	6.4
6	Pneumonia	8.8	Trauma	5.1
7	Ageing and weakness	8.5	Tuberculosis	3.9
8	Meningeal tuberculosis	7.6	Neurological disease	3.8
9	Kidney	7.4	Kidney	2.5
10	Respiratory disease	6.5	Rheumatism	1.6

Basic factors (*e.g.* age, sex, locality *etc.*) were used in the compilation of the table - the age ranges between the two dates are similar.

The effects of environmental pollutants on the human body are characterized by multifactorial causes, multiple sources, low dose, and long period of exposure. The effect of the environment on health is very complex, since physical, chemical, and biological factors in the environment can all influence health status.

Among the environmental factors mentioned above, chemical factors have a particularly important impact. Relevant data indicate that over 100,000 chemical products are marketed in the world today. Every year over 1000 new chemical products appear on the market as indicated by the United Nations Environment Programme/ International Register of Potentially Toxic Chemicals. Common chemical pollutants, such as sulphur dioxide, nitrogen dioxide, photochemical smog, heavy metals (*e.g.* lead, mercury, and cadmium), agricultural herbicides and pesticides, chemical carcinogens *e.g.* asbestos, chromium, polyaromatic hydrocarbons, *N*-nitrosamines, polychlorinated biphenyls (PCBs), and many other environmental pollutants have the potential to cause acute or chronic toxicity. Various chemical residues can be found in the environment and accumulated in the body. By concentration in the food chain, they could be harmful to health. Therefore, evaluation of chemical hazards in the environment is very important. In recent years, such research has indicated that certain chemical pollutants are carcinogenic and cause other diseases. This has attracted more attention to the important problem of preventive medicine.

In China, the harmful health effects that environmental pollution may produce are receiving increasing attention. In 1983, in order to strengthen preventive medicine, China established, within the Chinese Academy of Preventive Medicine, the Institutes of Environmental Health Monitoring, Environmental Health and Engineering, Nutrition and Food Hygiene, Food Safety Control and Inspection, and Occupational Health and Diseases. Risk assessment of many types of chemical agents (including indoor air pollutants, air pollutants in work areas, and chemicals in water, soil, and food) is carried out in these Institutes. These Institutes also conduct epidemiological studies in many areas. Several medical colleges in China have established departments of public health or hygiene to train public health doctors. Some of the graduates of these departments will take on the responsibility of assessing environmental quality and health effects. Sanitary and anti-epidemic stations have been established at the provincial, city, and county levels, In these stations, scientists are responsible for carrying out risk assessments of chemical products and epidemiological surveys on environmental factors. Because much foreign and domestic literature indicates that the environmental factors may be crucial causative factors for cancer and many other diseases, Chinese scientists have put much emphasis on the risk assessment of chemical contaminants in the environment.

2 Toxicological Studies and Epidemiological Investigations are Important Methods for the Risk Assessment of Environmental Chemicals

As a result of the daily increase in the quantity and variety of chemical products, people are becoming increasingly concerned about the safety of new chemical products. Toxicological research has become an important approach for conducting such assessment.

In China, the goals of toxicological research are generally categorized as follows:

i) To assess the toxicity of chemical substances; for example, to study the toxicity and to determine thresholds of lead, benzene, and organophosphates. This also includes studies on special toxic effects, such as carcinogenicity and mutagenicity.

ii) To study chemical substances in the body and how organs, tissues, and cells react to and detoxify such substances.

iii) To integrate the toxicological findings with epidemiological data on exposure to occupational toxicants and thus establish scientifically based standards for safe levels of exposure. In addition, studies on effective detoxification of drugs by experimental therapeutic approaches are also carried out.[2]

It is important to emphasize the very important role that epidemiological studies play in assessing the risks of chemicals in the environment in China. Approximately 800 million of China's 1000 million inhabitants reside in the rural areas and their residential living patterns are relatively stable, *i.e.* people live for a long time in one place. This provides ideal conditions for conducting epidemiological studies on the effect of long-term exposure to chemicals on the human body. Because of this, great importance is attached to methods of epidemiological investigation and analysis.

In conducting risk assessment of chemicals, attention is paid to evidence from laboratory studies on animals, as well as long-term, chronic effects on the human body. In most situations, if a conflict appears between toxicological and epidemiological findings, more importance is attached to the conclusions based upon epidemiological studies on human population.

3 The Application of Epidemiological Investigations in Risk Assessment

Epidemiological investigations are very important in the realm of risk assessment of chemicals. As is commonly known, epidemiology is the study of disease in human populations, particularly the studies on the distribution and cause of disease.

When analysing and assessing the possible etiological roles that environmental chemicals and factors play in the occurrence of cancer and other diseases, it is extremely important to use an epidemiological approach.[3]

3.1 Objectives

(1) To determine whether data obtained from epidemiological studies and clinical observations are consistent with disease causation hypotheses generated from laboratory experiments.

(2) To provide evidence and data necessary to formulate health standards, to develop preventive measures and to provide health services.

3.2 Methods

(1) Investigations into the distribution of population-specific diseases among various human populations (including distribution according to age, genetic characteristics, and sex), different geographical conditions (including natural and social environments), and time relationships (including variation according to year, season, or shorter time interval), and investigations into factors which cause specific diseases, as well as factors which affect causative agents, and also the study of disease incidence and mortality rates, and, in addition, monitoring of the trends of their occurrence and changes.

(2) Investigations and inspections of environmental factors and conditions which cause diseases, including the presence of pollutants and various naturally occurring micro-organisms in the air, water, soil, and food and their environmental burden, changes according to time and place, and the levels of human exposure.

(3) Analysis of data used to evaluate the extent and severity of pollution, health effects, the relationship between exposure and effects (dose and response). According to this information, studies on the threshold of chemical burden, hygienic standards, and chemical measures are also conducted.

(4) Survey of the occurrence and development of diseases and disorders in population, presentation of biological, microbiological, chemical, physical, genetic, and other scientific evidence and explanation of the causes of diseases.

(5) Comprehensive analysis and investigation of epidemiological and laboratory data, and the presentation of hypotheses and evidence of causes of disease.[4]

As mentioned above, the epidemiological approach is used widely to interpret causes of disease, according to the background data; this provides scientific evidence for developing preventive measures and establishing health standards.

4 Examples

4.1 Indoor Air Pollution and Lung Cancer. In general, most people spend about 33 to 50% of their life in houses and buildings.

In Xuanwei county, in Yuannan Province of China, scientists discovered a very significant phenomenon. In this county there was an unusually high incidence of female lung cancer: the annual age-adjusted mortality rate for lung cancer from 1973 to 1979 was 27.7 per 100,000 in males (among China's highest) and 25.3 in females (Chine's highest) (see Table 2). Since tobacco smoking is common in males (≥40%), but rare in females (<0.2%), indoor air pollution is suspected to be the major risk factor for female lung cancer.

Table 2 *Annual lung cancer mortality rates in China*

		Mortality rates (per 100,000)				
Areas	*Years*	*Unadjusted*			*Age-adjusted 1964 China population*	
		Males	*Females*	*Total*	*Males*	*Females*
China	1973-75			5.0	6.8	3.2
Yuannan province	1970			2.8	4.3	1.5
Xuanwei county	1973-79	27.0	24.5		27.7	25.3
High-risk areas in Xuanwei	1973-79	114.4	120.6		118.0	125.6
Low-risk areas in Xuanwei	1973-79	4.0	2.8		4.3	3.1

Xuanwei county has a population of about 1 million people, of which 90% are farmers. Local residents customarily use one of the three domestic fuels: 'smoky coal', 'smokeless coal', and 'charcoal'. Fuel burning in shallow grates or hearths without chimneys has resulted in serious indoor air pollution. Women spend most of their time in preparing and cooking food for the family, while men generally are working in the field.

In order to investigate the potential effect of indoor environmental pollutants on lung cancer, the scientists studied the air in high- and low-risk areas in Xuanwei county. The results show that the indoor and outdoor concentrations of benz[a]pyrene (BP) in high-risk areas were 626.9 $\mu g\ 100\ m^{-3}$ and 130.0 $\mu g\ 100\ m^{-3}$, respectively, whereas in low-risk areas they were only 45.7 and 0.2 $\mu g\ 100\ m^{-3}$, respectively. The concentrations of sulphur oxides (as sulphur dioxide) were 0.44 mg m^{-3} and 0.01 mg m^{-3}, respectively in the high-risk areas, and 0.03 and 0.01 mg m^{-3}, respectively in the low-risk areas. In addition, the concentrations of total suspended particles (TSP) in the high-risk areas were 10.45 and 0.24 mg m^{-3} (indoor and outdoor), and in the low-risk areas 2.57 and 0.20 mg m^{-3}, respectively. From the above data, it is considered that the indoor air pollution was more serious in the high-risk areas of lung cancer than in the low lung cancer mortality areas, and that indoor air pollution is a potential etiological factor for the high lung cancer mortality in Xuanwei county of China.[5,6]

In this project, Chinese scientists have completed the following research work relating to Xuanwei county:

Comparative studies on lung cancer of mice induced by subcutaneous injection with soot extract from coal and wood burning.

Studies on lung cancer of mice induced by soot extract by intratracheal injection.

Heritable translocation test with the extract of soot collected from indoor air in high lung cancer incidence areas.

Identification of mutagenicity in human urine from high and low lung cancer incidence areas.

Studies on the induction of lung cancer in animals inhaling coal and wood smoke.

Organotrophy observations with the extract of soot collected from high lung cancer incidence areas.

Cytogenetic changes of animals exposed to emission from coal stoves in mice and rats.

Studies on the mutagenicity of extracts of respirable particles collected from indoor air.[7]

These studies principally support the hypothesis that the indoor burning of 'smoky coal' is an important factor in the etiology of lung cancer in Xuanwei county.

4.2 *N*-Nitroso-compounds and Stomach Cancer. Stomach cancer is the leading cause of death among all malignancies in China and accounts for 23.03% of total cancer death with no sign of decline. Its cause is not known, but some evidence shows that environmental factors may play a very important role in the etiology of stomach cancer.

From the results obtained from epidemiological and laboratory studies carried out in high- and low-risk areas of stomach cancer in China, the *N*-nitroso-compounds (NNO) seem to be one of the most important potential risk factors.

In order to evaluate the role of NNO in the etiology of stomach cancer, the concentrations of NNO in drinking water,[8] pickled vegetable, fish sauce and gastric juice in high- and low-risk areas of stomach cancer were determined. These results are as follows:

NNO in drinking water: Putian county in Fujian province is one of the high-risk areas of stomach cancer in China with an age-adjusted mortality rate of 61.19 $100,000^{-1}$ in males. An epidemiological survey showed that this high mortality was related to the severe contamination of sources of drinking water. In 33 water samples collected from different sites along the river in Putian county, the following six carcinogenic NNOs have been detected: *N*-nitrosodimethylamine (NDMA), *N*-nitrosodiethylamine (NDEA), *N*-nitrosopiperidine (NPIP), *N*-nitrosodipropylamine (NDPA), *N*-nitrosodibutylamine (NDBA), and *N*-nitrosomorpholine (NMOR).

Home-made pickled vegetables are very common in the high-risk areas, *e.g.* Wuwei county, where stomach cancer prevails. In winter and spring, fresh vegetables are not available, and pickled vegetable becomes a major component of the local diet. Two kinds of home-made pickled vegetable (pickled vegetable with flour-paste and clear-soup pickled vegetables) were collected from Wuwei county in Gansu province and analysed for *N*-nitroso-compounds. Six volatile *N*-nitrosamines were found in pickled vegetables with flour-paste and the highest content of NNO was 54.21 p.p.b. In the clear-soup pickled vegetables, eight *N*-nitrosamines were detected and the highest content of NNO was as high as 162.27 p.p.b. It was also found that elevation of environmental temperature and increase of storage time markedly increased the content and type of *N*-nitrosamines.

NNO in fish sauce: home-made fish sauce was collected from Changle county in Fujian province, another high-risk area for stomach cancer. Seven volatile carcinogenic *N*-nitrosamines were found; the highest content was of NDMA. Compared with the data obtained from commercial fish sauces in low-risk cities, the concentrations of cyclic *N*-nitrosamines [*N*-nitrosopyrrolidine (WPYR) and NMOR] in home-made fish sauce were significantly higher.

The NNO concentration in gastric juice collected from chronic gastritis patients was correlated positively with stomach cancer mortality.

In vitro mutagenicity tests for gastric juice were carried out. The result showed that the rates of mutagenicity for sister chromatid exchange (SCE) of human peripheral lymphocytes and micronuclei test of Chinese hamster ovary cells were 52.1% and 85.7% in the high-risk area and 7.4% and 65.7% in the low-risk area, respectively.[9-12]

From the above epidemiological data, it is concluded that NNOs are closely related to the stomach cancer. However, the etiological role of NNOs in the cause of human stomach cancer is still unknown and more research work is required.

4.3 Selenium and Keshan Disease. Keshan disease is an endemic myocardial disease spreading in a long, narrow strip of land covering 309 counties from the north-east to the south-west of mainland China. The disease usually has a sudden on-set and high case fatality. The most susceptible populations are young children and women of childbearing age.[13]

Medical records show that Keshan disease has been endemic in China for about 100 years. In the winter of 1935, there was an outbreak of the disease in Keshan county, Heilogiang province. Many hypotheses have been proposed concerning the etiology of this disease; the selenium deficiency hypothesis has been studied the greatest.

The epidemiological relationship of this disease to selenium occurrence in susceptible populations was studied. The major evidence obtained so far is as follows:

(1) Blood samples with an average selenium content of <0.02 p.p.m. were observed in Keshan disease-affected areas, whereas all the samples from the non-affected areas contained >0.02 p.p.m. selenium. The average blood selenium content was 0.021 ± 0.001 p.p.m. from the affected areas and 0.095 ± 0.088 p.p.m. from non-affected areas.

(2) Average hair selenium concentrations in the affected areas were <0.12 p.p.m., whereas those from non-affected areas were >0.20 p.p.m. Hair selenium concentrations of residents at different sites of affected areas averaged 0.074 ± 0.050 p.p.m. and those of non-affected areas averaged 0.343 ± 0.173 p.p.m.

(3) The average urinary excretion of selenium (12 h) by rural children was 0.59 ± 0.18 μg in affected areas and 1.50 ± 0.13 μg in non-affected areas.

(4) Blood glutathione peroxidase (SeGSHpx) activities of children in the affected areas were significantly lower than those in the non-affected areas. Significant correlations were found between the selenium concentrations in blood and hair and blood GSHpx activity (r=0.57 and 0.64, respectively).

(5) Average selenium concentrations in maize produced in affected and non-affected areas were 0.005 ± 0.002 and 0.035 ± 0.056 p.p.m., respectively, while those of rice were 0.007 ± 0.003 and 0.024 ± 0.038 p.p.m., respectively. These differences are highly significant (P<0.01).

These data clearly indicate that the poor selenium availability for residents in the Keshan disease area, as shown by the low selenium levels in their blood, hair, and urine, and their low GSHpx activity, is caused by the low selenium levels in local foods.

The above findings show that the residents of Keshan disease areas have low selenium availability, as compared to those in non-affected areas. Among these residents, susceptible groups (*i.e.* children in rural farming families) have an even lower selenium availability than non-susceptible groups (*i.e.* adults in the same area). Selenium (in the form of sodium selenite) intervention has been proved to be very effective in the prophylaxis of Keshan disease. The oral supplementation of sodium selenite in children (1 mg week^{-1} for 6-9 years-old and 0.5 mg week^{-1} for 1-5 years-old) of high-risk areas resulted in a more than 90% decrease in the incidence of this disease. This result has been confirmed by different investigators in various endemic areas. Different methods of intervention (*i.e.* sodium selenite tablets or fortified tablet salt) are used. Thus, it is very likely that selenium deficiency plays an important role in the etiology of this disease. However, other factors such as molybdenum, magnesium, Vitamin E, and thiamin insufficiency, or barium and nitrite excess, or even viruses or mycotoxins may be also involved in the pathogenesis of Keshan disease. Present data do not permit the ruling out of any one of these factors acting as a co-factor with selenium deficiency in the etiology of Keshan disease.[14]

4.4 Methylmercury and Methylmercury Poisoning. Methylmercury is a highly toxic chemical, which caused the Minamata disease in the 1950s. Therefore, it is important to study methylmercury pollution in surface water (*e.g.* river and lake) and to evaluate its health effect.

For the purpose of evaluating methylmercury pollution risks in Songhua river in North-east China, Chinese scientists have carried out the following research work:

(i) To investigate the sources of methylmercury pollution and to monitor the level of pollution and the concentration of methylmercury in river and sludge.

(ii) To determine the concentration of mercury in finless fishes and water plants.

(iii) To investigate the history of fish consumption in fishermen, the quantity of fish eaten, and to measure the accumulation level of mercury in the hair, blood, and urine of fishermen.

(iv) To examine treated subjects (patients) and control subjects, including field of vision, perception of sound, and the co-ordinative function of sensing, *etc.*

The results of the investigation on Songhua river show that the acetic acid and the polyethylene production plants were the major sources for methylmercury pollution. The acetaldehyde production area of the acetic acid plant was a particularly serious source for methylmercury. The results are presented in Table 3.

Table 3 *Concentrations (p.p.m.) of mercury in the acetaldehyde production area*

No.	*Samples*	*Sampling site*	*Total mercury*	*Methylmercury*
1	Reaction liquid	Reaction pot	50.00	4.64
2	Crude acetaldehyde	Condensing unit	45.60	3.76
3	Effluent	Distillation tower	6.35	0.43

The mercury contents of fish (catfish, silver carp, crucian, and carp) are shown in Table 4. The data in Table 4 show that all the mean values of mercury in Songhua river fish exceeded the National and WHO Guideline limit for mercury in fish (0.3 and 0.5 p.p.m., respectively), but the trend also shows that the level of mercury pollution in Songhua river has been gradually reduced.

Table 4 *Concentrations (p.p.m.) of mercury in fish at different periods in Songhua river*

Years	*Mean values*	*Maximum value*
1973-1974	0.88	3.50
1975-1976	0.74	3.24
1978	0.66	1.58
1980	0.44	0.77
1982	0.56	0.98

The content of mercury in water plants in Songhua river was 1.69 p.p.m. and in the feathers of *ichthyophogous* birds it was 9.15 p.p.m., values close to the corresponding mercury levels reported in Canadian polluted areas.

The concentration of mercury in the organs of autotrophic cats was also determined. It was found that the maximum content of mercury was in liver, followed by kidney, brain, and blood. There was significantly more mercury in the internal organs of fisherman's cats than in the control cats.

The above data indicated that the Songhua river has been polluted by industrial waste (including mercury), and an abnormal content of mercury has been found in finless fish, water plants, and autotrophic cats.

The manner in which mercury enters the human body is via the food chain. The fish and oysters have the ability to accumulate mercury from the mercury-containing industrial waste water discharged into surface water. Studies carried out on fishermen showed that 90% of mercury consumed by fishermen came from mercury-polluted river fish (Table 5).

Table 5 *The consumption of fish and intake of mercury*

Food and water	*Food consumption* (g day^{-1})	*Mercury intake* (mg day^{-1})	% *of total intake*
Fish	500-900	0.548-0.914	92.51
Grain	570-890	0.045-0.073	7.37
Water	2.5 (l)	0.001	0.12

The half-life of methylmercury in the human body is about 70 days and its target organ is the brain. When the intake level of mercury exceeds its excretion level, it will cause damage to the nervous system, and finally result in chronic mercury poisoning.

Some data on the content of mercury in hair of different populations are shown in Table 6.

Blood samples collected from fishermen in polluted areas had a mercury content of 53.6-228.0 p.p.b. and the content of methylmercury ranged from 15.5 to 59.2 p.p.b. It is generally recognized that the minimum harmful concentration in human blood is 200 p.p.b.[15]

Table 6 *Contents of mercury in hair of different populations*

Populations	*No. of subjects*	*Hair mercury (mean, p.p.m.)*
Fishermen (hospital patients)	86	10.35-43.70
Fishermen (healthy control)	241	3.51-4.32
Residents (urban)	839	0.75
Residents (rural)	156	0.26

The above epidemiological data serve not only as the basis for comprehensive risk assessment of methylmercury contamination in the environment, but also it was demonstrated that the major symptoms and signs caused by methylmercury pollution are nerve terminal hypoesthesia, tunnel vision, and neural deafness. However, it must be emphasized that the symptoms and signs shown by the residents along the Songhua river were not all serious, so that patients could be diagnosed as suffering from mild chronic methylmercury poisoning.

In order to facilitate the control of methylmercury poisoning, 'Diagnostic Criteria and Treatment Guidelines for Chronic Methylmercury Poisoning Caused by Water Pollution' were established. This was based on the extensive research data obtained by the Chinese scientists and the available international literature. In recent years, the government has placed a strong emphasis on the control of Songhua river pollution. As the result, the environmental quality of Songhua river has been greatly improved, leading to a significant reduction in the potential hazard to residents living along the Songhua river.

5 Conclusions

The risk assessment of environmental chemicals is an important topic. It is also a new field, and there are many problems that merit further study. Obviously, epidemiological studies need to be strengthened, especially in the survey methodologies and criteria for risk assessment. It is believed that international information exchange will be very helpful in improving the quality of environmental chemical risk assessment, thus making a greater contribution to environmental protection and human health protection.

6 References

1 G. Sun and S. Wang, 'Environmental Medicine', Science and Technology Publishing House, Tianjing, 1987, Chapter 1, p.6.

2 S. Yin and S. Wang, 'Fundamentals and Advances in Toxicology', Institute of Health, Chinese Academy of Preventive Medicine, Beijing, China.

3. See P.C. Elwood, in 'Toxic Hazard Assessment of Chemicals', ed. M.L. Richardson, The Royal Society of Chemistry, London 1986, p.188.

4. G. Sun and S. Wang, 'Environmental Medicine', Science and Technology Publishing House, Tianjing, 1987, Chapter 12, p.308.

5. J.L. Mumford and X.Z. He, *Science*, 1987, **235**, 217.

6. X. He and S. Cao, 'Etiological Research on Lung Cancer in Xuanwei, China', Institute of Environmental Health and Engineering, Chinese Academy of Preventive Medicine, Beijing, 1986, p.1.

7. S. Li, 'A Brief Introduction on Environmental Toxicology in the People's Republic of China', Proceedings of the Regional Workshop on Environmental Toxicity and Carcinogenesis, Organized by Mahidol University, Bangkok, Thailand, January 15-17 1986, p.257.

8. M.L. Richardson, K.S. Webb, and T.A. Gough, *Ecotox. Env. Safety,* 1980, **4**, 207.

9. C. Chen and L. Yu, *Chin. J. Epidemiol.*, in press.

10. R. Zhang and L. Yu, *China Environ. Sci.,* 1986, **6**, 77.

11. R. Zhang and C. Chen, *Chin. J. Oncol.,* in press.

12. R. Zhang and Z. Zhang, *Chin. J. Oncol.*, 1986, **8**, 427.

13. G. Sun and S. Wang, 'Environmental Medicine', Science and Technology Publishing House, Tianjin, 1987, Chapter 6, p.197.

14. G. Yang and J. Chen, 'Advances in Nutritional Research', 1984, 6, 203.

15. X. Lin and S. Hou, 'Research Papers Compilation on Environmental Medicine', Bethune Medical University, Changchun, 1984, p.1.

13
Solvent Exposures and Risk Assessment

W.A. Temple and D.G. Ferry

NATIONAL TOXICOLOGY GROUP, UNIVERSITY OF OTAGO MEDICAL SCHOOL, PO BOX 913, DUNEDIN, NEW ZEALAND

1 Introduction

A solvent may be defined as a substance (usually liquid) having the power of dissolving other substances in it. Everyone is exposed to solvents. Each year millions of tonnes of organic solvents are manufactured and used throughout the world. Although industrial users constitute the major group of consumers of organic solvents, many products used in our homes contain these substances. For example, adhesives, aerosols, cosmetics, deodorants, fuels, paints, paint thinners, paint removers, and window cleaners are common household products which contain solvents. Exposure to solvents therefore may take place in the home or by occupational exposure in the workplace. An employee of a paint manufacturing plant may be exposed to solvents whilst at work and perhaps later at home through using some of the products mentioned earlier, thereby extending his or her exposure to solvents.

Exposure at work may be minimal as a result of adequate controls, but outside exposure may be significant. It is therefore essential in the evaluation of risk and hazard to consider the total sum of exposure at work, in the home, or in the general environment.

The risk posed by a specific solvent depends on the nature of the hazard presented and the probability of its occurrence. The probability of a given effect occurring under a specified dosing regimen depends on the potency of the toxicant, the susceptibility of the exposed individuals, and the level of exposure to which the target organism is subjected.

Adverse or toxic effects in individuals are not produced by a solvent unless that substance or its biotransformation products reach susceptible sites in the body at a concentration and for a sufficient length of time to produce a toxic response.

Animal toxicity studies provide a database that can be used as a starting point to assess the risk (or evaluate the hazard) to humans associated with a situation in which the solvent, the subject and the exposure are defined. There is often a wide variety of animal toxicity data available for a particular chemical. These data may include acute or chronic toxicity effects, teratogenic, reproductive, mutagenic, and carcinogenic effects, sensitization, and skin and eye irritation results. Invariably, however, all these tests suffer from difficulties inherent in systems employing animals as surrogates for humans. Sometimes epidemiological studies of humans are available but this is more the exception than the rule. With new chemicals epidemiological studies are never available. Much valuable human data may, however, be available from reported cases of poisoning. The widespread use of solvents, both in commercial and household environments, provides the potential for adverse human exposure, either in acute or in chronic, long-term situations. These may include circumstances such as solvent abuse, suicidal ingestion, or overt accidental exposure, occupational exposure, and adventitious low-level intake such as trihalomethanes in water. Many exposures of the above nature are frequently referred to Poisons Information Centres for comment. These centres will either have their own, or have access to, extensive databases of information[1] concerning the health effects of chemicals to humans and/or animals. Information received by the centres may add much valuable data to the health profile of a particular chemical. Through its observations and experience, a Poisons Information Centre is able to contribute to the prevention of poisonings through the identification of high-risk circumstances for poisonings in the community and by calling alerts to potential emergency situations involving poisonings. This is an essential function of Poison Information Centres and has become known as 'toxicovigilance'.[2]

Various types of solvent exposures are now illustrated by use of actual case histories recorded by the New Zealand Poisons Information Centre.

2 Solvent Abuse

Case 1. At approximately 10 p.m. on a Saturday evening, A, an 18 year old work skills trainee and three of his friends were in a picnic reserve area. They had with them a fire extinguisher containing bromochlorodifluoromethane pressurized with dry nitrogen. Each of the four was squirting the contents of the fire extinguisher into the air and inhaling the fumes with the intention of becoming intoxicated or getting a 'high feeling'. Each of them had two turns at this when A, after having his second turn, rolled onto his stomach and began coughing and spluttering. He was then rolled over onto his back by one of his friends and at this stage it appeared as if he had choked; his eyes and mouth were open and he had stopped breathing. Two of his friends started cardiopulmonary resuscitation treatment while the third went to call an ambulance. On the arrival of the ambulance, at approximately 10.40 p.m., a resuscitation unit was set up. However, at approximately 11.30 p.m. a physician examined him in the ambulance at the scene and pronounced him dead.

Case 2. B, a 14 year old student, presented complaining of tiredness, trouble with concentration and memory, and a decline in examination results. Since her liver was enlarged and results of urinalysis were abnormal, tests for liver and kidney function were obtained. The results of these tests were markedly abnormal. Although she had denied substance abuse, when presented with these results B admitted to increasing sniffing of adhesives and typing corrective fluid over the previous two years. She said that at school she and other students would often pour the typing corrective fluid solvents onto their collars, cuffs, handkerchiefs, or neck scarves so that they could sniff the substance unobtrusively during class time. Her family agreed to further clinical monitoring and were willing to ensure her abstinence from solvents and undertook a family counselling programme.

The practice of voluntarily inhaling vapours for the euphoric or intoxicating effect is not a new phenomenon. For example, in the 19th century nitrous oxide (laughing gas) was used for pleasure and later ether[3] and chloroform vapours[4] were inhaled. During the present century solvent abuse among adolescents was reported in the USA as early as the 1950s. During the 1960s the practice spread throughout the USA, on to Western Europe and Scandinavia, and by the 1970s solvent sniffing had reached Australia and New Zealand. Table 1 contains a representative list of some of the products abused and their active ingredients.

Table 1 *Some agents commonly involved in solvent abuse*

Common name	*Chemical constituents*
Cleaning fluid	Trichloroethane, trichloroethylene, toluene
Petrol (gasoline)	Hydrocarbons including paraffins, olefins, naphthenes, and other aromatics
Typing corrective fluid	1,1,1-Trichloroethane, trichloroethylene

Table 1 (continued)

Fingernail polish remover	Acetone, ethyl acetate, methylcellosolve acetate
Lighter fluid	Kerosene, naphtha, and various proportions of aliphatic hydrocarbons
Paint and paint thinners	Toluene, ethyl acetate, methylcellosolve acetate, and alcohols
Adhesives	Toluene, acetone
Aerosol propellants	Freons, including trichlorofluoromethane, dichlorodifluoromethane, propane and butane
Paint remover	Dichloromethane
Pure solvents	Toluene, acetone, ethyl ether, chloroform

2.1 Methods of Abuse. The method of inhalation depends to a large extent on the nature of the substance being abused and the type of container it is sold in. Fumes of the substance are inhaled deeply through the nose or mouth till the sniffer experiences a euphoric feeling or hallucinations. The substance to be used may be placed on a rag, piece of gauze, or cloth which is held over the mouth or nose during inhalation. Alternatively the substance may be placed directly into a paper or plastic bag and the opening of the bag is held tightly over the mouth or nose. Aerosol products are often simply sprayed into the bag and then inhaled or the substances sprayed directly into the mouth. Sniffing sessions can last from a few minutes to several hours, depending on the desire of the sniffer and the vapours being inhaled.

2.2 Effects of Inhalation. The inhalation of solvents can have physiological effects on the sniffer depending on the age of the sniffer, personality, physique, the vapour being inhaled, the quantity inhaled, and the length of sniffing history. Inhaled vapours of these substances pass rapidly from the alveolae into the blood and from the blood into the brain.[5] The initial effects may be experienced within seconds. The solvent inhalants cause stimulation of the central nervous system followed by depression. Effects range from mild excitement to euphoria to loss of self-control to hallucinations to stupor to unconsciousness and sometimes death.[6,7] The sniffer may suffer from headaches, dizziness, nausea, vomiting, disturbed vision, irritation round eyes, nose, and throat, and drowsiness.

2.3 Toxic Effects. Most disturbances resulting from short-term sniffing have been found to be temporary. Studies on the effects of long-term sniffing vary in their findings. Some have found damaged brain, liver, kidneys, bone marrow, and lungs but the findings are not always consistent. The most consistent findings have been urinary abnormalities which appear to be transient, and elevated blood lead levels in habitual petrol sniffers. Chronic encephalopathies have been reported from lead-containing petrol and from toluene. Most of the nephrotoxic effects of solvent inhalation have been reported for the chlorinated hydrocarbons. The effects of solvent inhalation on liver function appear to parallel closely the effects of solvents on kidney function. The chlorinated hydrocarbons are hepatotoxic whilst some hydrocarbons, particularly toluene, are implicated also in liver damage. Concurrent alcohol abuse may exacerbate organ damage, especially in the liver. Sudden death in abusers can occur by a variety of mechanisms, including asphyxiation resulting from using a bag over the head to inhale the solvent, or by aspiration of gastric contents. Other mechanisms include cardiac arrhythmia (fluorocarbons or chlorinated hydrocarbons),[8,9] respiratory arrest, and liver failure.

<u>2.4 Individuals at Risk</u>. There are two different kinds of abuser, the chronic abuser who sniffs on a regular basis and the social user who sniffs only occasionally. Populations at risk for solvent abuse include those whose work brings them into contact with solvents. Medical workers for more than a century have been at risk from anaesthetic agents including ether and chloroform. Occupations at risk for solvent abuse include floor layers, printers, cabinet makers, shoe makers, painters, hairdressers, laboratory workers, dry cleaners, and petrol station attendants, as they are more at risk from accidental inhalation.[10]

Children and adolescents are at risk in any community but are most at risk in communities with widespread unemployment, disrupted homes, minority ethnic backgrounds, and a lower socio-economic group. Studies both in New Zealand and overseas have found that sniffing is concentrated in the adolescent years from 12 to 18.[11] Sniffers of younger ages (4 to 7) have been reported but it is not as prevalent as in the adolescent years.[12] The practice tends to fall away with increasing age. Most research has found the practice to be more common amongst males than females. In the USA a high proportion of solvent abusers are of Spanish/American and Mexican/American cultural backgrounds in comparison with Caucasians and Blacks. In New Zealand research has indicated a high proportion of sniffing Polynesians in comparison with the Caucasian population. In all countries where research has been carried out the highest incidence of abuse appears amongst the most impoverished groups of society who are often experiencing culturative stress.

Frequently mentioned in studies is a high incidence of family disorganization, alcoholism among adult members, and ineffective parenting. Because of the young age of the sniffers, the home environment is very influential in determining solvent abuse. It is widely agreed that many children try sniffing as an experiment or because of peer pressure but quickly cease the practice. The abusers who cause greatest concern are those who have become psychologically dependent on the habit and have encouraged others to try. A general picture of the habitual sniffer is not the healthy, happy adolescent from a stable, caring home, but is the adolescent experiencing problems at home, at school, or within the society and the sniffing is a form of escapism.

2.5 Identification of At-Risk People. Identification of solvent abusers may be made through clinical evaluation, in particular looking for damage to the central nervous system (CNS) peripheral nervous system, kidneys, liver, lungs, heart, and bone marrow. Children and adolescents may manifest declining school performance, declining interest in sports activities, or incorrigibility. Workers are less productive, become irritable with their fellow workers, and may injure themselves. Inflammation or erosions of nasal membranes may be observed. Nutritional deficiency and respiratory tract infections may occur from poor self-care and lack of judgement. Specific biological tests primarily developed for exposure of workers in occupational settings such as urine monitoring may be employed and will be discussed later under 'Occupational Exposure'. Various neuro-sensory function tests such as electromyography and nerve conduction tests should also be considered when evaluating possible solvent exposure.

2.6 Treatment. Most liver, kidney, bone marrow, lung, and even central nervous system damage from solvent abuse is repaired with discontinuance and time alone. Chronic, heavy solvent abusers may, however, suffer from withdrawal symptoms which may require sedation (diazepam). If lead poisoning is present through inhalation of petrol, chelating agents such as penicillamine may be necessary to hasten excretion. Assistance to abusers to establish and maintain abstinence may involve environmental changes, such as changing occupations, neighbourhoods, schools, friends, or playmates.[13,14]

3 Accidental or Intentional Exposure

Case 3. An obese lady, Mrs. C, presented to an Accident and Emergency Department having drunk approximately 250 ml of diethyl ether in a suicide attempt approximately one and half hours prior to admission. She had vomited once before admission. On arrival she was drowsy but fully rousable and co-operative and, apart from the strong smell of ether, physical examination was unremarkable. She underwent gastric lavage with removal of ether-smelling gastric contents and six packs of activated charcoal were instilled after lavage. She was transferred from Accident and Emergency Department to a medical ward for observation where her conscious level improved to normal over 24 hours. An ether smell persisted on her person for about the same time. She made an uneventful recovery and was discharged and referred to a psychiatric unit for further assessment.

Case 4. A male, Mr. D, was found unconscious by his wife after having ingested possibly 50-100 ml of chloroform in an unknown number of vodka drinks. Medical examination revealed that he had possibly inhaled vomitus at the time of ingestion and was cyanosed. His airway was cleared, ventillation was established and he was admitted to hospital where it was thought the initial vomiting had saved him from what would otherwise have been a lethal dose of chloroform. In hospital he was lavaged and given antibiotics to cover the aspiration. He had only a minimal rise in liver enzymes and after six days in hospital was discharged well.

Case 5. Following a period of methanol ingestion, Mr. E presented to the Accident and Emergency Department at approximately 10.30 p.m. one evening where he appeared confused and was answering questions in a sluggish manner. He was admitted to the hospital. Apparently at midday on the day prior to admission he had complained of headache, soreness behind the eyes, and photophobia. Later in the same day he started vomiting and this continued throughout the day of admission. Further questioning revealed that he was a laboratory trainee in a private pathology laboratory but no other helpful information was elicited, either from Mr. E or from his girlfriend who accompanied him. Initial examination showed him to have widely dilated but reacting pupils and acidotic breathing. His fundi were normal to examination, his abdomen was soft and non-tender and his chest was clear to auscultation and percussion. Fifteen minutes after the initial examination he appeared more confused and blood gases showed pH 7.03 and pCO_2 of 9 mm Hg and bicarbonate 2 mmol l^{-1}. He was given intravenous sodium bicarbonate but 15 minutes later he was totally non-responsive and a series of *grand mal* convulsions occurred which were treated with intravenous valium. By one hour after admission he was having prolonged apnoeic episodes for which he was intubated and ventilated initially by ambubag and later by a ventilator. Blood gases at the time of intubation were pH 6.76, pCO_2 92 mm Hg, bicarbonate 6 mmol l^{-1}. At this point he was deeply unconscious and totally unresponsive to any painful stimuli and there were no spontaneous respirations. On the morning of the next day he was still deeply unconscious and areflexic with large non-reacting pupils and no spontaneous respiration. His optic discs showed the pallor of early atrophy. Serum methanol level was reported at 30 mmol l^{-1}. This was treated with nasogastric ethanol and haemodialysis. He later developed hyperglycaemia with ketosis and his serum amylase was reported at 2020 units l^{-1} By the morning of the second day after admission no methanol was detectable in his bloodstream and he was still totally unresponsive to any form of stimuli and lacked any spontaneous respiration. At this time it was noted that Mr. E. was putting out large amounts of dilute urine and further investigation showed he had *diabetes insipidus*. On the fourth morning Mr. E suddenly went into ventricular fibrillation and rapidly proceeded into asystole and died shortly afterwards.

Toxic effects from solvents are more the domain of occupational exposure or solvent abuse via the inhalation route; such adverse effects are rarely the subject of ingestion. However, when recorded, cases may provide many valuable data in assessing the risk to future poisoning victims who have ingested these substances. Reference to such data will provide guidance with the four essentials of overdose management that should be considered for every patient. These are: (a) supportive care, (b) prevention of further absorption, (c) enhancement of excretion (if possible), and (d) administration of an antidote (if available). Supportive care involves the maintenance of airways, breathing, and circulation.[15,16]

Prevention of further absorption of ingested poisons is generally achieved by emesis, usually with the use of Syrup of Ipecac or by gastric lavage followed by the use of activated charcoal and a cathartic. Syrup of Ipecac is an emetic which is commonly used with poisonings except in cases of corrosive agents or with volatile solvents which may aspirate the stomach contents into the respiratory tract. (Ipecac is the dried root of *Cephaelis ipecacuanha*, which is found in Brazil. The principal alkaloids of ipecac are emetine and cephaline.) Gastric lavage may also be used for removing substances from the stomach. Activated charcoal is used as a safe, effective, and inexpensive gastro-intestinal absorbent and is ideal for use with ingested solvents. Although many good controlled studies have demonstrated the effectiveness of activated charcoal in decreasing serum levels of many toxins, the use of a cathartic (*e.g.* magnesium citrate or sulphate) is an additional method of achieving this by decreasing gastro-intestinal transit time of a toxin and is commonly used. Although there are various methods for enhancing the excretion of drugs and toxins, their use in treatment of poisoning is limited. Procedures intended to enhance the excretion of the ingested agents from the body include forced diuresis, alterations of urine pH, dialysis, haemoperfusion, and interruption of the enterohepatic circulation.[17,18]

An understanding of the pharmacokinetics and toxicokinetics of a substance can greatly aid in the treatment of an overdose of it. Toxicokinetics are concerned with levels of a substance or its metabolites in the body following overdose situations. For example, in Case 5, of methanol poisoning, although some of the methanol is eliminated by the lungs and expired air, the main route of metabolism is through oxidation by alcohol dehydrogenase. Whereas ethanol is metabolized to carbon dioxide and water by alcohol dehydrogenase, 40% of the dose of methanol is metabolized by the same enzyme with zero-order kinetics to produce formic acid via formaldehyde. The metabolism of methanol proceeds five times more slowly than that of ethanol, and approximately 30% of a dose of methanol remains unchanged in the body up to 48 hours after ingestion. Formic acid, which is six times more toxic than ethanol, is completely responsible for metabolic acidosis and ocular toxicity[19,20] as exhibited in Case 5. Treatment of methanol poisoning is directed first towards correcting the metabolic acidosis, then inhibiting the oxidation of the parent compound, and finally removing circulating amounts of the parent compound and its toxic metabolites.[17,18] Intravenous alkali therapy by the use of large amounts of sodium bicarbonate may be used to correct metabolic acidosis. Ethanol is given in the management of methanol ingestion because it competes with alcohol dehydrogenase which is the enzyme responsible for the critical first step in the metabolism. Ethanol saturates this enzyme, since it has a greater attraction towards ethanol than methanol. The administration of ethanol will therefore avoid a build-up of the toxic products in metabolism and allows for an increased excretion of unchanged parent compound through the kidneys.

From the many accumulated human case reports of methanol poisoning a good database is now available from which the risks of patients surviving acute poisoning from high doses can be eradicated. It is known that the fatal dose of methanol to humans varies widely from 15 to 240 ml (0.2 to 3.0 g kg^{-1}), but usually more than 70 ml is required to cause death (certainly if left untreated).[21] It has also been stated that about 25% of patients with severe methanol poisoning and a blood bicarbonate level less than 20 mmol l^{-1} will die of respiratory failure.[18] Patients with blood methanol levels greater than 0.5 mg ml^{-1} should be considered for dialysis therapy.[18] The chance of surviving acute poisoning by a high dose of methanol is therefore likely to be low unless prompt clinical management is commenced as soon as possible after the poisoning event.

4 Occupational Exposure

An important objective of experimental and clinical investigation in the field of toxicology is the proposal of 'safe' levels of exposure. The traditional approach to establishing 'safe' levels of human exposure is to apply a suitable factor to that dose level observed to produce no adverse effects in toxicological studies. The magnitude of the safety factor depends on scientific judgement, and is intended to allow for potential differences in sensitivity between animal and man.

Occupational exposure to solvents occurs primarily by the inhalational route although percutaneous absorption, and to a lesser extent ingestion, should not be overlooked. Many different kinds of guides for exposure to airborne contaminants have been proposed and some of them have been used for years. The guides that have gradually become the most widely accepted ones are those issued annually by the American Conference of Government Industrial Hygienists (ACGIH), and are termed 'threshold limit values' (TLVs)[22] (see also Glossary of Terms). TLVs refer to airborne concentrations of substances and represent conditions under which it is believed nearly all workers may be repeatedly exposed, day after day, without adverse effects. However, because of wide variations in individual's susceptibility, a small percentage of workers may experience some discomfort from some substances at or below the threshold limit; a smaller percentage may be affected more seriously by the aggravation of a pre-existing condition or by the development of an occupational illness. With the accumulation of new information on the toxicity of these chemicals, the proposed permissible levels must be re-evaluated at regular intervals. It should also be made clear that these levels are guides only and should not replace close medical surveillance of the workers if significant exposure can occur.

TLVs are therefore not 'safe' doses but represent instead control levels at which exposure is currently believed to be acceptable. They can of course be altered, usually downwards, either as a result of new evidence of risks and potential hazards or because recent technological changes allow for reduced exposures. If TLVs are to be useful, however, safety practicioners must be ever vigilant of the potential for solvent exposure in the workplace as is graphically illustrated by the following fatal case.

Case 6. A fatal case of poisoning occurred at a dry cleaning establishment where a young and healthy woman was employed to iron clothes subjected previously to trichloroethylene cleaning. The trichloroethylene concentration in the work place air considerably exceeded allowable limits. The first symptoms were dizziness and strong dyspnoea, and afterwards heavy cyanosis and loss of consciousness. Her clinical course was consistent with pulmonary emphysema leading to death. Pathological and histological examinations revealed no other changes at *post mortem*. On the basis of data obtained it is suspected that phosgene (from trichloroethylene) formed on the hot surface of the iron.

Another analogous approach to risk management with solvents in the occupational setting is the use of biological exposure limits. These represent the maximum permissible quantity (or concentration) of a chemical compound (or its metabolites) in biological fluids (urine, blood, alveolar or expired air) or tissues of exposed workers. Biological monitoring assesses the health risk by evaluating the internal dose and provides an additional means of protection. ACGIH has therefore recently introduced the Biological Exposure Indices (BEIs)[23] which represent warning levels of biological response to the chemical or warning levels of the chemical or its metabolic product(s) in tissue fluids or exhaled air of exposed workers, regardless of whether the chemical was inhaled, ingested, or absorbed through the skin.

Introduction of the BEI is a step in the evolution of the concept of TLVs. The BEI provides the health professional with an additional tool to aid in the assessment of workers' safety. TLVs are a measure of the composition of the external environment surrounding the workers. BEIs are a measure of the amount of chemical absorbed into the body. Recommended values of BEIs are based on data obtained in epidemiological and field studies or determined as bioequivalent to a TLV by means of pharmacokinetic analysis of data from controlled human studies. Because elimination of chemicals and their metabolic products as well as biological changes induced by exposure to the chemical are kinetic events the list of BEIs is strictly related to 8 hour exposures and to the specified timing of the collection of biological samples. Other factors to be considered when the BEI is applied are changes induced by strenuous physical activity, changes induced by environmental conditions, changes induced by water intake and by physiological functions induced by pre-

Table 2 *Biological exposure indices (recommended) and threshold limit values established by ACGIH for some common solvents*

Solvent	*Indices*	*Time*	*Recommended BEI*	*TLV-TWA (p.p.m.)*
Toluene	Hippuric acid in urine	End of shift	2.5 g g^{-1} creatinine	100
Trichloroethylene	Trichloroacetic acid in urine	End of work week	100 mg l^{-1} or below	50
	Trichloroacetic acid and trichloroethanol in urine	End of shift and end of work week	300 mg l^{-1} or below 320 mg g^{-1} creatinine	
	Free trichloroethanol in blood	End of shift and end of work week	4 mg l^{-1}	
Methyl ethyl ketone	MEK in urine	End of shift	2 mg l^{-1}	200

existing disease or congenital variation, changes in metabolism induced by congenital variation of metabolic pathway, and changes in metabolic pathway induced by simultaneous administration of another chemical. Table 2 lists the TLVs and BEIs for three common solvents.

The ACGIH TLV booklet[23] contains lists of both environmental and biological exposure limits. The former includes some 650 compounds, the latter only ten substances (of which seven are solvents). Although the present lack of BEIs constitutes a serious hindrance to biological monitoring they nevertheless remain as another useful risk management technique for solvent exposure.

5 Solvent Neurotoxicity

The first two steps in the risk management process (hazard identification and risk estimation) require collection and analysis of data and are essentially scientific in nature. Scientific research often focuses on newly identified hazards and in this respect it is considered worthwhile that the reader should be aware of some somewhat controversial research in the area of solvent neurotoxicity.

Recent reports,[24–26] particularly from Scandinavian countries, have drawn attention to the potential neurotoxic effects of prolonged exposure to organic solvents. These studies have shown that prolonged occupational exposure to solvent mixtures of certain types may result in irreversible structural damage to the nervous system, manifest as changes in behaviour and neurological functions. Although many organic solvents are, in sufficient doses, capable of causing reversible central nervous system effects, few unequivocally induce chronic or long-lasting or irreversible changes. Two industrial solvents, n-hexane and methyl n-butyl ketone, both of which are biotransformed to hexane-2,5-dione, have caused many cases of occupational neuropathy. These solvents have been used in glues and cleaning fluids in the manufacturing of shoes and printed fabrics. Initial case reports included Japanese[27] and Italian factory workers.[21] Carbon disulphide has also been demonstrated to be neurotoxic and there are strong suggestions that trichloroethylene and toluene can cause damage to the nervous system.[29] Evidence for toluene has come from reports concerning solvent abusers who have been repeatedly inhaling either pure toluene or toluene-based paints. Chronic neuro-behavioural disorders have not been demonstrated in workers exposed to toluene at levels close to the TLV. This, however, cannot be ruled out since few studies have been performed and these have not focused on the early signs of chronic neurological dysfunction which include decreased auditory response, memory disturbances, impaired verbal abilities, and impaired psychomotive function. Relatively few animal studies have examined the neurotoxic effects of toluene. Research is required to develop an animal model to determine the threshold for nervous system alterations. Other proposed human neurotoxicants[24] include diethyl ether, ethylene chloride, nitrobenzene, pyridine, styrene, tetrachloroethane, xylene, and white spirit. Various mixed solvents have also been proposed as possible human neurotoxic agents and a considerable amount of research in this area has been carried out by Scandinavian researchers.[30] They have performed a large number of studies in which various solvent-exposed groups have been evaluated with a variety of neuro-physiological and neuro-behavioural tests. Occupations monitored included painters, cabinet makers, boat builders, and floor layers. Electrophysiological evaluation of workers exposed to solvent mixtures using both standard techniques for evaluating peripheral nerve function and electro-encephalography have noted mild changes in neuro-physiological parameters. Behavioural studies noted an excess of subjective symptoms, abnormalities of psychomotor performance, memory deficits, impairment in verbal concept formation, and disturbances of mood when compared with control groups. This accumulated evidence supports the occurrence of a syndrome of toxic encephalopathy caused by excessive exposure to some organic solvents. This syndrome[31] has been given a number of different names, such as psycho-organic syndrome, chronic encephalopathy, pre-senile dementia, painters' syndrome, and neurasthenic syndrome. This syndrome is characterized by memory disturbances, impaired psychomotor function, impaired verbal abilities and disturbances of mood with onset during periods of excessive exposure to solvents and persistance after exposure has ceased. Early non-specific manifestations of this condition are tiredness, irritability, depression, and episodes of anxiety. Although these symptoms may, of course, occur because of a variety of other causes it is important to consider the possibility of such a syndrome in individuals who may be occupationally exposed to solvents. Although it is clear that toxic encephalopathy may occur in individuals heavily exposed to some solvents over a period of months to years, the exact exposure levels that lead to irreversible neuro-behavioural changes remain to be determined. Should these levels prove to be lower than the TLVs currently established for the particular solvents in question then as part of the risk management procedure careful consideration should be given to lowering the existing TLVs in light of these new data. Testimony to the controversial nature of this research into neurotoxicity alluded earlier to at the beginning of this section is evident by the fact that a WHO working group in 1985[32] distinguished between an organic effective syndrome and a chronic toxic encephalopathy that may be mild or severe. However, four months later at a meeting held in North Carolina, another classification was suggested, implying that the condition is still not as well defined or understood as one would like. Obviously this is a complex problem but it is a particularly important one since the exposure of human populations in the environment is not usually to a single chemical. Therefore, more knowledge must be gained concerning the interactive effects of solvents such as potentiation, synergism, and additive or, indeed, antagonistic effects. Without this necessary body of knowledge one can only make use of data on the toxicology of individual solvents when considering risk assessment procedure. However, one should be ever cognizant of the possibility of interactive effects.

6 Trihalomethanes in Drinking Water

In the early 1970s it was discovered that chloroform, other trihalomethanes (THMs), and higher molecular weight halogenated compounds are produced during the chlorination of water supplies.[33,34] (THMs are believed to be formed from the interaction of chlorine with certain organic acids such as humic and fulvic acids which are usually present in coloured surface waters.) This information, together with the knowledge that chloroform can cause liver damage to humans, and the United States Cancer Institute rodent study[35] which reported a dose-related incidence of malignant tumours in the kidneys of male rats and the liver of mice of both sexes, in 1976, raised questions of safety with respect to the chlorination by-products in drinking water. Additional stimulation to evaluate the risk was provided by an Environmental Defence Fund study[36] which in 1976 reported elevated cancer rates in Louisiana parishes served by the Mississippi River.

In 1979 the US Environmental Protection Agency advised that the maximum permitted level of THMs in water delivered to the consumer should be 100 $\mu g\ l^{-1}$. Other countries recognizing health risks in THMs have set guideline values between 25 and 350 $\mu g\ l^{-1}$ for total THMs[37] (see also Chapter 8).

Since the initial finding of THMs in drinking water many systematic THM data gathering exercises have been performed, particularly in the USA, as well as many epidemiological studies to examine the question of organics in drinking water and cancer. These latter studies have used many different exposure and effect measures and applied various epidemiological and statistical techniques. Several early ecological or demographic correlation studies were performed.[36-39] These studies indicated that cancer mortality was higher in geographical areas with relatively poor quality drinking water or with water measurements having recently shown high levels of THMs. Because of the lack of sensitivity and specificity no estimate of quantitative risk could be determined but the results were helpful in suggesting tenable hypotheses. A number of case control studies[40,41] based on death certificate data were therefore undertaken to determine whether, in fact, it was exposure to chlorinated by-products that accounted for the apparent relationship. Data evaluated included source characteristics of water supply to the last residence of decedents as well as treatment methodology and THM levels. These studies collectively suggested that bladder, rectal, and colon cancer were associated with exposure to chlorine-treated municipal water supplies particularly from surface or shallow ground sources. Later generation case control studies[42,43] have either confirmed or failed to detect such an association. Several other studies are in progress. It is hoped that these will be able to provide more satisfactory answers to this important question than are now available.

There remains a possible association between certain types of cancer and THMs in water. Other THMs, often formed when bromide ions are present, are only now being tested for weak animal carcinogenicity in bioassays similar to those used to show the chloroform is a carcinogen under test conditions. These other THMs are, however, known to be more active than chloroform in the Ames salmonella test for mutagenicity.[44]

Extrapolation from the rodent studies using the linear one-hit statistical model gives us some idea of the reality of the cancer risk from drinking water containing the recommended maximum acceptable concentration of chloroform. This lies somewhere between no risk and a maximum of 1.6 cases per million population per year.[45] No experimental tools at present can define the true rate between these points with accuracy. What is clear is that we should be taking effective steps to reduce THMs in drinking water. A high THM concentration could not arise if the water contained small amounts of precursors or if they were removed by treatment before chlorination. Care must be taken, however, not to compromise the microbiological quality of the water.

7 Conclusions

The process of risk assessment may involve a series of distinct stages, ranging from hazard identification and risk estimation to the selection and implementation of an appropriate risk management strategy. This process is in essence identical to the toxicovigilance programmes of Poison Information Centres which was referred to earlier in this chapter. In the instance of solvent abuse, Poison Information Centres by carefully recording and evaluating case data have been able to identify at risk populations in the community and have played a prominent role in formulating measures to reduce the risks that this presents. Some risks are relatively easy to evaluate such as the chances of surviving acute poisoning by high doses of a toxic chemical, where the survival chances are low. In the case of determining the risk of THMs in water this evaluation is extremely complex involving many animal tests and epidemiological studies (see Chapter 11). No matter how much toxicological testing with animals has been done, however, it can never be proven that a chemical is absolutely safe for humans. Humans react to some chemicals quite differently to the way in which laboratory animals do and there is no way in which this can consistently be predicted. Animal testing does, however, remain the primary source of information in assessing the safety of chemicals. However, we have seen that a better assessment may be achieved after assessing human exposure data and that Poison Information Centres are a valuable source of such data.

Once the scientific evaluation of a particular risk has been defined, then comes the issue of deciding on the acceptable risk. This is a most difficult step, moving from the world of facts to a world of values. In making a decision there must be a balanced risk and benefit. Such a decision will necessarily be multi-faceted. In the case of chlorination it will include such factors as the benefits gained from the use of chlorine, the availability and adequacy of possible alternatives, effects on water quality, as well as economic and employment considerations. Unfortunately our conception of risks often correlates poorly with the scientific information and values may differ widely between individuals. However, it is society which must ultimately decide what is acceptable as far as risk in concerned (see also Chapter 6).

8 References

1. 'Toxic Hazard Assessment of Chemicals', ed. M.L. Richardson, The Royal Society of Chemistry, London, 1986, p.15.

2. Toxicovigilance in 'Collection de Medecine Legale et de Toxicologie Medicale', ed. L. Roche, Lyon, 1978, No. 110.

3. Beluze, *Ann. Hyg. Publique Med. Leg.*, 1886-1887, 15, 539.

4. S. Kornfield and G. Bikeles, *Wein. Klin. Wochenschr.*, 1893, **6**, 64.

5. I. Astrand, *Br. J. Ind. Med.*, 1985, **42**, 217.

6. H.R. Anderson, B. Dick, R.S. Macnair, J.C. Palmer, and J.D. Ramsey, *Hum. Toxicol.*, 1982, **1**, 207.

7. H.R. Anderson, R.S. Macnair, and J.D. Ramsey, *Br. Med. J.*, 1985, **290**, 304.

8. C. Steadman, L.C. Dorrington, P. Kay, and H. Stephens, *Med. J. Aust.*, 1984, **111**, 115.

9. G.S. King, J.E. Smialek, and W.G. Troutman, *J. Am. Med. Assoc.*, 1985, **253**, 1604.

10. G.E. Barnes, *Int. J. Addict.*, 1979, **14**, 1.

11. J.M. Watson, *Practitioner*, 1984, **228**, 487.

12. J.M. Watson, *Br. J. Addict.*, 1980, **75**, 27.

13. I. Sourindhrin and J.A. Baird, *Br. J. Addict.*, 1984, **79**, 227.

14. G. Masterton, *J. Adolesc.*, 1979, **26**, 65.

15. T.A. Gossell and J.D. Bricker, 'Principles of Clinical Toxicology', Raven Press, New York, 1984.

16. P.D. Byrson, 'Comprehensive Review in Toxicology', Aspen Systems Corpn., Rockwell, Maryland, 1986.

17. L.M. Haddad and J.F. Winchester, 'Clinical Management of Poisoning and Drug Overdose', W.B. Saunders Company, Philadelphia, 1983.

18. R.H. Dreisbach and W.O. Robertson, 'Handbook of Poisoning', Appleton and Lange, Los Altos, California, 1987.

19. K. McMartin, J. Ambre, and T. Tephly, *Am. J. Med.*, 1980, **68**, 414.

20. M. Smith, *Ear, Nose, Throat J.*, 1983, **62**, 126.

21. C.J. Polson, M.A. Green, and M.R. Lee, 'Clinical Toxicology', 3rd Edn., Pitman Books Ltd., London, 1983.

22. American Conference of Governmental Industrial Hygienists, 'Documentation of the Threshold Limit Values', 5th Edn., ACGIH Inc., Cincinnati, Ohio, 1986.

23. 'Threshold Limit Values and Biological Exposure Indices for 1985-86', American Conference of Governmental Industrial Hygienists, Cincinnati, Ohio.

24. P.S. Spencer and H.H. Schaumburg, *Scand. J. Work Environ. Health,* 1985, **11**, (Suppl. 1), 53.

25. A.M. Seppalaien, *Scand. J. Work Environ. Health*, 1985, **11**, (Suppl. 1), 61.

26. K. Ekberg, L. Barregard, S. Hagberg, and G. Sallsten, *Br. J. Ind. Med.*, 1986, **43**, 101.

27. A. Yamada, *Jpn. J. Ind. Health*, 1964, **6**, 192.

28. N. Frontali, A.M. Guarani, A. Spagnol, and M.C. Amatine, *Ann. Inst. Super. Sanita.*, 1979, **15**,, 273.

29. J.B. Cavanagh, *Br. J. Ind. Med.*, 1985, **42**, 433.

30. E.L. Baker Jr., T.J. Smith, and P.J. Landrigan, *Am. J. Ind. Med.*, 1985, **8**, 207.

31. C. van Vliet, G.M.H. Swaen, J.J.M. Slangen, Tj. de Boorder, and F. Sturmans, *Int. Arch. Occup. Environ. Health*, 1987, **59**, 493.

32. WHO Environmental Health, Vol. 6, 'Chronic Effects of Organic Solvents on the Central Nervous System', WHO, Copenhagen, 1985.

33. T.A. Bellar, J.J. Lichtenberg, and R.C. Kroner, *J. Am. Water Works. Assoc.*, 1974, **66**, 703.

34. J.J. Rook, *J. Soc. Water Treat. Exam.*, 1974, **23**, 234.

35. N.P. Page and U. Saffiotti, 'Report on carcinogenesis bioassay of chloroform', National Cancer Institute, Division of Cancer Cause and Prevention, MD, Bethesda, 1976.

36. T. Page, R.H. Harris, and S.S. Epstein, *Science*, 1976, **193**, 55.

37. 'Drinking water standards for New Zealand', A report prepared for the Board of Health by the Department of Health, Wellington, 1984.

38. R.J. Kuzma, C.M. Kuzma, and C.R. Buncher, *Am. J. Public Health*, 1977, **67**, 725.

39. K.P. Cantor, R. Hoover, T.H. Mason, and L.J. McCabe, *J. Nat. Cancer Inst.*, 1978, **61**, 979.

40. T.B. Young, M.S. Kanarek, and A.A. Tsiatis, *J. Nat. Cancer Inst.*, 1981, **67**, 1191.

41. M.S. Gottlieb, J.K. Carr, and D.T. Morris, *Int. J. Epidemiol.*, 1981, **10**, 117.

42. T.B. Young, D.A. Wolf, and M.S. Kanarek, *Int. J. Epidemiol.*, 1987, **16**, 190.

43. I. Cech, A.H. Holguin, A.S. Littell, J.P. Henry, and J.O'Connell, *Int. J. Epidemiol.*, 1987, **16**, 198.

44. J.C. Loper, *Mutat. Res.*, 1980, **76**, 241.

45. R.G. Tardiff, *J. Am. Water Works Assoc.*, 1977, **69**, 658.

14

Background Data for Risk Assessment: An Alternative to the Traditional LD_{50} Study?

M.J. van den Heuvel

DEPARTMENT OF HEALTH AND SOCIAL SECURITY, HANNIBAL HOUSE, ELEPHANT AND CASTLE, LONDON SE1 6TF, UK

1 Introduction

The starting point for the health or environmental risk assessment of a substance or preparation has customarily been an estimation of its acute toxicity, often in the form of a traditional LD_{50} study and for many of the chemical compounds and mixtures currently marketed in the United Kingdom that estimation is the only laboratory-generated toxicity data available to risk assessors. In the past a variety of animal species were used to evaluate acute toxicity and, for solids and liquids, the oral route of dosing was considered the most appropriate.

2 Accepted Procedures for Acute Oral Toxicity Assessment

The generally accepted methods for assessing acute oral toxicity are based upon graduated dose procedures in which groups of laboratory animals are exposed to a range of dose levels with the top dose chosen to produce a high level of mortality. These procedures were codified in the internationally agreed guideline published by the Organization for Economic Co-operation and Development (OECD) in 1981.[1] This guideline, which is summarized in Table 1, suggested the exposure of rats to at least three dose levels (five male and five female rats at each level), with the dose levels being chosen to span the 'expected' LD_{50} dose. The expected LD_{50} dose was to be estimated by analogy with similar compounds or from a pilot or sighting study and, where it was anticipated that no compound-related deaths would occur at a dose level of 5000 mg kg^{-1} body weight, the study could be reduced to an examination of the toxic effects at that single dose level (the so-called 'limit' test).

Table 1 *OECD guideline for acute oral toxicity testing (Guideline* 401, 1981)

Principle:	Test substance administered to rats in graduated doses
Dose levels:	Three dose levels selected to span expected LD_{50}. (Expected LD_{50} established by structure-activity relationships (SAR) or range-finding study)
Observation period:	14 days following administration
No. of animals used:	3 groups of 10 rats (5 male and 5 female)
Total No. of animals:	30 rats
Total time for study:	14 days minimum
Limit dose study:	5 male and 5 female rats Dose 5000 mg kg^{-1} body wt. If compound mortality occurs, full study is required

When properly conducted, this type of acute toxicity investigation - often referred to as an LD_{50} study - provides data not just upon what doses of the test material may prove lethal but, more importantly, on the dose response to the test substance including information on the nature, time to onset, and outcome of signs of toxicity. This latter information is essential for effective human health and environmental risk assessment. In many studies it is also possible to estimate statistically the LD_{50} dose and the shape (slope) of the dose-response curve. These estimates, in spite of their recognized limitations, are considered important by regulatory authorities and used by them particularly for classification and labelling purposes.

For some years, the rationale behind the methodology used to generate LD_{50} figures has been called into question by workers in the field of toxicology both on scientific grounds and ethically in relation to the humaneness of the procedure.[2-5] At the same time, toxicologists accepted that for the foreseeable future there would be a continuing need to investigate acute toxicity of materials *in vivo* not only for the purpose of estimating possible lethal dose levels for man and other mammals but also to obtain the more important data on likely nature and time to onset of significant signs of toxicity and on the probable rate and extent of recovery from sub-lethal doses. Risk managers, also, recognized that such information would be required for the foreseeable future for regulatory purposes.

3 Acute Toxicity Testing

In 1986, the OECD reviewed its guidelines for the investigation of acute toxicity and published revised texts.[6] The major changes introduced were a reduction in the number of rats used in a study. The new guidelines (Table 2) proposed use of five rats of one sex at each of three dose levels with a cross-check on the susceptibility of the other sex by application of a single dose to five animals. The dose to be used in the 'limit' test was reduced from 5000 to 2000 mg kg^{-1} body weight. It is not certain, yet, how well these changes will be received by industry and by regulatory authorities, particularly the proposal that only one sex should be used in the first stage of the test. From a regulatory point of view, a major problem could be the absence of any indication as to how the dose level for the test on the second sex is to be chosen. This is not surprising, perhaps, in view of the small number of animals used in the first stage of the test and the known unreliability of the statistically estimated LD_{50} figure even when larger numbers of animals are used. It seems likely that difficulties will be experienced by toxicologists in deciding upon the dose level for the test on the second sex and that frequently the result of the second sex test will be uninterpretable. The sequential nature of the tests on the two sexes will also increase time needed to carry out the study and, therefore, the costs for industry. More important for the long term, however, is that despite the welcome reduction in the number of animals used which will result from introduction of the new guidelines, the study remains unsatisfactory both scientifically and ethically: scientifically because the LD_{50} figure derived remains suspect in that even within one laboratory widely varying figures can be obtained in separate studies with the same compound; and ethically because the dose levels are still to be specifically chosen with a view to killing 50% of the dosed animals. This must result in animals experiencing severe pain and distress in a fair proportion of such studies.

An alternative procedure, which is scientifically and ethically more acceptable, and which would continue to provide the information needed by risk assessors and managers, is thus urgently required. Several new approaches to acute toxicity testing have, indeed, been proposed and of these, three have received particular attention.

Table 2 *OECD guideline for acute oral toxicity testing (Guideline* 401 *Revised* 1987*)*

Principle:	Test substance administered to rats in graduated doses
Dose levels:	Three dose levels selected to span expected LD_{50} (Expected LD_{50} established by structure activity relationships (SAR) or range-finding test)
Observation period:	14 days following administration
No. of rats used:	3 groups of 5 rats, one sex only. Assessment of toxicity in second sex determined by separate study using 5 rats
Total no. of animals:	20 rats
Total time for study:	28 days minimum (if both sexes tested)
Limit dose study:	5 males and 5 female rats Dose 2000 mg kg^{-1} body wt. If compound mortality occurs full study is required.

4 Step-wise Procedure

Of these, one is a step-wise procedure (Table 3) put forward in 1985 by the German Ministry of Health (BGA). This procedure is currently undergoing validation tests in Germany but would appear to provide an acceptably sound basis for evaluating signs of toxicity. It is also designed to generate a 'number', similar and not far removed from the concept of an LD_{50} and would thus permit ready 'classification' of a tested substance. It is difficult to assess, in the absence of the results of the validation study, whether the procedure represents a scientific advance on the traditional acute toxicity study. One must suspect that the removal of the choice of dose levels from the toxicologists conducting the study together with the reduction in number of animals used is likely to reduce still further any shred of confidence which might still remain in the LD_{50} figure which results. What is clear is that the full procedure requires 18 rats (two fewer than the current OECD guideline). Its 2-stage sequential nature will tend to make it unattractive to industry, and furthermore, the doses used in the second stage are, intentionally, in the acutely toxic range so that a high proportion of all the animals used in a full study will die or develop severe signs of toxicity. The method will, therefore not commend itself to those concerned for animal welfare and it will attract the same ethical objections as the traditional acute toxicity test procedure.

Table 3 *Acute toxicity study - BGA method,* 1985

Principle:	Step-wise procedure - Phase 1 - Test substance administered at one or two pre-set dose levels. Level to be selected should produce some mortality within 14 day observation period. Phase 2 - Test substance administered at two dose levels chosen (based) on results of Phase 1 test to define the toxic range.
Dose levels:	Phase 1 - One or two dose levels chosen from 25, 200, or 2000 mg kg^{-1} body wt. Level(s) chosen on basis of SAR or range-finding study. Phase 2 - Two dose levels chosen to span Phase 1 estimated median lethal dose.
Observation period:	14 days following each dose level administered.
No. of animals used:	Phase 1: 6 to 12 rats (male and female) Phase 2: 6 rats all male or all female (more vulnerable sex based on Phase 1 result).
Total No. of animals used:	12 - 18 rats.
Total time for study:	28 days minimum (up to 42 days in some cases).
Limit dose study:	3 male and 3 female Dose - 2000 mg kg^{-1} body wt. If compound related mortality occurs, full study is required.

5 Up and Down Test

In 1985, an approach to acute toxicity testing which has become known as the Up and Down Test (Table 4) was described by Bruce.[7]

Table 4 *The Up and Down Acute Toxicity Test (Bruce[7])*

Principle:	Sequential administration to female rats.
Dose levels:	Initial dose administered to one female rat based upon expected LD_{50} dose. If animal dies within 48 h, a lower dose is administered to the second animal. If animal survives for 48 h, a higher dose is administered to the second animal. Doses are lowered or raised by a factor of 1.3. Once an administered dose causes the death of an animal, dosing continues on the above basis until a further five animals have been treated. A follow-up study using six males dosed at female LD_{50} dose to define sex difference in sensitivity is carried out.
Observation period:	Two days between doses. All surviving animals kept under observation for 14 days. If delayed deaths occur (*i.e.* more than 48 h after dosing) an alternative test procedure is indicated.
No of animals used:	6 - 8 female rats; 6 male rats.
Total time for study:	38 days minimum (if both sexes are tested).
Limit dose study:	Not required. An appropriate maximum dose level (*ca.* 2000 mg kg^{-1} body wt.) can be prescribed.

This ingenious procedure is fundamentally attractive in that it permits the calculation of a reasonably precise and apparently reproducible measure of lethality and provides adequate information on signs of toxicity. It uses relatively few (6 to 14) animals and although sequential testing is involved, this is so arranged that a study is not excessively protracted as compared to a traditional acute study unless both sexes need to be investigated. The method has been validated to some extent but does have some severe limitations. If it is the nature of the test compound to produce delayed deaths - and this is not an unusual event where the doses used are close to lethal levels - the method is recognized as being inappropriate and resort must be made to some alternative procedure. In addition, all the animals in a study receive a dose which is either lethal or not far removed from a lethal dose. All may be expected to show severe signs of toxicity and the intention is that 50% of them should die. Thus, whilst the method is simple to carry out, it does not appear to be applicable to all types of compound, and even though very few animals are used, all those used are likely to experience pain and distress.

6 British Toxicology Society Approach

In 1984, a rather different approach was suggested by the British Toxicology Society (BTS).[8]

Table 5 *The BTS acute toxicity study (Revised 1987)*

Principle:	Test substance administered to rats at one dose level.
Dose levels:	Dose level is selected from pre-set levels (which equate with classification/ranking systems). Dose level selected should produce toxicity but no mortality (based on SAR or range-finding study). If mortality occurs, re-testing at a lower dose level may be required: if no signs of toxicity are elicited, re-testing at a higher dose level may be required.
Observation period:	14 days following administration.
No of animals used:	Initially, one group of 5 male and 5 female rats.
Total No of animals used:	10 (20 if first dose level selected proves to be incorrect.
Total time for study:	14 days minimum (16 - 28 days if a second dose level has to be investigated)
Limit dose study:	Not appropriate. Maximum pre-set at 2000 mg kg^{-1} body wt.

The basis of this BTS procedure (Table 5) is that it avoids using death of animals as an end-point, and relies instead on the observation of clear signs of toxicity developed at one of a series of fixed dose levels. The dose level at which toxic signs are detected is used to rank or classify the test materials, and those originally suggested by the BTS were chosen to equate with the EEC system of classification.[9] The procedure is such, however, that additional dose levels could be introduced to meet other national classification systems.

The basis for the BTS procedure of examining acute toxicity is relatively simple. A retrospective examination of traditional LD_{50} studies revealed that between 80% and 90% of those compounds which elicited clear signs of toxicity but no deaths at a dose level of 5, 50, or 500 mg kg^{-1} body weight had estimated LD_{50} figures (derived from the same studies) of >25, 25-200, or 200-2000 mg kg^{-1} respectively [the European Economic Community (EEC) classification banding for Very Toxic, Toxic, and Harmful].[10] Similarly, where no clear signs of toxicity were elicited by a dose of 500 mg kg^{-1} body weight, 80% of the time the LD_{50} was estimated as being greater than 2000 mg kg^{-1} body weight (*i.e.* the compound would be unclassified on the basis of the EEC system). It was suggested, therefore, that a study designed to examine a group of rats at a dose level at which signs of toxicity would be seen but which would not result in any deaths in that group should enable the likely upper and lower limits, between which the LD_{50} of the test compound must lie, to be estimated with an adequate degree of certainty (~80%).

The results of such a study would thus both provide for ranking or classification and, because signs of toxicity would necessarily have been observed, all the information required for adequate risk assessment would also become available. The study would have other advantages. Clearly, because deaths would not determine classification, any animal thought to be suffering unduly could be sacrificed without affecting the results of the study, and the lower sub-lethal doses prescribed by the method would reduce the severity of toxic signs. Both these latter points would make the study more acceptable from the ethical point of view. Carefully conducted, the study should also reduce the number of animals used to as few as 10 rats and rarely more than 20 rats per study. Although a dose-response curve would not be generated, compounds with high LD_{50} figures but which nevertheless produced toxic effects at low doses (*i.e.* compounds with shallow dose-response curves) would rightly attract a more severe toxicity classification than the LD_{50} figure alone would have allocated. Lastly, there are reasons to believe that less inter-laboratory variation of classification would result from use of the method than is the case with current procedures which rely on notoriously unreliable estimated LD_{50}s.

7 Validation of Procedures

The BTS procedure has recently been subjected to a validation study in the UK.[10] Five commercial/industrial laboratories engaged on a day-to-day basis in the toxicity testing of substances and preparations examined a total of 41 materials (substances and preparations) and compared the results obtained using the BTS procedure with those obtained from a conventional acute toxicity study (OECD) in respect of both the generation of the scientific data on the compounds and in relation to the allocation of classification. Inter-laboratory variation was not evaluated in this exercise. The results were extremely encouraging. The exercise showed that an adequate ranking for classification purposes was possible using the new procedure (Table 6). Although studies using the BTS procedure did not provide all of the information on signs of toxicity that was provided by the conventional studies, it was considered that small modifications to the BTS procedure could largely rectify this. For example, whilst the top dose level of 500 mg kg^{-1} body weight prescribed in the procedure was adequate for classification purposes, compounds which poduce no significant toxicity at that level should normally be investigated at higher levels if a full risk assessment is to be made. The procedure thus requires the addition of a fourth dose level, at ~2000 mg kg^{-1} body weight, so that those low toxicity materials which develop signs of toxicity between 500 mg kg^{-1} and 2000 mg kg^{-1} can be identified and information on the onset, nature, and duration of those signs obtained. From the animal welfare point of view, the trial provided considerable evidence of advantage over the conventional procedure in that fewer animals were required and those developing signs of toxicity suffered less pain and distress (Table 7).

Table 6 *Effects of procedures on classification of test compounds*[11]

Classification status	*No. of compounds*
Classification unchanged	31
Classification changed to a less toxic class by BTS procedure	8
Classification changed to a more toxic class by BTS procedure	2

Table 7 *Number of animals used and number dying or humanely killed during OECD and BTS studies on 41 compounds*[11]

*No. of animals used**		*No of animals dying or humanely killed during studies*	
OECD	BTS	OECD	BTS
1705 (1230)	610 (630)	705	92

*Figures in brackets indicate the number of animals which would have been used had the OECD and BTS protocols always been followed.

8 Conclusions

On the basis of the UK validation study, it is possible to state that an alternative to the conventional acute toxicity study now appears to exist. It would be unrealistic, however, to expect that in spite of its advantages it could be immediately adopted by laboratories, given the international nature of the chemical industry and the differing national requirements of regulatory authorities. Certainly, a further and this time international validation study will need to be carried out. Such a study will have to examine again, and hopefully confirm the aspects looked at in the UK exercise but additionally will need to look at the inter-laboratory variation. Also, adaptation of the procedure to take account of the various international classification systems will be necessary. It is to be hoped that an international validation of the procedure will not now be long delayed. If successful, the confidence gained by regulatory authorities in results obtained from tests using the BTS approach should then enable its rapid international adoption for the purpose of human health and environmental risk assessment.

9 References

1. 'OECD Guidelines for the Testing of Chemicals - Guideline 401'. OECD, Paris, 1981, Decision of the Council, C(81)30 Final.

2. G. Zbinden and M. Flury-Roversi, *Arch. Toxicol.*, 1981, **47**, 79.

3. M.L. Tattersall, *Arch. Toxicol. Suppl.*, 1982, 5, 267.

4. E. Schütz and H. Fuchs, *Arch. Toxicol.*, 1982, 51, 197.

5. R. Bass et al., *Drug Res.,* 1983, 33, 81.

6. 'OECD Guidelines for the Testing of Chemicals', OECD, Paris, 1987, Third addendum - Revised Guideline 401.

7. R.D. Bruce, *Fundam. Appl. Toxicol.*, 1985, 5, 151.

8. *Human Toxicol.*, 1984, 3, 85.

9. 'Council Directive 79/831/EEC amending for the sixth time Council Directive 67/548/EEC on the approximation of laws, regulations and administrative provisions relating to the classification, packaging, and labelling of dangerous substances'. *Official Journal of the European Community*, 1979, L259: 20 (Annex VI).

10. See H.P.A. Illing, in 'Toxic Hazard Assessment of Chemicals', ed. M.L. Richardson, The Royal Society of Chemistry, London, 1986, p.137.

11. M.J. van den Heuvel, A.D. Dayan, and R.O.Shillaker, *Human Toxicol.*, 1987, 6, 279.

15

Risk Assessment: A Physiologically Based Pharmacokinetic Approach

R.B. Drawbaugh, H.J. Clewell, and M.E. Andersen

TOXIC HAZARDS DIVISION, ARMSTRONG AEROSPACE MEDICAL RESEARCH LABORATORY, WRIGHT-PATTERSON AIR FORCE BASE, OH 45433, USA

1 Introduction

The major responsibility of toxicologists is to predict chemical safety for man. An important aspect in achieving this goal is to understand clearly the test regimen used in the laboratory assessment of toxicity and how to extrapolate these findings to predict possible human risks. In general, there are two approaches to extrapolation of animal toxicity data to man: (1) qualitative extrapolation, attempting to determine the utility of an animal model of toxicity as a possible representative for man, and (2) quantitative extrapolation, using physiologically based mathematical modelling as a tool for scaling-up laboratory animal data to man.

It is hoped to convince the reader that despite a disproportionate emphasis on species differences, physiologically based pharmacokinetic (PB-PK) modelling of test animals can provide an attractive approach to predict target tissue dose and ultimately an overall assessment of toxicity in man. This chapter describes physiological modelling and the application of PB-PK models to risk estimation of exposed humans.

2 Physiological Modelling

The subject of classical pharmacokinetics was developed to describe the flow of chemicals and their metabolites through discrete compartments of model systems by fitting experimental data on chemical levels to mathematical equations. These compartments may be purely theoretical concepts that are postulated solely to fit observed kinetic values which are not easily related to the underlying anatomy and physiology. Classical pharmacokinetic models have been used extensively in clinical medicine and provide useful information in determining required dose and schedule of maintenance therapy to obtain an efficacious level of drug in the body. Linear one- and two-compartment open models (Figure 1) are used most frequently.

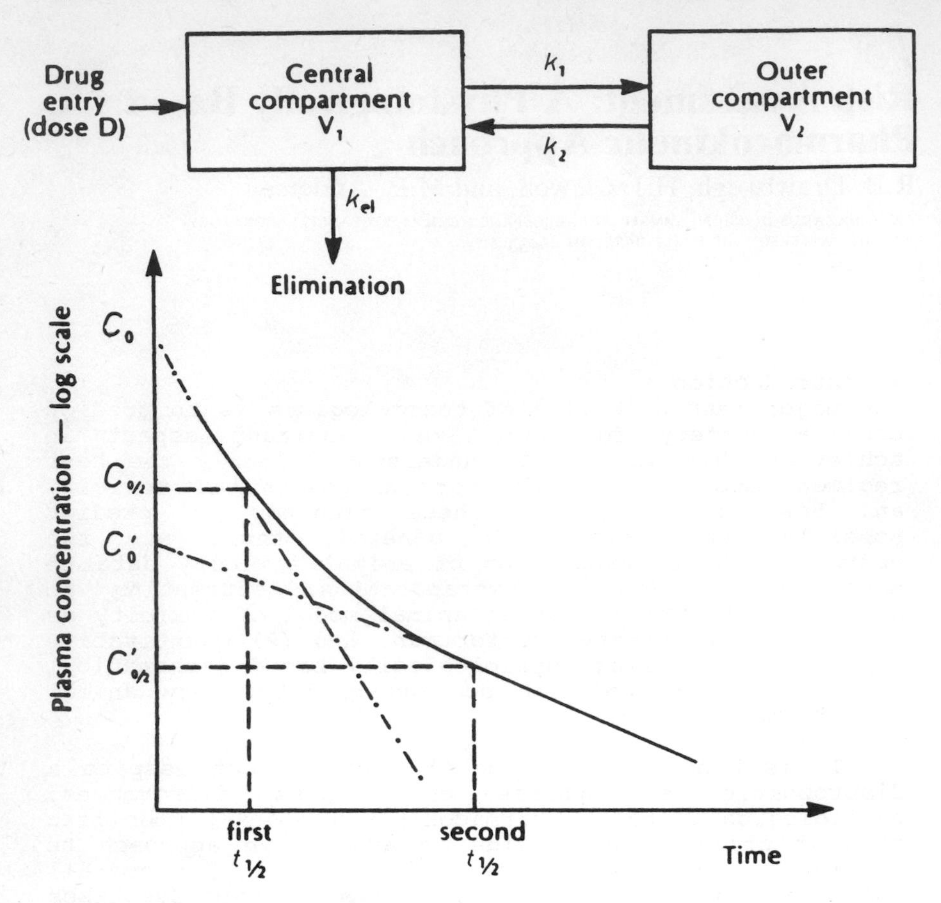

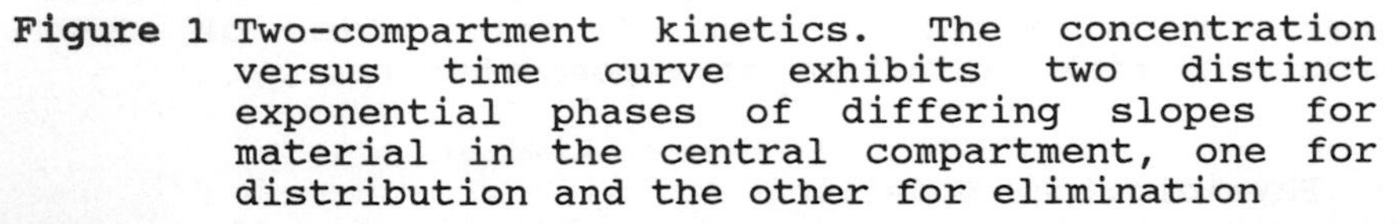
Figure 1 Two-compartment kinetics. The concentration versus time curve exhibits two distinct exponential phases of differing slopes for material in the central compartment, one for distribution and the other for elimination

Physiologically based pharmacokinetic models differ considerably from the classical compartmental models in that they assume that distribution and disposition of a xenobiotic results from a complex set of physiological processes and biochemical interactions (Figure 2). Several review papers have been written describing physiologically based pharmacokinetic modelling (PB-PK).[1–4] A PB-PK model utilizes known physiological and anatomical functions (Table 1) as a basis for the model which was proposed over 50 years ago by Teorell.[5,6] In addition, actual physico-chemical and biochemical constants are obtained to assist in describing the uptake, distribution, and disposition of a foreign chemical in the body. After gathering this

Table 1 *Kinetic constants and model parameters used in the physiologically based pharmacokinetic model for dichloromethane*

	B6C3F1 Mice[a]	F344 Rats[b]	Hamsters[b]	Humans
Weights (kg)				
Body	0.0345	0.233	0.140	70.0
Lungs (x 10^3)	0.410	2.72	1.64	772.0
	Percentage of body weight			
Liver	4.0	4.0	4.0	3.14
Rapidly perfused	5.0	5.0	5.0	3.71
Slowly perfused	78.0	75.0	75.0	62.1
Fat	4.0	7.0	7.0	23.1
Flows (l h^{-1})				
Alveolar ventilation	2.32	5.10	3.50	348.0
Cardiac output	2.32	5.10	3.50	348.0
	Percentage of cardiac output			
Liver	0.24	0.24	0.20	0.24
Rapidly perfused	0.52	0.52	0.56	0.52
Slowly perfused	0.19	0.19	0.19	0.19
Fat	0.05	0.05	0.05	0.05
Partition coefficients				
Blood/air	8.29	19.4	22.5	9.7
Liver/blood	1.71	0.732	0.840	1.46
Lung/blood	1.71	0.732	0.840	1.46
Rapidly perfused/blood	1.71	0.732	0.840	1.46
Slowly perfused/blood	0.960	0.408	1.196	0.82
Fat/blood	14.5	6.19	6.00	12.4
Metabolic constants				
V_{max} (mg h^{-1})	1.054	1.50	2.047	118.9
K_m (mg l^{-1})	0.396	0.771	0.649	0.580
KF (h^{-1})	4.017	2.21	1.513	0.53
A1	0.416	0.136	0.0638	0.00143
A2	0.137	0.0558	0.0774	0.0473

[a] Parameters correspond to the average body weight of B6C3F1 mice in the NTP bioassay (NTP, 1985).

[b] Parameters correspond to the average body weight in gas uptake studies.

information, a physiological model is developed which conceptually expresses the behaviour of the chemical in the animal.[7] In this approach the various time-dependent biological processes are described by a set of mass-balance equations. The resultant mathematical model can easily be written and executed using available IBM-compatible hardware and software.[8]

The physiological model may appear similar to that of a classical compartment model, but the physiological models are clearly more involved and require a larger number of model parameters. The complexity should not be feared, but rather looked on as a small price to pay for their greater predictive power that can be used to great advantage in the risk assessment process.

3 Example of a Physiological Model

Methylene chloride (dichloromethane, DCM) is an important industrial chemical, and its pharmacokinetic behaviour has recently been described using a physiological model (see also Chapter 9).[9] This study applied a PB-PK model of DCM disposition to risk assessment in humans exposed to low concentrations of DCM. This effort was undertaken jointly by members of the Toxic Hazards Division of the Armstrong Aerospace Medical Research Laboratory and Dow Chemical, Midland, ML, USA. Initial work on constructing a physiological model for DCM had already been accomplished,[10] and it was felt that such a model could have a significant impact on the regulatory rulemaking process.[11]

An important step in the pharmacokinetic risk assessment is to determine the dose level to be used in extrapolating from the tumour incidence observed in rodent to tumour incidence expected for human exposure. The initial modelling effort[10] had to be expanded so that metabolism in both target tissues (lung and liver) could be included, and to estimate the amount of DCM oxidation and conjugation expected in species studied.[12,13] A PB-PK description for DCM could now be used to: (1) estimate tissue exposure to DCM, (2) identify tissue exposure to metabolites of the oxidative pathway, and (3) estimate tissue levels of metabolites from the glutathione dependent pathway during the inhalation exposures. It was concluded that unmetabolized DCM was not the tumourigenic agent (based on mechanistic considerations).

In the DCM risk assessment initially proposed by the US EPA[12,13] the relationship between administered dose and target tissue dose was assumed to be linear. Andersen et al.[9] developed a PB-PK tissue dose extrapolation which was clearly non-linear (Figure 3).

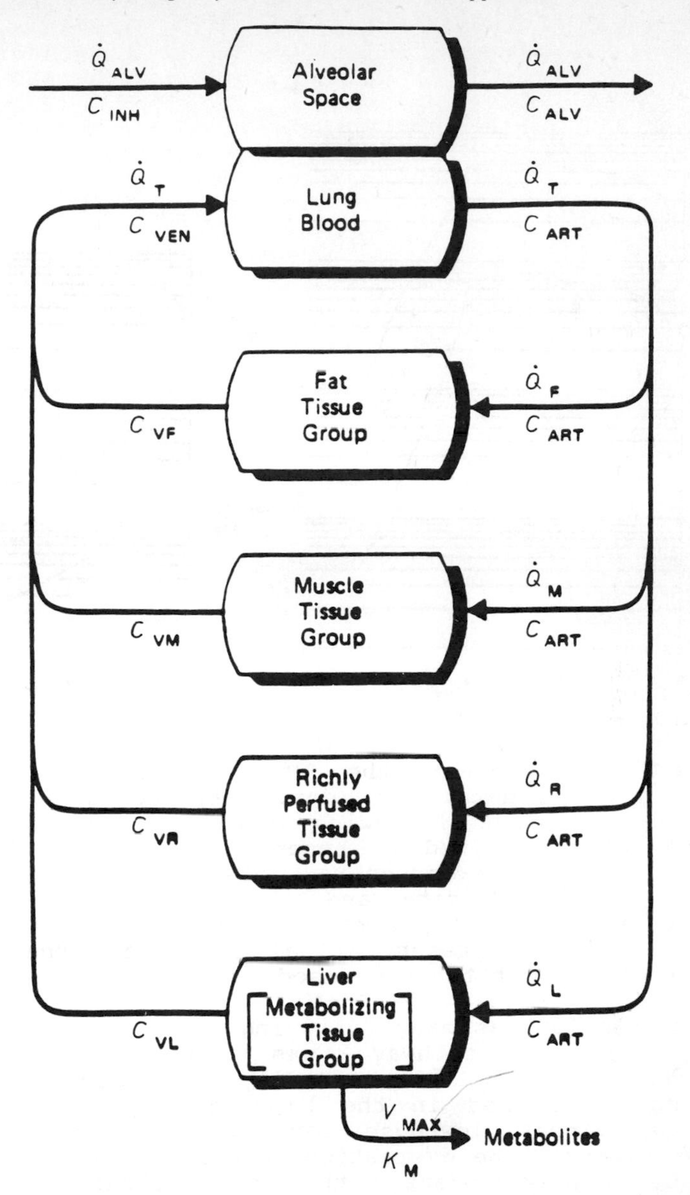

Figure 2 Flow diagram of a generic physiologically based pharmacokinetic model for volatile organic chemicals. In this description, organs or groups of organs are defined with respect to their volumes, blood flows *(Q)*, and partition coefficients for the chemical. The uptake of vapour is determined by the alveolar ventilation *(Q_{alv})*, cardiac output *(Q_t)* blood: air partition coefficient, and the concentration gradient between arterial and venous pulmonary blood *(C_{art} and C_{ven})*

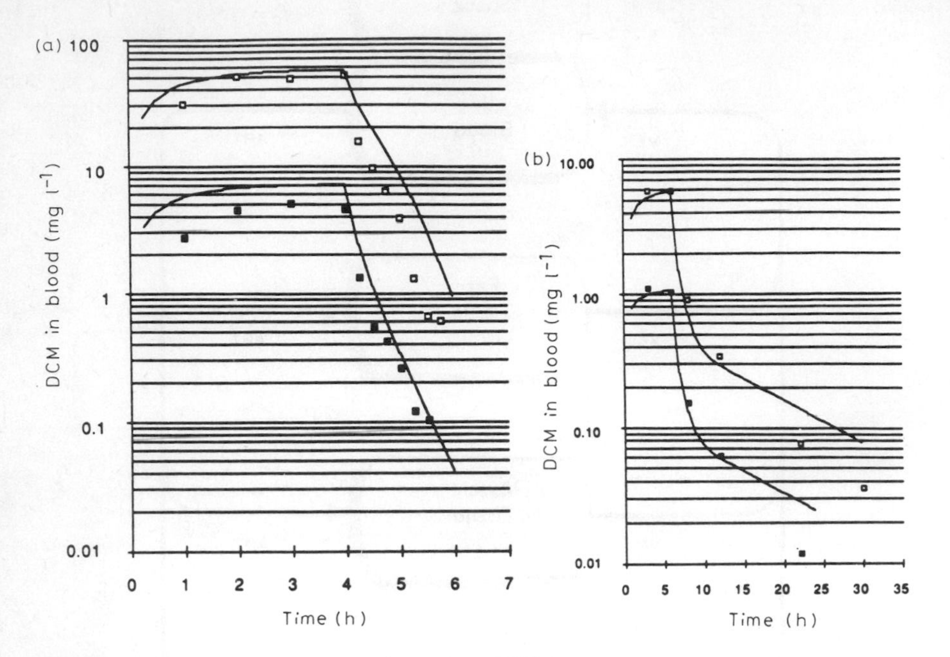

Figure 3 Validation of the PB-PK model with experimental data:[9] (a) data obtained in F344 rats during and following inhalation exposure; (b) data obtained in humans during and following inhalation exposure. The solid lines are simulated data and the closed/open symbols represent experimental determinations. The points represent actual data while the curves represent the exercised model

This conclusion was made following the observation that the oxidation pathway was favoured at low concentrations, but is readily saturated at the concentrations used in the bioassay exposures. Thus, using a linear low-dose extrapolation, the target tissue dose will be overestimated. If the model is then used to compare rodent with man it clearly suggests that man is less sensitive than mouse, traditionally a view which is contrary to US EPA thinking. Overall, conventional models predicted an acceptable exposure limit which is approximately 50 to 100 fold lower than the predictions made using physiologically based modelling approaches.

This example shows how a physiologically based pharmacokinetic model for DCM which addresses the previous failures to consider the effect of enzyme saturation at dose levels used by the National Toxicology Program bioassay and consideration to man's lower level of metabolic enzymes, thus being less sensitive to DCM than rodents, has been used to strengthen the scientific bases of risk assessment. It is extremely important to understand that the approach (Figure 4) is not uniquely applicable to DCM but can be and has been used for volatiles, pesticides, and other environmental toxicants. Additional models developed by this laboratory include styrene,[7] bromomethane,[14] and hydrazine.[15]

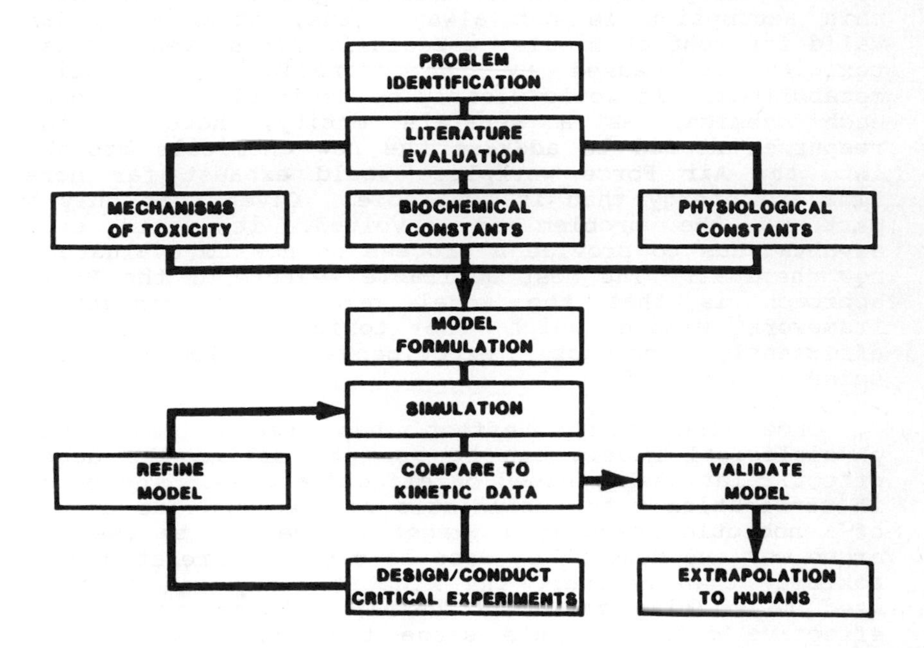

Figure 4 Flow-chart of the approach to chemical hazard assessment using physiologically based pharmacokinetic modelling. The points represent actual data while the curves represent the exercised model

4 Other Applications for Physiological Models

Physiologically based models can be used to reduce the number of animals required for an *in vivo* toxicological assessment, as a front-end analysis for pharmacodynamic modelling, and to allow for retrospective analysis of data used to develop an exposure limit for the workplace.

5 Conclusions

The intense political pressure on regulatory agencies to react quickly on public health issues, and the limited resources available, support the requirement for a reasonably inexpensive, but scientifically based approach to risk assessment extrapolation. It is in this context that a physiologically based pharmacokinetic approach is presented to satisfy partially the requirement for a cookbook-like approach to human carcinogenic risk assessment.[16] The use of predictive physiological kinetic models in risk assessment is predicated on a very simple premise: an effective target organ dose in one species is expected to be equally effective in another species.[17] Although this assumption is not always true, it does appear valid for most chemicals, especially for solvents whose toxicity is caused by the formation of reactive metabolites. It would clearly be desirable to approach each chemical as a separate entity; however, the resources needed to address the new chemicals brought into the Air Force workplace would exhaust far more time and energy than is available. Given that only a part of the problem is involved, it would seem advantageous to provide a **process** on how to evaluate a new chemical. The most desirable feature of the PB-PK approach is that the model provides a conceptual framework within which the toxicologist can more efficiently conduct experiments following the scientific method.

Once the toxic effect has been identified, physiological models can be used to define the dose-effect relationship based on effective dose. The model in combination with mechanistic studies and the amount of xenobiotic reaching a target tissue can be used to argue why one measure of dose is better correlated with toxicity than another. The kinetic study can then be used to clarify the relationship between effect and effective dose. At this stage the major task is to relate this effective dose in an experimental animal to an effect in the human.

Using the interspecies extrapolative capability of physiological models, a researcher can outline a potential approach to this problem. The physiological model can be exercised using appropriate human exposure conditions to predict the environmental levels of a substance that will produce the same effective dose in man as that observed in the laboratory species used for the study. Generally speaking, physiological models show significant promise as an analytical tool for enhancing the scientific basis of risk assessment for a number of industrially important chemicals.

6 References

1. K.B. Bischoff, 'Physiological pharmacokinetics', *Bull. Math. Biol.*, 1986, **48**, 309-322.

2. L.E. Gerlowski and R.K. Jain, 'Physiologically based pharmacokinetic modeling: principles and applications', *J. Pharm. Sci.*, 1983, **72**, 1103-1126.

3. J.M. Collins and R.L. Dedrick, 'Pharmacokinetics of anticancer drugs', 1982, in 'Pharmacologic Principles of Cancer Treatment', ed. B.A. Chabner, Saunders, Philadelphia, pp. 77-99.

4. K.J. Himmelstein and R.J. Lutz, 'A review of the applications of physiologically based pharmacokinetic modeling', *J. Pharmacokinet. Biopharm.*, 1979, **7**, 127-145.

5. T. Teorell, 'Kinetics of distribution of substances administered in the body. I. The extravascular modes of administration', *Arch. Int. Pharmacodyn.*, 1937, **57**, 205-225.

6. T. Teorell, 'Kinetics of distribution of substances administered to the body. II. The intravascular mode of administration', *Arch. Int. Pharmacodyn.*, 1937, **57**, 266-240.

7. J.C. Ramsey and M.E. Andersen, 'A physiological model for the inhalation pharmacokinetics of inhaled styrene in rats and humans', *Toxicol. Appl. Pharmacol.*, 1984, **73**, 159-175.

8. H.J. Clewell III and M.E. Andersen, 1986, 'A multiple dose-route physiological pharmacokinetic model for volatile chemicals using ACSL/PC', in 'Languages for Continuous System Simulation', ed. F.D. Cellier, Society for Computer Simulation, San Diego, pp. 85-101.

9. M.E. Andersen, H.J. Clewell III, M.L. Gargas, F.A. Smith, and R.H. Reitz, 'Physiologically based pharmacokinetics and the risk assessment process for methylene chloride', *Toxicol. Appl. Pharmacol.*, 1987, **87**, 185-205.

10. M.L. Gargas, H.J. Clewell III, and M.E. Andersen, 'Metabolism of inhaled dihalomethanes *in vivo*: differentiation of kinetic constants for two independent pathways', *Toxicol. Appl. Pharmacol.*, 1986, **82**, 211-223.

11. P. Shabecoff, 'EPA reassesses the cancer risks of many chemicals', *New York Times*, 4 Jan. 1988.

12. US Environmental Protection Agency, Health assessment document for dichloromethane (methylene chloride). Final report, EPA-600/8-82-004F, 1985.

13. US Environmental Protection Agency, Addendum to the health assessment document for dichloromethane (methylene chloride): Updated carcinogenicity assessment of dichloromethane (methylene chloride), EPA/600/8-82-004FF, 1985.

14. J.N. McDougal, M.E. George, H.J. Clewell III, and M.E. Andersen, 'Dermal Absorption of Hydrazine Vapors', Proc. 17th Annual Conference on Environmental Toxicology, October 1986, pp 88-94.

15. J.N. McDougal, G.W. Jepson, H.J. Clewell III, M.G. MacNaughton, and M.E. Andersen, 'A Physiological Pharmacokinetic Model for Dermal Absorption of Vapors in the Rat', *Toxicol. Appl. Pharmacol.*, 1986, **85**, 286-294.

16. US Environmental Protection Agency, 'Proposed guidelines for carcinogenic risk assessment', *Fed. Reg.*, 1984, **49**, 4294-4301.

17. H.J. Clewell III and M.E. Andersen, 'Risk assessment extrapolations and physiological modeling', *Toxicol. Ind. Health*, 1985, **1**, 111-131.

Section 3: Incidental Emissions: Air and Water

16
Decision Making in Control of Air Pollutants Posing Health Risks

D.H. Trønnes

CENTRE FOR INDUSTRIAL RESEARCH, PO BOX124, BLINDERN 0134, OSLO 3, NORWAY

and H.M. Seip

DEPARTMENT OF CHEMISTRY, UNIVERSITY OF OSLO, PO BOX 1033, BLINDERN 0315, OSLO 3, NORWAY

1 Introduction

Air pollution is nothing new in the history of mankind. No doubt our ancestors when living in caves were exposed to heavy smoke with high concentrations of many harmful compounds. Coal smoke caused the major air pollution problems from about the thirteenth century and in many areas this is the situation even today. Concern about effects of pollutants on health and environment has been expressed by a few foresighted individuals long before the problems were taken seriously by most people. Thus John Evelyn discussed smog problems in London in his famous 'Fumifugium' published in 1661.

We have recently witnessed catastrophes such as the accident in the nuclear reactor in Chernobyl and the spread of deadly methyl isocyanate in Bhopal (see Chapter 7). In these cases the cause of the damage is obvious. However, considerably lower air concentrations of many compounds may also cause harm to health or environment. Thus the frequency of lung cancer is usually higher in cities than in the rural environment; it seems likely that this in part is related to the air quality.[1] In such cases the causal relationship is more uncertain and it may be extremely difficult to find the most effective countermeasures.

A decision maker dealing with air pollution problems is therefore often in a very difficult situation. On the basis of a huge amount of data, which may be interpreted in different ways by experts, she/he must try to reach a rational decision (to do nothing is also a decision that may have serious consequences). To assist the decision maker formal methods have been developed. So far all the methods have considerable weaknesses. Often the process of applying the methods is more useful than the results obtained. The decision maker is forced to structure the problem and, as far as possible, quantify the consequences of various options. The process also gives valuable opportunities for communication between the various interest groups and the decision makers.

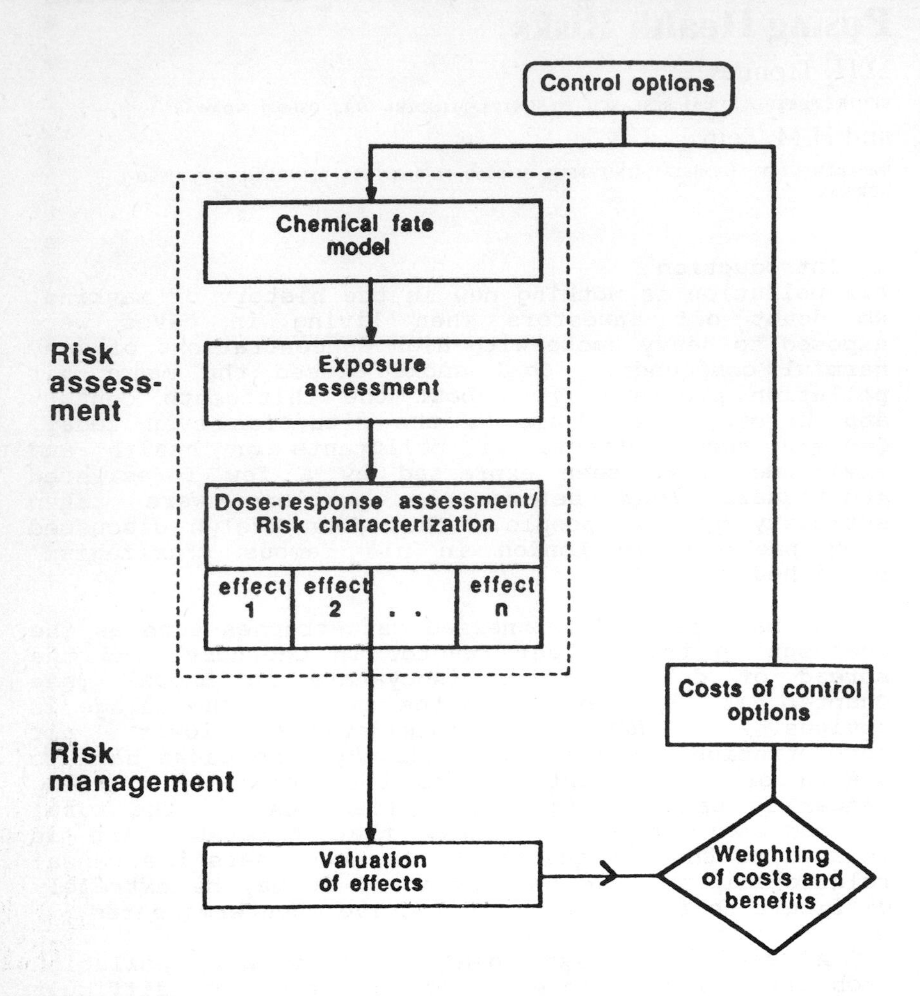

Figure 1 Components of an analysis of pollution control

Important steps in a decision analysis of pollution control are illustrated in Figure 1. The steps depicted in the upper part of the left-hand side are often denoted risk assessment, while the lower part and the right-hand side are denoted risk management (*cf.* Section 2). Models for dispersion and fate of pollutants and for exposure, as well as quantitative toxicity data, are required in a response model for effects. One control measure may affect a number of pollutants, and each compound may affect human health and the environment in several ways. To make the situation even more complicated for the decision maker, synergistic or antagonistic interactions may increase or decrease the effects compared with those expected by adding the individual effects. In most cases the costs of various measures can be estimated with reasonable accuracy. The uncertainties are generally much larger in the estimates of exposure and, in particular, in the relationships between dose and response (*e.g.* per cent mortality). It is of utmost importance that the uncertainties originating from all steps in the analysis are included in the presentation of the results.[2]

Though the uncertainties are often large, both exposure doses and related effects can in principle be determined objectively. This is not the case for valuation of effects which is necessarily subjective. We are often dealing with commodities for which there are no market prices. There is, for instance, no absolute answer to the question of the value of human life[3] or that of an antique building. Even if market prices do exist, their usefulness may be questioned. The problems were elegantly illustrated by Oscar Wilde when he defined a cynic as a man who knows the price of everything and the value of nothing.

In this chapter, a very brief review will firstly indicate some quantitative decision analysis techniques; later two examples will be given. The first one is indoor exposure to radon. This case is relatively simple, *i.e.* there is just one substance and the effects are fairly well known. In the study the quantification of uncertainties plays an important role. The second case deals with effects of air pollutants in Oslo. This is a much more complicated problem. A huge number of compounds should actually be considered, but concentration data are available for only a few. The number of possible effects is also large. In this case drastic simplifications are necessary, but a formal analysis is still of value.

2 Quantitative Decision Analysis Techniques

Decision analysis is used, here, as a general term for a series of methods. Others use this term only for techniques designed to take uncertainties explicitly into account. Detailed discussions of various techniques for decision analysis are beyond the scope of this chapter; the emphasis is on the two case studies described in the later sections. However, some key methods in decision analysis are mentioned below, with emphasis on problems involving human health. For further details see for example Zeleny[4] or Morgan.[5]

2.1 Risk Assessment. The first step in determining risk to human health caused by a chemical is normally to carry out tests to establish whether the chemical has a potential to cause health damage. If these tests show the chemical to be non-toxic, or that the chances of exposure of man (and environment) are small, a full risk assessment is unnecessary. If the chemical represents a potential danger two additional tasks must be carried out (see Figure 1). First, an assessment has to be made of how strongly the population is exposed to the chemical considered, and secondly the connection between the exposure (concentration and duration) and the probability of health damage must be established.

2.1.1 Exposure assessment. Exposure under present conditions may be determined by direct measurement. However, prediction of exposure under future conditions (*e.g.* after construction of a new plant or after implementing measures to reduce emissions from an existing one) involves, in general, the use of one or more mathematical models describing how the chemical is transported and transformed on its way to the receptor. The approach tends to depend strongly on the particular case investigated. It will therefore not be discussed further here, but reference should be made to the case studies described in the Sections 3 and 4.

2.1.2 Dose-response relationships. The goal of a dose-response assessment is to obtain a quantitative relationship between duration and level of exposure and the probability of the occurrence of a certain health damage. There are two main approaches: epidemiological studies and laboratory studies on living organisms. The latter may be acute or long-term tests on animals (often rodents) or short-term tests for mutagenicity using bacteria or cell cultures. Unfortunately the uncertainty in the dose-response relationships tends to be large. Because of the lack of controlled exposure conditions in epidemiological studies one should include in the analysis all factors that could contribute significantly to the observed effect. Some of these factors may not be known. Moreover, important information on those factors known to play a role, such as data on ambient concentrations of the chemical investigated, may be scarce or unavailable.

In animal bioassays the scientist can have full control over exposure conditions. However, the uncertainty introduced in transferring the results to humans may be of several orders of magnitude.

Generally speaking the uncertainties in quantification of health risk are of three different kinds: (i) the processes involved are inherently stochastic in nature, or at least so complex that it is infeasible to build and apply precise deterministic models; (ii) if plausible functional relationships for important variables have been established, the uncertainties in the parameters describing the relationships must be determined; (iii) lack of information may prevent formulation of reliable mathematical models.

The first type of uncertainty represents, in principle, no real problem for the decision maker, since probabilistic models can be applied. In the case study on effects of radon (Section 3) a functional relationship between radon dose and probability of lung cancer has been used. In the case study described in Section 4, no model has been established, and a simpler approach was chosen.

2.2 Risk Management.

2.2.1 Cost-benefit analysis. This is a procedure for determining whether the expected benefits from a proposed action outweigh the expected costs. It is assumed that all costs and benefits can be expressed in monetary units. For some benefits market prices may be obtained. In other cases it may be possible to find a 'surrogate market'. Alternatively, people's willingness to pay may be obtained from interviews. The approach taken in evaluating the health effects of radon (Section 3) may be regarded as part of a cost-benefit analysis (or of one of the modified versions described below). We have, however, not yet carried through all the steps.

2.2.2 Contingent cost-benefit analysis. In this variant unit prices are estimated for some of the benefits as in a normal cost-benefit analysis. For those benefits that are more difficult to express in monetary terms, the unit prices are varied to find the smallest values that make the net benefit positive. Alternatively, one may find the range in monetary units that makes a certain control option the best one.[6] This procedure does not solve the difficult problem of valuing commodities for which there are no market prices. However, in many cases it may be rather easy to agree that the price should be above a certain value or fall in a given range.

2.2.3 Risk-benefit analysis. This may be viewed as a simplified or incomplete cost-benefit analysis. Only some of the benefits are expressed in monetary terms. Those that are most difficult to quantify in this way (*e.g.* decreased mortality), are expressed in other, more natural units. However, the final balancing of costs and benefits, which is left to the decision maker, will usually implicitly imply monetary values for all benefits.

2.2.4 Cost-effectiveness analysis. This is an examination of the costs of alternative means to achieve a given goal (*e.g.* a certain pollution level).

2.2.5 Multiobjective decision making techniques. This term is often reserved for techniques that are not variations of cost-benefit analysis; comparison of benefits are made without expressing them in monetary units. A number of such techniques have been developed.[4]

Multiattribute utility analysis (MUA) may be derived from a set of axioms on how people make decisions under uncertainty.[7] In MUA costs and benefits are brought onto a common scale by assigning so called utilities to the consequences considered. This technique thus makes it possible to balance incommensurate consequences against each other without having to express everything in monetary terms. The utility of each consequence and the relative importance of the different types of consequence are derived from answers given by individuals or groups to certain importance-ranking questions. A special advantage of MUA is that people's attitude towards risk may be taken explicitly into account.

Simplified versions of MUA have been developed, *e.g.* Simplified Multiattribute Rating Technique (SMART).[8] The approach used to valuate countermeasures against air pollutants in Oslo (Section 4) is also a simplified form of MUA.

3 Health Risks Caused by Indoor Radon Exposure

This section describes one approach for quantification of health risks to the Norwegian population caused by indoor exposure to radon, ^{222}Rn. Similar methods may be used for other environmental toxicants. Radon is a noble gas produced by radioactive decay chains starting with uranium or thorium. There are three different radon isotopes of which ^{222}Rn is the most important one and the only radon isotope considered in this chapter. ^{222}Rn is produced by decay of ^{238}U (uranium) via ^{226}Ra (radium). Among the sources of indoor radon are soil, tap water, and building materials. The largest source of radon in Norwegian dwellings is radon influx from the underlying soil. Radon is radioactive and the α-particles emitted in the decay process can contribute to lung cancer when the decay process takes place in the lungs. The radon gas enters houses through cracks and openings in the structure. The radon concentrations in dwellings in Norway, Sweden, and Finland are very high with mean values roughly two times the concentrations in USA, Canada, UK, Germany, Italy, and the Netherlands.[9] It has been estimated that today's exposure to radon may cause 120-360 deaths by lung cancer in Norway annually. This is approximately 10-30% of the total number of deaths by lung cancer.[10] Since there are reasons to believe that new buildings tend to have higher radon concentrations than old ones, and since there is a long time period between exposure and death by lung cancer, the numbers given above are probably overestimates of deaths due to radon exposure. Rather, the numbers should be regarded as predictions of future lung cancer incidence in Norway caused by radon, if no countermeasures are taken.

To estimate the lung cancer incidence, models describing the exposure situation and the dose-response relationship are needed. Large uncertainties are generally involved in estimates of exposure data and dose-response relationships. Furthermore, there may be different opinions among the experts on various assumptions in the models. It is therefore important to state explicitly the assumptions and to develop a model in which the uncertainties can be accounted for. When the decision makers are considering measures to reduce exposure to indoor radon, it is important for them to know the uncertainties in the benefit in terms of reduced lung cancer incidence expected for a given measure.

3.1 Exposure Data. Nero et al.[11] have summarized information from a number of studies on indoor concentrations of radon in the United States. They found that various lognormal distributions can be used to fit the exposure distributions in the USA, both on a national scale and in smaller areas such as individual states. Knowledge about exposure to radon in Norwegian homes is mainly based on a survey of concentration levels in 1500 houses.[10] Houses in all counties in Norway were represented in the survey, but they were not chosen completely at random. Nearly all measurements were carried out in houses consisting of one or two storeys. The measurements seem to be lognormally distributed.[10] Some aggregate results from the measurements are shown in Table 1.

Table 1 *Distribution of annual average concentrations of radon-222* (Bq m^{-3}) *in Norwegian Homes*

Mean	*Percentage of houses with concentrations above*		
	200	400	800
100	10	3	1

Using the arithmetic mean of the data and the fact that 10% of the houses have radon concentrations above 200 Bq m^{-3} (Table 1), the parameters in the lognormal distribution that fit these data can be found. The parameters are ν = 4.3 and τ = 0.78 (Figure 2).

However, in this distribution the percentage of houses with concentrations above 400 Bq m^{-3} is 1.5%. This is less than the 3% suggested by the measurements shown in Table 1. By requiring an arithmetic mean of 100 Bq m^{-3} and that the percentage of houses with concentrations above 400 Bq m^{-3} should be 3%, a lognormal distribution with parameters ν = 4.08 and τ = 1.016 is obtained. This distribution has a heavier tail than the first distribution (Figure 2).

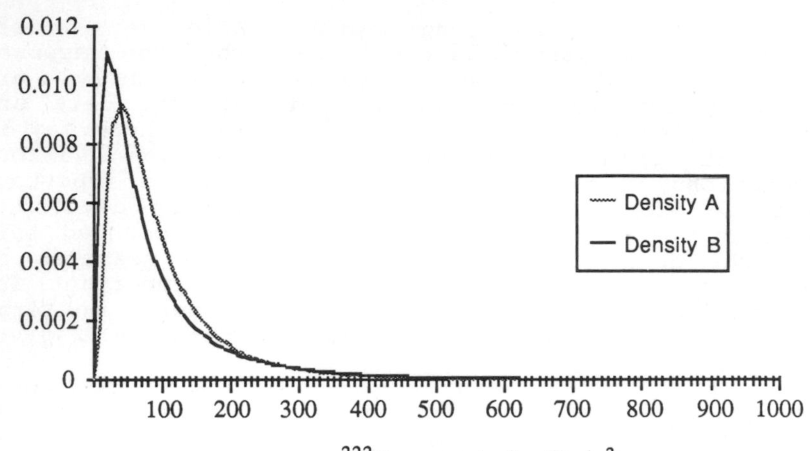

Figure 2 Two probability densities describing the distribution of indoor radon concentrations in houses with one or two storeys in Norway (the numbers on the y-axis are scaled so that the areas under the curves are equal to 1). Density A is lognormal with parameters ν = 4.3 and τ = 0.78 (which gives an arithmetic mean of 100 Bq m^{-3} and a probability of 10% for concentrations above 200 Bq m^{-3}). Density B is lognormal with parameters ν = 4.08 and τ = 1.106 (which gives an arithmetic mean of 100 Bq m^{-3} and a probability of 3% for concentrations above 400 Bq m^{-3})

The concentration of radon in a building varies very much between the storeys. The radon concentration at the first floor is about 30% of the concentration at the ground floor, and at higher floors the concentrations are considerably lower. It is therefore important to account for the number of persons living in apartments on different floors when evaluating the exposure distribution in the Norwegian population. Sanner et al.[12] assumed that 75% of the population live in houses with one or two storeys, that 10% live in apartments on the first floor, and that 15% live in apartments on the second or higher floors. In this chapter the same assumptions are made. Further it is assumed that the concentration distributions in Norwegian homes, given in Figure 2, are representative only for those living in houses with one or two storeys. The concentration for persons living on the first floor is 30% and for those living on the second or higher floors the concentration is assumed to be 10% of ground floor concentrations.

When considering average exposure, account is also taken of how much time people spend in their homes. According to data from the Central Bureau of Statistics[13] people in Norway on average spend roughly 16 hours indoors in their homes and 8 hours elsewhere, *e.g.* (outdoors, at work, *etc.*) during a 24 hour period. There are, however, some variations with sex, age, and between urban and rural districts that will not be taken into account in this chapter. In accordance with Sanner et al.[12] it is assumed that the concentrations outside the home are 10% of the concentrations they are exposed to at home.

<u>3.2 Dose-Response Relationship</u>. Data on the relationship between exposure to radon and the probability of developing lung cancer are primarily based on epidemiological studies on radon-exposed underground miners. The results of some of these studies are summarized in a report from the International Commission on Radiological Protection.[9] Several models have been developed to describe the dose-response relationships. Among these models the relative risk projection model gives the best description of the epidemiological data.[9] Use is made of this model to describe the risk of getting lung cancer. The relative risk projection model is based on the assumptions that the age-specific lung cancer rate attributable to radon, $\lambda_r(t)$, is proportional to the normal lung cancer rate, $\lambda_0(t)$, and to the accumulated radon dose up to age t minus a constant lag time, τ. In this chapter, exposure to constant concentrations is assumed. The lung cancer rate attributable to radon can then be expressed as:

$$\lambda_r(t)=r\cdot C_{Rn}\cdot(t-\tau)\cdot\lambda_0(t) \quad ; \; t\geq\tau \qquad (1)$$

$$(\lambda_r(t)=0 \text{ for } t<\tau)$$

and the total lung cancer rate as:

$$\lambda(t)=\lambda_0(t)+\lambda_r(t)=\lambda_0(t)\cdot[1+r\cdot C_{Rn}\cdot(t-\tau)] \qquad (2)$$

where r is the relative excess risk coefficient that expresses the impact of a given dose of radon on lung cancer rates. C_{Rn} is the concentration of ^{222}Rn in Bq m^{-3}. The lung cancer rates, $\lambda(t)$, $\lambda_0(t)$, and λ_r are expressed in terms of the number of lung cancer deaths per year per 100,000 persons in age group t.

The normal lung cancer rates in a population differ considerably between males and females (Figure 3), and between smokers and non-smokers.

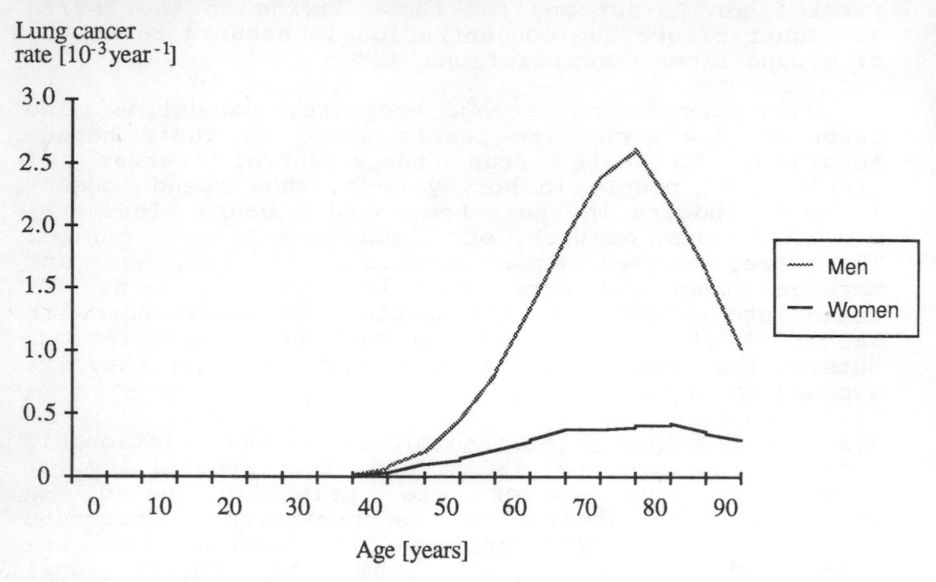

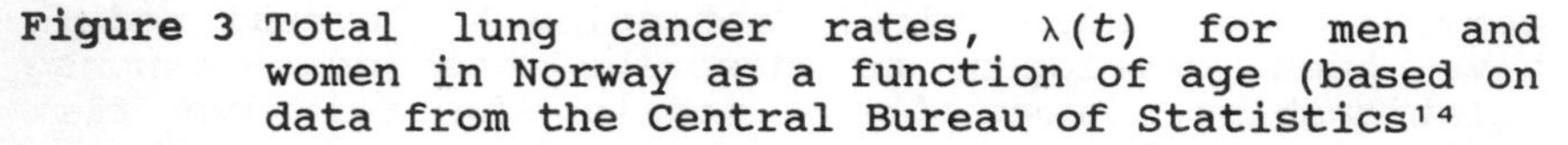
Figure 3 Total lung cancer rates, $\lambda(t)$ for men and women in Norway as a function of age (based on data from the Central Bureau of Statistics[14]

It is of interest to find a more aggregated quantity describing the lung cancer risk than the age-specific rates. The population-related annual lung cancer frequency attributable to radon exposure, F_r, is such a quantity. The age distribution varies among different populations, so to find F_r we have to weight the lung cancer rates attributable to radon exposure, $\lambda_r(t)$, with the age distribution of the population in question:

$$F_r = \sum_{t=0}^{\infty} v(t) \cdot \lambda_r(t) = \sum_{t \geq \tau}^{\infty} v(t) \cdot r \cdot C_{Rn} \cdot (t-\tau) \cdot \lambda_0(t) \tag{3}$$

$$= C_{Rn} \cdot r \cdot \sum_{t \geq \tau}^{\infty} v(t) \cdot (t-\tau) \cdot \lambda_0(t)$$

where $v(t)=n(t)/N$ is the fraction of the population in age group t, $n(t)$ is the number of persons in age group t, and N is the number of persons in the total population.

Using data for the Norwegian population[14] and τ=10, the sum in the last expression above equals 54.8 h year^{-1} for women and 221.4 h year^{-1} for men. Further, the relative excess risk coefficient, r, is estimated to be 0.8 x 10^{-8} (m^3 Bq^{-1} h^{-1}) for concentrations given in Bq m^{-3} of ^{222}radon.[9] (r is found to be slightly age dependent and will decrease slightly at exposure to extremely high radon concentrations. However, to simplify the description, it was chosen to treat r as a constant in this chapter.) Using these values the annual lung cancer frequencies attributable to radon exposure for men and women become:

$$F_{r,men} = 1.78 \times 10^{-6} \cdot C_{Rn} \qquad (4)$$

and

$$F_{r,women} = 0.44 \times 10^{-6} \cdot C_{Rn} \qquad (5)$$

The unit for F_r is year^{-1}

3.3 Estimating Human Health Risks. The number of annual lung cancer deaths attributable to radon exposure can now be obtained by integrating over all possible radon concentrations:

$$R = N_{women} \cdot k \cdot l \cdot \int_0^{\infty} F_{r,women} \cdot f(C_{Rn}) \cdot d(C_{Rn}) + N_{men} \cdot k \cdot l \cdot \int_0^{\infty} F_{r,men} \cdot f(C_{Rn}) \cdot d(C_{Rn}) \qquad (6)$$

where N_{women} = 2074000 and N_{men} = 2033000 are the total number of women and men, respectively, $f(C_{Rn})$ is the frequency distribution of concentrations in houses with one or two storeys (*cf.* Section 3.1), and k and l are the occupancy and residence factors, respectively. The factors k and l are applied to adjust for the fact that not all time is spent indoors in people's homes and that not all people live in one or two storeyed houses. The numerical values obtained from the assumptions given previously are k=0.7 and l=0.8.

In Figure 2 it is seen that the number of persons exposed to radon decreases for increasing concentrations (above a certain concentration). On the other hand the probability of developing lung cancer, given by F_r, increases proportionally with radon concentrations (as can be seen from equations (4) and (5)). The products of these quantities and the factors k, l, and N for men and women (*cf.* equation (6)), are plotted as a function of ^{222}Rn concentrations in Figure 4.

The functions in Figure 4 are useful for example when considering how large reductions in lung cancer incidence that can be achieved by implementing counter-measures in all houses with ^{222}Rn concentrations above certain levels. Calculations based on the figure show that the majority of the deaths (69%) will be among persons exposed to concentrations below 200 Bq m^{-3}, which is the action level proposed by the International Commission on Radiological Protection.

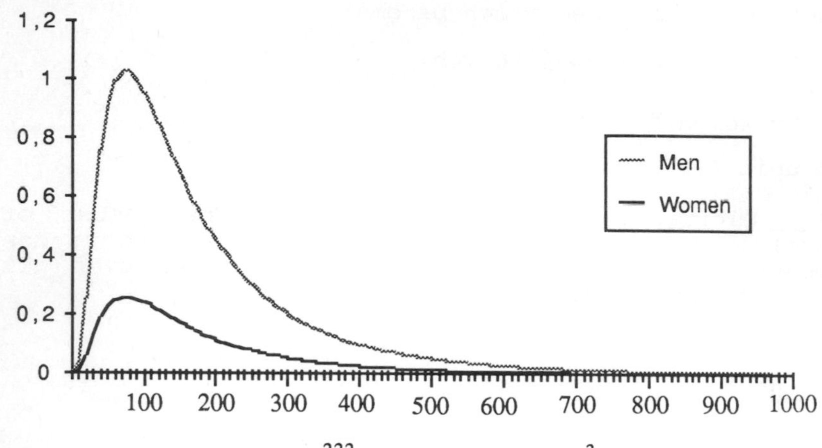

Figure 4 Distribution of lung cancer incidence caused by radon exposure as a function of ^{222}Rn concentrations for men and women in Norway (based on exposure density A, Figure 2). The areas under the curves between two concentration values give the number of lung cancer cases per year among men and women, respectively, exposed to concentrations in this interval.

<u>3.4 Assessing Uncertainties in Lung Cancer Incidences Caused by Radon Exposure</u>. In Section 2 it was pointed out that there will generally be large uncertainties involved in health risk assessments. This is due to uncertainties both in the exposure data and in the dose-response relationship. In this section it will be shown how the uncertainties in the exposure and dose-response functions can be quantified and how these uncertainties can be propagated through the effect model to give the resulting uncertainties in the effect estimates.

<u>3.4.1 Modelling uncertainties in the exposure data</u>. Uncertainties in the exposure data will be modelled by

ascribing uncertainties to the parameters in the lognormal distribution (see Figure 2).

For the purpose of performing Monte Carlo simulations a distribution function is needed for the parameter ν. It is reasonable to choose this function so that the 95% confidence interval comprises the ν values for the two functions shown in Figure 2, *i.e.* 4.08 and 4.30. Somewhat arbitrarily a normal distribution with mean 4.15 and standard deviation 0.1 was chosen. The corresponding 95% confidence interval is [3.95, 4.35]. In accordance with the data in Table 1, the second parameter in the exposure distribution, τ, is given a value that keeps the arithmetic mean of the concentrations equal to 100 Bq m^{-3}.

<u>3.4.2 Modelling uncertainties in the dose-response function</u>. The dose-response relationship is given by equation (3). In this expression there is uncertainty in the relative excess risk coefficient, r, which is estimated mainly from epidemiological studies on underground miners. There are also uncertainties in the estimates of the lung cancer rates, $\lambda_0(t)$, and in the length of the latency period, τ. Further, there is model uncertainty caused by the assumptions in equation (1), on which the relative risk projection model is based.

The uncertainty in the relative excess risk coefficient, r, has been estimated to be ±50%.[9] This is interpreted as meaning that $[0.5 \cdot \underline{r}, 1.5 \cdot \underline{r}]$ constitutes a 95% confidence interval for r, with $\underline{r}$ representing its mean value. For the purpose of performing Monte Carlo simulations on the effect model one needs to specify a probability distribution for r. A normal distribution with mean $\underline{r}$ and standard deviation $0.25 \cdot \underline{r}$ would approximately satisfy the given confidence interval, but it would give negative values of r with some positive probability. Therefore, a lognormal distribution was chosen for r, and parameters were calculated to satisfy the following two conditions: the arithmetic mean of the distribution is $\underline{r}$ and the probability for r-values to be greater than $1.5 \cdot \underline{r}$ is 2.5% (*cf.* the confidence interval). This distribution is not very different from the 'corresponding' normal distribution as shown in Figure 5.

It is very difficult to estimate the uncertainty in the model assumptions. One could try different models and perform an analysis of each alternative. Uncertainties in the cancer rates, $\lambda_0(t)$, which are relatively small, and the latency period, τ, can be estimated from available data. No attempt was made here to estimate these uncertainties. Instead, a subjective estimate of the total uncertainty was made by treating the sum in the last expression in equation

(3) as a stochastic variable. Normal distributions with mean values of 54.8 and 221.4 for women and men, respectively (as given above), and with a standard deviation of 0.125 times the mean values, were obtained.

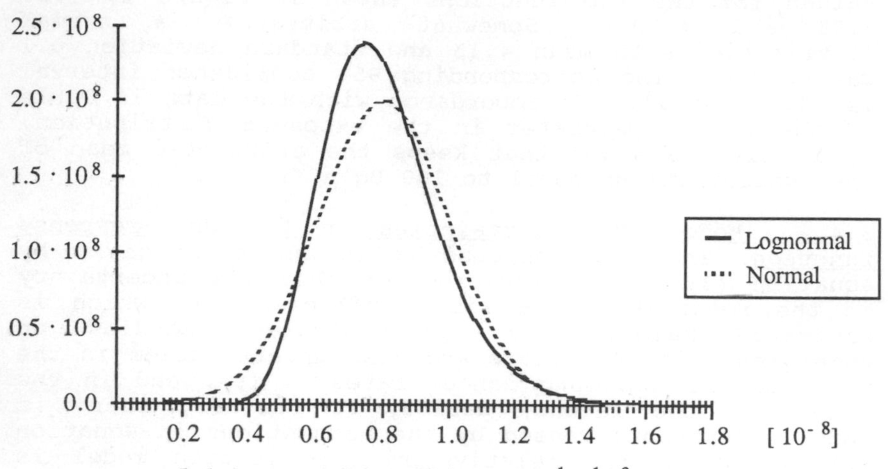

Figure 5 Two probability densities describing the uncertainty in the relative excess risk coefficient, *r* (the numbers on the y-axis are scaled so that the areas under the curves are equal to 1). Lognormal probability density (parameters - 18.67 and 0.22), compared to a normal probability density with mean $0.8 \cdot 10^{-8}$ and standard deviation $0.2 \cdot 10^{-8}$

<u>3.5 Monte Carlo Simulations</u>. To find how the uncertainties in exposure and dose-response functions affect the uncertainties in our estimates of the number of lung cancer cases per year in the Norwegian population attributable to radon exposure, a method is needed to propagate these uncertainties through the effect model. Stochastic simulation or Monte Carlo simulation is such a technique.

The calculations are carried out by drawing values for the uncertain parameters according to their probability distribution as specified above. Using one set, the resulting number of lung cancer cases, R_1, is obtained from equation (6). It is known that R is an uncertain quantity and this first value can be considered as an 'observation' of this quantity. To get a better estimate of R, more 'observations' of the quantity are needed. A new observation, R_2, is obtained by drawing new values for the uncertain parameters and calculating the resulting value of *R*. This procedure is replicated a large number of times. The mean value, standard deviation, and various fractiles of *R* can be obtained together with a plot of its frequency distribution.

To calculate the uncertainties in the estimates of the number of lung cancer cases and to get an impression of the sensitivity to the uncertainties in the various parameters, four different Monte Carlo simulations have been carried out.

Simulation 1. In this simulation it is assumed that the only uncertain parameter is *r*. The parameters in the concentration distribution are ν=4.3 and τ=0.78.

Simulation 2. This simulation is based on the same assumptions as in simulation 1, except that the parameters in the concentration distribution are: ν=4.08 and τ=1.016.

Simulation 3. The same assumption with respect to the dose-response function as in simulations 1 and 2 is applied. Uncertainty in the concentration distribution (as explained above) is also taken into account.

Simulation 4. Uncertainties in the dose-response function are accounted for by uncertainties in the parameter *r* and in the weighted sum of lung cancer rates, [equation (3)]. No uncertainty in the exposure distribution is taken into account and the parameters in the concentration distribution are ν=4.15 and τ=0.954.

3.6 Discussion of the Results. Some results from the simulations are given in Table 2.

Table 2 *Characteristics of frequency distributions for lung cancer caused by radon exposure*

	Number of replications	*Mean*	*S.D.*	*P(X>350)*
Simulation 1	1000	249.4	54.4	0.05
Simulation 2	1500	243.1	53.3	0.04
Simulation 3	5000	247.5	55.5	0.05
Simulation 4	6000	248.8	63.7	0.07

Comparison of simulations 1 and 2 shows that the concentration distribution with the heavier tail (simulation 2) gives a lower mean value for the number of lung cancer cases. This might seem surprising, but is explained by the fact that the probability of concentrations in the range 40-270 Bq m^{-3} is larger in distribution 1 than in distribution 2 (*cf.* Figure 2) to compensate for the heavier tail since the arithmetic means of the two distributions are the same. However, when considering the effect of implementing countermeasures in all houses with radon concentrations exceeding a given guideline value (*e.g.* 200 Bq m^{-3}), the concentration distribution with the heavier tail used in simulation 2 may give a higher estimate of the benefit.

In simulation 3 the uncertainty in the concentration distribution is taken into account, keeping, however, the arithmetic mean of the concentrations unchanged. As can be seen from the last column in Table 2 the uncertainty in the concentration distribution has very little effect on the uncertainty in the estimates of the number of lung cancer cases.

In simulation 4 the uncertainties with respect to the model assumption in equation (1), to the lung cancer rates and to the length of the latency period are included. The uncertainties in the effect estimates increase considerably in this simulation. As can be seen from the last column in Table 2, the probability of a large number of lung cancer cases (*e.g.* >350) becomes greater.

The frequency distribution, obtained from simulation 4, of the number of lung cancer deaths per year attributable to indoor radon exposure in Norway is plotted in Figure 6.

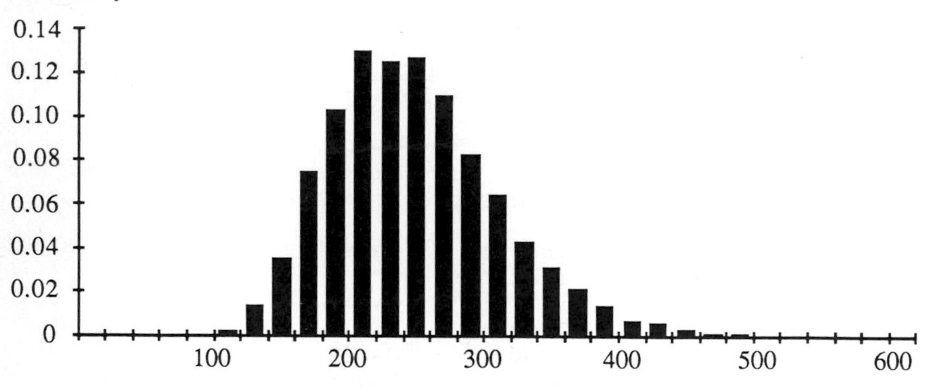

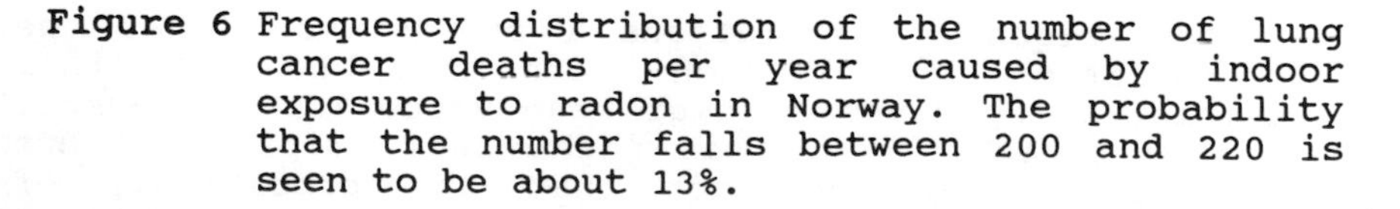
Figure 6 Frequency distribution of the number of lung cancer deaths per year caused by indoor exposure to radon in Norway. The probability that the number falls between 200 and 220 is seen to be about 13%.

<u>3.7 Evaluating Countermeasures</u>. The effect model described above has been developed for the purpose of estimating the reduction in the number of lung cancer cases due to countermeasures. One way to define a countermeasure is to specify a guideline concentration and implement the necessary measures to eliminate all exposure to concentrations above this guideline. To obtain the effect, in terms of reductions in the number of lung cancer cases, of eliminating exposure to concentrations above the guideline, it is usually not sufficient to integrate from the guideline concentration in the effect model [equation (6)], since the countermeasures will generally not eliminate radon completely. The results for concentration distributions before and after implementing the measures should be compared.

To evaluate measures to reduce indoor exposure to radon, reductions in the number of lung cancer cases have to be weighted against the costs and possible side effects. Many possible countermeasures have been suggested including better sealing of the structure, reduction of the chimney effect that actively sucks radon into the houses, and devices to pump radon out from the ground under the structure. The most suitable measure depends on how the house is built and on the magnitude of the indoor radon concentrations, and on whether it shall be implemented in an existing house or in one under construction. In general, measures in houses under construction and in existing houses with moderate concentrations are fairly cheap. When considering costs of reducing radon exposure in existing houses, the costs to identify the houses with concentrations above the guidelines must also be taken into account.

Possible side effects from the countermeasures are better indoor climate also with respect to components other than radon due to better ventilation, better insulated and less humid cellars, *etc.*, but in some cases increased heating costs.

4 Abatement of Air Pollution in Oslo

In this section an analysis is described that was carried out to find the best pollution control measures to reduce air pollution in Oslo. The work was carried out in collaboration with the Norwegian State Pollution Control Authority (SFT). A more complete description is given by Trønnes.[15] The analysis of the air pollution in Oslo is much more complex than the radon case described in Section 3. There are many objectives to be pursued and many possible means to achieve these objectives. In this situation formal methods to structure the objectives and to manage the risks become a very important part of the analysis. In addition to the larger complexity of the analysis, knowledge of dose-response relationships describing the effects on human health, ecology, *etc.* caused by air pollution is much poorer than for radon. The risk assessment will therefore be more primitive in this analysis.

4.1 Objectives and Attributes. Before starting the analysis it is important to define the problem and to structure the objectives. The overall objective of abatement of air pollution in Oslo is to find the best pollution control action. The primary motivation for the considered abatement measures is to reduce human exposure to air pollution. However, to get a complete picture, all relevant impacts of the countermeasures and the costs must be considered. In total, five general concerns have been identified: exposure, well-being, acidification, side effects, and costs, as shown in the second level of the objectives hierarchy in Figure 7.

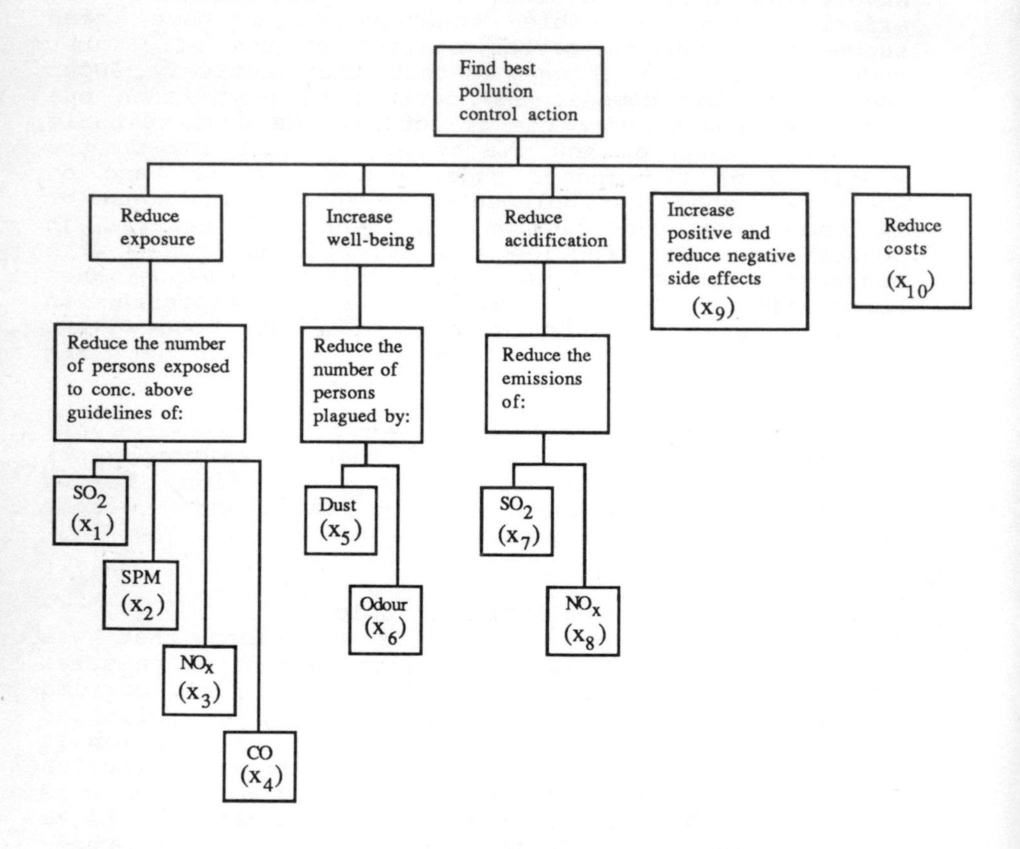

Figure 7 Objectives hierarchy for the abatement of air pollution in Oslo

The meaning of each general concern is defined by the objectives given below it in the hierarchy. The objectives listed in the hierarchy also represent the limits of the problem. Concerns that are not included here will not be taken into account in the formal analysis. Acceptance of the objectives hierarchy from the decision makers and the interest groups involved is a prerequisite for acceptance of the results from the analysis. In the following we give a brief description of the five general concerns considered in this analysis.

4.1.1 Reduce exposure. Exposure to air pollutants can contribute to long-term health effects like lung cancer and other lung and heart diseases. An obvious goal for measures aimed at improving air quality is therefore to reduce harmful health effects. In the first version of the objectives hierarchy 'reduce health effects' was one of the five general concerns. However, to relate exposure to health effects, dose-response models are needed. Because satisfactory models are not available, it was considered to be too speculative and too controversial to quantify health risks. Therefore, use was made of exposure assessment in the analysis and the objectives hierarchy had to be changed accordingly. In particular, the objective 'reduce exposure' replaced 'reduce health effects'. The objective 'reduce exposure' is divided into reducing the number of persons exposed to sulphur dioxide (SO_2), suspended particulate matter (SPM), nitrogen oxides (NO_X) and carbon monoxide (CO) in concentrations above established guidelines. The guidelines used are those suggested by SFT:[16] 100 $\mu g\ m^{-3}$ with 24 h averaging time for SO_2, SPM, and NO_X and 10 mg m^{-3} with 8 h averaging time for CO. In the analysis it is implicitly assumed that exposure to concentrations above the guidelines can contribute to long-term health effects and therefore should be reduced. Since quantification of health effects has been omitted, the responsibility for assessing the severity of exposure is left to the decision makers in the sense that they will have to make value trade-offs between exposure and other objectives.

4.1.2 Increase well-being. About 50% of the population in Oslo report that they are afflicted by dust and unpleasant odour. The levels were subjectively estimated through interviews. Since levels of 'dust' are obtained very differently from the SPM concentrations (Section 4.1.1), these two parameters are treated separately. This approach is supported by the fact that 'dust' generally consists of larger particles than SPM. Other factors that may affect the well-being are included under side effects.

4.1.3 Reduce acidification. Thousands of lakes in southern Norway have lost their fish populations as a result of acidification.[17] There is also concern about acidification of soil and effects on vegetation. Emissions of SO_2 and NO_x in Oslo contribute to these damages and it is therefore desirable to reduce these emissions.

4.1.4 Side effects. Most countermeasures will, in addition to reducing air pollution, have other effects that should be taken into account in estimating total benefits. The objective 'increase positive and reduce negative side effect' is included in the objectives hierarchy to account for any effects that may result from the countermeasures other than those explicitly mentioned in the hierarchy. Examples of side effects are: changes in travel times, changes in the number of traffic accidents, changes in the number of persons afflicted by noise, changes in heating expenses. The side effects are directly converted into monetary terms.

4.1.5 Costs. Implementation of countermeasures necessarily will result in costs. Some investments and running expenses will be covered by public budgets and some will be covered by the private sector. Naturally, reducing the costs of implementing the countermeasures is an important objective.

4.1.6 Attributes. To each of the lowest-level objectives variables or attributes were assigned. By means of these attributes the impacts of an abatement measure can be described quantitatively. The attributes employed are shown in Table 3.

Table 3 *Attributes for the abatement of air pollution in Oslo*

Exposure

x_1 = Number of persons exposed to SO_2 concentrations above guidelines

x_2 = Number of persons exposed to SPM concentrations above guidelines

x_3 = Number of persons exposed to NO_x concentrations above guidelines

x_4 = Number of persons exposed to CO concentrations above guidelines

Table 3 (continued)

Well-being

x_5 = Number of persons afflicted by dust

x_6 = Number of persons afflicted by odour

Acidification

x_7 = Tons of SO_2 emitted per year

x_8 = Tons of NO_X emitted per year

Side effects

x_9 = Annualized present value of net change in side effects

Costs

x_{10} = Annualized present value of costs of implementation

4.2 Estimating the Impacts of the Countermeasures.

4.2.1 Countermeasures. Road traffic and space heating are the most important sources of air pollution in Oslo, but industry and sewage incinerators also contribute significantly. There are many possible approaches to reducing emissions from each source, resulting in a large number of countermeasures. In the present study SFT identified 38 measures to be analysed. In this chapter, however, results for only 13 measures are presented (see Table 6). A more complete list is given by Trønnes.[15]

The impact assessments started with detailed specifications of each abatement measure. Then costs, reduction in emissions, and side effects were estimated. Most of this work has been carried out by institutions and organizations working in the respective fields. For example, the Institute of Transport Economics, the Oslo transportation Company, and the municipality of Oslo have been working on the traffic regulating measures, and Oslo Electricity Works have been working on some of the measures related to space heating.

4.2.2 The reference state. To measure the impact of abatement measures one has to compare with some reference state. Here, the scenario to be expected if none of the abatement measures considered in this analysis are implemented, defines the reference state. The reference state can be denoted by the vector of attributes values, x^0:

$$x^0 = (x^0{}_1, \; x^0{}_2, \ldots\ldots, x^0{}_8, \; 0,0)$$

The last two attributes will take the value 0, because there are no side effects or costs involved if no abatement measures are implemented.

The reference scenario is not fully defined before deciding on a specific year, because of expected changes in energy consumption, road traffic, *etc.* Also some abatement measures are already approved for implementation and will affect the pollution in the future independently of the measures considered here. Catalytic converters installed in new cars from 1989 is an example of a measure that will be implemented and will contribute to considerable reductions in exposure to NO_x and CO. This abatement measure is expected to reach full effect in year 2000. It is therefore convenient to define the reference state to be the pollution situation in year 2000 if none of the measures considered in this analysis are implemented. The values of the attributes for the reference state are shown in the last row of Table 4. The low number of persons exposed to concentrations above guidelines for NO_x and CO are to a large extent a result of the decision to install catalytic converters in new cars from 1989.

4.2.3 Exposure assessment and impact assessment. The aim of the impact assessment is to quantify the effects of the abatement measures in terms of the attributes or performance measures defined in Table 4. The impacts of an abatement measure i will be described in terms of differences in the attribute values in the year 2000 if the measure i is carried out and the reference state. The vector x^i describes these changes.

To obtain estimates of the number of people exposed to excessive concentrations of the pollution components for several scenarios, simulation models must be used. Dispersion models contain equations describing the fate of air pollution components under different meteorological conditions like wind direction and velocity, whether it is inversion or not, *etc.* Inputs to these models are the amount and location of emissions. The output can be obtained in terms of 'concentration maps' which give the concentration of each component for every square kilometre cell on a map. In Oslo road traffic is an important source of air pollution so the highest concentrations often occur in streets with heavy traffic. The Norwegian Institute for Air Research (NILU) has developed dispersion models that describe dispersion in streets and more global models for Oslo.[18] Combined with data on the population's place of residence, working place, and travelling habits it is possible to obtain estimates of the number of people exposed to various concentrations of the pollutants. In this study NILU performed most of the work on exposure assessments.

The assessment of the number of persons afflicted by dust and unpleasant odour is based on a survey in which a representative sample of the population in Oslo was interviewed.[19] In the interviews people were also asked about what sources contributed most to their nuisances. The effects of the abatement measures were then estimated by SFT, on the basis of how much each measure would contribute to reductions in the relevant sources in various parts of the city.

The results of the impact assessments are given in Table 4. The uncertainties in these estimates, which in some cases are large, are discussed later.

4.2.4 Correction factors. The costs and benefits are not equally distributed through the lifetime of the abatement measures. To be able to compare the costs of the measures, annualized present values (annual costs) have been computed. In these calculations a discount rate of 7% p.a. and a lifetime of 30 years have been assumed. In Table 4 the costs and side effects are given in annualized present values.

For the abatement measures correction factors are introduced for the attributes x_1 - x_8 to account for the fact that some of the measures will not have full effect immediately. When the full effect estimate for year 2000 (which is given in Table 4) is multiplied with the correction factor, one gets the average effect per year, or the 'annualized' values, for these attributes over the 30 year lifetime. The values of the correction factors for the abatement measures are given in Table 4.

4.3 Valuation of Impacts of Countermeasures. To evaluate and rank the countermeasures some criterion or evaluation method is needed. These methods vary in complexity and need for subjective value judgements. In the present analysis a relatively simple evaluation model was wanted that would be easy to communicate to the decision makers and would be easily accepted by them. It should not be too demanding for the decision makers to make the necessary value judgements and trade-offs between the objectives needed to arrive at a conclusion.

Table 4 *Attribute values[a,b] for the abatement measures and for the situation in 1985 and in year 2000 (reference state)*

	Attributes										
Abatement measure	SO_2 (x_1)	SPM (x_2)	NO_x (x_3)	CO (x_4)	Dust (x_5)	Odour (x_6)	SO_2 (x_7)	NO_x (x_8)	Corr factor[c]	Side eff.[e] (x_9)	Costs[e] (x_{10})
A	2630	14630	9000	0	40000	7500	23	500	0.8	750	251
B.1	0	19900	0	0	3500	18000	0	0	1	1.5	5
B.2[d]	0	32020	2020	0	8390	62500	0	768	0.84	0.13	30
C[d]	0	0	0	0	800000	0	0	0	1	473	534
D	0	2750	1080	0	480	2380	0	100	0.8	0	0.41
E	109750	10500	0	0	0	0	-116	0	0.8	227	308
F	2440	0	0	0	0	0	31	0	0.82	7.4	4.3
G.1	0	40810	0	0	7700	5540	0	0	0.74	0	23
G.2	16130	1750	0	0	2750	1250	200	0	0.8	46	20
G.3	0	0	0	0	22530	25060	0	0	0.87	0	7.5
H.1[d]	180000	0	0	0	0	0	16780	0	1	55	101
H.2	47100	6500	0	0	180	960	380	11	1	0.4	7.1
I	0	0	0	0	46880	0	0	0	0.8	0	20.4
Situation in 1985	30000	130000	100000	1000	248000	216000	2700	8600			
Reference state	184000	150000	12000	0	306000	230000	5160	7800			

[a] See Table 3 and Table 6 for definition of the attributes and the measures

[b] In the table the attributes x_1 - x_8 are given in terms of improvements in the attribute values

[c] See text for explanation

[d] The abatement measure will be implemented on a national basis, and the improvements may therefore be greater than the attribute values for the reference state which refers to the situation in Oslo only

[e] Attribute values are given in 10^6 NOK (1 NOK = $ 0.15)

<u>4.3.1 The evaluation model</u>. A commonly used evaluation model in multiattribute utility analysis is the additive form of the utility function U.[20,21]

$$U(\mathbf{x}) = \sum_{i=1}^{n} k_i u_i(x_i), \tag{7}$$

where the k_i s are scaling factors representing value trade-offs between units of the corresponding attributes and the u_is are the utility functions for the attributes. The use of equation (7) requires participation from the decision maker in determining both the utility functions and the scaling factors.

The evaluation model used in this analysis, shown in equation (8), differs from equation (7) in two aspects: the attribute values are used directly instead of the utility functions for each attribute (this simplification means that we assume a linear relationship between the attribute values and the utility), and the benefit/cost ratios, R, have been used to evaluate the abatement measures instead of utility.

$$R = B/C = \left(\sum_{i=1}^{9} k_i x_i\right) / k_{10} x_{10} \tag{8}$$

Here R is the benefit/cost ratio, and B and C are the benefits and costs respectively, measured in some common unit. The k_i s (i=1,2,...10) are weights representing a valuation of the attributes into a common unit, and the x_i s are the attribute values. Alternatively the net benefit, N, given by:

$$N = B - C \tag{9}$$

can be used to evaluate the abatement measures.

The benefit/cost ratio, R, is used to rank the abatement measures. R>1 indicates that the benefit for this measure is greater than the cost and consequently that it might be useful to implement the measure.

<u>4.3.2 Valuations and trade-offs</u>. Before the evaluation model can be used the weights, k_i, must be determined. This is the subjective part of the analysis (*cf.* Figure 1), and one expects different interest groups to come up with the different sets of weights, reflecting their preferences. Even if this part of the analysis is subjective, one will find some guidance in the literature and in earlier practice with similar decisions. SFT worked out a set of weights based on their preferences and references to earlier works.[22]

This list of weights are given in Table 5.

Table 5 *SFT's Attribute weights pr. unit of the attributes*

General concern	*Attribute*		*Unit*[a]	*Attribute weight*[b]
Exposure	SO_2	(x_1)	persons exposed to SO_2	1.89
	SPM	(x_2)	persons exposed to SPM	4.15
	NO_x	(x_3)	persons exposed to NO_x	3.77
	CO	(x_4)	persons exposed to CO	5.66
Well-being	Dust	(x_5)	persons afflicted by dust	0.23
	Odour	(x_6)	persons afflicted by odour	0.23
Acidification				
	SO_2	(x_7)	tons of SO_2 emitted year^{-1}	2.45
	NO_x	(x_8)	tons of NO_x emitted year^{-1}	0.25
Side-effects		(x_9)	10^3 NOK year^{-1}	1
Costs		(x_{10})	10^3 NOK year^{-1}	1

[a]By 'persons exposed' in this Table we mean exposure to concentrations above the guidelines.
[b]Attribute weights have been normalized to give 10^3 NOK year^{-1} the weight 1.

The attribute weights in Table 5 have been normalized so that 1000 NOK year^{-1} have been given the weight 1. The weight corresponding to attribute x_1 is 1.89. This means that it is considered to be 1.89 times more important to obtain a reduction of one in the number of persons exposed to excessive sulphur dioxide (for one year), as it is to save one thousand NOK. In other words, one is willing to pay 1890 NOK to have one person less exposed to excessive sulphur dioxide for one year. The other attribute weights in Table 5 can be given similar interpretations.

The weights for NO_x and CO exposure in Table 5 are two and three times as large as for SO_2 exposure, respectively. Syversen[22] argues that a greater portion of those exposed to NO_x and CO in Oslo are exposed to concentrations much higher than the guidelines than is the case for sulphur dioxide. In addition, exposure to concentrations just above the guidelines for sulphur dioxide and NO_x will affect only the most sensitive groups of the population, while for exposure to carbon monoxide all persons will be affected. Further, it is argued that exposure to suspended particulate matter (SPM) should be given a high weight relative to sulphur dioxide because it is assumed that SPM contains hydrocarbons and other organic compounds that contribute to elevated risks for lung cancer.

4.4 Results. Using Tables 4 and 5 it is possible to calculate the total damage, in monetary terms, caused by air pollution in Oslo. Without implementing any of the abatement measures considered here, the annual total damages, in year 2000 will be 1.15 x 10^9 NOK. In a similar manner the benefit/cost ratios of each abatement measure can be calculated using equation (8). Results from these calculations for some of the measures are shown in Table 6; a more complete list is included by Trønnes.[15]

Table 6 *Benefit/cost ratio and net benefit of the abatement measures*

Rank No.[a]	*Abatement measure No.*	*Description*	*Benefit/cost ratio*	*Net benefit*[b]
1	D	Buses, better maintenance to reduce exhaust emissions	52.6	21.6
2	B1	Diesel vehicles, better maintenance	29.2	141.0
3	H2	Usage of low-sulphur oil under inversions	16.4	109.3
4	B2	Diesel trucks, reduced exhaust emissions	7.2	186.0
5	G1	Reduced usage of wood for space heating	5.5	103.5
6	H1	Reduced sulphur content in low sulphur oil	4.7	373.7
7	G2	Oil-based heating installations, better maintenance	3.9	58.0
8	A	Reduced private car usage	3.3	577.3
9	F	Energy saving measures	2.6	6.9
10	E	Remote heating, oil-based	1.4	123.2
11	C	Prohibition on usage of studded tyres	1.4	213.6
12	G3	Reduced emissions from industry	1.2	1.5
13	I	Vegetation screens	0.4	-12.2

[a]The measures are ranked according to their benefit/cost ratio.
[b]The net benefits are given in 10^6 NOK.

When an abatement measure is implemented it will in general affect the benefit/cost ratios of the other measures. For example, the implementation of certain measures may reduce the concentrations in one part of the city to levels below the guidelines. Implementation of another measure will then not contribute to a further improvement of the situation, and the benefit/ cost ratio will be reduced accordingly. Table 6 gives the benefit/cost ratios and net benefit for each measure when no other measures have been implemented.

<u>4.5 Dealing with Uncertainties</u>. Uncertainties in the effect attributes of the abatement measures will result in uncertainties in the benefit/cost ratios given in Table 6. These uncertainties have to be quantified when considering whether the benefits are significantly greater than the costs of an abatement measure, or whether one measure is significantly better than another. In this analysis no detailed quantification of the uncertainties in the effects estimates of the abatement measures has been carried out. At the start of the project, however, there were plans to do so. There are several reasons why the uncertainty analysis has been simplified. In addition to lack of funding the most important reasons are: no quantitative data to estimate uncertainties statistically are available, many people involved in the quantification have little experience with probability theory, quantifying uncertainties means more complicated data handling.

However, it is important to get some idea of the magnitude of the uncertainties in the benefit/cost ratio and of what factors contribute to the uncertainties. In many cases the only method to quantify the uncertainties is by subjective expert assessments.

To get an impression of the uncertainties in the benefit/cost ratios of the abatement measures considered, measure G2, 'better maintenance of oil-based heating installations', was selected for further investigation. The uncertainties in the effects of this abatement measure were assessed subjectively in terms of fractiles in their probability distributions. In the assessment the ten attributes were grouped into five different categories, one for each of the general objectives: exposure, well-being, acidification, side effects and costs. The relative uncertainty was assumed to be the same in each category. Table 7 shows subjectively estimated relative fractiles in the probability distribution for these five categories.

In Table 7 the fractiles are given as percentages of the median values. To find the actual fractiles for a given attribute one has to multiply its median value given in Table 4 by the relative fractile value from Table 7.

Exposure data are assumed to have the highest degree of uncertainty since uncertainties are introduced at many steps in the estimation of these data.

Table 7 *Relative fractiles of the probability distributions, given as percentages of the median value, for the five uncertainty categories.*

Uncertainty category	*Attributes*	*Fractile* 0%	5%	20%	50%	80%	95%	100%
Exposure	$x_1 - x_4$	25	50	70	100	130	150	195
Well-being	$x_5 - x_6$	40	60	80	100	125	145	180
Acidification	$x_7 - x_8$	60	80	90	100	110	120	140
Side effects	x_9	50	70	85	100	115	130	150
Costs	x_{10}	70	85	90	100	120	140	170

4.5.1 Monte Carlo simulation. The uncertainties in the benefit/cost ratio resulting from the uncertainty distributions given above can be estimated using Monte Carlo simulations as explained in Section 3. The 'observations' of the benefit/cost ratio, obtained by replicate runs of the simulation model were sampled into intervals of length 0.2. Using the results from this sampling, one can plot the frequency distribution of the benefit/cost ratio. The resulting plot, shown in Figure 8, is based on 5000 replications.

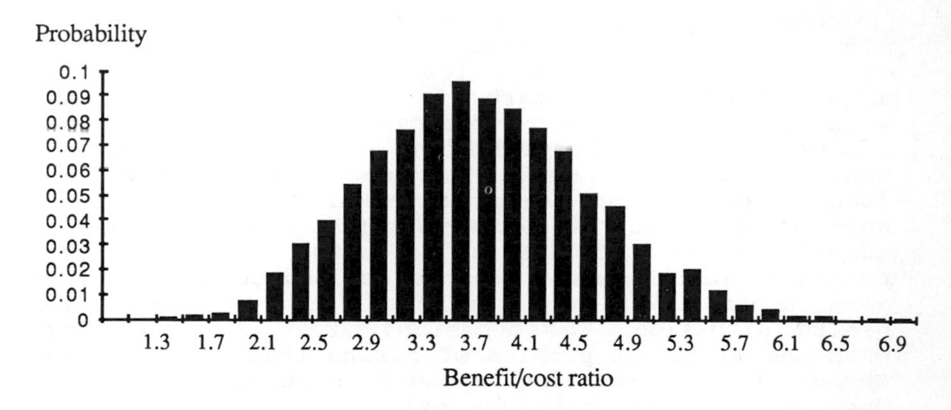

Figure 8 Frequency distribution for the benefit/cost ratio of abatement measure G.2. The probability that the ratio falls between 3.5 and 3.7 is seen to be about 10%

The simulated mean value of the benefit/cost ratio for abatement measure G.2 is 3.78 and the simulated standard deviation is 0.86. The probability is about

0.1 that the benefit/cost ratio is less than 2.7, and about 0.9 that the benefit/cost ratio is less than 4.9. An 80% confidence interval for the benefit/cost ratio of abatement measure G.2 is therefore approximately [2.7, 4.9]. In Table 6 three of the abatement measures have benefit/cost ratios within this interval (in the more complete study this was the case for 8 measures[15]).

5 Summary and Conclusions

This chapter briefly describes some methods developed to assist decision makers involved in pollution control. The importance of including the uncertainties in all steps in the analysis is emphasized. The results will then generally be given in terms of probability distributions of effects. In practice this may turn out to be difficult to achieve. It is a paradox that when 'experts' feel that a quantity is very uncertain they are often reluctant to quantify the uncertainty and only a point estimate is obtained.

The studies discussed represent cases of very different complexity. The results for health effects of radon are preliminary. The study will be improved by taking synergism between effects of smoking and exposure to radon into account and by obtaining more reliable estimates for some of the uncertainties. Furthermore, the risk management aspects, *e.g.* costs and effects of countermeasures, must be included. The approach taken, however, seems satisfactory.

Limited resources enforced us to make drastic simplifications in the study of air pollution in Oslo. We ended up with a much less comprehensive analysis than planned at the start of the project. One of the main motivations for performing this analysis was indications that air pollution in Oslo causes adverse health effects in humans like lung cancer, obstructive lung diseases, and angina attacks. However, lack of good dose-effect and dose-response models made the quantification of these effects difficult and controversial. It was therefore decided to use exposure assessments as final attributes for describing the health situation. The decision makers are then left with the difficult problem of making trade-offs between objectives like reducing exposure to sulphur dioxide on one hand and to nitrogen oxides on the other, and between reducing exposure versus changes in other attributes. In our opinion it would be better to utilize medical expertise as far as possible in these evaluations than to leave the question totally to the decision makers. It is also important to identify optimal combinations of abatement measures and consider these combinations of measures as the final decision alternatives. However, to be able to identify optimal combinations and calculate their benefit/cost ratios, one has to describe dependencies between the measures. In this project the available resources did not permit the establishment of a model for these dependencies, and the combination of measures had to be done on a rather *ad hoc* basis.

In spite of the difficulties discussed for these case studies we strongly advocate the use of such methods. The Norwegian State Pollution Control Authority felt that the analysis of air pollution in Oslo was useful in giving priorities to counter-measures. Only by carrying out a number of case studies will it be possible to improve the methods so that more powerful tools for decision analysis can be developed. This is of great importance to enhance the chances of making the right decisions in the many complicated pollution control questions to be settled in the future.

6 Acknowledgements

This work could not have been carried out without the support of a number of people. In particular we want to thank Anders Heiberg for his contribution to Section 2 and Terje Kronen who acted as co-ordinator in the project described in Section 4.

7 References

1. C. Davies and T. Sanner, 'Air Pollution and Lung Cancer', (in Norwegian), The Norwegian Radium Hospital, Oslo, Norway, 1983.

2. G.W. Suter, L.W. Barnthouse, and R.V. O'Neill, *Environ. Manage.*, 1987, **11**, 295.

3. D. Føllesdal, 'Risk: Philosophical and Ethical Aspects', 'Risk and Reason. Risk Assessment in Relation to Environmental Mutagens and Carcinogens', ed. P. Oftedal and A. Brøgger, Alan R. Liss, Inc., New York, 1986, p.41.

4. M. Zeleny, 'Multiple Criteria Decision-Making', McGraw-Hill, New York, 1982.

5. M.G. Morgan, 'The Role of Decision Analysis in the Implementation of Environmental Policies', Organization for Economic Co-operation and Development, ENV/CHEM/CM 83.5, Annex, Geneva, 1983.

6. 'Air-Borne Sulphur Pollution. Effects and Control', Air Pollution Studies No. 1, Economic Commission for Europe, United Nations, New York, 1984.

7. R.L. Keeney and H.F. Raiffa, 'Decisions with Multiple Objectives: Preferences and Value Trade-offs', John Wiley & Sons, 1976.

8. W. Edwards, 'Use of Multiattribute Measurement for Social Decision Making', 'Conflicting Objectives in Decisions', ed. B.E. Bell, R.L. Keeney, and H.F. Raiffa, Wiley, New York, 1977, p.247.

9. 'Lung Cancer Risks from Indoor Exposures to Radon Daughters', The International Commission on Radiological Protection, Pergamon Press, Oxford 1986.

10. E. Stranden, 'Radon-222 in Norwegian Dwellings', 'Radon and its Decay Products. Occurrence, Properties, and Health Effects', ed. P.K. Hopke, ACS Symposium Series No.331, American Chemical Society, Washington DC, 1987, p.70.

11. A.V. Nero, M.B. Schwehr, W.W. Nazaroff, and K.L. Revzan, *Science*, 1986, **234**, p.992.

12. T. Sanner, E. Dybing, and E. Stranden, 'Lung Cancer Risks from Indoor Exposures to Radon', (in Norwegian with English summary), National Institute of Radiation Hygiene, PO Box 55, N-1345 Østerås, Norway, Report 1988: 3.

13. 'Tidsnyttingsundersøkelsen 1980-1981', Central Bureau of Statistics, Oslo, 1983.

14. 'Statistical Yearbook 1983', Central Bureau of Statistics, Oslo, 1983.

15. D.H. Trønnes, 'Abatement of Air Pollution on Oslo', in 'Risk Management of Chemicals in the Environment', ed. H.M. Seip and A. Heiberg, Plenum Press, in press.

16. 'Air Pollution. Health and Environmental Effects', (in Norwegian), SFT-report No.38, The Norwegian State Pollution Control Authority, Oslo, 1982.

17. L.N. Overrein, H.M. Seip, and A. Tollan, 'Acid Precipitation - Effects on Forest and Fish', Final report of the SNSF-project 1972-1980, NIVA, Oslo, 1980.

18. K.E. Grønskei, F. Gram, and S. Larssen, 'Calculation of Dispersion and Exposure for Some Air Pollution Components in Oslo' (in Norwegian), NILU/OR 8/82, The Norwegian Institute for Air Research, Lillestrøm, 1982.

19. 'Inquiry about Air Pollution Nuisance in Oslo', (in Norwegian), Norsk Opinionsinstitutt, Oslo, 1986.

20. G. Anadalingam, *Eur. J. Op. Res.*, 1987, 336.

21. M.W. Merkhofer and R.L. Keeney, *Risk Anal.*, 1987, 7, 173.

22. T. Syversen, 'Comparing the Utility of Various Impacts of Measures to Reduce Air Pollution', (in Norwegian), The Norwegian State Pollution Control Authority, Oslo, 1987.

17
Acid Rain — Modelling Its Risks to the European Environment

L. Hordijk*

NATIONAL INSTITUTE FOR PUBLIC HEALTH AND ENVIRONMENT PROTECTION, PO BOX 1, 3720 BA BILTHOVEN, THE NETHERLANDS

1 Introduction

As the public debate on acid deposition escalates, governments and industry are hard pressed to decide whether to install additional controls on power plants and clean up other potential sources of pollution; to take steps to mitigate possible effects of acid deposition (*e.g.* liming of waterways and soils, development of resistant species of biota); or to wait perhaps five or ten years until there is more conclusive information about the complex relationship between emissions and environmental effects. Acting now to reduce emissions carries the risk that large expenditures will be made with little or no advantages. Yet, hesitation poses the serious threat of irreversible ecological damage that might have been prevented by prompt action. For example, a common policy being implemented in Europe for controlling acidification impacts is a 30% reduction of sulphur dioxide emissions by 1993 relative to the 1980 level. Although such a policy might be costly for many countries, the benefits to the natural environment are not well defined.

Research aimed at improving scientific understanding of the acidification problem has increased drastically over the past few years. But augmenting scientific information will not necessarily lead to the identification of suitable policies for controlling acidification of the environment.

*Former leader of the Acid Rain Project,
International Institute for Applied Systems Analysis,
A 2361 Laxenburg,
Austria.

This chapter also appeared as an article in the March 1988 issue of the journal *Environment* and appears here by permission of Heldref Publications, Washington DC.

This information must be structured in a form that can be used for decision-making based on available scientific evidence and credible judgements about the probability of future events.

At the International Institute for Applied Systems Analysis (IIASA) in Austria the Regional Acidification Information and Simulation (RAINS) model has been developed to synthesize the vast amount of unstructured information about acidification and for dealing with the many crucial uncertainties associated with pollution emissions and their environmental effects. In close collaboration with the UN Economic Commission for Europe and many European and North American policy advisors, RAINS has been set up as a tool to assist decision-making. As far as possible in the model construction, existing models have been used and linked in one single framework. Scientists from many countries have served in the model's review process.

2 The RAINS Model

The emphasis of RAINS is on the transboundary aspect of air pollution. Hence the spatial coverage is all of Europe including the European part of the USSR, and the time period is from 1960 up to 2040. The model is currently sulphur-based because of the principal role of sulphur dioxide as a precursor of acid deposition. However, the model is being expanded to include emissions of oxides of nitrogen and ammonia.[1-4] RAINS consists of three linked compartments: pollutant generation, atmospheric processes, and environmental impacts (see Figure 1).

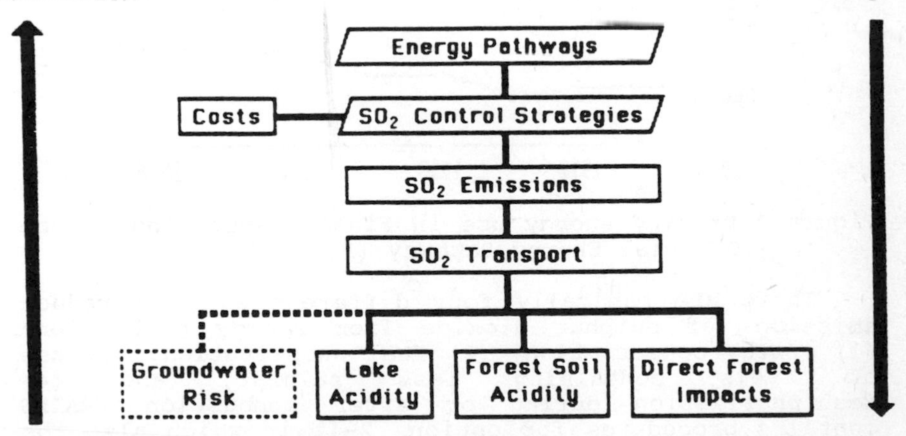

Figure 1 Scheme of the RAINS model

As a starting point the model user has a choice of three possible energy pathways for each of 27 European countries considered (including the European part of the USSR). The first pathway, the 'Official Energy Pathway', is based on recent IIASA studies and international data.[5,6] The 'Maximum Natural Gas Utilization Pathway' investigates the possibility of increased introduction of natural gas, and the 'Nuclear Phase Out Pathway' is based on the assumption that no nuclear power plants are built in Western Europe after 1990. In an interactive fashion a user of RAINS can select one of these pathways and further investigate the potential emission reductions. The energy and emissions submodel accounts for five emissions producing sectors: conversion (*e.g.* refineries), power plants, domestic, industry, and transportation. Moreover eight fuels are considered: brown coal, derived coal (*e.g.* briquettes and coke), light oil, medium distillates, heavy oil, gas, and 'other fuels'. The gas and 'other fuels' sectors are assumed to produce no sulphur dioxide emissions. Figure 2 shows typical output from RAINS energy pathways.

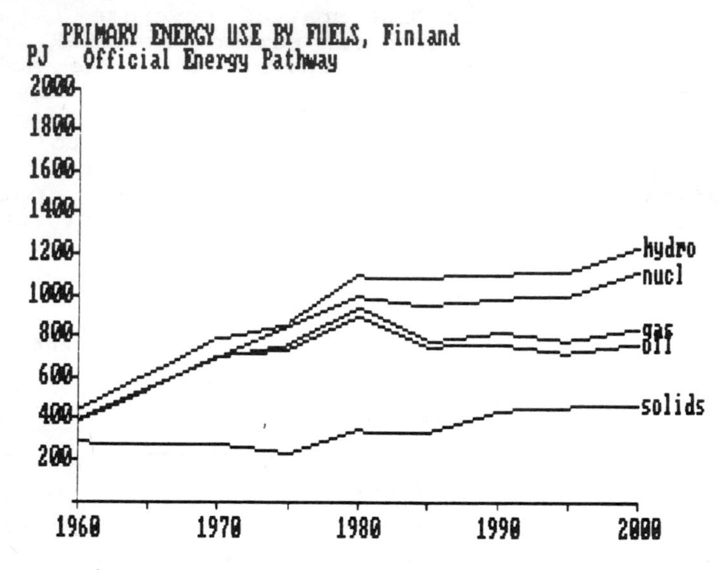

Figure 2 Primary energy use in Finland according to the Official Energy Pathway (PJ)

There are basically four different ways to reduce emissions of sulphur dioxide from energy combustion: (1) energy conservation, (2) fuel substitution, (3) use of fuels containing less sulphur, and (4) desulphurization during or after combustion. RAINS contains procedures for options 2-4, in which also the costs of the policies are estimated. Fuel substitution (option 2) can be performed within ranges which are derived from the differences between the energy pathways and consistency of the energy balance is preserved. Desulphurization during or after combustion (option 4) is described by three technologies: combustion modification, flue gas desulphurization (FGD), and regenerative processes.

Table 1 *Sulphur dioxide emissions (kt sulphur)*

Country	*1980*	*30% reduction*	*Major sulphur controls*
Albania	39	27	15
Austria	159	111	89
Belgium	432	303	142
Bulgaria	508	355	363
Czechoslovakia	1832	1282	592
Denmark	226	158	77
Finland	294	206	100
France	1657	1160	448
West Germany	1602	1121	464
East Germany	2415	1691	996
Greece	345	242	226
Hungary	813	569	352
Ireland	119	83	71
Italy	1898	1328	640
Luxembourg	20	14	12
Netherlands	243	170	155
Norway	72	51	43
Poland	1741	1219	841
Portugal	130	91	91
Romania	757	530	566
Spain	1879	1315	966
Sweden	243	170	100
Switzerland	67	47	38
Turkey	497	348	779
United Kingdom	2342	1639	967
USSR	8588	6012	2878
Yugoslavia	837	586	446
Total Europe	29752	20826	12455
% reduction	-	30	58

Source: International Institute for Applied Systems Analysis, Laxenburg, Austria.

As a numerical illustration of RAINS, Table 1 provides the 1980 emissions of sulphur dioxide together with emissions for the year 2000, assuming that Major Sulphur Controls take place in all European countries. In this scenario FGD with 90% efficiency is applied to 90% of the energy used in the conversion, power plants, and industry sectors. In the domestic and transport sectors fuels containing sulphur are replaced by fuels that contain only half as much sulphur. Together these measures reduce the total European emissions by 58%. To illustrate current policies, Table 1 also shows the resulting emissions after a 30% roll-back in all European countries.

Emissions of sulphur dioxide in one country are often deposited in another country. In RAINS the atmospheric transport submodel computes sulphur dioxide air concentration and sulphur deposition in Europe due to the emissions in each country and then sums the contributions from each country with a background contribution to compute the total deposition and/or concentration in each of the 150 x 150 km grid cells into which Europe has been subdivided. The submodel consists of a transfer matrix based on a Lagrangian model of long-range transport of air pollutants. The model[7,8] on which the transfer matrix is based accounts for the effects of winds, precipitation, and other meteorological and chemical variables on sulphur deposition and air concentration. In Figure 3 deposition patterns of total sulphur in Europe in 1980 and 2000 are shown, for the latter year assuming the Major Sulphur Control scenario. Figure 4 depicts another output mode of the RAINS model, a three-dimensional picture of deposition in 1980 is presented.

3 Soil Acidification

Soil acidification is an important link between air pollution and effects on the terrestrial and aquatic environment. The RAINS soil submodel deals only with forest soils and its working is based on the soil's ability to buffer acid deposition. Soil acidification has been defined as the decrease in acid neutralizing capacity of the soil.[9] The submodel calculates the soil acidity in the 50 cm top layer from the acid load and the buffering characteristics of the soil. The buffering characteristics are divided into 'buffer capacity', the total reservoir of buffering compounds in the soil, and 'buffer rate', the maximum potential rate of the reaction between buffering compounds and acid load.[10-12] Figure 5 presents an example of soil model output in the form of a map of the country-by-country status of soil acidity in the year 2000 resulting from the Major Sulphur Controls scenario and compared with an assumed 30% reduction of sulphur dioxide emissions. The user of RAINS has the freedom to select not only the year but also the pH values. In the example, pH values between 4.0 and 4.3, respectively were chosen. These values reflect a doubling of the hydrogen-ion concentration. As can be noticed in the Figure, forest soils in Central Europe are mainly in lower pH classes. But the Major Sulphur Controls scenario results in higher pH levels in this area compared with the 30% reduction scenario. In Figure 6 it can be noticed that on a longer time horizon the difference between the two scenarios becomes greater.

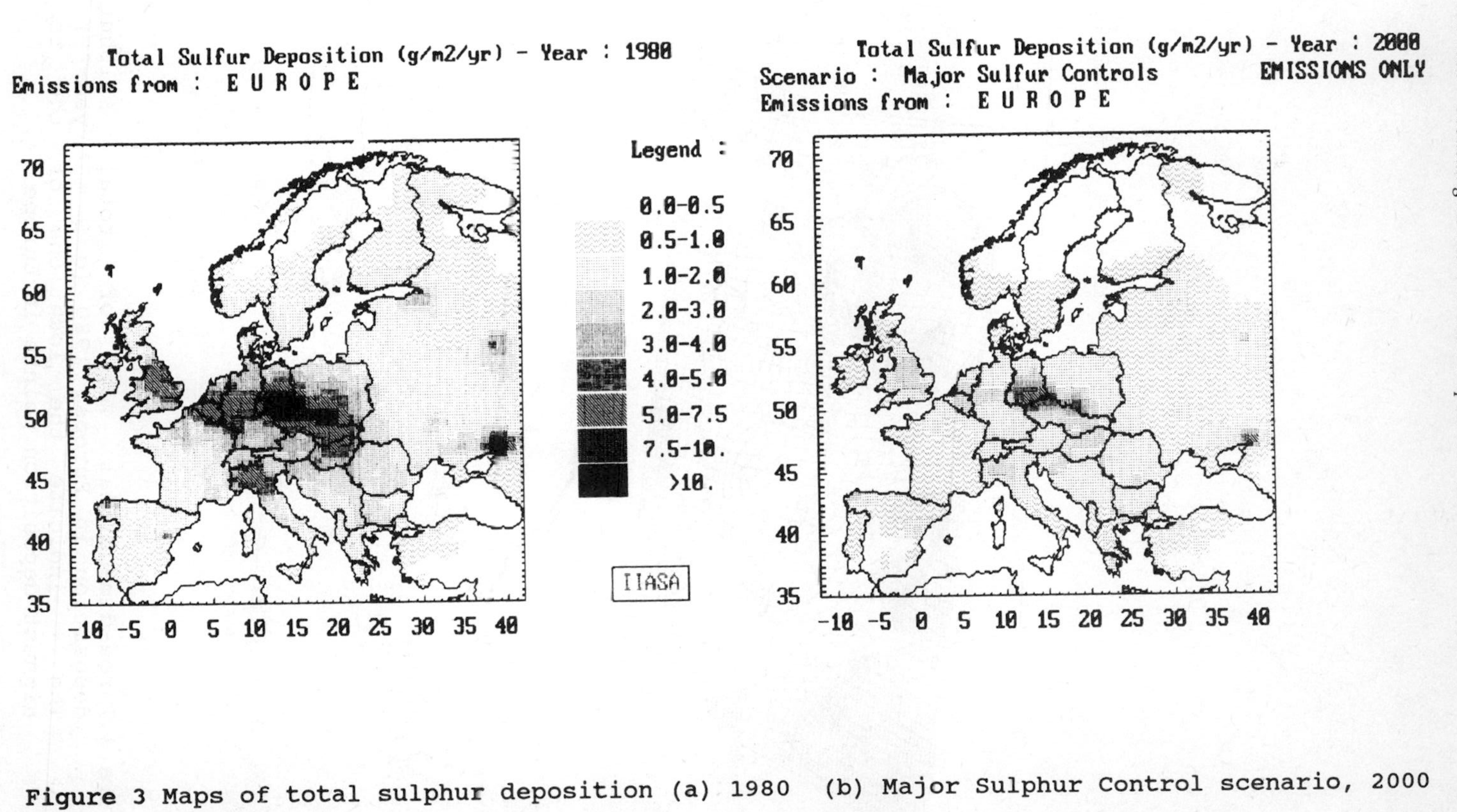

Figure 3 Maps of total sulphur deposition (a) 1980 (b) Major Sulphur Control scenario, 2000

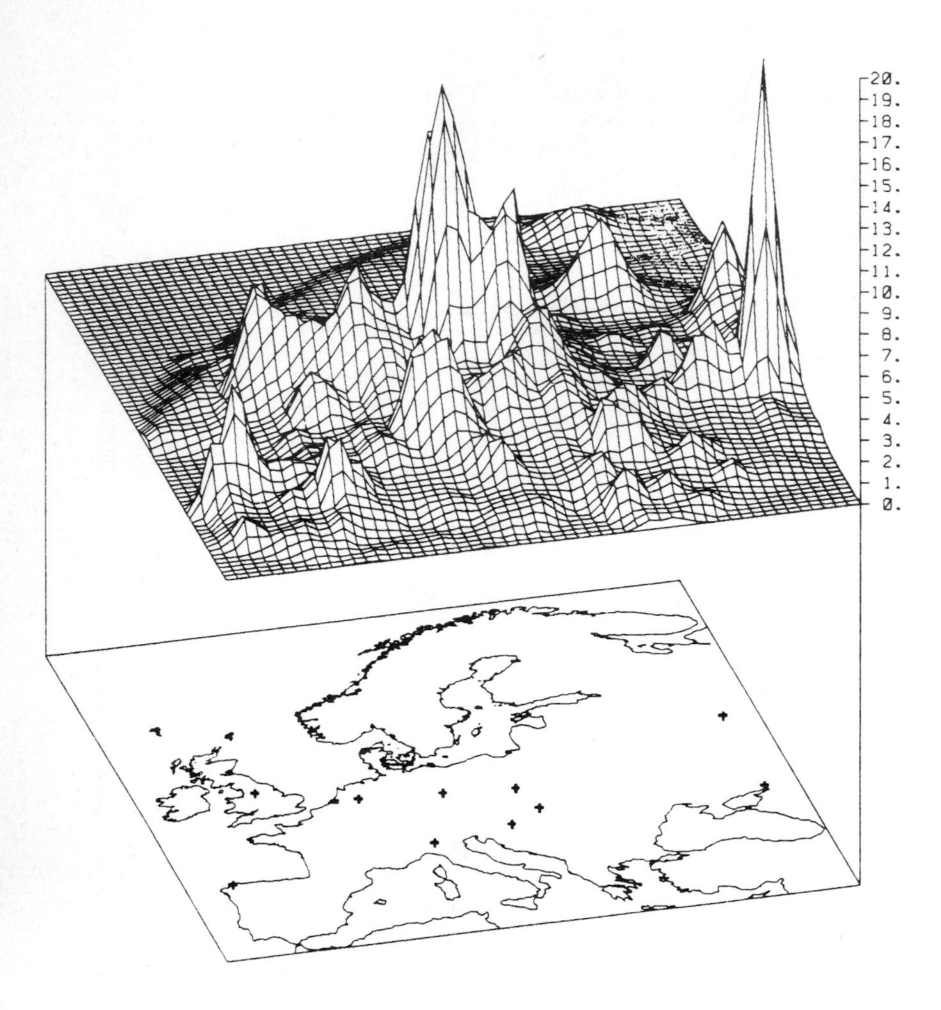

Figure 4 Three-dimensional map of total sulphur deposition in Europe, 1980 (g S m^{-2} $year^{-1}$). The + indicate the locations of the ten highest deposition points in Europe

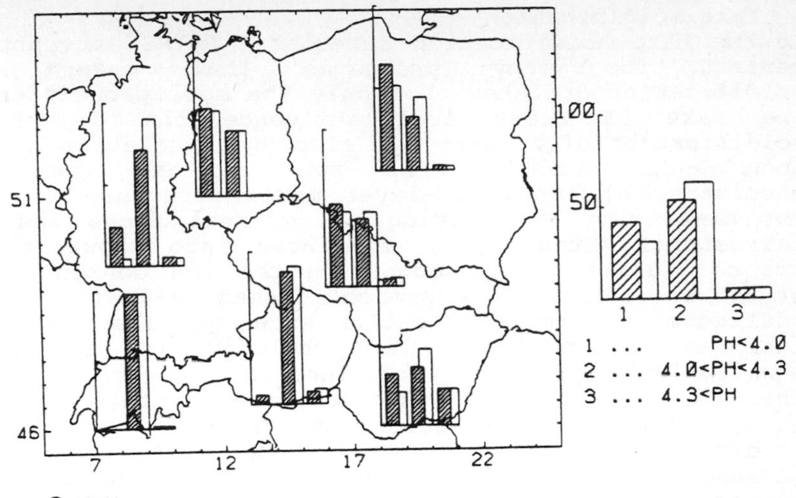

Figure 5 Distribution of Central European forest soils in pH classes for 30% reduction (left bars) and Major Sulphur Controls (right bars) scenarios in the year 2000. The bar chart at the right-hand side gives the aggregated distribution for all countries shown

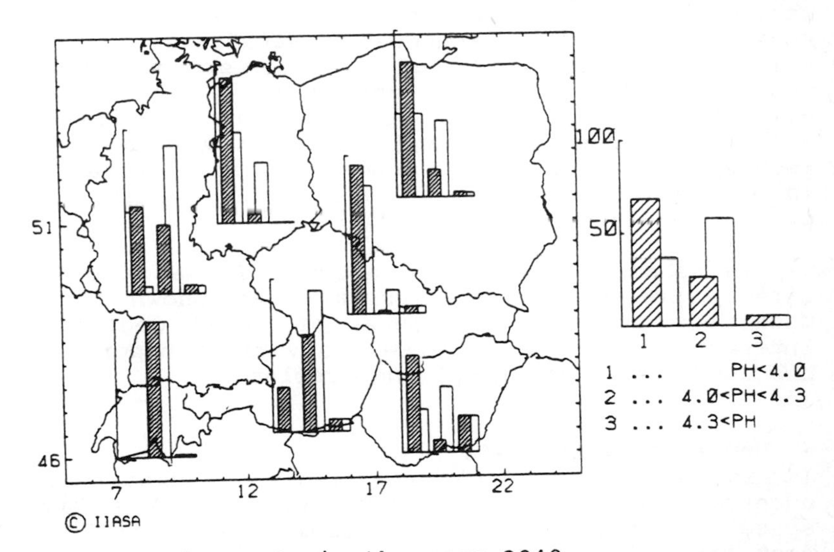

Figure 6 As Figure 5, in the year 2040

4 Lake Acidification

In the Lake Acidification submodel[13] RAINS attempts to reflect the many processes that govern the acidification of lakes. Not only the soil properties of the lake catchment area influence the degree of acidification of a lake, but also such variables as the amount of snowmelt, the run-off, and the lake chemistry. A simple two-layer model structure is used for simulating the routing of internal flows and the terrestrial catchment is segmented into snowpack and two soil layers. For computing the ion concentration of the flows the same approach is used as in the forest soil submodel including cation exchange. The change in lake water chemistry is predicted by means of equilibrium expressions for inorganic carbon species. The ion loads to the lake are mixed within a layer which depends on location and season. The lake acidification submodel has been implemented for Finland, Norway, and Sweden. The output of this submodel is similar to that of the soil model in that for a variable number of lake areas the distribution of the lakes into pH classes (of which the boundaries are set by the user) is shown on a map of Scandinavia.

5 Other Impact Submodels

The reduction of the soils' buffering capacity may also lead to acidification of groundwater. Evidence of this comes from measurements of both wells and surface waters fed by groundwater.[14] In the RAINS model we have implemented a groundwater sensitivity map which produces European maps of aquifer susceptibility to acidification.[15] Input factors to this map include soil type, soil depth, soil texture, aquifer size, mineral composition, and water availability for recharge. This submodel needs further elaboration and is not yet at a stage where solid results can be presented.

The same can be stated about a RAINS submodel for direct forest impacts. Prinz[16] has shown the large uncertainties about the causes of forest decline. The interested reader is referred to various papers showing the approach to the modelling problem.[17,18]

6 Uncertainty Analysis

It is very important to try to evaluate the many uncertainties in RAINS. However, the large time and space scales treated in the model make it difficult to test the models rigorously against field data. So far an extensive uncertainty analysis has been carried out on the atmospheric submodel, results of which have been reported elsewhere,[19,20] and work has started on similar analysis for the emissions submodel of RAINS.[21]

As an example of uncertainty, some results from the atmospheric transport model are presented. Uncertainty comes from various sources including inter-annual meteorological variability, model structure, climate change, and parameter estimation. Together these uncertainties add up to total uncertainty of this submodel that varies over space. As an example of how information about uncertainty is presented to the model user, Figure 7 depicts the uncertainty in the 2 g m^{-2} deposition isoline caused by a ±25% error in computing total deposition. Note that the effect of a constant uncertainty has a very strong spatial variability.

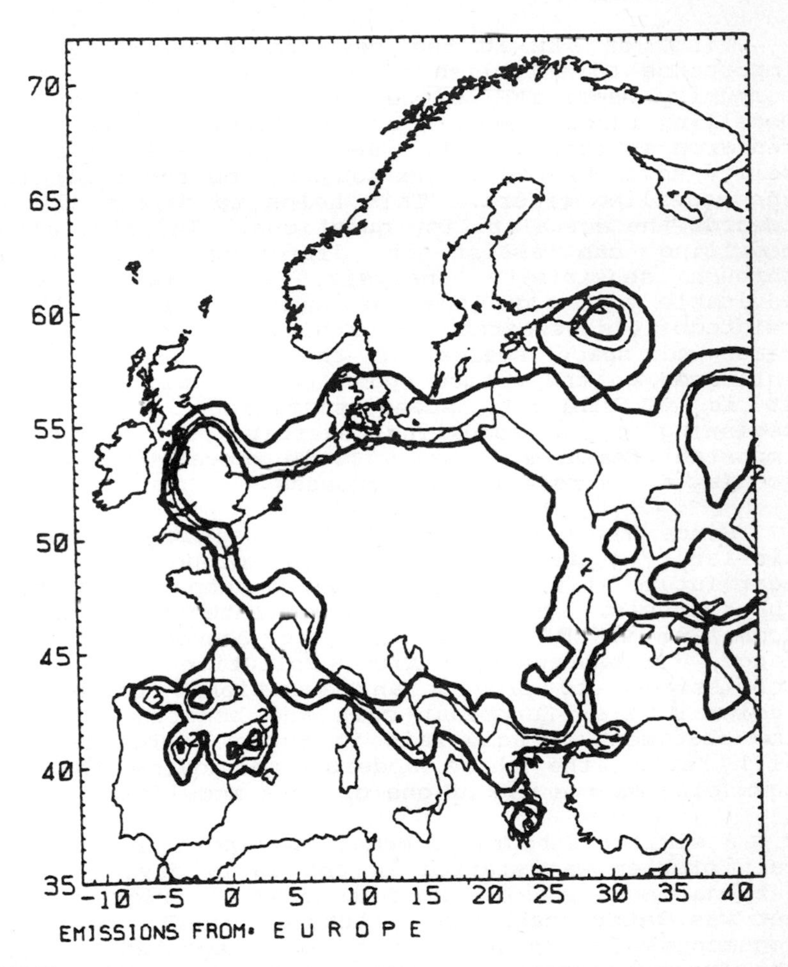

Figure 7 Computed 2 g S m^{-2} $year^{-1}$ isoline of total sulphur deposition (light line) with ±25% uncertainty lines (heavy lines) assuming a 30% reduction scenario (year 2000)

Tentative conclusions from the not yet finished uncertainty analysis are that in many cases model errors seem to compensate and that it is more important to know the range of the parameter uncertainty than to know the type of their probability distribution.

7 Concluding Remarks

In this chapter an attempt has been made to illustrate a model approach to a large-scale environmental problem. In these concluding remarks, some important points are emphasized resulting from modelling acid rain in Europe.

First of all it has been shown to be of utmost importance to establish close links with the research community especially since (at least in IIASA's case) modelling takes place in an institution where no field research is carried out. Secondly, the input from the policy advisors has to be sought from the beginning of the modelling effort. This helps to direct the model towards the actual policy questions. Thirdly, although modelling can assist in directing field research through sensitivity analysis,[22] it is in general advisable that modelling does not move faster than the related field research. Fourthly, the selection of the time and space scales of the model bears a major influence on the entire model construction. Therefore it is advisable to spend sufficient effort in the beginning of a modelling exercise to define these important features. Again, field researchers will have to play a key role in this process.

Since the beginning of the project at IIASA two similar projects have started in Europe.[23,24] In the near future it will be possible to compare results from these models. The Task Force on Integrated Assessment Modelling which comes under the Convention on Long-Range Transboundary Air Pollution[25] has been commissioned to produce an operational framework for regional (*i.e.* European) cost and benefit analysis of the abatement of acid rain in Europe. The Task Force will review the three models and advise the policy community on the use of one or more model(s).

A similar integrated modelling approach to the acid rain problem was started in the early days of the USA National Acid Precipitation Assessment Program (NAPAP), but was later abandoned. Although NAPAP's most recent documents[26,27] do not refer to integrated modelling, the European experience might help to reinforce these very useful activities in the USA. In addition, ongoing modelling work in various European countries including the Netherlands, Norway, Finland, and the German Democratic Republic might help to catalyse modelling elsewhere.

8 Acknowledgements

The author is indebted to his former IIASA colleagues, who have spent so much painstaking effort to build the RAINS model. Colaboration from EMEP, WMO and ECE is gratefully acknowledged, as are the various research grants made available to IIASA's Acid Rain project by organizations in Finland, France, the Federal Republic of Germany, Netherlands, Norway, Sweden, and the USA. Errors and misinterpretations are the author's sole responsibility.

9 References

1. See B. Lübkert, 'A Model for Estimating Nitrogen Oxide Emissions in Europe', IIASA Working Paper WP-87-122, IIASA, Laxenburg, Austria, 1987.

2. R.G. Derwent and K. Nodop, 'Long-range Transport and Deposition of Acidic Nitrogen Species in North-west Europe', *Nature (London)*, 1986, **324**, 356-358.

3. E. Buijsman, H.F.M. Maas, and W.A.H. Asman, 'Anthropogenic NH_3 Emissions in Europe', *Atmos. Environ.*, 1987, **21**, 1009-1022.

4. W.A.H. Asman and A.J. Janssen, 'A Long-range Transport Model for Ammonia and Ammonium for Europe and Some Model Experiments', Insitute for Meteorology and Oceanography, University of Utrecht (Netherlands), Report R-86-6, 1986.

5. UN Economic Commission for Europe, *Energy Data Bank*, Geneva, 1986.

6. International Energy Agency, *Coal Information Report*, Paris, 1986.

7. A. Eliassen and J. Saltbones, 'Modelling of Long-range Transport of Sulphur over Europe: a Two-year Model Run and Some Model Experiments', *Atmos. Environ.*, 1983, **17**, 1457-1473.

8. J. Lehmhaus, J. Saltbones, and A. Eliassen, 'A Modified Sulphur Budget for Europe for 1980', EMEP/MSC-W Report 1/86, Norwegian Meteorological Institute, Oslo, 1986.

9. N. van Breeman, C.T. Driscoll, and J. Mulder, 'Acid Deposition and Internal Proton Sources in Acidification of Soils and Waters, *Nature (London)*, 1984, **307**, 599-604.

10. J. Alcamo, M. Amann, J.-P. Hettelingh, M. Holmberg, L. Hordijk, J. Kämäri, L. Kauppi, P. Kauppi, G. Kornai, and A. Mäkelä, 'Acidification in Europe: a

simulation model for evaluating control strategies', *Ambio*, 1987, **16**, 232-245.

11. P. Kauppi, J. Kämäri, M. Posch, L. Kauppi, and E. Matzner, 'Acidification of Forest Soils: Model Development and Application for Analysing Impacts of Acidic Deposition in Europe', *Ecol. Modelling*, 1986, **33**, 231-253.

12. M. Posch, L. Kauppi, and J. Kämäri, 'Sensitivity Analysis of a Regional Scale Soil Acidification Model', IIASA Working Paper WP-85-45. Laxenburg, Austria, 1985.

13. J. Kämäri and M. Posch, 'Regional application of a simple lake acidification model to Northern Europe', in 'Systems Analysis in Water Quality Management', ed. H.B. Beck, 1986, pp. 73-84.

14. U. von Brömssen, 'Acidification and Drinking Water - Groundwater', in 'Acidification and Its Policy Implications', ed. T. Schneider, Elsevier, Amsterdam, 1986, pp. 251-266.

15. M. Holmberg, J. Johnston, and L. Maxe, 'Assessing Aquifer Sensitivity to Acid Deposition', presented at the International Conference on the Vulnerability of Soil and Groundwater to Pollutants, Noordwijk aan Zee, the Netherlands, 1987.

16. B. Prinz, 'Causes of forest damage in Europe. Major hypotheses and factors', *Environment*, 1987, **29**, 11-15.

17. A. Mäkelä, J. Materna, and W. Schöpp, 'Direct Effects of Sulphur on Forest in Europe - a Regional Model of Risks', IIASA Working Paper WP-87-57, Laxenburg, Austria, 1987.

18. A. Mäkelä and S. Huttunen, 'Cuticular Erosion and Winter Drought in Polluted Environments - a Model Analysis', IIASA Working Paper WP-87-48, Laxenhurg, Austria, 1987.

19. 'Atmospheric Computations to Assess Acidification in Europe: Work in Progress', ed. J. Alcamo and J. Bartnicki, IIASA Research Report RR-86-5, Laxenburg, Austria, 1986.

20. J. Alcamo and J. Bartnicki, 'A Framework for Error Analysis of a Long-range Transport Model with Emphasis on Parameter Uncertainty', *Atmos. Environ.*, 1987, **21**, 2121-2131.

21. J. Alcamo, 'Uncertainty of Forecasted Sulfur Deposition Due to Uncertain Spatial Distribution of SO_2 Emissions', Preprints of the Sixteenth NATO/CCMS International Technical Meeting on Air Pollution Modelling and its Applications, Lindau, FRG, 1987.

22. A sensitivity analysis of the forest soil model has shown that the model is particularly sensitive to base saturation, silicate buffer rate, and a filtering factor representing the fact that forests catch more air pollutants than open land does. For these results see ref. 12.

23. S.R. Watson, 'Modelling Acid Deposition for Policy Analysis', *J. Operations Res. Soc.*, 1986, **37**, 893-900.

24. M.J. Chadwick, 'Co-ordinating Economic and Ecological Goals', in 'The Assessment of Environmental Problems', ed. G.R. Conway, Imperial College, London, 1986, pp.41-49.

25. P.H. Sand, 'Air Pollution in Europe. International Policy Responses', *Environment*, 1987, **29**, 16-20.

26. NAPAP, Annual Report 1986 to the President and Congress, Washington, DC, 1987.

27. NAPAP, 'Interim Assessment. The Causes and Effects of Acidic Deposition', Washington, DC, 1987.

18
Nitrogen Oxides as an Environmental Problem

I. Pollo

TECHNICAL UNIVERSITY OF LUBLIN, UL. HIRSZFELDA 23/6, 20-092 LUBLIN, POLAND

1 Introduction

Among the subjects discussed within inorganic chemistry there is a large group of nitrogen-oxygen compounds with a nitrogen atom in oxidation states from +1 to +6. All of these compounds can appear in the atmosphere as pollutants and most are present in the environment as natural components, although mainly in trace quantities.

2 Environmental Chemistry

However, environmental chemistry often treats mixtures of nitrogen monoxide (NO) and nitrogen dioxide (NO_2) (in equilibrium with N_2O_4) as 'nitrogen oxides, NO_x'. The majority of nitrogen oxides are water soluble, producing nitrous and nitric acids when dissolved. The appearance of nitrogen oxides in nature (*e.g.* during thunderstorms) has always been considered to be a useful phenomenon as it provides a source of soil enrichment. In contrast, the increasing nitrogen oxides concentration from man-made sources is regarded as a cause of serious damage in the biosphere and in the corrosion of building materials. This differentiation indicates that the excess of nitrogen compounds is disadvantageous. As the UN Economic and Social Council data show, the stimulation of plant growth in crops should be expected only when the concentration of nitrogen dioxide does not exceed 10-20 p.p.b. v/v. At concentrations >50 p.p.b. v/v a significant inhibition of growth was observed. Similar investigations have shown that the fertilization of coniferous forests gives the best results only when the nitrogen dose does not exceed 15-20 kg N ha^{-1} per year. In the region where serious damage to forests caused by nitrogen oxides has been observed, the yearly nitrogen deposition exceeds 40 kg N ha^{-1}. Man and animals are not able to assimilate inorganic nitrogen compounds. It seems reasonable to assume that all nitro-derivatives act disadvantageously and, furthermore, it has been proved that several *N*-nitroso-compounds have carcinogenic properties. Nitrogen oxides and the products of their reaction with water, especially nitrous acid and its salts, are agents causing nitration, nitrosation, oxidation, and acidulation. The most dangerous aspect for man and animals is their ability to form nitroso-haemoglobin and methaemoglobin. Infants are the most sensitive in this respect.

The most important man-made source of nitrogen oxides is the combustion of fuels. In some countries the emission of nitrogen oxides comes mainly from mobile sources (cars, aircraft, and also from power plants. Recently, the fraction of NO_x emitted by mobile sources has been shown to be increasing. Apart from nitrogen suboxide (N_2O) which is of relatively low toxicity and not very reactive, nitrogen monoxide, nitrogen dioxide, and dinitrogen tetraoxide constitute the group of N-O compounds which is the best known environmentally. In the atmosphere this group constitutes a very reactive system which is capable of chemical transformation. The presence of water vapour and organic compounds increases the possibility of their participation in different processes which result in the formation of even more toxic substances than nitrogen oxides. This fact should be taken into account when estimating the danger which nitrogen oxides create for man, animals, and plants. On the other hand, each of the nitrogen oxides has its particular chemical and toxic properties. It is significant that man-made sources emit almost entirely nitrogen monoxide (NO). Only after a long period is it oxidized to NO_2, mostly by oxygen but also by ozone and by excited singlet oxygen. On the other hand, nitrogen dioxide (NO_2) easily undergoes photochemical decomposition into NO and an oxygen atom. This process requires near-UV light of wavelength 395 nm. Therefore, even after a long time there is a mixture of different nitrogen oxides in the lower layers of the atmosphere.

On the other hand the oxidation of nitrogen monoxide to nitrogen dioxide by ozone is a very fast reaction: hence the presence of ozone in the air indicates nitrogen oxides pollution and that almost all the NO_x is in the form of NO_2. In the presence of water vapour nitrous and nitric acids are formed.

3 Technical and Legal Limits

The limitation of nitrogen oxides concentration in the atmosphere is the object of technological and legal actions. Formal regulations determine limit values of acceptable nitrogen oxides concentrations in the workplace and in atmospheric air. The comparison of limit values officially accepted in different countries indicates that there is no agreement as to the problem of the highest nitrogen oxides and nitric acid concentrations in the atmosphere which are considered safe for humans.

In the USA nitrogen monoxide (limit value 30 mg m^{-3}) and nitrogen dioxide (6 mg m^{-3}) are differentiated, with the recommendation of 2.5% of these levels (0.75 and 0.15 mg m^{-1} respectively) for urban air.

In the Soviet Union a limit value of 0.085 mg m^{-3} for both oxides is officially accepted for inhabited regions. Polish regulations also do not differentiate between NO and NO_2 and determine both compounds together with a limit value of 0.15 mg m^{-3} for 24 hours average time and 0.022 mg m^{-3} for one year average time.

4 Conclusions

Taking into account that the determination of limit values is connected with the comprehension of 'safe' concentration for man, the absence of any agreement concerning this problem is of concern. Since the atmosphere constitutes a global system, there is a requirement for uniform criteria for regulations concerning acceptable quantities of nitrogen oxides emissions.

Questions concerning natural water contamination by nitrates and nitrites formed during the deposition processes from nitrogen oxide present in the atmospheric air are separately regulated, but include the use of nitrates as fertilizers. However, they belong to the general problem of the nitrogen oxides presence in the environment and their solution is connected with the limitation of nitrogen oxides emission from man-made sources.

19
River Danube Pollution and Its Risk Assessment

P. Benedek
BENECZÚR U. 28, H-1068 BUDAPEST, HUNGARY

1 Introduction

The primary objective of this study is to describe the pollution of a truly international river intersecting eight countries of different political and economic characters. As a secondary objective, efforts have been made to define similarities or even coincidences between water pollution control (often called, rather unfortunately, water quality management) and risk management.

Besides incidental risks, attention is given to the 'accidental' water pollution in three riparian countries (Austria, Czechoslovakia, Hungary), and to the 'intentional' risk problems among the other environmental impact assessments of major hydraulic structures built or under construction in Austria, Czechoslovakia (see Chapter 18), Hungary, Yugoslavia, and Romania. None of these can be appreciated without being acquainted with the major geographical, climatic, and hydrological characteristics of the river.

2 Geography, Morphology, Hydrology, and Hydraulic Engineering Developments

Figures 1 and 2 attempt to give information on the major geographical, morphological, and hydrological characteristics of the second largest river of Europe (only the Volga is larger). The Danube crosses eight countries: Federal Republic of Germany, State of Bavaria, population of the Bavarian catchment: 7 million), Austria (7 million), Czechoslovakia (7 million), Hungary (10 million), Yugoslavia (13 million), Romania (19 million), Bulgaria (4 million), USSR (Ukraine, 1 million); altogether 67 million people live in the catchment area of 817,000 km². The total length of the river is 2857 km, the width varies between 50 and 1200 m, and the depth between 0.8 and 27 m.

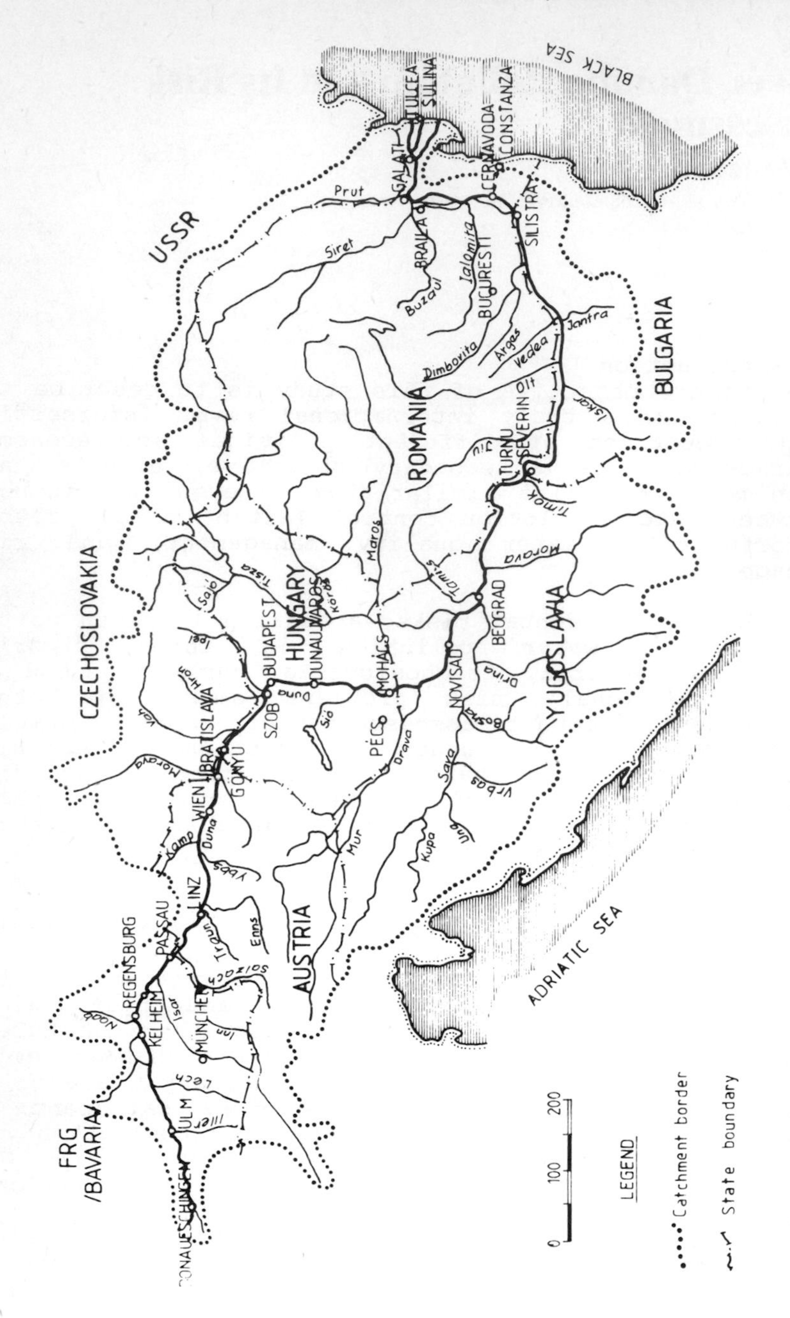

Figure 1 The Danube catchment (817,000 km^2)

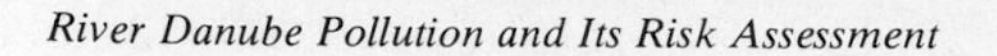

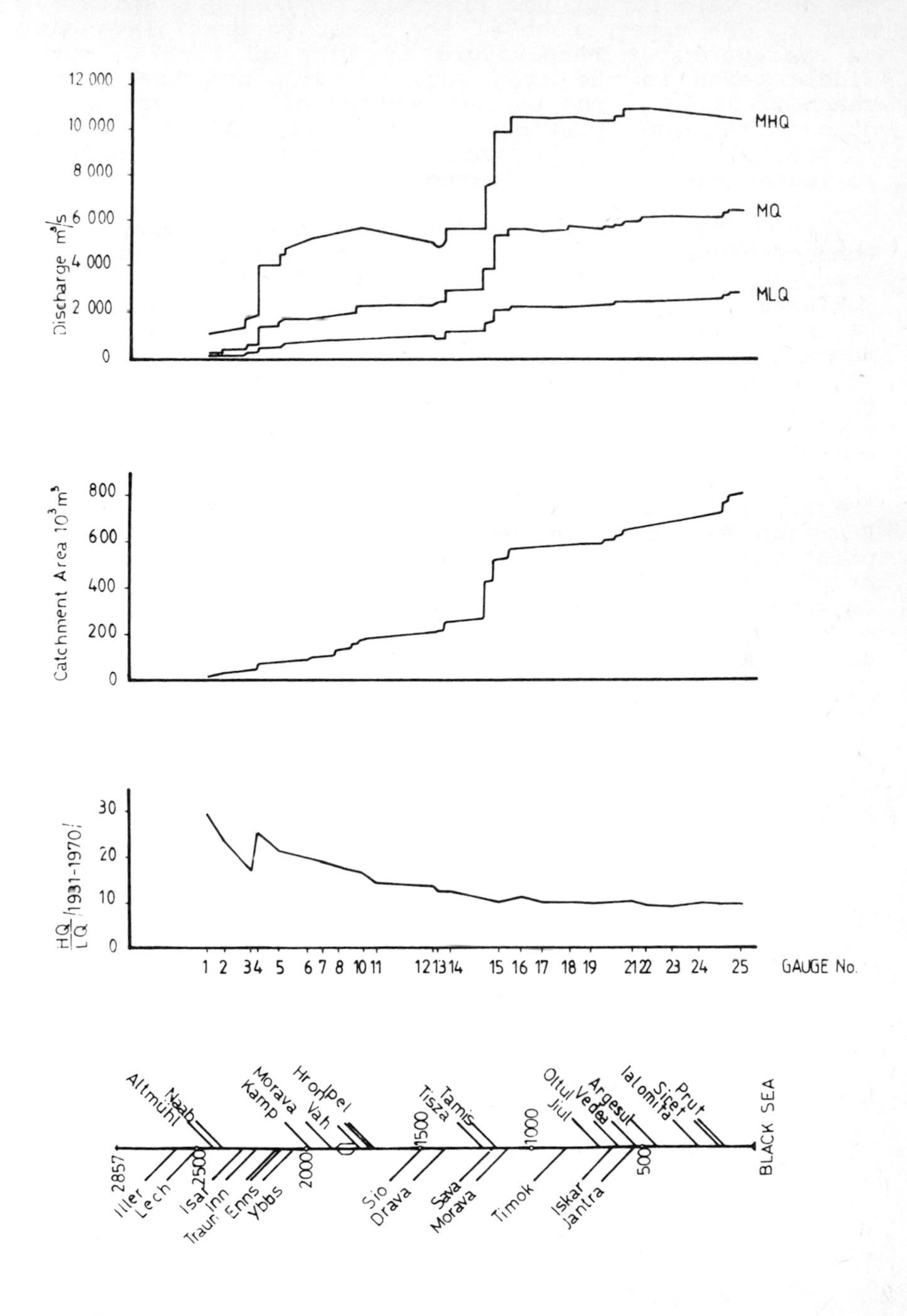

Figure 2 Hydrologic profile of the Danube[1]

The mean velocity of the river is between 0.6 and 2.65 $m\ s^{-1}$. The upper reach of the river to Bratislava has an average water temperature in July of 16-17°C, the middle reach to the Iron Gate 18-22°C, and the lower reach 22-23.5°C. The average period of ice cover is 7 days in the upper, 24 days in the middle, and 28 days in the Delta region, however, these figures are subject to recent changes as indicated in Section 7.

In the Federal Republic of Germany at Donaueschingen, two minor rivers the Brigach and the Breg, originating from the Schwarzwald (Black Forest) coalesce and thence flow across Bavaria as the river Danube (Donau). This confluence is situated 670 m above and 2780 km from the Black Sea. The greatest morphological influence on the river is exercised by the Alpine tributaries. These have enormous bed-load, sediment transport building up alternately flat and steep river-bed sections. Further, the morphology is characterized by narrow channels, where the river traverses steep sided mountain valleys, such as the Bavarian Forest at the German-Austrian border. At this point the river Inn joins the Danube doubling its flow. At the Austrian-Czechoslovakian border the Carpathian tributaries represent a major turning point in the morphological character of the river somewhat downstream from Bratislava. There is a considerable change in the slope of the bottom of the riverbed (the bottom slope decreases from 0.35 to 0.10 $^{0}/_{00}$). Upstream from here a fall of 150 m occurs in about 350 km of the river's length, whereas downstream the same fall of 150 m takes about 1600 km.

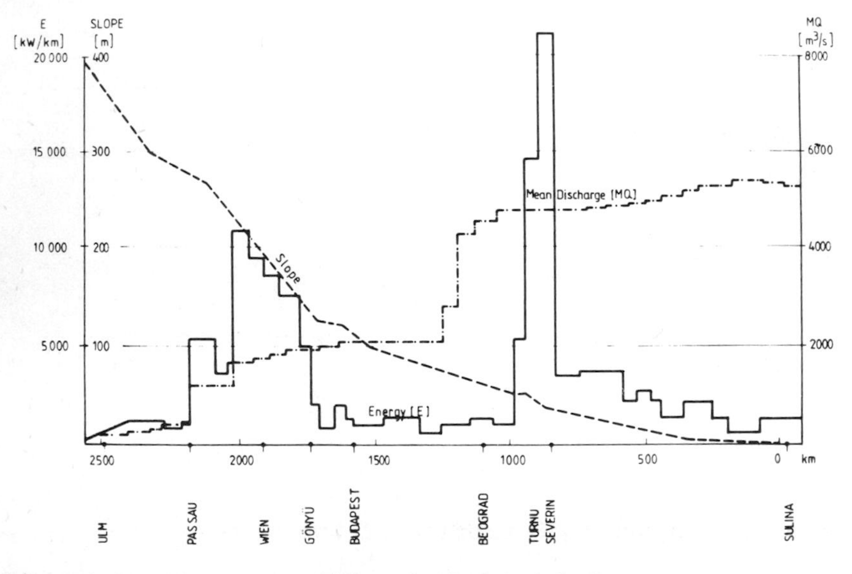

Figure 3 Energy profile of the Danube[2]

There is one more sudden change in the river's slope: along the Iron Gate reach. In Figure 3 the 'energy' line (product of slope and discharge) is shown, underlining the importance of the reaches mentioned for energy potential, river management and navigation.[2] It is important to mention the bed-load and sediment transport. Whereas at Bratislava bed-load transport is 650,000 m^3 $year^{-1}$, above Budapest (Szob) it is only 13,000 m^3 $year^{-1}$. As a consequence of the enormous sediment accumulation downstream from Bratislava, the river has several branches (see later) and even the principal channel is hardly navigable.

Downstream from Budapest, the bottom slope of the river-bed decreases to 0.07-0.04 $^0/_{00}$ while meandering across the Hungarian plain. In Yugoslavia, the largest tributaries, the Drava (stemming from the South-East Alps), the Tisza (a tributary with the greatest subcatchment area), and at Belgrade the Sava, double the mean flow to 5700 m^3 s^{-1}. The greatest breakthrough section is situated across the South Carpathian mountains with roughly 117 km of cataract formation. In the past, the mean width of the river decreased in this section from 800 to 150 m and the average slope was 0.245 $^0/_{00}$. The maximum water velocity was around 5 m s^{-1}. Today, this section is impounded by the Iron Gate (Djerdap) barrage system with a mean velocity of 0.17 m s^{-1}.

From the Iron Gate, the Danube flows across the wide flatland to the Black Sea. Along the common Romanian-Bulgarian section the Romanian bank is flat, wide, forested, and periodically flooded while the Bulgarian bank is steep owing to the adjoining chalk plateau (~ 200m high). The lower section of the Danube is sediment-rich in spite of the impounded section before the Iron Gate. From Sinistra the river turns to the North-East, diverted by the Dobrudsha plateau and the depression zone south from Braila. Here the bed-slope decreases to 0.004 $^0/_{00}$. At Galati, the two great left side tributaries, the Siret and Prut have their confluence, and the Danube reaches the Black Sea through a characteristic Delta formation and with a mean flow of 6500 m^3 s^{-1}. Above Tulcea there are two main branches, the northern Kilija-branch with 62%, and the Tulcea branch with 38% of the total discharge. The latter branch bifurcates. The so-formed middle branch is called the Sulina branch and this is the major navigation route. The Delta has an area of 5640 km^2 below sea level) with many canals and islands, including floating islands, and amphibious land (called 'Plauri') with an enormous nature reserve with rare fauna and flora. The large sediment transport of the river forces the Delta area increasingly into the Black Sea where the land is being reshaped by its marine streams.

The hydraulic engineering activity along the Danube river started in Roman times. Firstly, Emperor Traianus ordered (in 100 AD) the necessary measures (cutting a channel in the narrow gorge of Kazan) along the Iron Gate cataracts to enable (upstream) navigation. Generally speaking almost all the engineering works (river management) in past centuries were aimed at the tasks both of navigation and of flood control. The latter were also dedicated to alleviating excessive ice formation.

Typical river engineering includes transverse dykes built along the convex shore of the river, narrowing down the river-bed and driving the current towards the concave shore. By its silting effect it also contributes to the formation of the convex shore.

3 Water Uses

Since 1856, for the requirement of free navigation, the Danube countries have enacted several conventions and in 1948 the 'Danube Commission' was formed at the Belgrade meeting of the riparian countries. The Danube Commission has its seat in Budapest and deals with all the hydraulic engineering and water resources management problems related to the navigation.

Whereas in 1976, the shipping capacity of the Danube countries was a very modest 3,372,589 tonnes,[2] the estimated capacity in the year 2000 will be 52 x 10^6 tonnes.[3] The Danube Commission prescribes the dimensions of the river-bed and of the ship locks and the barrages to be guaranteed by the States participating in that Commission. It should be mentioned that the Danube Commission is taking important steps for ensuring the interests of international water quality protection concerned with navigation.[4] Among accidental water pollution incidents, oil plays an important role (see later).

Since 1959, a new Main-Danube Canal has been under construction with a bottom width of 31 m and water depth of 4-4.25 m. The sluice-gates are of such a size that ships of 1500 tonnes can navigate them. The section of 204 km between Regensburg and Bamberg (Northern Bavaria) will be supplied with 18 locks. This enormous construction work probably will be finished in the next decade. When in operation 15 m^3 s^{-1} discharge of the Danube will be lost in the Main catchment. In 1984, Romania opened the Danube-Black Sea Canal, 64 km in length, between Cernavoda and Constanza. This canal is 90 m wide and 7.5 m deep, and it decreases the navigation route by 370 km, simultaneously providing the irrigation of agricultural fields of 700,000 ha.

Since 1950, an increasing number of barrages have been built (altogether 49 have been built or are planned) from Ulm (Bavaria) to Cernavoda (Romania).[5] All of these hydraulic engineering works are of multi-purpose character aiming at flood control, low-flow augmentation, ice-free transport, navigation, hydroelectric power production, and in several cases irrigation and drinking water supply. In Bavaria up to 1984 between Ulm and Kelheim a chain of 20 impoundments had been built. Two others in the Regensburg area help the extension of the navigable river section as far as Kelheim. The total hydro-power capacity of these amounts to 250 MW. Between Regensburg and the Austrian border there are six more barrages (partly under construction), but only three of them produce energy, with a total of 100 MW capacity. The last one is situated at the FRG-Austrian border (Jochenstein) with bilateral exploitation of the 130 MW energy potential by the two countries.

The considerable energy potential of the Austrian Danube reach (see Figure 3) whose major part has been utilized by Austria[6] can be seen. From the 350 km Austrian Danube reach, 250 km has been impounded by the construction of eight barrages and at least two more have been planned in the vicinity of Vienna (Wien). The energy-producing capacity of these sections are up to 1923 MW and for the projected total construction 2574 MW.

Figure 4 shows the layout sketch of the Czechoslovak-Hungarian hydropower system Gabcikovo-Nagymaros (GNV) to be completed in the mid-1990s. The capacity of the upper plant is 720 MW and that of the Nagymaros plant 160 MW. The raised water level necessary for the upper plant is ensured by the Dunakiliti transverse barrage (river km 1842) behind which the flow velocity decreases considerably. According to the plans of the hydropower system there will be a 'pulse release operation' which already has been realized at the Iron Gate hydropower station (see later) and it is in an experimental stage along the Austrian Danube reach. The pulse release operation means that the hydropower plant operates within the requirements of the energy consumption. Only the Gabcikovo hydropower plant will operate in this way, while the Nagymaros plant will mitigate these effects.

Considering navigation, the chronic difficulties of this particular Danube reach will be eliminated by the GNV system. The continuous dredging (besides being excessively expensive) did not always overcome these difficulties, even though navigation is continuously increasing.

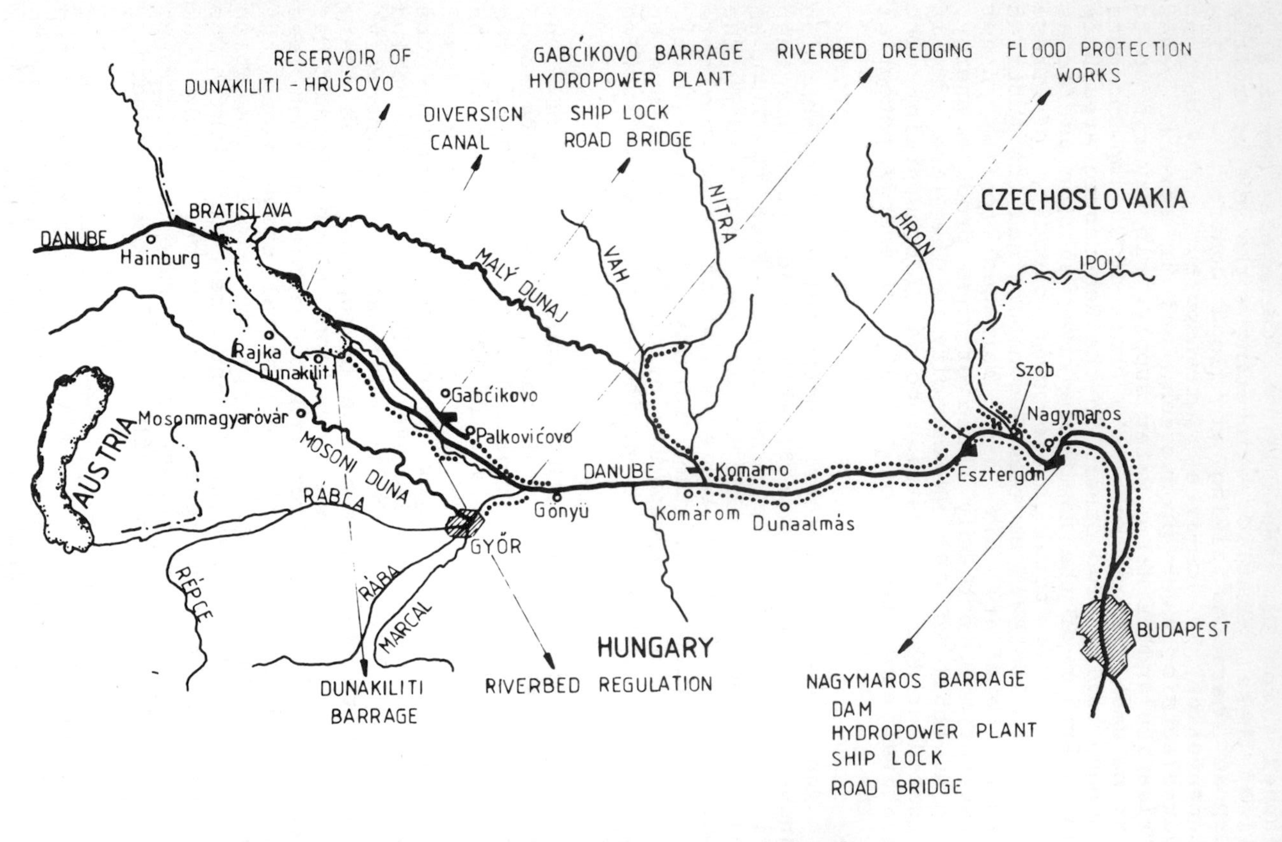

Figure 4 Layout of the Gabcikovo-Nagymaros barrage system[7]

This barrage system will be the most complex along the river. Its by-pass canal of 25 km length (see Figure 4) is planned to discharge 5200 $m^3 s^{-1}$. Apart from the navigational significance and energy production, it has an important task in flood control. It should be kept in mind that in Hungary, a quarter of the total population is living, and a third of the railway system and 20% of the road network are situated in a land potentially inundated by flood; this is defended by dykes of a total length of 4183 km, of which 1350 km are in the Danube valley.

The outstanding energy potential of the river (see Figure 3) is utilized by the hydropower plants Iron Gate Nos. I and II, both constructed by joint efforts of Yugoslavia and Romania.

The hydroelectric plant Iron Gate No. I (Djerdap or Portile de Fier I) at river km 942 is one of the biggest river power stations of Europe. Its maximum discharge is 22,300 $m^3 s^{-1}$, its maximum impounding head is 33.5 m, and its hydropower capacity is 2100 MW, *i.e.* the annual production is 11.3 x 10^8 kWh. The impoundment is about 270 km long (reaching almost the mouth of the Tisza river).[8]

The second Iron Gate plant (Gruia or Portile de Fier II) at river km 863 commenced operation in 1984. Its impoundment abuts Iron Gate I and it works as a balancing reservoir against the pulse release operation of Iron Gate I. It has a hydropower capacity of 364 MW.

Liepolt[9,10] described the Danube's ecosystem and referred to the most ancient water use: fishing. According to his estimate, the annual commercial fish yield from the river is about 4400 t, three-quarters of which is taken in the lower river reach. A further 45000 t $year^{-1}$ of fish is bred in the lakes of the flood plain and in the Delta area. The Author has no reason to discount these figures, but he feels that data from the past decade should be reassessed.

The **drinking and industrial water supply** from the river will be dealt with largely on the basis of a WHO report on a working group meeting held in Budapest.[11] The principal water users, both for supply and discharge of water, are listed in Table 1.

The major part of the drinking water along the Danube is produced from wells sunk into the alluvial sandy gravel deposit of the river. This is the so called 'bank side well filtration' water supply system. Some water works and the industries situated as isolated establishments some distance from the riparian cities use direct water intake. The majority of the

Table 1 *Major municipal and industrial water users in the Danube basin*

*Country (Water consumption)**	*Danube river users (water supply and waste water discharge)*		*Primary tributaries in the catchment polluting the Danube*		*Remarks References*
	Cities with major industries†	*Characteristic industries*	*Rivers*	*Polluting cities +industries+secondary tributaries*	
FRG (170)	Ulm	Pulp+paper, sugar	Lech	Augsburg	Grundremmingen atomic power, 10
	Ingolstadt	Oil refinery			
	Kelheim	Oil refinery, chemicals, sugar	Naab		
	Regensburg	Miscellaneous	Isar		
	Straubing				
Austria (360)	Linz	Iron & steel, chemicals	Traun	Pulp & paper and others	
	Tulln	Sugar			
	Vienna	Misc.	Ybbs		
	Schwechat	Oil refinery			
Czechoslovakia (2200)	Bratislava	Misc. oil refinery	Morava	Sugar	
			Vah	Misc.	
			Hron	Misc.	
			Ipel	Pulp & paper	

Hungary (5400)	*Győr*	Iron&steel,	Sió	Chemicals	Paks atomic power
	Budapest	Misc.			
	Százhalombatta	Oil refinery			
	Dunaujváros	Pulp & paper, Iron & steel			
Yugoslavia (1500)	*Novi Sad*	Misc.	Tisza	Szeged (H), Mures (RO)	Vinca atomic power
	Belgrade	Oil refinery, petrochemical	Sava	Zagreb, Tuzla	Krsko atomic power, 10
		Misc.	Morava	Krusevac, Svetozarevo, Nis	
Romania (14000)	Giurgiu	Not known	Jiu	Craiova, chemicals	
	Braila	Not known	Arges	Pitesti, Bucuresti	
	Galati	Not known	Jalomita	Ploesti, petrochem.	
			Siret	Misc.	
Bulgaria (6002)	*Ruse*	Not known	Iskar	Sofia	Kozlodni 1760 MW atomic power
USSR (1129)	Izmail	Not known	Prut	Not known	Only guessed by author, more information needed

* Estimated water consumption from the Danube in limits of 10^6 m^3 $year^{-1}$ in 1980[5]

† Cities over 100,000 population are printed in *italic type*
Cities over 500,000 population are underlined
Cities over 1,000,000 population are printed in **bold type**

direct industrial water intake is for use as cooling water. This is particularly large in the case of atomic power plants. The greatest user of the Danube water for drinking and industrial purposes is probably Hungary. Almost half of the total water requirement of the country comes from the Danube, 80-90% through bank side filtered wells (for Budapest alone ~1 x 10^6 m^3 day^{-1}).

Though the majority of the bank-wells yield good quality water currently, and usually there is no other treatment than routine chlorination, several case studies collected by a WHO Working Group indicated an ever increasing pollution threat to the future use of this water resource.[12]

A survey of the quality problems of water abstraction from bank-filtered wells is summarized in Table 2. Many problems are attributed to the pollution of the bottom deposit. In an anaerobic environment not only iron and manganese, but the micropollutants adsorbed on to the suspended solids can be remobilized, and although as yet routine monitoring does not indicate any problems in the river's water, health-damaging pollutants might appear in the drinking water.[13,14] At present, this is only a local problem along the Danube arising particularly at places where there is a sewage outfall into the river section above the wells. The fast settlement taking place within the sewage plume along the shore is only increased by the 'transverse dykes' for river management. The investigation of this phenomenon itself is important, but along the impounded river reaches it also requires a more intensive analysis of the efficiency of sewage treatment.

Along the Danube, there are also several surface water intakes for drinking water supply purposes. The largest one is at Ulm (FRG) supplying Württemberg province with 2.4 m^3 s^{-1} water.[10] In Hungary, the Budapest Metropolitan Water Works has a direct intake of 2.3 m^3 s^{-1} capacity and there is another one near the Yugoslav border (Mohács) supplying Pécs (180,000 inhabitants) with water. These surface water intake systems face several problems. Because of ammonia pollution (mainly in winter) and the algae content of the river, its water has to be chlorinated continuously prior to the sedimentation-filtration unit. As a consequence, there is an increased formation of trihalomethanes (THM). In the Danube, there are other THM-precursors, for instance, the lignin-sulphonic acid pollution by the pulp and paper industry of the Austrian and the Slovak Danube reaches. This pollution appears in the chemical oxygen demand (COD), but, simultaneously, has to be considered as a THM-precursor. Considering other industrial pollutants,

Table 2 *Water quality problems of water supply from bank-filtered water along the Danube*[11]

Source and type of pollution	*Pollution from the river side*	*Pollution from the background*
1. Polluted organic bottom deposit and re-mobilization of heavy metals (River bed engineering)	Water works of Csepel (down stream Budapest) Szentendre Island (upstream Budapest) (H)*	
2. Dissolved organic pollutant	Nussdorf (A) 1976 Lobau (A), 1981/82 Bratislava (Cz)	Chemical waste deposit, Vác, 1981 (H)
3. Nitrate (downstream Budapest)		Water Works of Csepel pollution of agricultural origin (H)
4. Sediment settlement on impounded river reaches (similar effect to point 1 because of decreased dissolved O_2)	Goldwörth near Linz (A) 1974, Belgrade (Yu)	
5. Excessive dredging of river bed (similar effect to point 2, because of decreased filter capacity	Szentendre Island (H)	

*A = Austria, Cz = Czechoslovakia; H = Hungary; Yu = Yugoslavia the traditional water treatment technology of Danube water satisfies less and less the hygienic requirements; hence expensive active carbon treatment or slow sand filtering and groundwater recharge has to be introduced.

Significant water use has resulted in the disposal of municipal and industrial wastewater into the river. As a carrier of sewage, the Danube - at least from the confluence with the Inn, *i.e.* from the FRG/Austrian border - has an impressive assimilative capacity. This relates mainly to the biodegradability of waste constituents (see Table 3 below). The problem is that the riparian states have abused this 'power' (it is also known as 'dilution power') of the river and have discharged raw sewage into it. In the past decade, several treatment plants have been built (Linz, Vienna, and a part of Budapest) but in the case of industries (mainly pulp-paper mills, iron and steel works, chemical plants, sugar factories, *etc.*: see Table 1) the situation is not improving. The tributaries are usually more polluted than the Danube and carry considerable waste load. The following parameters of the wastewater discharges are characteristic.[15,16]

(1) Wastewaters (and polluted tributaries) are not able to mix with the bulk of the flow, and this plume effect can, sometimes, be observed 100 km downstream.

(2) Raw wastewaters may cause enormous suspended solids deposition and bacterial contamination locally and immediately downstream of the discharge points.

(3) Organic and inorganic micropollutants (*e.g.* chlorinated hydrocarbons, heavy metals, *etc.*), mainly of industrial origin are generally increasing in the river water, and are usually not included in the routine monitoring procedures.

(4) Many accidental pollutions (oil, oil derivatives, toxic wastes, *etc.*) are registered, menacing water intake works which are without any automatic early warning system.

As far as irrigation is concerned, this is a general water use along the entire length of the river with increasing significance towards the lower reaches which is plausible when considering the climatic conditions of that part of Europe.

In the FRG irrigation from the river is only utilized to a minor extent (a few hundred hectares for vegetable growth). In Austria the irrigated land is at present 50,000 ha, with a projected increase to 130,000 ha. It should be mentioned that in almost all the riparian countries, irrigation is mostly exercised from the tributaries of canals established directly from this purpose (*e.g.* the Vojvodina of Yugoslavia) and as a direct intake from the river.

Another important agricultural water use of the Danube (and its tributaries) is the drainage of inland water, from precipitation but particularly from snow melting. This is itself a significant source of non-point-river-pollution, which greatly contributes to the increasing trend of nitrate and phosphate content of the river water (see below).

In the Slovakian part of Czechoslovakia 800,000 ha is irrigated, in Hungary 437,000 ha (one third directly from the Danube, the remainder from Tisza). In the latter country, a lowland, drainage is also an important river water use. In Yugoslavia (Vojvodina), the situation is similar: 760,000 ha is drained while 360,000 ha is irrigated (data from 1972). The possible extension of the irrigated area is one million ha.

In Bulgaria irrigation is a historical tradition. From the Danube, 128,000 ha is directly irrigated by pumped water. In some cases the water has to be pumped to an elevation of 100 m. In Romania the irrigated lowland along the Danube is about 700,000 ha.

Much of the above information, unless otherwise indicated, was abstracted by the Author from the monograph 'Die Donau und ihr Einzugsgebeit',[1] which is the result of recent co-operation of the riparian countries.

4 General Water Quality Evaluation from Available Publications

As a tentative approach to the general water quality characterization of the Danube, Table 3 represents the most recent data collected from different sources as indicated in that Table. These are consequently inhomogeneous in their character because of the different years of observation and the different methods in sampling and analysis.

The dissolved oxygen content is generally near to the saturation value with a few exceptions as indicated in Table 3. The COD values are traditionally determined by the potassium permanganate method, though in most countries the dichromate method is used.

It should be mentioned that downstream of the major polluters there are drastic short-term deteriorations in river quality not reflected in Table 3, bacterial contamination being particularly acute.

Table 3 *Water quality characteristics of the Danube river*

Country	*River* km	*City*	*Year of data*	*Conductivity* $\mu S\ cm^{-1}$	Cl^- $mg\ dm^{-3}$	*Dissolved* O_2 $mg\ dm^{-3}$	BOD_5 $mg\ dm^{-3}$
FR Germany	2588	Ulm	1979–1985	200–400	25		
	2510	Donauwörth	1984				2 - 5
	2460	Ingolstadt	1984				
	2398	Bad Abbach	1983	475	24.5	10.9	3.6
	2203	Jochenstein	1983	347	17	10.9	3.8
Austria	2131	Linz	1985	369	17	10.4	2.1
	2060	Ybbs	1985	368	17	10.4	2.5
	1920	Vienna	1983	370	17	9.9	3.2
	1873	Wolfstahl	1985	378	10.4		3.9
Czechoslovakia	1873	Karlova Ves	1985	407		10.3	4.5
	1806	Medvedov	1977–1986				3.5
Hungary	1848	Rajka	1981–1986	430	21	10.4	4.4
	1708	Szob	1981–1986			10.3	
	1654	Budapest N.	1981–1986	430	27	10.7	5.3
	1629	Budapest S.	1981–1986			10.5	6.3
	1480	Baja	1981–1986	480	29	10.3	6.4
Yugoslavia	1424	Bezdan	1974–1980			9.94	4.73
	1173	Belgrade	1984			6.5–10.7	2.4–9.0
	943	Iron Gate	1974	323–383		6.1–12.4	3.1–7.7
	861	Prahova	1974–1980			8.90	2.49

Bulgaria	845-375	Bulg.reach	1959-1963	*	9.5-26.5	5.0-12.0	
	845	Novo Selo	1975-1983			6.3	
		Nikopol	1975-1983			6.9	
	495	Rusze	1979			8.2-9.5	2.28
	375	Silistra	1975-1983			7.5	4.38
Romania	1073						
	694						
	244						
	150	Galati	1982		47	9.96-11	2.2-2.5
	176	Braila	1967		10-24		
	2	Sulina	1967				
USSR		Kilija	1979-1980			6.1-11.9	
		Kilija	1984				0.4-7.0

*The conductivity fluctuates in the range 300-400 $\mu S\ cm^{-1}$ to the estuary of the river

Table 3 continued

Country	*River*	$KMnO_4^-$	NH_4^+	NO_3^-	PO_4^{3-}	*Saprobity index*	*Reference*
	km	mg dm^{-3}	mg dm^{-3}	mg dm^{-3}	mg dm^{-3}		
FR Germany	2588			17	0.2		17†
	2510					1.81-2.8	18†
	2460					2.71-3.2	18
	2398	19.2	0.25	12		2.3	19†
	2203	18.4	0.30	10		2.4	19
Austria	2131	19	0.29	12.1	0.44	1.81-2.3	20†
	2060	18	0.38	11.2	0.59	1.81-2.3	20
	1920	21	0.37	10.8	0.69	2.31-2.7	20
	1873	18	0.41	10.2	0.76	2.31-2.7	20
Czechoslovakia	1873	25	0.51	11.6		2.31-2.7	20
	1806	25	0.5	10	-	2.4-2.6	21
Hungary	1848	22	0.62	9.5	0.58	2.7	22†
	1708	28.8	0.57	10.2	0.44		22
	1654	27.6	0.47	9.8	0.41	2.6	22
	1629	28.8	0.54	10.2	0.44	2.7	22
	1480	27.6	1.00	11.5	0.59	2.6	22
Yugoslavia	1424	29.4					23
	1173	16-32	0.1-0.2	3-12			24
	943			1.0-7.3	0.1		9
	861	21.8					23

Bulgaria	845-375	18-57		1-10	0.10		25
	845						26**
							26**
	495						26,27**
	375						26,27**
Romania	1073		0.36	3.4	0.31		28
	694		0.30	3.7	0.35		28
	244		0.38	4.3	0.50		28
	10		0.32	3.3	0.42		28
	150	19.3-22.8	0.4-0.6				29
	176	18-47		0.4-4			**
	2	17-53		0.5-2.9			9
USSR						1.85-2.48	
		10-40	0.1-1.5	0.1-2.0	0.01-0.29		30

† Yearly averages
** Random data

In Czechoslovakia and Hungary, the Gabcikovo-Sturovo common Czech-Hungarian Danube section was investigated from the viewpoint of bacterial contamination.[31,32] Mycobacteria indicated that the agricultural non-point source pollution originated from large-scale animal farms. While the chemical parameters showed a first/second quality class on the basis of bacterial contamination, this section of the Danube is in the third/fourth quality class. This pertains to the influence of the Bratislava and Győr wastewater discharges and that from the more polluted tributaries.

Very limited data exists on viral contamination, though this should be noticeable below large wastewater discharge points. Small amounts of *Coxsackie* viruses were detected in the Czechoslovakian Danube section.[33]

Beyond the routine measurements of bacterial contamination of the river there are several studies dealing with local problems, *e.g.* the *Salmonella* incidence in the Austrian impounded river sections.[34] Whereas in the water there were positive samples in only 2.3% of the total number, *Salmonella* could be isolated in 27.4% of bottom sediment samples. These originate partly from raw wastewater, but also from treated samples and from aquatic birds. In the bottom sediment *Salmonellae* are of long living strains and during the turbulent flood period they might be repeatedly introduced to the water body.

At Galati in Romania during a sampling period in 1982 the *Escherichia. coli* bacteria were in the order of magnitude of hundred thousands per litre and some samples were *Salmonella* positive.[29]

The general water quality picture depends on the flow and climatic conditions. The Danube is largely a river of Alpine character with discharge variations of more or less similar yearly pattern due to snow melt and regular rainy seasons. This has enormous influence on the suspended solid load and oxygen regime of the river. In winter conditions, with ice cover, there is a definite effect on biogenic components, nitrogen compounds, favouring enhanced NH_4^+ content (above 1 mg dm^{-3} in Hungary during the winter season); as a consequence the mean annual values in Table 3 are subject to marked seasonal variations.

5 Heavy Metals and Hazardous Organics

Publications in the riparian countries ranging from the FRG to the Soviet Union indicate the presence of heavy metals in the water and in the sediment. Table 4 shows the incidence and concentrations of several metals in various Danube sections. It can be seen that, apart

Table 4 *Heavy metal content of the Danube river*

Country	*River* km	*City*	*Water concentrations* $\mu g\ dm^{-3}$ Cd	Cu	Hg	Pb	Zn	*Remarks,*	*References*
			<0.1	0.5-3	0.03	0.2-1	5-20	Background values,	35
FR Germany	2423	Weltenburg	<0.2	3.1	<0.3	<3	32	1984-85 mean values,	35
	2412	Kelheim	<0.2	5.2	<0.5	<3	102	1984-85 mean values,	35
	2240	Passau	<0.3	4.4	<0.3	<3	39	1984-85 mean values,	35
	2203	Jockenstein	<0.1	3	0.4	1	<10	1983 mean values,	19
Austria	2203-1873	Austrian river reach	0.63		0.07	3.66	34.4	1976-79 mean values,	36
			0.13		0.06	1.23	12.8	1984 mean values	
Hungary	1708-1430	Hungarian river reach	0.1-0.5	3-5	0.1-0.3	1-2	20-40		13
									13
Yugo-slavia	1116	Smederevo	1-1.9		0.6-0.8				37
	955	Tekija	0.3-0.6		0.2-0.3				37
Bulgaria	845-375	Bulgarian reach		0.01		0.003-	0.02	Data of 1959-63	9
				0.03		0.012	0.13		
Romania	1073				0.72		0.37		28
	694				0.57		0.25		28
	244				0.27		0.23		28
	150	Galati			0.24-3.86				29
	10				0.39		0.10		28
USSR	0-117	Kilija arm of Danube			0.03-0.06			1979-84	38

from local problems downstream from major polluters, generally there is no problem in the water body of the river. However, in the above-mentioned sections it is quite obvious that the bottom sediment is heavily contaminated.[39] Part of the contaminated sediment is transported further and dissipated in the lower sections by major floods resulting in the metals in the sediment forming a permanent sink (Table 5 and Figure 5).

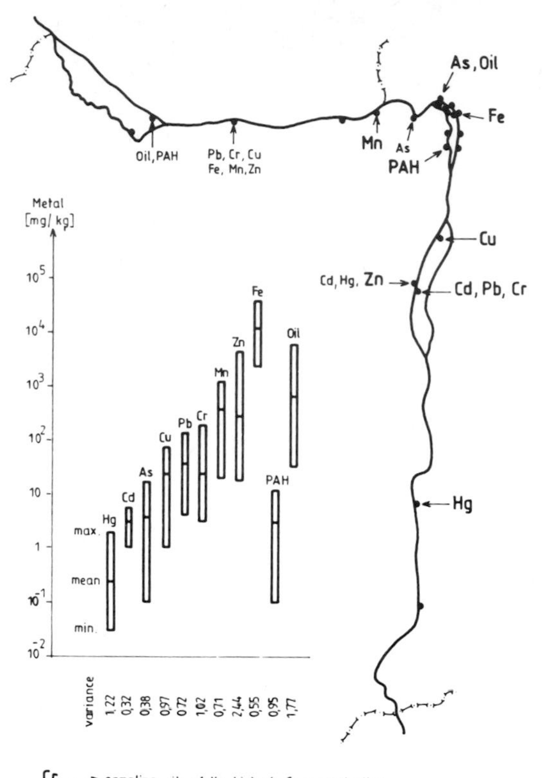

Figure 5 Characteristic values of different pollutants in the bottom sediment along the Hungarian river reach of the Danube[13]

It is interesting to know what is the final fate of the heavy metals in the liquid and solid phases at the Delta of the Danube, and what is transported to the Black Sea. Kozlova presented a very interesting report

Table 5 *Heavy metal content in the sediment of the Danube river*

Country	*River* km	*City*	*Sediment concentration* mg kg^{-1} Cd	Cu	Hg	Pb	Zn	*Remarks*	*References*
FR Germany	2423	Weltenburg	0.40	28	0.55	16	106	1984-85 mean values	35
	2412	Kelheim	0.30	19	0.61	15	129		
	2240	Passau	0.17	19	0.23	14	109		
Austria	2203-1873	The entire Austrian reach	0.68		0.39	68	246	1984 mean values	36
Hungary	1708-1480	See Figure 5							
Yugo-slavia	1116-940	Iron Gate impoundment	1.5	72	0.4	106	325	1972-77 mean values	40
			2.0-4.8		0.8-1.9		244-484		41
USSR	0-117	Kilija arm of Danube			0.2-0.7* 0.7-1.2†			1979-84* in the Delta †in the zone of hydrofront (sea)	38

on the fate of mercury in the Kilija Delta.[38] The majority of mercury, *i.e.* 60-95%, is in suspension and the amount changes with the seasons (flood). The yearly mercury transport of the Danube to the Kilija Delta is 40 t, out of which 5.4 t is in dissolved form and 34.6 t bound to suspended solids. Of the bound portion 6.0 t remains in the Delta bottom sediment and 28.6 t reaches the sea. However, owing to the drastic salt content change in the hydrofront, the majority of the mercury accumulates in this hydrofront zone (see also Table 5).

Table 6 and Figure 5 indicate a relatively low level of organic micropollutants.

From the papers collected by the Author at the Bavarian Institute of Water Research (Bayerisches Landesanstalt für Wasserforschung, München) in 1985 it was realized that in 1984 in the Bavarian Danube the concentration of volatile chlorinated hydrocarbons fluctuated between 1 and 15 $\mu g\ dm^{-3}$ and that of poly-aromatic hydrocarbons (PAH) compounds between 10 and 510 ng dm^{-3}. The accumulation factor of PAH in sediment was 10^4-10^6 (see also Table 6).

In the Czechoslovakian Danube Section, mostly downstream from Bratislava, chlorinated insecticides were identified in the river and in the adjacent groundwater in the order of magnitude 10^{-7}-10^{-9} g dm^{-3}. The concentration of those pesticides has not changed during the past ten years.[33]

Further, it might be added to Table 6 that in the Yugoslavian Danube during the observation period 1979-1982 the pesticides Lindane, Aldrin, Dieldrin, and DDT were found in >40% of samples.

The lack of information on organic micropollutants can be mainly regarded to be due to a lack of the necessary instrumentation, most of the laboratories being unable to detect harmful substances in low concentrations.

6 The Multiple Effects of Barrages on the River Ecosystem

The biota of the river have been investigated in all of the riparian countries, mainly with the co-operation of the Working Association of Danube Researchers (IAD) seated in Vienna.[9] It is not feasible to review the enormous amount of work done in the IAD; however, the following points reflect the effects that the barrages may have on the biota.

Table 6 *Organic micropollutants in the Danube river*

Country	*River km*	*City*	*Year of observation*	*Occurrence of contamination*					*Remarks*	*References*
FR Germany	2203	Jockenstein	1983	$CHCl_3$	CCl_4	C_2HCl_3	C_2Cl_4	$C_2H_3Cl_3$		
				0.4	0.04	0.5	0.5	0.1*	*Mean values in μg dm^{-3}	19
				1.8	0.70	0.9	0.7	0.8†	†90% duration values in μg dm^{-3}	
Austria	2203-1873		1973-80	Extractable material 0.7-1.2 mg dm^{-3}						42
Czecho-slovakia	1840	Dobrohost	1976-84	Petroleum hydrocarbons in tenths of mg dm^{-3}, Chlorinated hydrocarbon insecticides 10 μg dm^{-3}, *o,m,p*-xylenes, ethylbenzene, toluene, naphthalenes 0.10 - 10 μg dm^{-3}						11
				Phenols	ANA-detergents	Mineral oil				
Hungary	1848	Rajka	1984	0.004	0.074	0.36			in mg dm^{-3}	22
	1659	Budapest	1984	0.004	0.12	0.14			For sediment contents	
	1452	Mohács	1984	0.003	0.15	0.34			See also Figure 4	
Romania	150	Galati	1982	Pesticides: HCH 0.145 μg dm^{-3} DDT 0.365 μg dm^{-3}						29
	1073-10		1981-85	Phenols: 0.02-0.03 μg dm^{-3} DDT 0.239-358 μg dm^{-3}						28

The multidisciplinary character of a river's ecology is very dependent on the construction of a barrage across the river, creating an impoundment. Anthropogenic effects are now abundant: river regulation, flood control, water uses, change in water temperature, pollution, *etc.* However, all of these changes develop successively and then the living world of the river has an opportunity to adapt. When a barrage blocks the old river bed, it takes decades for a new ecosystem to develop.

The basic problem in the case of a barrage is of hydraulic character, *i.e.* the drastic decrease of flow velocity in the impoundment. It has many consequences. First there is the higher rate of settling of suspended solids and blocking of the bed-load transport. This results in a greater transparency of the water and greater extremes in water temperature, both between the various seasons and through the bulk of the impounded water body. Longer detention time, greater transparency (more light), and possible remobilization of nutrients in the bottom sediment (anaerobic conditions) might have a consequence in the trophic character, as well as saprobic conditions. Benthos and biotekton (live coating of the surface of bottom formations) find completely new environments in the more abundant and finer bottom mud.

It should be borne in mind that all of the above phenomena are influenced by the fluctuating discharge conditions of the river in the seasonal flood and low-flow periods. Further, below the barrage, the old river bed is also subjected to changes mainly because of the reduction in the bed-load and in part due to the suspended solids. This results in an eroded (sinking) river bed, which has been observed in many cases.

In the following sections, the Danube will be observed from the viewpoint of the effects of barrages: first the existing impoundments of the Austrian 'Altenwörth' and the Yugoslavian-Romanian 'Iron Gate No. I and No. II', and then a common Czechoslovakian-Hungarian hydrocomplex under construction, *i.e.* the 'Gabcikovo-Nagymaros' (GNV) barrage-system.

6.1 Altenwörth. The Austrian Academy of Sciences, with the support of the Ministry of Science and Research, and the company responsible for the construction of the power stations 'Donaukraftwerke', in 1984 launched a five year research project (Man and Biosphere/MAB) project) for the better understanding of the multiple effects of the Danube barrages in Austria.[43] The model area investigated is an impounded river section upstream from the Altenwörth Barrage (river km 1981, commissioned in 1976). The first results of this large-scale research activity have been recently published, a short review of these is presented.[34,44-48]

In the Austrian river section (see also Figure 3) all the barrages constructed are situated above Vienna on the high graded slope of the river bed. The bed-load transport of this river reach was very large (10^6 t year^{-1} at Vienna) before the construction of the barrages. The present load is estimated to be about half that value. The suspended sediment transport at Vienna is fluctuating between 1 and 7 x 10^6 t year^{-1} which is also considerably less than previous conditions. Consequently, it has been found that the river bed is sinking by ~1-2 cm year^{-1} below Vienna owing to erosion and decreased sediment transport.

The investigation of the Altenwörth impoundment indicated that the bed-load movement stopped at the upstream end of the impounded river section, while in the vicinity of the barrage fine sediment accumulated with significant effects on the benthos conditions, which will be mentioned later. The granular composition of the sediment changes with the distance from the barrage, and floods result in major displacement of the previously settled sediment driving it through the weirs of the barrage.

The Altenwörth barrage has a total length of 24 km. The maximum, mean, and minimum discharges are approximately 5500, 1800, and 800 m s^{-1} respectively. At mean discharge the average depth at river km 2005 is ~4.6-5.0 m and at the barrage 17 m. The width of the channel-like impoundment is fairly constant and the flow velocity (0.9-0.3 m s^{-1}) is a function of river depth (hydraulic radius) and discharge. The average detention time is 10-12 h in the impounded section, which might be close to one day at minimum discharge. The total suspended solids deposition in the impoundment was 2.9 x 10^6 t. Particulate phosphate attached to the fine sediment is accumulated near the barrage. The highest figures for organic carbon, nitrogen, and phosphorus content of the sediment are 1650, 120, and 120 mg 100 g^{-1} respectively.

It was observed in the last 5 km stretch of the river (nearest to the barrage) that, owing to increased transparency and higher nutrient yield, the biological activity was about 20% higher than in the previous impounded stretch. The oxygen saturation value in the bulk of water is highly dependent on the primary production and in July at 18.5 °C water temperature (at low primary production) the oxygen saturation fell to 64%, whereas it grew to 120% at peak production time, in October.

All of these values are also a function of flow velocity, the low value resulting in greater nutrient recycling and hence higher primary production.

In the upstream river section, it was interesting to observe both the generally decreasing trend of NH_4^+ due to higher output of wastewater treatment, and also the increase of nitrate in winter due to agricultural run-off.

The organic load of the impoundment is fairly modest. The COD value is between 7.1-15.2 mg dm^{-3}, the $KMnO_4$ value 18.7-30 mg dm^{-3} (expressed O_2 consumption = 4.7-7.6 mg dm^{-3}), and BOD_5 1.3-4.1 mg dm^{-3}.

The distribution of the heavy metal content of the sediment also deserves attention. These metals together with the organic carbon are attached to the finer sediment particles (smaller than 60 μm) and in consequence they are found in greater abundance near the barrage (see sediment composition above). Compared with the geochemical average values a moderate external load is illustrated by zinc and lead.

The sediment accumulation in the vicinity of the barrage influences the groundwater quality supplied by the river. This has been observed in a bankwell for drinking water supply (1 km upstream from the barrage and 500 m from the river bank). Oxygen has disappeared from the well water, and high NH_4^+ concentration, increasing COD, and iron and manganese contents are typically caused by fine, organically polluted bottom sediment settled over the sandy, gravelly river bed in the vicinity of wells. In these cases, thorough treatment of the well-water cannot be avoided.

In summing up the results of the observations in the first two years of the project it can be said that no detrimental effects were observed on the water quality of the river water; on the contrary, a slight improvement in almost all parameters is characteristic. This is due to increased self-purification capacity within the half- to one-day detention time. This fact together with decreased reaeration rate (increasing water depth) and seasonal variation of primary production results in widely varying O_2-saturation (120-60%). The effect of fine-grained sediment accumulation has been noticed in deteriorating groundwater quality in the adjacent water-bearing strata. Further research is needed, in the Author's opinion, for better understanding of the development in the bottom layers and benthos population.

<u>6.2 Iron Gate No.I and No.II</u>. The effect of the Iron Gate barrages on the suspended solids condition is well characterized by Bulgarian observations downstream of the two barrages (No.I was commissioned in 1971, No.II in 1985). The impounded water volume is 3.1 x 10^9 m^3

(Iron Gate No.I = 2.5, No.II = 0.7 x 10^9 m^3). Immediately downstream of Iron Gate No.II at the Bulgarian town of Novo Selo (river km 834) the reduction of suspended solids transport due to the barrage is greater than 50%. This loss is more moderate at downstream stations (40%) as a consequence of remobilized bottom sediment. Figure 6 gives an interesting picture of the changing suspended solids load at river km 554 during the past 25 years.[49] It would be interesting to know what had caused the loss in solids in the years before Iron Gate No.I was commissioned.

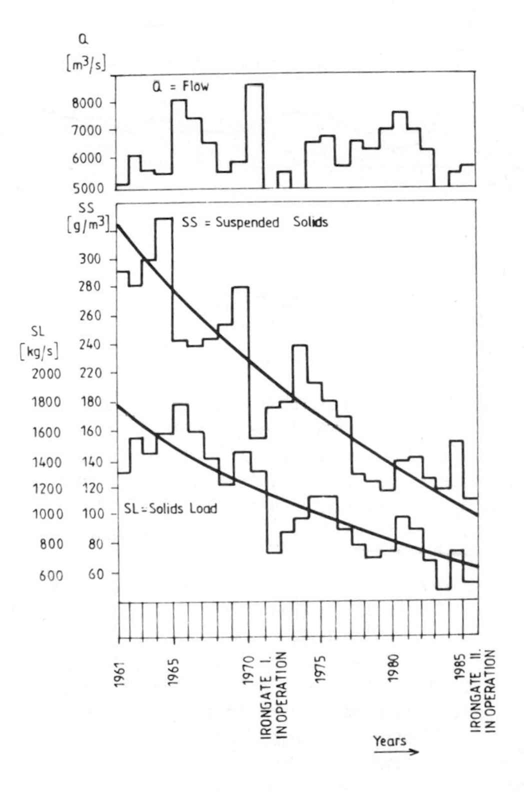

Figure 6 Changes in flow-suspended solids- solids load at Swishtow (river km 554) in Bulgaria in the period 1961-1985[49]

In reality, the observations in the Iron Gate No.I impoundment between 1971 and 1984 show an enormous difference between inflow- and outflow-sediment transport (see Figure 7).

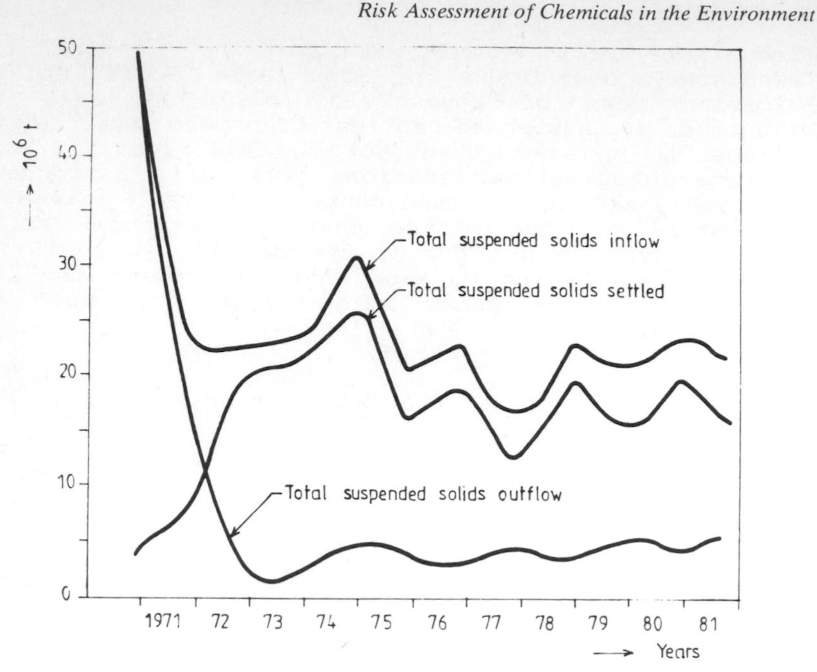

Figure 7 Sedimentation conditions in the Iron Gate No.I impoundment in the period 1971-81 (after Perisic et al.[40])

The effects of Iron Gate No.I at river km 943 (and partly that of No.II) on the biota, the saprobity and trophity, and also on the self-purification and oxygen conditions have been recently investigated by Yugoslavian and Romanian scientists.[37,40,50-52] The summary of their findings is as follows.

Taking into account 160 km of the effective length of the investigated impoundment behind Iron Gate No.I (1116-957 river km) and assuming ~2.0 x 10^9 m^3 impounded water volume for this specific reach, and 10,000, 5000, and 2000 m^3 s^{-1} discharges at maximum, medium, and low flow conditions respectively (see Figure 2), the flow-through times in the impoundment should be 2.5, 5, and 12 days, and the flow velocities 0.8, 0.4, and 0.16 s^{-1} respectively. Considerable pollution load arrives at the impoundment, partly from the Sava river, partly from the Beograd (Belgrade) area. At low flow conditions (1800-2400 m^3 s^{-1}) in the autumn of 1986 the BOD removal was 75%. At the same time a considerable amount of organic degradable material settled together with metals; *e.g.* the influent concentrations of cadmium and mercury were 1.6 and 1.8 μg dm^{-3}, decreasing to 0.4 and 0.2 μg dm^{-3}, in the outflow during the period investigated in 1986.

The O_2-deficit is somewhat reduced again in the second impounded section above Iron Gate No.II (about 850 river km). Photosynthesis, however, does not contribute to the O_2-pool of the impoundment because of intensive sedimentation of the plankton. On the contrary, the accumulation of the sedimented phytoplankton with the organically polluted solids leads to a considerable O_2-consuming activity of the bottom layer with the potential menace of remobilizing heavy metals from the bottom sediment. Altogether it can be said that O_2-consumption and restoration in the Iron Gate impounded sections show a negative balance, at least at low-flow conditions, when reaeration rate is unfavourable due to little flow-velocity, comparatively narrow width (~1.8 km), and considerable water depth of up to 70 m. It is interesting to note that no eutrophication has been observed, though the total phosphorus content is >0.2 mg dm^{-3} and sometimes is 1 mg dm^{-3}.

The sedimentation and self-purification effects of the two Iron Gate impoundments positively influence the saprobic conditions of the downstream section of the river. Observation in the period 1973-1985 indicated a definite improvement in the saprobity index, shifting it from the β-meso-polysaprobic range towards the α-mesosaprobic. Simultaneously, owing to the sedimentation of finer solids in the impoundment, the benthos-zoocenosis also changed.

In summing up of the results it should be noted that the quality of the water leaving the impounded sections is basically improved in comparison with that arriving. This can be related to the high self-purification capacity, sedimentation, and high nitrification rates. The reaeration rate is low, however; in the second impounded section (of Iron Gate No.II) the oxygen balance is more or less restored.

<u>6.3 Gabcikovo-Nagymaros</u>. As shown in Figure 4, this is a barrage-hydrocomplex built partly in Czechoslovakia and partly in Hungary. Extensive research has been devoted to determination of the ecological and water quality consequences of the complex. There are three main problem areas: (a) the impounded river section behind the Dunakiliti reservoir (see Figure 4), (b) the abandoned Danube bed between Dunakiliti and Gabcikovo, and (c) the impounded reach upstream from Nagymaros. At present the whole complex is under construction and will be finished in 1992.

Along the hydrocomplex the variation of the river's discharge is roughly between 7600 and 1000 m^3 s^{-1} (observed greatest and smallest discharges at Nagymaros 8180 and 590 m^3 s^{-1}) with an average of 2350 m^3 s^{-1}). The Dunakiliti reservoir has an approximate capacity of 200 x 10^6 m^3 and a surface area of 54 x 10^6 m^2. The figures for the Nagymaros impoundment are calculated at 550 x 10^6 m^3 and 96 x 10^6 m^2. This means that the average flow-through times are one and three days in Dunakiliti and Nagymaros, respectively, the extreme values being 2 - 3 times greater or less than these values.

The bed-load and sediment transport towards the Dunakiliti impoundment will depend on the gravel excavation from the river bed in the Bratislava area and on the effect of the planned barrages downstream from Vienna.[53]

For the Nagymaros barrage it was demonstrated by model simulations that sedimentation with an expected rate of 0.04-0.05 m $year^{-1}$ will occur, mainly along the river banks.[54]

Comparative studies have been launched along the common Czechoslovakian/Hungarian Danube reach from Rajka to Gönyü both in the main river and in its branches having lower velocity or stagnant water to forecast the expected ecological effects of the GNV-barrage system.

Besides the possible shift in trophic character of the impounded river water, there are some problems with the bacterial contamination of this river section as reviewed in Section 4. This cannot be eliminated unless the present pollution load of the dischargers in the area is greatly reduced.

As far as heavy metal pollution is concerned it is interesting that the suspended sediment is much more polluted with metal than the bottom sediment.[13] This reflects the well known fact that metals are attached to the finest portion of the suspended solids. There is a question as to what extent the barrages will alter this present day condition. It was experienced in connection with bankwells situated in the vicinity of silted-up river bed that anaerobic conditions in the filtration zone increased the iron content of the well water tenfold to that found in the river, whereas NO_3^- concentration decreased by an order of magnitude.[55] These observations concur with the Austrian experiences (see above).

In summary, the results of these investigations seem to strengthen the previous forecasts that caution should be exercised in connection with the construction of a barrage system as to the likely tendencies in increasing trophity (the planktonic production capacity of water) level and various consequences of enhanced settling properties in the impoundments.[3,56,57] The only possible way to counterbalance these effects is wastewater load reduction. This has been foreseen and calculated in the costs of the hydropower system of GNV.[7]

6.4 Delta Region. In the Delta area, especially in the Kilija arm (Soviet Danube section) a complex hydrobiological investigation was carried out over the period 1981-85.[30] The saprobity of the river corresponded to the β-mesosaprobic and in winter conditions to the α-mesosaprobic stage. The mean value of phytoplanktonic algae fluctuated in the range 2-3 g m^{-3}. At the same time the number of heterotrophic bacteria varied in the range 3.7-26.7 x 10^6 cm^{-3}. The maximum number of *E. coli* was 10,000 dm^{-3} and the average 3000-7000 dm^{-3}. The biomass of the zooplankton fluctuated in the range 0.4-1.3 g m^{-3} The higher biomass in spring time was regarded as a consequence of the barrage impoundment 'Shelesnije Worota' built in 1970. The benthos biomass was usually 0.1-0.3 g m^{-2}, which at flood periods might fully disappear, while in favourable conditions the biomass might reach 10.7 g m^{-2}.

6.5 Bavarian Danube. Little is known (by the Author) of the effects of the barrages built in the Bavarian river section. It seems, however, that heavy metals are a topic of concern. In this connection, the problem of possible accumulation of heavy metals in the food chain deserves attention. From Wachs's work it is known that in the Bavarian Danube between river km 2240 (Passau) and river km 2423 (Weltenburg) the levels of various heavy metals in fish muscle were low, *e.g.* cadmium <0.05, mercury 0.24, and zinc 6.0 mg kg^{-1}.[35] Similar values were found by Ebner and Gams in the Austrian Danube, *e.g.* the average mercury concentration in fish muscle was 0.32 mg kg^{-1}, and the maximum cadmium value was 0.014 and zinc 10 mg kg^{-1}.[58] These are well below the criteria of food safety (Cd = 0.05, Hg = 0.5 mg kg^{-1} in fish muscle) as indicated by Wachs.[35]

7 Risk Assessment and Future Trends in the Water Quality of the River Danube

As far as the pollution risk of the river is concerned three main categories of risk (and risk management problems) might be defined: (1) Incidental, (2) Accidental, and (3) Intentional risks. All of these in the case of the Danube are complicated by their possible international character, *i.e.* the effects exported to the downstream neighbours.

The routine procedure of risk management (identification, evaluation, and reduction) roughly equates with normal water pollution control. In the latter case 'identification' is called 'monitoring and data processing', 'evaluation' is the same in both cases involving model development, and 'reduction' is often called 'decision-making'. It should be noted here that water quality criteria are laid down by legislation in all countries and that this is a co-operative effort of health- and water-management agencies. The problem with the criteria is that they differ both in concept and figures from country to country.

In the following, the incidental and mainly the intentional risks stemming from river pollution (and at the same time influenced by river regulation, hydropower development, *etc.*) will be characterized through the observation of water quality trends.

As may be seen from previous sections and from Tables 3-6 the water quality of the Danube is acceptable at present, even if it is considered that the river basin covers more than 800,000 km^2 of catchment area with a total population of ~67 million. The peculiarity of the basin is that the upstream countries are far more industrialized than the downstream ones which means that the smaller the river's mean discharge the larger is the wastewater load, in absolute figures. Upstream from river km 1900, *i.e.* in the territories of Austria and the Bavarian State of the Federal Republic of Germany for a length of about 1000 km the river is flowing through a densely populated, highly industrialized area with a subcatchment of about 130,000 km^2. From the old water quality maps of Bavaria it is seen that some sections of the river were overloaded with untreated wastewater. The situation is, however, improving. The latest quality maps, based mainly on the degree of saprobic conditions, indicate that in 1986 roughly two-thirds of the Bavarian river reach is already in the state of 'moderately loaded', and only a few km below Kelheim (around river km 2400) are still fairly polluted; between Regensburg and Deggendorf it is, however, critically loaded.[18]

The same improvement can be seen in the Austrian reach if the water quality picture of Liepolt[9] from 1967 is compared with the data of 1985 published by Weber et al.[20] Before 1967 the river sections downstream of Linz and Vienna (Wien) were in the heavily polluted category in contrast to the present state after the completion of the wastewater treatment plants of those large cities (the second water quality category deteriorates slightly below Vienna). Moreover, according to the evaluation work made by Fleckseder, the metal content of the river and its sediment at Vienna has also decreased during the past decade.[36] Whereas in the period 1976-79, the mercury value was 0.066 μg dm^{-3} in the water, in 1984 it was only 0.057 μg dm^{-3}; the same data for cadmium were 0.63 and 0.13, for lead 3.66 and 1.23, and for zinc 34.4 and 12.8 μg dm^{-3}, respectively. Even more impressive is the decrease of the saprophytic bacteria downstream from Vienna. Whereas in 1978-80 the number per cm^3 was 617,000, in 1981-83 it was 50,000. For *E. coli* 2800 and 248 cm^{-3} respectively were reported.[59]

For the percentage changes in the water quality parameters of the Austrian, the common Czechoslovakian-Hungarian, and the Hungarian Danube sections Table 7 indicates the figures based on ten years observations. Nitrate concentration has increased, as shown in Table 8, in the Hungarian Danube over the period 1971-85 together with the increase in use of nitrogen fertilizers. Though the concentration of nitrogen compounds in the river should be significantly influenced by wastewater discharges, it is interesting to note also the obvious correlation with nitrogen fertilizer use in agricultural areas (run-off water).[25]

The lower Danube, from the confluence of the Tisza and Sava, roughly for the last 1200 km of its length towards the Black Sea, has such an increased mean flow (5000-6000 m^3 s^{-1}) that the river's self-purification capacity must be enormous. This is promoted by the two barrages of Iron Gate Nos.I and II. As a consequence, it can be seen from Tables 3-6 that the quality conditions are satisfactory with the exceptions of short reaches downstream from Belgrade, Rusze, the Arges river confluence, *etc.* (see Table 1) which are heavily and directly polluted with untreated wastewaters.[24,61] Simultaneously it should be borne in mind that the industrial power of this sub-basin of the river, though at present modest in comparison with the upstream sections, is rapidly developing together with the agricultural activities. An ever increasing ion content of the river has been observed during the past 30 years by Romanian and Bulgarian researchers.[62,63] The total ion concentration has grown since 1954 by 30% from 300 to 420 g m^{-3}, and if the total ion load is investigated by decades it can be seen that it was 2000 kg s^{-1} in the first decade, 2160 in the second, and 2420 in the last, indicating an increasing rate of salination of the river. Maybe it is more important than the change in the composition of the ions. The sulphate and chloride contents were 10.6% and 7.5% in 1959; these had grown by 1982 to levels of 16 and 11%, respectively.

The ecological impact of the barrages poses an intentional risk. As indicated in Section 6, the

Table 7 *Tendencies in the change of water quality parameters in the Austrian (1978-85), the Czechoslovak-Hungarian, and Hungarian (1976-85) Danube sections*[60]

City	*River* km	COD*	*Percentage change of components* BOD_5	Dissolved O_2	Ammonia	Ortho phosphate	Conductivity†
Felsen-Hütt	2200	-2.5	+1.0	+0.4	+4.4	+0.2	+0.2
Linz	2130	-3.9	+3.1	+0.6	+0.5	+0.2	+0.1
Ybbs-Persenbeug	2060	-2.6	+2.6	+0.6	+3.6	+0.1	+0.3
Vienna-Nussdorf	1920	-3.1	+5.9	+0.9	+4.4	+2.5	+0.4
Wolfsthal	1983	-1.6	+3.4	+1.7	+3.9	+1.7	+0.4
Rajka	1848	-1.9	-2.6	-0.1	+0.6	-0.4	+0.4
Komárom	1760	-0.5	-1.9	-0.2	-0.1	-1.0	+0.2
Szob	1708	-0.7	+0.1	+0.5	-0.7	+0.0	+0.4
Budapest	1654	-1.7	+1.3	+0.8	-3.4	+1.7	+0.1

Changes: - improves, + deteriorates

* Permanganate method in the Austrian Sector
† Total dissolved solids in the Hungarian Sector

Table 8 *Development in the NO_3^- content of the Hungarian Danube section between 1971 and 1985*[25]

Parameter (and dimension)	*Place and river km*		*Five year periods*		
			1971-75	*1976-80*	*1981-85*
Nitrogen fertilizer used (in 1000 tonnes)	In Hungary altogether		479	556	603
NO_3^- average concentrations (mg dm^{-3})	Rajka	1848	7.06	8.58	9.53
NO_3^- average concentrations (mg dm^{-3})	Baja	1480	6.94	8.72	8.76
NO_3^- concentration of 95% duration (mg dm^{-3})	Rajka	1848	9.75	13.56	14.61
NO_3^- concentration of 95% duration (mg dm^{-3})	Baja	1480	11.10	14.86	12.78

ecological transformation in the impounded river sections could not have been evaluated owing to the short time span after construction; only partial results of investigations were presented. The trends, however, can be deduced with high probability, as was attempted in Section 6. Water management can do a lot to decrease the detrimental effects and enhance the positive effects.

Another group of intentional risks arises from the restricted capability of water management to curb the anthropogenic effects of a chemical, physical, or biological nature. The limited efficiency of wastewater treatment (risk reduction) is well known. The objective of the management is to exploit the dilution and self-purification capacity of the river, *e.g.* in the case of such pollutants as wasted heat, salts, and refractory organics.

Hydrological phenomena are interrelated with biochemical and chemical ones as will be seen in the following.

With low flow conditions and low water temperature during the winter season, the resulting ice-cover of the river plays a significant role in self-purification and generally in almost every important water quality component. Therefore, it is interesting to observe the trend in ice formation during the past decades. Investigations show that power stations (atomic or fossil fuel) of 1000 MW capacity with their concentrated heat load (thermal production) produce +0.8 °C increase in river water temperature at 80-100 km distance from the discharge point. Evaluating the heat output of anthropogenic origin and water temperature data of the Danube (at a gauging station near Vienna) since 1900 has proved an increasing trend of those related to identical aggregated negative air temperatures.[64] Comparing the period of 1901-1941 with that of 1942-1985 the increase of heat output is 104-114% and that of water temperature is 102-108% related to 100-500 °C aggregated (summed up) air temperatures. This anthropogenic heat output is more or less consumed by melting the floating ice floes, resulting in a decreasing number of 'icy days'. Thus, in the period 1877-1950 the average number of icy days (with ice drifting) was 18, but during the period 1951-1980 it decreased to 10.

Another interesting observation is that during the past decade a higher daily 'critical' air temperature has been required for the formation of solid ice cover than in previous decades. The difference is about 30%. This is attributed to the changing chemical composition of the river water: *e.g.* at Budapest during the past five decades the dissolved solids concentration of the water has grown from 210 to 360 g m^{-3}. This leads to 12% increase in the time span required for ice cover formation. The trend in salt content increase in the Lower Danube has shown similar effects.

Simultaneously, there has been a drastic change of opposite character in the ice conditions of the impounded river sections. In spite of the above trends due to increasing salt concentration, the reduced flow velocity (winter temperature coincides with low discharge) and low level of turbulence favour solid ice cover formation, particularly in Austria. Behind the barrages solid ice covers the impoundments and this again reduces the downstream ice-drifting development. Considering the ecological effects first, oxygen saturation, reaeration, and self-purification should be mentioned. Whereas all of these are favourably influenced in the ice-free river sections, the opposite occurs in the impounded sections, the ice cover contributing to the poor reaeration conditions indicated in Section 6.

8 Accidental Risks

There is quite a significant history of accidental pollutions in Czechoslovakian-Hungarian Danube and these can be seen in Table 9 and Figure 8.

Table 9 *Causes of accidental water pollutions in the Hungarian Danube catchment between 1976 and 1986*[60]

Type of pollution	*All accidents in 10 years*
Municipal sewage release	90
Oil and oil derivatives	912
Organic industrial wastes	206
Inorganic industrial wastes	179
Manure (liquid)	70
Pesticides	39
Fertilizers	11
Putrescible organic waste of unknown origin	65
Ammonium ion	25
Others of unknown origin	510
Total number	2107
Number of accidents originating from abroad	319

The predominant sources of accidents are industrial plants and oil, which are released at a high frequency. In Austria, *e.g.*, during the eleven year period 1976-87

22 oil accidents occurred on the Danube downstream from river km 2000 (Krems) of which only four were caused by ships. However, in the catchment area (in the tributaries) pertaining to this river stretch 30 accidents were registered each year, of which 55-65% were related to oil spills.[65] The majority of the accidents happening on the tributaries do not effect the main river.

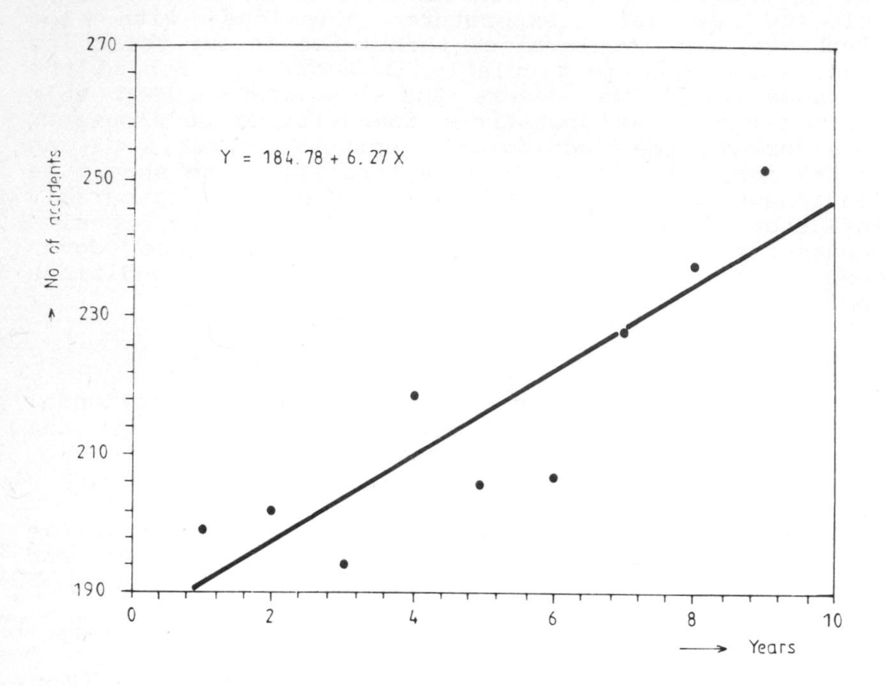

Figure 8 Trend analysis of accidental water pollutions in Hungary on the Danube and its tributaries during the past ten years. (About 20% of all accidental pollutions had their origin in the upstream countries)[60]

In the past decades, significant accidents occurred in the Middle Danube, *i.e.* in Czechoslovakia at the Bratislava oil refinery, in Hungary at Vác with chemical dumping, and at Dunaujváros (see map, Figure 1) with heavy oil release. In the Bratislava refinery raw petrol leaked from the bottom of oil tanks and contaminated the groundwater resources along the river supplying Bratislava with drinking water. At Vác (north of Budapest) the leakage from the chemicals in a nearby depot also contaminated the bankwell filtration zone. The waterworks has been out of operation since 1981. In Dunaujváros (south of Budapest) more than 100 tonnes of heavy heating oil escaped incidentally in winter time from a storage reservoir and the frozen oil slicks were washed off along the river bank for more than 100 km downstream of the discharge point.

9 Conclusions

The catchment area of the Danube and the geographic borders of the riparian countries can be seen in Figure 1. As a major water resource and traditional communication route (sometimes a natural border line) it has yielded excellent opportunities for urban, agricultural, and industrial development from Roman times to the present day. Its significance has grown since the second world war as a link between East and West.

Figure 2 presents information on the most important features of the hydrology of the river and Figure 3 completes, thus indicating river-sections of high hydro-energy potential. During the past decades tremendous development in the catchment area, urbanization, industrialization, hydropower exploitation, and intensified agriculture led to changes in the morphology, hydrology, and biogeochemical, and ecological characteristics of the river, such as increasing flood potential, decreasing icy periods, and decreasing bed-load and sediment transport, all leading to a deteriorating trend in water quality. Major water uses can be seen in Table 1 and an approximative picture of the water quality indicators is presented in Tables 3-6. These show that in comparison with other watercourses the quality of the river is still fairly good with the exception of relatively short stretches downstream from major wastewater discharges and polluted tributary confluences.

Turning to risk assessment problems, it was pointed out that apart from incidental and accidental risks (see also Table 9 and Figure 8) intentional risks appear as significant factors in water management. Though management can control the major effects of wastewater discharges, it has to tolerate (intentionally) some residual pollutants in the water due to technical and mainly economical (cost-benefit) considerations. The same is true for the run-off water from agricultural areas. Specific changes in the ecosystem are expected as a result of exploiting the hydropower potential of the river (see Section 6).

All of these result in certain trends as summarized in Table 10. Some of these might be favourable, such as reduced ice formation due to increased salt content and water temperature in free-flowing river sections, or improving saprobity in the impounded sections.

Table 10 *Summary of trends in the ecosystem of the River Danube*

Causes	*Hydrology*	*Physical*	*Chemical*	*Biological*
		Water quality parameters		
Inorganic pollutants (residuals of waste water treatment and agricultural run-off)	Ice formation (+) Sediment accumulation (-)	Transparency (-) Suspended solids composition (-)	pH (-) Salts, ions (-) Nutrients (-)	Toxicity* (-) Halobity* (-) Trophity* (-) Benthon composition (-)
Organic pollutants (residuals of wastewater treatment)	Sediment accumulation (-)	Colouring (-) Transparency (-) Suspended solids composition (-) Scum formation (-)	Taste-, odour-deterioration(-) THM-precursors (-) Carcinogenic, mutagenic agents (-)	Toxicity* (-) Benthon (-) composition (-) Fish-breeding (-)
River impoundment (hydropower plants)	Ice formation (-) Bed-load reduction (?) Sediment accumulation (-)	Temperature (-) Transparency (+)	O_2-saturation(-) Organic decomposition (+)	Trophity* (-) Saprobity* (+) Benthon and biotekton composition (?) Fish-breeding (?)

Changes: (+) Improvement; (-) deterioration; (?) questionable
* Lump parameter of hydrobiological characterization of the water body

Certain changes, however, cannot be evaluated either in a positive or a negative sense, at least not in the short term. Only time can decide their effect.

Risk analysis of a living world such as that of a large international river is very complicated and it poses a challenging task for scientists in the future. They have to understand the responses originating from anthropogenic interference with nature. Monitoring of 'pollution' is not enough; the entire web of effects and their interdependence must be evaluated.

10 References

1. 'Die Donau und ihr Einzugsgebiet. Eine hydrologische Monographie', Regionale Zusammenarbeit der Donauländer, ed. Bayerisches Landesamt für Wasserwirtschaft. München, 1986.

2. Gy. Fekete, *Water Quality Bull.*, Canada Centre for Inland Waters, Burlington, Canada, 1979, **4**, 56.

3. VITUKI, 'The effect of the GNV-barrage system to the water quality of the Danube' (in Hungarian), Research Report Summary, Budapest, 1985.

4. Commission Du Danube, Réunion d'experts pour l'examen du project de Recommendations relatives a la prévention de la prévention de la pollution des eaux du Danube par la navigation, Manuscript, 1985.

5. B. Hock and G. Kovács, A large international river: the Danube. Summary of hydrological conditions and water management problems in the Danube Basin', International Institute for Applied Systems Analysis, Manuscript, Laxenburg, Austria, 1987.

6. 'Donaustrom', Verlag A.F. Koska, Wien, 1984.

7. VIZITERV, 'Ecological impact study of the Gabcikova-Nagymaros Barrage System', (in Hungarian), Manuscript, Budapest, 1985.

8. B. Jankó, *Közlekedéstudományi Szemle,* Budapest, 1978, **XXVIII**. No.1.

9. 'Limnologie der Donau', ed. R. Liepolt, Scheizerbartische Verl. Stuttgart, 1967.

10. R. Liepolt, *Water Quality Bull.*, Canada Centre for Inland Waters, Burlington, Canada, 1979, **4**, 78.

11. WHO, Regional Office for Europe, 'River bankwell filtration for potable water supply', Report on Working Group. Copenhagen, 1984.

12. F. László *et al.*, 'Vizügyi Közlemények', Budapest, 1987, Vol.69 (in press).

13. F. László, 'Potential release of pollutants in riverbank filtration systems along the river Danube, Hungary', Proceedings of the 3rd International Symposium on Interaction between Sediments and Water, CEP Consultants, Edinburgh, UK, 1984, p.264.

14. P. Literáthy and F. László, 'Uptake and release of heavy metals in the bottom silt of recipients', in 'Interaction Between Sediments and Fresh Water', ed. H.L. Golterman, WB. Junk Publishers, The Hague, 1977, p. 403.

15. P. Benedek, 'New trends in water quality management'. IAWPR/Pergamon Press, *Water Sci. Technol. (London)*, 1982, **14**, p.47.

16. UNESCO/UNEP/WHO/WMO Workshop on the Assessment of Particulate Matter Contamination in Rivers and Lakes, Manuscript, VITUKI, Budapest, 1978.

17. D. Backhaus, 'Ergebnisse chemischer und biologischer Langzeituntersuchungen an der oberen Donau bis Ulm', 26. Arbeitstagung der IAD, Passau, 1987, p.5.

18. Gewassergütekarte Bayern, Bayerisches, Staatsministerium des Innern, München, FRG, 1984, 1987.

19. Bayerisches, Landesamt für Wasserwirtschaft. Auszug aus Jahrbuch für Gütepegel Jochenstein, 1986.

20. E. Weber et al., 'Ergebnisse der Monatlichen Gewässergüteuntersuchungen der österreichischen Donaustrecke', in 'Wasser und Abwasser', Technische Universität, Wien, 1986, Vol. 30, p. 541.

21. J. Ardó, 'Jahreszeitliche Veränderungen der Wasserqualität im tschechoslowakischen Donauabschnitt', 26. Arbeitstagung der IAD, Passau, 1987, p.1.

22. VITUKI, 'Data Inventory on Water Quality of Surface Waters of the Research Institute for Water Resources Development', Budapest, 1986.

23. WHO/EURO correspondence, 'Information note on the quality of the waters of the Danube river in the territory of Yugoslavia', Excerpts from the

Hydrological Almanacs, Federal Hydrometeorological Bureau, Belgrade, 1985.

24. G. Petrovic and Zs.T. Dvihally, 'Sauerstoff-production in der Mittleren Donau', 25. Arbeitstagung der IAD, Bratislava, 1985, p. 135.

25. B. Hock, Development in the nitrogen content of surface waters between 1971-85 (in Hungarian), Manuscript, VITUKI, Budapest 1987.

26. K. Zankov *et al.*, 'Prognose und Beurteilung der Sauerstoffverhältnisse im bulgarischen Donau-abschnitt', 24. Arbeitstagung der IAD, Szentendre/Ungarn, 1984, Vol. I, p. 41.

27. K. Zankov and K. Ivanov, 'Verunreinigung der Donau mit organischen Stoffen im bulgarischen Abschnitt', 22. Arbeitstagung der IAD, Basel, 1981, p. 53.

28. R. Antoniu, The River Danube water quality in the Romanian Section. Planning Meeting on the River Danube Water Quality Protection Project. World Health Organization Reg. Office for Europe, Report. Laxenburg, Austria, 1987.

29. Gh. Zamfir *et al.*, 'Die hygienisch-sanitare Erforschung des Donaustroms in seinem Endabschnitt', 24. Arbeitstagung der IAD, Szentendre/Ungarn, 1984, Vol. II, p. 67.

30. 'Gesamtbericht der IAD über die wissenschaftliche Tätigkeit in den einzelnen Donauländern', ed. E. Weber, Wien, 1985.

31. B. Trzilova and L. Miklosovicova, 'Die Qualität des Donauwassers', 25. Arbeitstagung der IAD, Bratislava, 1985, p. 180.

32. B. Trzilova *et al.*, 'Die Wasserqualität des Tschechoslovakisch-ungarischen Donauabschnittes Gabukovo-Nagymaros', 26. Arbeitstagung der IAD, Passau, 1987, p.65.

33. L. Rosival, 'The river Danube water quality (Presentation from Czechoslovakia). Planning Meeting on the River Danube Water Quality Protection Project', World Health Organization Reg. Office for Europe, Manuscript, Laxenburg, Austria, 1987.

34. G. Kasimir, 'Bakterien im Stauraum Altenwörth', 26. Arbeitstagung der IAD, Passau, 1987, 485.

35. B. Wachs, 'Ökologisches Verhalten Umweltrelevanter Schwermetalle', in *Münchener Beiträge zur Abwasser-, Fischerei- und Fluszbiologie*, München, 1986, **40**, p.460.

36. H. Fleckseder, 'Schwermetalle in der österreichischen Donau', in 'Wasser und Abwasser', Technische Universität, Wien, 1986, Vol. 30, p.483.

37. M. Perisic et al., 'Einige Aspekte der Qualitäts-änderung des Donauwassers in Speicherbecken "Derdap I" während der Niedrigwasserperiode', 26. Arbeitstagung der IAD, Passau, 1987, p. 170.

38. S.I. Kozlova, 'Die Raum und Zeitveränderlichkeit des Queksilbergehalts des Kilijadeltas der Donau', 25. Arbeitstagung der IAD, Bratislava, 1985, p.34.

39. WHO, 'Micropollutants in river sediments', Report on a WHO working group. Copenhagen, 1982, EURO Reports and Studies, No. 61.

40. M. Perisic *et al.*, 'Some aspects of the river Danube self-purification in the Iron Gate I reservoir', Jaroslav Cerni Institute, Manuscript, Belgrade, 1987.

41. G. Petrovic and V. Schleicherte, *Arch. Hydrobiol. Suppl.*, 1981, **52**, (4), 323.

42. B. Gunselius, 'Verunreinigungen der Österreichischen Donau durch Mineralöl', 22. Arbeitstagung der IAD, Basel, 1981, p.21.

43. Ökosystemstudie Donaustau-Altenwörth', A study of the ecosystem of the impounded Donaube-reach at Altenwörth (in German), Österreichische Akademie der Wissenschaften, 1984.

44. R. Belocky and M. Sager, 'Zur Korngrössen-abhängigkeit der Zusammensetzung von Feinsedimenten aus Stauräumen der Donau', 26. Arbeitstagung der IAD, Passau, 1987, p.98.

45. G. Bretschko and M. Leichtfried, 'Organischer Kohlenstoff, Gesamtstickstoff und Gesamtphosphor in den Weichsedimenten des Stauraumes Altenwörth', 26. Arbeitstagung der IAD, Passau, 1987, p.109.

46. A. Herzig, et al., 'Quantitative Zoobenthos-untersuchungen im Stauraum Altenwörth, 26. Arbeitstagung der IAD, Passau, 1987, p.127.

47. M. Nausch, 'Qualitative und quantitative Analyse des Phytoplanktons der Donau im Stauraum Altenwörth 26. Arbeitstagung der IAD, Passau, 1987, p. 161.

48. F. Zibuschka, 'Beeinträchtigung der Grundwasserqualität durch Stauhaltung', Arbeitstagung der IAD, Passau, 1987, p. 517.

49. D. Petschinov, 'Veränderung des Schwebstoffabflusses der Donau im Bulgarischen Abschnitt unter dem Einfluss von anthropogenen Faktoren', 26. Arbeitstagung der IAD, Passau, 1987, p. 186.

50. E. Gruia and S. Marcoci, 'Zur Sauerstoffbilanz der Stauseen Eisernes Tor I und Eisernes Tor II', 26. Arbeitstagung der IAD, Passau, 1987, p.524.

51. S. Marcoci, 'Uberblick über die Saprobität des Donauwassers', 26. Arbeitstagung der IAD, Passau, 1987, p. 547.

52. V. Marinescu, 'Die Struktur der Benthos-Zoozönosen bei Mraconia in den Jahren 1981-82.
Die Entwicklung der Benthos. Zoozönose in der Donau in Bereich des Stausees Eisernes Tor II', 26. Arbeitstagung der IAD, Passau, 1987, p. 147.

53. J. Rotschein, 'Qualitative prognosis of the effects of the barrage', (in Czech.), VUVH Report, Manuscript, Bratislava, 1976.

54. L. Rákóczi, 'The effect of the GNV-system to the sediment transport' (in Hungarian), VITUKI Report, Manuscript, Budapest, 1985.

55. F. László and Zs. Homonnay, 'Study of effects determining the quality of bank-filtered well water', Conjunctive Water Use. IAHS, Publ. 1986, 181.

56. P. Benedek and F. László, 'A large international river: the Danube', IAWPR/Pergamon Press, *Progr. Water Technol.*, Cincinnati, 1980, 13, 61.

57. B. Hock, 'The expected water quality impacts of the barrage system Gabcikovo-Nagymaros (in Hungarian), Research report, Manuscript, Budapest, 1978.

58. F. Ebner and H. Gams, 'Schwermetalluntersuchungen in der Donau im Zeitraum 1976-1984', in 'Wasser und Abwasser', Technische Universität, Wien, 1984, Vol. 28, p.105.

59. H. Fleckseder, OZE, Wien, 1986, *Jg.39* (7/8), 141.

60. VITUKI, 'Evaluation of accidental pollution' (in Hungarian), Research Report, Manuscript, Budapest, 1987.

61. G. Petrovic, 'Die Ergebnisse der hydrochemischen Forschung im Stauraum Djerdap', 25. Arbeitstagung der IAD, Bratislava, 1985, p. 53.

62. F. Gruia, 'Einige Probleme bezüglich der Mineralisierung des Donauwassers in der Zeit vom 1958 bis 1982', 26. Arbeitstagung der IAD, Passau, 1987, p.19.

63. K. Ivanov and D. Petschinov, 'Veränderung des Ionen-gehaltes der Donau bei der Stadt Russe unter dem Einfluss der Wirtschaftstätigkeit', 26. Arbeitstagung der IAD, Passau, 1987, p.22.

64. J. Déri, 'Hydrothermal and hydrochemical effects to the ice conditions of the Danube', IAHR Symposium on Ice, Iowa City, USA, 1986.

65. H. Frischherz, Personal communications, 1987.

20
Total Index of Environmental Quality as Applied to Water Resources

J. Riha

FACULTY OF CIVIL ENGINEERING, CZECH TECHNICAL UNIVERSITY, 16629 PRAGUE, THAKUROVA 7, CZECHOSLOVAKIA

1 Introduction

In advanced countries the pollution of the hydrosphere is a complex problem. The combined causes of pollution may be divided into two groups, *i.e.* point sources of pollution (discharges of polluted waste waters from industrial production processes, domestic sewerage systems without waste water treatment plants, *etc.*) and non-point pollution sources (agricultural production, herbicides, pesticides, *etc.*), which may affect large areas. The negative impacts of point sources of pollution may be technically eliminated or diminished by the construction of treatment plants; this is not so with non-point sources which are a universal world problem, *e.g.* eutrophication.

In a number of countries, accidents of various kinds are a serious problem because they cause random and unexpected discharges of dangerous, toxic substances into the environment. The danger level following from accidents and the release of chemical substances depends on many factors, namely on their type and character, amount, place, hydrogeological substratum, local topography, the human factor, *etc.*

Experience gained in Czechoslovakia and in other countries has shown that crude oil accidents caused primarily by the failure of technical equipment and the human factor are the main cause of accidents resulting in the pollution of the hydrosphere with chemical substances, marine pollution, in the case of oil, having by far the greater consequences. Underground pipelines are generally considered to be the most satisfactory installations for the transport of crude oil and other liquid chemical compounds. Considerable care is devoted to their construction but there are still accidents involving crude oil pipelines causing subsequent release of crude oil into hydrogeological structures and dramatic contaminations of surface and ground waters.

The probability that a pipeline accident will not take place with the observance of n effects may be expressed by the relationship

$$P = \prod_{i=1}^{n} (n - p_i) \qquad (1)$$

Where p_i is the probability that failure will take place owing to the i-th cause and $(1 - p_i)$ is the probability that an accident will not take place owing to the i-th cause.

As an example let us consider the following list of events (causes) which may contribute to the cumulative assessment of risk:

p_1 - the chosen route of the pipeline with regard to the geological substratum, natural obstacles, population density, *etc.*,

p_2 - the quality of structural and technological implementation, including the technical standards for pipelines, pumping plants, and other installations and equipment,

p_3 - the efficiency of the human factor, *i.e.* qualification and reliability of operators,

p_4 - the efficiency of automatic control and safety equipment for long-distance crude oil transport (sensors, electronic devices, dispatch facilities, *etc.*),

p_5 - the efficiency of the surface and ground water quality monitoring systems, *e.g.* systems for control and identification of pollutants in aquifers.

p_i values may be determined by expert assessment. In comparative studies of several variants the highest value of P corresponds to the smallest risk. In any case it must be compared with the value of the acceptable measures of risk for the given case and society.

2 Cumulative Environmental Impact Analysis (CEIA) and the Decision-making Strategy

Let us consider the pollution of water caused by accidents to structural and technological installations. During the planning and design of a new structure including its technology and siting in a certain area, we are concerned with a decision-making problem: the determination of the optimal choice of variant with regard to the set of many criteria.

There are four main factors in the theory of decision-making, *i.e.* decision-making under conditions of: (1) certainty, (2) risk, (3) uncertainty, (4) unknown factors. In decision-making (1) all future states are known and may unequivocally be determined. Decision-making under factor (2) assumes that subsequent states are known with definable probability which is known or is determined by expert estimation. In decision-making under conditions of uncertainty (3) subsequent states may be known but there is no knowledge of their probability. In decision-making under conditions (4) subsequent states and their probability are unknown factors. In the traditional concept, risk is therefore linked with probability. Uncertainty and indefiniteness are by their nature unique and unprecedented; decision-making cannot be based on empirical experience. The theory of FUZZY sets (FUZZY set models are models focusing on decision criteria or constraints that are not sharply defined, so that the decision area is not demarcated in an unambiguous manner) is (to a certain extent) applied as the instrument for resolving the problem of uncertainty and indefiniteness.

In principle, any decision-making may be considered as being the result of a thinking process in which the main role is played by the value preference of the decision-maker. It is mainly his attitude to indefiniteness and the measure of importance which the decision-maker in his value hierarchy attaches to various aspects which affect the final decision.

The explicit expression of the value of the decision-maker may be implemented through the multi-dimensional functions. This may be implemented such that its form will consider all important values and preferences of the decision-maker. Such preferences may be identified through a special dialogue between the decision-maker and the so-called analyst who constructs the benefit function system.

The first step of the solution will be to choose between the application of the so-called real functions of partial benefit systems, *e.g.* from ecological relations and standards and monotonous functions expressing the tendency to risk. The summary solutions of both approaches have been described in the literature.[1]

The latter approach has been theoretically clarified.[2,3] Mathematical proofs were presented for the claim that decision-making:

(a) has a tendency to risk when the function is convex,

(b) has an aversion to risk when the function is concave,

(c) has a neutral tendency to risk precisely when the function is linear.

Mathematical proofs are taken from the theory of chance and the problem of the unacceptability of risk may be explained illustratively by the so-called indifference curve.

According to reference 1 there is the similarity between:

(a) ecologically optimistic evaluation and unacceptability of risk (concave shape of curve),

(b) ecologically pessimistic evaluation and preference for risk (convex shape of function).

Thus, for instance, for the rising concave exponential function $f(x) = -e^{-cx}$ the value of the coefficient c is the measure of resistance to acceptance of risk.

3 Method of the Total Index of Environmental Quality

The application of the multicriterial axiomatic multiattribute utility theory[4] proceeds from the philosophical approach and from the assumption that the total quality of the environment for the given region is determined by essential properties of the individual components of the environment whose quality may be assessed using available analytical and diagnostic indicators. These indicators of quality will form a catalogue of criteria whose values are determined either exactly (analytically) or by estimation by experts. Utility can be defined as representing an overall evaluation of benefit and costs. Utility models are based on the assumption that the whole vector of relevant objectives can be translated by means of a weighting procedure into the master control of one unambiguous utility function. This assumption of explicit and known trade-offs between objectives is essential in multicriterial axiomatic multiattribute utility theory. The variety of properties usually prevents the conversion to a common value measure which is made possible by the following working procedure.

Assuming that:

V_i is the variant solution of projects for $i = 1,2,\ldots,m$, where m is the total number of constructed variants;

$\bar{P}_y$ is the cardinal parameter of environmental quality which may be used as the criterion for the qualitative evaluation of anthropogenic impact at $y = 1,2,\ldots,z$, where z is the total number of selected criteria;

$P_j(Y)$ is the indicator of the criterion as the value of the analytically ascertained, possibly estimated parameter in objective or subjective units as the j-th partial consequence V_j;

$\bar{P}$ is the vector of parameters for which $\bar{P} = \bar{P}_1 \ldots \bar{P}_z$;

P is the total consequence V_i for which $P = P_1 \ldots P_n$;

w_j is the weight multiplier, *i.e.* the relative importance of the investigated $P_j(Y)$ within the framework of the whole set $j = 1,2,\ldots,n(Y)$;

U_j is the partial function as the quantitative multiplier having the character of the transformation function $f_j(P_j(Y))$ assuming the value at interval $0 < U_j = < 1$;

U is the total utility function.

It is simultaneously assumed that for the given region and set of variables j all values $P_j(Y)$ and U_j may be determined for which it applies that

$$U_j = F_j\,(P_j(Y)) \qquad (2)$$

which expresses the partial utility function in mathematical form. The total utility function U depends on the total impact P, and the partial utility function U_j is used for its construction. The preferential and utility independence of criteria and indicators is assumed. There also exists the indispensable assumption that, for the whole set V_i,

$$w_j = \text{constant} \qquad (3)$$

The total quality of the environment in the given region is given by the value of the total indicator of environmental quality

$$\mathrm{TIEQ} = U = \sum_{j=1}^{n} w_j U_j \qquad (4)$$

The said form of function U may be used in case it applies for the set w_j where

$$0 \leqslant w_j < 1, \; (j = 1,2,\ldots,n) \qquad (5)$$

and simultaneously

$$\sum_{j=1}^{n} w_j = 1 \qquad (6)$$

Equation (4) is the so-called additive form of function U. If the condition given by equation (6) does not apply, *i.e.* the sum of the w_js is not equal to one, it is necessary to use the so-called multiplicative form.

The individual values of the function U_j according to equation (2) are objective indicators of the quality of certain components of the environment only if they are related to stated standards of quality, possibly to foreign or world standards (for instance standards and limits recommended by the WHO). In exceptional cases it is possible to take into account scientific knowledge, regulations, prescriptions, and customary classifications of quality or non-quality *i.e.* measurements of pollution of the environment. In such cases the value of TIEQ assumes the character of the complex utility value of the region.

The procedure of general transformation is indicated in Figure 1 for direct dependence of U_j on $P_j(Y)$ and in Figure 2 for indirect dependence. It is evident from the graphs that the transformation function $U_j = f_j\,(P_j(Y))$ may assume different forms within the limits of envelopes A and B. The convex shape B expresses the risk preference and the concave shape A expresses aversion to the risk factor, *i.e.* the preference of utility.

The TIEQ method is based on the standard catalogue of selected indicators divided into six categories, *i.e.*

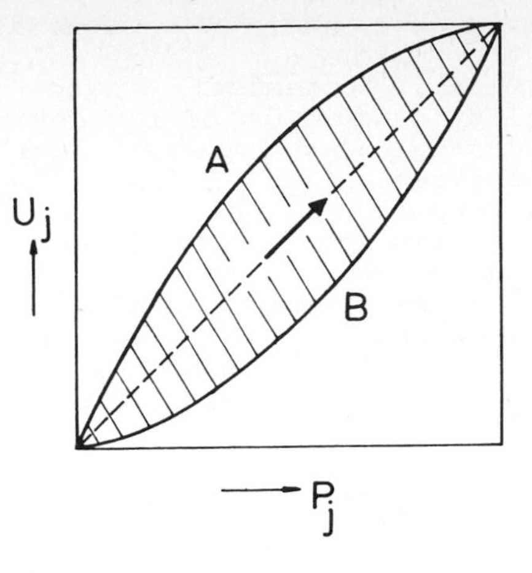

Figure 1 Direct dependence of transformation function $U_j = f_j(P_j(Y))$. The convex limiting curve B expresses the ecologically pessimistic evaluation, the concave curve A the ecologically optimistic evaluation

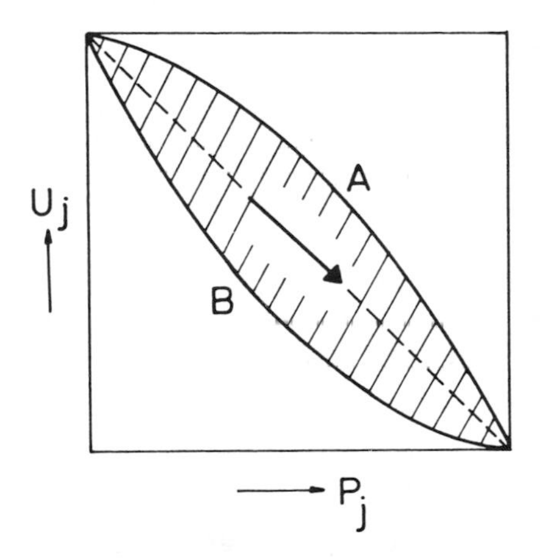

Figure 2 Indirect dependence of transformation function $U_j = f_j(P_j(Y))$. Explanation of the meaning of curves A and B is the same as for Figure 1.

Category A: Demographic and socio-economic phenomena in the region;

Category B: Physical, chemical, and biological quality indicators of the environment;

Category C: Psychological impacts and aesthetic properties of the geographical environment;

Category D: Structures of cultural and educational value in the region;

Category E: Ecological properties of the biophysical environment;

Category F: Investment and operating requirements, efficiency, and safety of the technical projects.

The individual categories of criteria involve diverse numbers of indicators.

In view of the fact that every $P_j(Y)$ in the given set $j = 1,2,\ldots,n$ has different significance, the weight parameter w_j is introduced whose finite value is the product of the output of the collaborating team approach as the standard value determined by selected experts using the Delphi method.

The socially maximally preferred (optimal) variant of the technical solution of the project V_i in relation to the total quality of the environment is reached when the function TIEQ reaches the highest value. Potential anthropogenic impacts of the individual V_i for $I = 1,2,\ldots,m$ on the ecosystem of the human environment should be assessed with regard to the time factor, *i.e.* the consequences of $P_j(Y)$ during the period of construction and during the period of the permanent full operation of the project.

Values U_j may be determined from auxiliary graphs or by computation and algorithm where the computer will determine the total value of TIEQ for every variant V_i.

4 Example of the Transformation Function as a Qualitative Multiplier

Part of the TIEQ method includes a scaling-weighting checklist with rating curves. Each criterion includes:

(a) Definition - designation of criterion by value of indicator P_j, the mathematical formulation of the relation $U_j = f_j(P_j(Y))$, the description of the applied analytical indicator $P_j(Y)$, the applied objective (technical) or subjective unit, and the description of the scale.

(b) Graph

(c) Discussion

(d) Method of determining the value of analytical indicator $P_j(Y)$

An example for solving the criterion of *biochemical oxygen demand* (organic pollution) is given for information.

(a) Definition

$j = 11$

$U_{11} = 1 - 0.004\ P_{11}^{2.322}$ for $P_{11} \in \langle 0;\ 8 \rangle$

$U_{11} = 0.004\ |P_{11} - 16|^{2.322}$ for $P_{11} \in \langle 8;\ 16 \rangle$

Indicator: Unit P_{11} for BOD_5 (mg l^{-1}).

(b) Graph: See Figure 3.

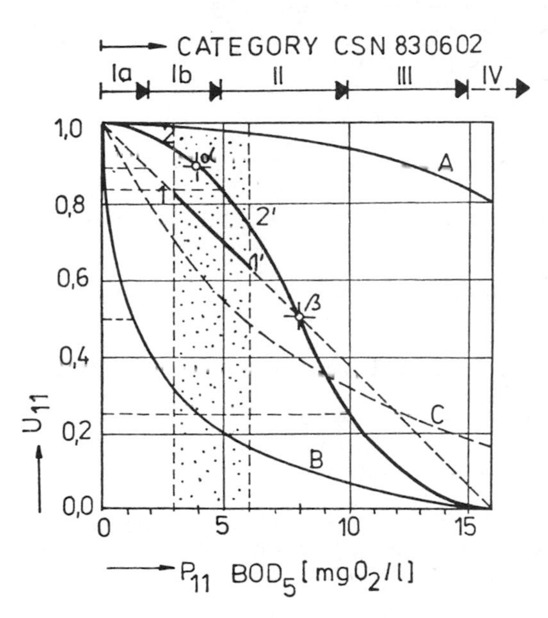

Figure 3 Transformation functions for parameter of biochemical oxygen demand in surface waters, indicator P_{11} = BOD_5 [mg O_2 l^{-1}]. Symbols: α, BOD_5(max) for water flows used for the water supply system; β, BOD_5(max) for other surface waters; 11' and 22', critical area according to ref. 9; curve A, for mountain water flow according to ref. 7; curve B for stagnant waters according to ref.7; curve C, according to ref.8.

(c) Discussion

Heterotrophic bacteria use organic matter for metabolic synthesis. Biochemical oxygen demand (BOD) is defined as the amount of oxygen consumed by micro-organisms for the degradation (mineralization) of organic substances under aerobic conditions. This amount of oxygen is proportional to the amount of available degradable organic substances. The level of BOD will therefore serve as the indicator of the level of biologically degradable organic pollution.

In view of the length of the standard method (20 days) the Czechoslovak State Standard[5] rules that a 5 day period is necessary for determining BOD. The value differs for different types of water and is dependent on water temperature. The following factors were taken into consideration for the choice of the transformation function and the construction of graphs:

(1) Initial rise of BOD_5 value at interval $P_{11} \in \langle 0;15 \rangle$ corresponds to clean and very clean water; see quality water category 1a and 1b according to reference 5. For the given interval the preference of economic utility was accepted and the mildly descending concave part of the parabola was chosen. The marginal domain of categories 1b and II corresponds to $P_{11} = 5$, *i.e.* $U_{11} = 0.83$. The value multiplier $U_{11} = 0.9$ was assigned to BOD_5(max), *i.e.* P_{11}(max) = 4 for water courses used for the water supply system according to reference 6.

(2) The critical area of deterioration of water quality corresponds to water quality category II, *i.e.* BOD_5 values within the region $P_{11} \in \langle 15;10 \rangle$, where for $P_{11} = 10$ is $U_{11} = 0.26$. For permissible BOD_5(max), *i.e.* P_{11}(max), = 8 for all surface waters the Czechoslovak Regulations[5] assign the value $U_{11} = 0.5$. All higher values of BOD_5 indicate below average water quality; the curve for $P_{11} > 8$ favours the ecological risk and has a highly convex shape.

(3) Values of $P_{11} > 15$ correspond to the fourth category of the Czechoslovak Regulations[5], *i.e.* such water is highly polluted. The value multiplier is $U_{11} = 0$.

(4) Transformation in the whole system $U_{11} \in \langle 0;1 \rangle$ is conducted by curve $U_{11} = f_{11}(P_{1'1})$, a descending S-shaped curve consisting of two parabolic branches with the inflexion point at $P_{11} = 8$, $U_{11} = 0.5$. In compliance with Czechoslovak regulations one curve was chosen for all types of surface waters using the alternative approach[7] discrimination of the behaviour and properties of flowing water, *i.e.* curve A, and stagnant water in reservoirs and lakes, *i.e.* curve B

gives explicit preference of risk. A compromise transformation is a curve provided by the National Sanitation Foundation (NSF).[8] The manual[9] classifies as conditionally suitable P_{11} ⟨3;6⟩ which for neutral (linear) transformation would correspond to points 11' to 22' on the S-shaped curve, *i.e.* both agree with the boundary between clean and polluted water quality.

(d) Method:

The determination of surface water is made in quality accordance to the BOD_5 indicator used under the Czechoslovak State Standards.[5]

5 Environmental Impact Assessment - the Recent Czechoslovak Case Study

The provision of electric power sources based on nuclear power in the Czechoslovak Socialist Republic is a complex problem, namely with regard to geographic conditions (seismicity, high population density, limited capacity of local water resources, *etc.*). The choice of suitable localities is always determined according to IAEA binding criteria. In view of the fact that new plants are planned with a capacity of 2000 MW or 4000 MW there is general interest that such plants should be sited and built with greatest social responsibility.

One method used as the basis for the final decision-making process was the multi-critical analysis proceeding from the axiomatic theory of ultimate benefit. The additive model was applied of the formal method of the Total Index of Environmental Quality, defined by equation (4). The partial function of benefit was constructed as the so-called real[10] monotonous increasing or decreasing function for interval $O \geqslant U_j \geqslant 1$ depending on the character of indicator P_j, *i.e.* depending on whether this is a direct dependence (according to the principle 'the higher → the better') or indirect dependence ('the lower → the better').

The additive nature of the model was maintained by observing the principle that for the set w_j it should apply equation (5) and simultaneously equation (6).

The task was to determine the order of implementation of five nuclear power plant variants:

V(1) - Opatovice 2000 MW (see also Table 3)
V(2) - Blahutovice 2000 MW
V(3) - Blahutovice 4000 MW
V(4) - Kecerovce 2000 MW
V(5) - Temelin 4000 MW

The Czech Power Works, which commissioned the study, set a catalogue of criteria for $n = 20$ and the input values for all variants (incidence matrix). The interaction with water resources was expressed by three criteria, criteria P_5, P_7, and P_9. The author of the case study determined the shapes of single-dimensional transformation function U_j.

Table 1 *The criteria used for water in the EIA procedure*

j	P_j	*Range*	w_j	*Dimension*
5	Average annual water consumption and losses	8	0.05816	$m^3\ s^{-1}$
7	Depth of underground	12	0.04737	m
9	Distance from water level	14	0.04588	km

Note: min $w_j = 0.02335$, max $w_j = 0.07450$.

Values for the set of quantitative multipliers w_j were determined using the team expertise method with a total of 39 individuals. The method of pair comparison and the modified Delphi method (single anonymous inquiry without subsequent correction) were used.

Values of the multi-dimensional vector U were determined using machine computation using the Hewlett Packard 9830A as follows:

Variant	$V(1)$	$V(2)$	$V(3)$	$V(4)$	$V(5)$
Value U	0.4575	0.5371	0.5885	0.4810	0.5779
Range	5	3	1	4	2

The analysis of the results of the solution has shown that priority was won by variants with designed double output, which significantly implements the priorities of the economic criterion to which most experts independently of one another attributed the highest importance.

6 Example - A Numerical Illustration

The use of the TIEQ method described in Sections 3 and 5 illustrated an additive form in accordance with equation (4). Use of the TIEQ method consists of obtaining baseline data on the 20 factors in Table 2 and, through use of their functional relationships, conversion into environmental quality scale values is possible. The partial functions of benefit were

Table 2 *Parameters used for comparative analysis and impact assessment*

j	*Parameter P_j*	*Dimension*
1	Settlement density < 5 km	population/km^2
2	Settlement density <10 km	population/km^2
3	Demolitions owing to construction	number of items
4	Demands on new flats	number of items
5	Average annual water consumption and losses	m^3/s
6	Thermal pollution - waste heat	MW
7	Depth of underground water level	m
8	Earthwork volume	10^6 m^3
9	Distance from water source	km
10	Health and safety - distance from reactor	km
11	Resettled population	population
12	Natural formations and protected areas	ha
13	Composite effect - architectural incorporation of technical structure in landscape	NP+/
14	Arable land seized	ha
15	Forest land seized	ha
16	Criterion of project electricity output	MW
17	Criterion of project heat output	MW
18	Criterion of costs	10^9 crowns/ 1000 MW
19	Criterion of complementary costs	10^9 crowns/ 1000 MW
20	Criterion of labour demand	number of persons

Note: NP+/ = Number of points according to quality descriptors.

Professional judgement must be exercised in the interpretation of the numerical results, with the focus being on comparative analyses rather than specific numerical values.

constructed according to equation (7):

$$U_j = \left[\frac{P_j - P_j(\min)}{P_j(\max) - P_j(\min)} \right]^k \tag{7}$$

for the so-called direct dependence and according to equation (8):

$$U_j = 1 - \left[\frac{P_j - P_j(\min)}{P_j(\max) - P_j(\min)} \right]^k \tag{8}$$

for indirect dependence.

The value of k has been derived from a set of real input data for all variants V_i, *e.g.* for parameter P_1 they are

V_1	V_2	V_3	V_4	V_5
39.3	109.6	110.6	330 (max)	30 (min)

According to equation (8) and for

$$\bar{P}_1 = \frac{39.3 + 109.6 + 110.6 + 330 + 30}{5} = 123.9$$

which corresponds to mean value of $U_1 = 0.5$, then $k = 0.597$. The resulting form of the first partial function of benefit is

$$U_1 = 1 - \left[\frac{P_1 - 30}{300} \right]^{0.597} \tag{9}$$

A similar procedure was used for each parameter.

These scale values, *e.g.* partial utility functions U_j, would then be multiplied by the appropriate weight and aggregated to arrive at a composite U_i score for each variant; see Table 3 for V_1. This numerical system provides an opportunity for displaying interchanges between the variants in terms of specific environmental parameters, impacts, *etc.* The socially maximum preferred optimal, *i.e.* the variant of the technical solution of projects V_i in relation to the total quality of the environment, is reached when the value of U = TIEQ reaches the highest value.

Table 3 *The value of the Total Index of Environmental Quality for the variant V(1) Opatovice 2000 MW*

j	w_j	P_j	U_j	$w_j U_j$
1	0.0649	39.3000	0.8743	0.0568
2	0.0641	297.8000	0.0050	0.0003
3	0.0234	126.0000	0.3644	0.0085
4	0.0327	131.0000	0.4000	0.0131
5	0.0582	1.3300	0.8569	0.0498
6	0.0478	3500.0000	0.8926	0.0426
7	0.0474	2.9000	0.0150	0.0007
8	0.0293	4.1000	0.8633	0.0253
9	0.0459	3.5000	0.9721	0.0446
10	0.0659	6.0000	0.0946	0.0062
11	0.0382	262.0000	0.6585	0.0251
12	0.0472	3.0000	0.0025	0.0001
13	0.0234	1.0000	0.0247	0.0006
14	0.0653	83.7000	0.7934	0.0518
15	0.0536	48.0000	0.7563	0.0405
16	0.0695	1962.0000	0.0673	0.0047
17	0.0615	1640.0000	0.5179	0.0319
18	0.0745	20.1800	0.2563	0.0191
19	0.0574	6.0840	0.5751	0.0330
20	0.0301	1.0180	0.0912	0.0027
Sum				0.4575

7 Conclusions

Despite all the efforts and objectives of these decision-making processes, with the aim of minimizing the value of accepted risk, one must deduce that there does not as yet exist a universal law which would make it possible to express mathematically a set of factors which are more or less significant.[11] In this concept the role of the subjective human factor is irreplaceable. The TIEQ method is only the first step for the next complex problem.

The total socio-economic and ecological effect of the application of the TIEQ method may contribute to the moderation of the negative environmental trends in Czechoslovakia. The method may be viewed as an important political tool which emphasizes the quality of the environment for the further development of society.

8 References

1. J. Riha, 'Multicriterial Assessment of the Investment Intents', SNTL, the State Techn. Publ. House, Prague, 1987 (Czech).

2. H. Raiffa, 'Decision Analysis, Introductory Lectures on Choices under Uncertainty', Addison-Wesley Publ. Co., Inc., Reading, Ontario, 1970.

3. R.L. Keeney and H. Raiffa, 'Decisions with Multiple Objectives: Preferences and Value Trade-Offs', J. Wiley & Son, New York, 1976.

4. P.C. Fishburn, 'Utility Theory for Decision-Making', J. Wiley & Son, New York, 1970.

5. 'Quality Inspection of Surface Water', Czechoslovak State Standard, CSN 83 0602.

6. 'Indexes of Permissible Water Pollution', Regulation of the Government of the CSR, No. 25/1975.

7. N. Dee et al., 'Environmental Evaluation System for Water Resource Planning', Battelle Columbus Laboratories, Columbus, Ohio, 1972 (Final Report).

8. R.M. Brown et al., 'A Water Quality Index - Do We Dare?', Water and Sewage Works, Washington, DC, 1970.

9. 'Guide for Environmental Regional Planning', Strojizdat, Moscow, 1980 (Russian).

10. J. Riha et al., 'Environmental Impact Assessment of Nuclear Power Plants' (Report prepared for the Czech Power Works, "Concern", Contract No.188/1986), Techn. Univ. of Prague, 1987 (Czech).

11. R.G.H. Turnbull, 'Risk and Hazard Assessment', in *Environmental Impact Assessment NATO ASI Series,* Martinus Nijhoff Publishers, The Hague, 1983, p.383.

21

Quantitative Structure-Activity Relationships and Toxicity Assessment in the Aquatic Environment

R.L. Lipnick

OFFICE OF TOXIC SUBSTANCES (TS-796), US ENVIRONMENTAL PROTECTION AGENCY, 401 M STREET, S.W., WASHINGTON, DC 20460, USA

1 Introduction

1.1 QSAR Models. Quantitative structure-activity relationships (QSAR) offer an attractive approach for systematizing the large literature of test data on the aquatic toxicity of organic compounds. The resulting QSAR models can provide a rational means of predicting the aquatic toxicity of large numbers of untested industrial organic chemicals. QSAR methodologies can also serve to identify random and systematic errors in such toxicity data. Only aquatic toxicity QSAR models will be discussed in this chapter; some corresponding models have been developed for toxicity to mammals via oral, dermal, inhalation, intraperitoneal, and intravenous administration[1,2] (see Section 1.5 and Chapter 4).

1.2 Toxic Substances Control Act (TSCA). Approximately 30,000 chemicals are now produced in commerce, and this number is rapidly growing each year.[3] Section 5 of the Toxic Substances Control Act (TSCA) provides the Office of Toxic Substances of the United States Environmental Protection Agency (EPA) with the authority to investigate the potential hazards to human health and to the environment posed by the commercial production of new chemicals,[4] which are being introduced at the rate of about 2000 per year.

1.3 Narcosis Mechanism: Baseline Minimum Toxicity. The EPA Environmental Research Laboratory in Duluth, Minnesota (ERL-Duluth), began an extensive programme in the late 1970s to test the toxicity of a large number of organic chemicals to the fathead minnow *(Pimephales promelas)* in a 96 h test, and to use the data in the development of QSAR models.[5] The development of a QSAR model generally requires the availability of test data on a group of similar chemicals which have been demonstrated or assumed to act by a common molecular mechanism of toxicity. Narcosis represents the most fundamental toxic mechanism, and is exhibited by all non-electrolyte organic compounds, irrespective of their chemical class. Narcosis action, however, will be masked in the case of compounds which also act by more specific mechanisms.

1.4 Log *P* Parameter. Charles Ernest Overton[6-8] and Hans Horst Meyer[9-11] independently reported at the turn of the century a correlation between the toxicity of chemicals acting by a narcosis mechanism and their olive oil/water partition coefficients. Such correlations are considered to reflect a mechanism in which a common molar concentration or molar volume of toxicant at the biophase site of action produces an equitoxic response. This narcosis site is regarded as residing within cellular membranes or critical proteins bound within such cell membranes. The alcohol n-octanol has been found to serve as a useful model for the physicochemical properties of this putative biophase site of action, and the n-octanol/water partition coefficient *(P)* is commonly employed as a model parameter to simulate the transport of a xenobiotic chemical from an aquatic environmental phase to the biophase site of action.

1.5 Bilinear Model. Veith and co-workers at ERL-Duluth reported that data on 60 non-electrolyte organic compounds consisting of alcohols, ketones, ethers, alkyl chlorides, and some substituted benzene derivatives were well fitted to a bilinear QSAR model as a function of log *P* [equation (1)], where LC_{50} is the 96 h 50% mortality to the fathead minnow.[3,12]

$$\text{Log } LC_{50} =$$

$$-0.94 \log P + 0.94 \log (0.000068\, P + 1) - 1.25 \qquad (1)$$

(n = number of compounds = 60)

The bilinear model[13] has found application in the field of drug design[14] and in the development of mammalian toxicity QSAR models for oral, dermal, intravenous, and intraperitoneal administration[1,2] in which the toxic dose reaching the biophase site of action is a function of both its intrinsic partitioning properties and the rate of transport to this site. The bilinear model reflects a similar consideration in the case of fish and other aquatic organisms. The maximum toxicity in a 96 h test was found for chemicals whose log *P* is approximately 4.5. Those chemicals in the data set having log *P* values larger than 4.5 can be regarded as not having achieved equilibrium partitioning within the studied 96 h test duration.

1.6 Linear Model. Könemann found that LC_{50} data on guppies *(Poecilia reticulata)* tested for 14 days follow a linear relationship up to a log *P* value of 5.8 [equation (2)],[15] providing support that non-equilibrium partitioning into the fish is responsible for the non-linear behaviour [equation (1)] observed with 96 h LC_{50} data.

$$\text{Log}\ (1/LC_{50}) = 0.871\ \log P - 4.87 \quad (2)$$

$(n = 50)\ (r = 0.988)\ (s = 0.237)$

By contrast, 24 h test data for the goldfish[16] follow a linear model up to a log *P* of 3, beyond which a bilinear model is needed to accommodate the data.

1.7 Comparative Toxicity. The slope and intercept of these models have been found to be very similar for a number of different species of freshwater fish,[16-18] and such data measured under equilibrium conditions can be treated in a combined fashion as a first approximation for chemical hazard assessment of chemicals acting by a narcosis mechanism. Under non-equilibrium conditions, the rate of uptake of chemicals will vary with species and environmental conditions, including temperature.

1.8 Water Solubility Cut-off. A non-electrolyte chemical will not produce lethality if the expected lethal concentration exceeds its water solubility. However, this is not strictly the case when testing is done in the presence of dispersing agents, which were not employed in generating the present test data. Chemicals non-toxic in a 24 or 96 h test can nevertheless exhibit toxicity in a longer test since, although the higher concentration required under non-equilibrium partitioning may exceed water solubility, this need not be the case under conditions of equilibrium partitioning.

2 QSAR Analysis of Aquatic Toxicology Data

2.1 Approach and Data. In order to illustrate the application of the above principles, 96 h LC_{50} data on the toxicity of 45 non-electrolyte organic compounds to the fathead minnow were selected from data contained in a three volume compilation of the test results from ERL-Duluth and the University of Wisconsin-Superior Centre for Great Lakes Studies research program.[19-21] An estimated 24 h LC_{50} was extracted where possible from graphs of LC_{50} as a function of time.

Log *P* values for all 45 chemicals could be estimated directly from chemical structure using the CLOGP3.3 computer program,[20] and used to predict baseline narcosis 96 h LC_{50} values based upon the above bilinear model [equation (1)]. Measured log *P* values, which were found[22] for a much smaller number of these chemicals, were in good agreement with the calculated values. Water solubility data were found for very few of the compounds in this study,[23] and this property was estimated using equation (3),[24] where S is the solubility in μmol l^{-1}, and MP is the melting point in °C. A nominal value of 25°C is used in equation (3) for solutes which are liquid at room temperature. Those chemicals in Table 1 which show no lethality or adverse effects in 96 h at water saturation have melting points in excess of 25°C.

$$\text{Log } P = 6.5 - 0.89 \log S - 0.015\, MP \qquad (3)$$

$(n = 27)$ $(r^2 = 0.92)$

2.2 Chemicals Showing No Toxicity in 96 Hours. Four of the chemicals in this study showed no mortality to fish at or near water saturation. These compounds are arranged in Table 1 in order of decreasing log P. Octachlorostyrene, with an extremely high log P of 7.94, is not expected to produce adverse effects within a 96 h test duration, although the LC_{50} at equilibrium [equation (2)] is predicted to be just within its water solubility. Toxicity for the remaining three compounds is water solubility limited. The predicted LC_{50} values for compounds (2) and (3) are within an order of magnitude of their water solubilities and these chemicals could potentially exhibit toxicity in a chronic test, for which QSAR studies indicate that effects take place at about an order of magnitude lower concentration.[25] Although compounds (2) and (3) are not individually toxic in the 96 h test, they would be expected to make a normal additive contribution[26] to the aquatic toxicity of other compounds acting by a narcosis mechanism if tested at a concentration sufficiently low so that their water solubility is not exceeded. Overton in his classic monograph 'Studien über die Narkose', observed some time ago that compounds whose lack of narcotic activity was previously unexplained, produced such additive effects.[7,8] The last chemical in this group, uracil, showed no toxicity at a very high concentration. Its narcotic toxicity is predicted to exceed its measured water solubility.

2.3 Toxicity Consistent with Baseline Narcosis QSAR Prediction. A comparison was made between the observed and QSAR-predicted toxicities of the remaining 41 chemicals, by calculating their excess toxicity, T_e,[16] according to equation (4),

$$T_e = LC_{50}(pred)/LC_{50}(obs) \qquad (4)$$

where $LC_{50}(pred)$ and $LC_{50}(obs)$ are the QSAR-predicted and observed toxicities, respectively. The T_e values were calculated using the bilinear QSAR model [equation (1)]. Veith and co-workers used this ratio to test for

Table 1 *Compounds not acutely toxic at the highest test concentration*

No.	Chemical name	CAS RN[a]	MW	Log P[b]	Pred LC_{50} (mg l^{-1}) 96 h[c]	MP (°C)[d]	Water Solubility (mg l^{-1})[e]	Comments[f]
1	Octachlorostyrene	29082-74-4	379.68	7.940	2.58 [0.003]	101	1.82×10^{-4} (6×10^{-3})[g]	No mortalities at 4.35 μg l^{-1} (70%saturation)
2	4-Bromo-2',4'-dinitrodiphenyl ether	72589-66-1	339.09	4.794	2.81 [1.68]	138.5	0.13 (0.2)[h]	No mortalities at saturation 0.2 mg l^{-1}
3	Dodecanamide	1120-16-7	199.34	4.056	2.96 [4.33]	110	1.55 (1.24)[i]	No deaths at 0.896 mg l^{-1} (72%saturation)
4	Uracil	66-22-8	112.09	-1.07	6.20×10^{4} [6.90×10^{4}]	>300	304 (3000)	No mortalities at 2500 mg l^{-1}

[a] Chemical Abstracts Registry Number
[b] Calculated using the CLOGP3.3 Computer program. Values in parenthesis are measured and are from the Pomona College Medchem 'starlist'.
[c] Calculated using bilinear QSAR [equation (1)]; values in brackets calculated using linear QSAR [equation (2)].
[d] Melting points are from the Aldrich Catalog Handbook of Fine Chemicals,[40] from Beilstein's Handbuch der Organischen Chemie, or Lange's Handbook of Chemistry,[48] unless otherwise indicated. L indicates liquid.
[e] Water solubilities calculated based upon equation (3); values in parenthesis are measured.
[f] From refs. 19-21.
[g] Calculated from 70% saturation concentration of 4.35 μg l^{-1} (ref. 21).
[h] From ref. 20.
[i] Calculated from 72% saturation concentration of 0.896 mg l^{-1} (ref. 20).

the statistical agreement between observed and predicted LC_{50}s for the 60 compounds used to derive equation (1).[12] This ratio was found to vary from 0.12 to 4.22 for 3,3-dimethylbutan-2-one and propan-2-ol, respectively. These two compounds belong to classes (saturated aliphatic monoketones and monohydric alcohols) well associated with a narcosis mechanism. Therefore, it is reasonable to ascribe the toxicity of any of the remaining 41 compounds belonging to the data set under consideration with T_e values less than 4.2 to solely a narcosis mechanism. T_e values have also been calculated for the chemicals in this data set using the linear QSAR model [equation (2)] to provide a comparison for those having a high partition coefficient, where the influence of pharmacokinetics becomes significant.

Prior studies of this type have demonstrated that no sharp dichotomy exists between non-electrolytes following a narcosis model and those showing a more specific mechanism of action.[16,17] To allow for possible problems in the assignment of calculated log *P* values for more complex molecules, compounds having a T_e value up to twice that of 3,3-dimethylbutan-2-one have been included in Table 2. A total of 25 compounds satisfy this criterion, and they are listed in order of increasing T_e value.

2.4 Chemicals Showing Excess Toxicity. The remaining 16 compounds (Table 3) have T_e values ranging from 8.8 to 1960. The discussion below provides a plausible explanation in each case for this excess toxicity based upon pharmacokinetics or mechanistic organic chemistry.

2.4.1 Pharmacokinetic effect. Two of these compounds, 2,6-diphenylpyridine (30) and 2,5-diphenylfuran (39) have log *P* values exceeding 4.5, beyond which equation (1) no longer predicts a linear relationship between log *P* and log LC_{50}. Comparison with the predictions of equation (2), which gives the linear prediction, yields T_e values of 4.7, and 4.8, respectively, consistent with a narcosis mechanism. Why these particular compounds undergo biological transport by the fish more rapidly than other non-electrolytes from which equation (1) was derived deserves further investigation.

2.4.2 Electrophilic toxicity: aldehydes and Schiff base formation. Table 3 contains test data on two aliphatic aldehydes and four aromatic aldehydes. In the case of the aliphatic aldehydes, butanal compound (40) and acetaldehyde (44), excess toxicity decreases with increasing partition coefficient. Hexanal (26), with a log *P* value of 1.89, shows an excess toxicity of only 6.7, which may be considered borderline between narcosis and a more specific mechanism. Overton reported a similar finding of an increasing tendency towards narcotic action within a series of increasing chain length.[7,8] Ferguson interpreted this behaviour in terms of a transition from chemical to physical toxicity.[28] Acetaldehyde and other aliphatic aldehydes have been found to form adducts with nucleic acids via Schiff base formation at exocyclic amino-groups.[29] A corresponding reaction with an ϵ-amino-group on a lysine residue within a critical enzyme is a likely event related to the observed excess toxicity of these aldehydes.

The four aromatic aldehydes tabulated are all ring-substituted benzaldehydes. 2,4,5-Trimethoxy-benzaldehyde (32) has a much lower log *P* value (1.38) than pentafluorobenzaldehyde (41), but the latter exhibits considerably greater excess toxicity (T_e = 51) than the former (T_e = 11). These compounds fall into two extremes in terms of the electronic effect of the ring substituents on the reactivity of the aromatic aldehyde moiety. The three methoxy-groups are all electron donating, whereas the fluoro substituents are electron withdrawing.[30] Perfluorobenzaldehyde has been found to be a useful reagent for derivatizing primary and secondary aliphatic amines to the corresponding Schiff bases for detection and separation using HPLC.[31,32] The ϵ-amino-groups in lysine residues in enzymes and other biological macromolecules containing such amino-groups offer a target site for Schiff base formation which would account for the toxicity of these aldehydes and the order observed in their excess toxicity, as illustrated in Scheme 1. *s*-Trioxane (19) is an acetal derived chemically from three molecules of formaldehyde. It shows only narcosis action, indicating that the required acetal cleavage does not take place enzymatically; chemical hydrolysis requires acid catalysis.

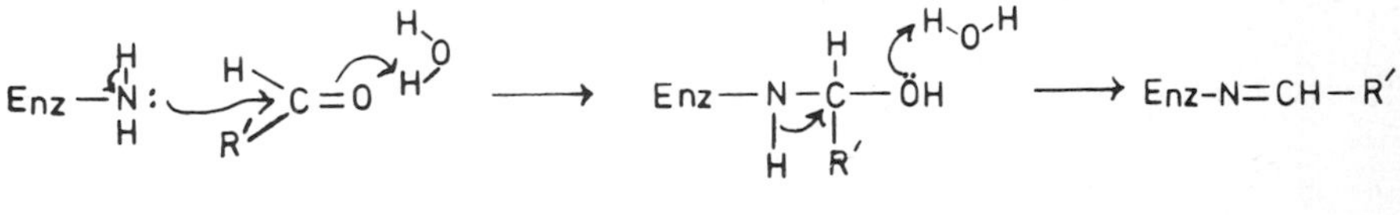

$$\mathrm{Enz{-}NH_2} + \mathrm{R'CH{=}O} \;(\mathrm{H_2O}) \longrightarrow \mathrm{Enz{-}NH{-}CH(R'){-}OH} \;(\mathrm{H_2O}) \longrightarrow \mathrm{Enz{-}N{=}CH{-}R'}$$

Scheme 1

<u>2.4.3 Other electrophiles: S_N2 chemical mechanism</u>. The excess toxicity of 3,4-dichlorobut-1-ene (31), phenyl disulphide (36), α,α'-dichloro-*p*-xylene (42), and chloroacetonitrile (45) is considered to reflect an

Table 2 *LC_{50} consistent with baseline narcosis prediction*[a]

No.	*Chemical name*	*CAS RN*	*MW*	*Log P*	LC_{50} (mg l^{-1})[b] 24 h	96 h	*R* (24/96)	*Pred* LC_{50} mg l^{-1}	T_e	*MP* (°C)	*Water solubility* (mg l^{-1})
5	2,4,5-Trimethyloxazole	20662-84-4	111.14	2.238	450	448	1.0	49.8	0.11	L	2.59×10^3
								[92.6]	[0.21]		
6	N,N,-Dimethyl-*p*-toluidine	99-97-8	135.21	2.986	52	52	1.1[d]	12.6	0.26[d]	L	455
					60	46		[25.1]	[0.51][d]		
7	4-Methyloxazole	693-93-6	83.09	0.940	2700	1390	1.9	611	0.44	L	5.56×10^4
								[935]	[0.67]		
8	*m*-Bromobenzamide	22726-00-7	200.04	2.481	100	92.7	1.1	53.4	0.57	155.3	15.8
								[102.4]	[1.1]		
9	2-Fluorotoluene	95-52-3	110.13	2.934	-	19.3	-	11.4	0.59	-62	424
								[22.7]	[1.18]		
10	n-Butyl sulphide	544-40-1	146.30	4.016	(3.5)[e]	3.58	(1.0)[f]	2.28	0.64	L	34.3
								[3.45]	[0.96]		
11	1,2-Dichloropropane	78-87-5	112.99	1.987	170	127	1.3	86.7	0.68	-100	5.04×10^3
								[156]	1.23		(2.7×10^3)
12	4,7-Dithiadecane	22037-97-4	178.35	3.517	-	752	-	5.99	0.80		
								[11.4]	[1.5]	L	152
13	Naphthalene	91-20-3	128.17	3.316	8	6.14	1.3	6.23	1.0	81-3	19.3
								[12.3]	(2.0)		(38.2)
14	1,3-Dichloropropane	142-28-9	112.99	1.707	130	131	1.0	158	1.2	-99	1.04×10^4
								[273]	[2.1]		(2.9×10^3)
15	4-Benzoylpyridine	14548-46-0	183.21	2.067	105	103	1.0	118	1.2	69-71	1.12×10^3
								[215]	[2.1]		
16	1,1,1,3,3,3-Hexafluoro-propan-2-ol	920-66-1	168.04	1.592	270	244	1.1	302	1.2	-4	2.08×10^4
								[511]	[2.1]		
17	t-Butyl disulphide	110-06-5	178.36	4.216	(1.5)[e]	1.37	(1.1)[f]	2.21	1.6	L	24.9
								[2.81]	[2.1]		

18	Butan-2-one oxime	96-29-7	87.12	0.361	1500	843	1.8	2240 [3130]	2.7 [3.7]	-29.5(L)	>10^5
19	s-Trioxane	110-88-3	90.08	-0.559	6500	5950	1.1	17000 [20500]	2.9 [3.5]	61-62.5	>10^5
20	Furan	110-00-9	68.08	1.348	(60)[e]	61	(1.0)[f]	207 [338]	3.4 [5.5]	-35	1.58×10^4
21	2,4-Dichlorobenzamide	2447-79-2	190.03	1.543	-	95.6	-	380 [638]	3.4 [6.7]	191-2	41.0
22	Pyrrole	109-97-7	67.09	0.758 (0.75)	220	210	1.1	732 [1088]	3.5 [5.2]	-23	7.19×10^4
23	Deca-1,9-diene	1647-16-1	138.25	4.896	0.38	0.29	1.3	1.10 [0.558]	3.8 [1.9]	L	3.32
24	2-Cyanopyridine	100-70-9	104.11	0.234 (0.50)	1050	726	1.5	3530 [4830]	4.9 [6.7]	78-80	5.12×10^4
25	2-Adamantanone	700-58-3	150.22	1.433	60	60.8	1.0	381 [629]	6.3 [10]	256-8	3.32
26	Hexanal	66-25-1	100.16	1.892	30	14.0	2.1	94.3 [167]	6.7 [12]	L(L)	5.71×10^3 (5.00×10^3)
27	Adamantane	281-23-2	136.24	3.982	(0.325)[e]	0.285	(1.1)[f]	2.22 [3.43]	7.8 [12]	205-210	0.027
28	1,4-Dicyanobutane	111-69-3	108.14	-0.424	2000	1930	1.0	1.52×10^4 [1.88×10^4]	7.9 [9.7]	1-3	>10^5
29	Chloromethylstyrene (60% *Para*, 40% *meta*)	1592-20-7[g] 39833-65-3[h]	152.62	3.978	1	0.31	3.2	2.50 [3.88]	8.1 [13]	3.5	39.4

[a] See footnotes in Table 1
[b] Toxicity to the fathead minnow (*Pimephales promelas*) from refs. 19-21.
[c] Ratio 24 h/96 h LC_{50}
[d] From average of two LC_{50} measurements
[e] 48 h LC_{50}
[f] Based upon 48 h LC_{50}
[g] *Para* isomer
[h] *Meta* isomer

Table 3 *Compounds showing excess toxicity*[a]

No.	*Chemical name*	*CAS RN*	*MW*	*Log P*	LC_{50} (mg l^{-1}) 24 h	96 h	*R* (24/96)	*Pred* LC_{50} mg l^{-1}	T_e	*MP* (°C)	*Water solubility* (mg l^{-1})
30	2,6-Diphenylpyridine	3558-69-8	231.30	4.861 (4.82)	0.45	0.212	2.1	1.87 [1.00]	8.8 [4.7]	74-76	0.841
31	3,4-Dichlorobut-1-ene	760-23-6	125.00	1.972	27	9.33	2.9	99.0 [178]	11 [19]	-61	5.80×10^3
32	2,4,5-Trimethoxybenzaldehyde	4460-86-0	196.20	1.381	80	49.5	1.6	556 [912]	11 [18]	112-14	1.32×10^3
33	4-Nitrobenzamide	619-80-7	166.14	0.830	-	133	-	1550 [2330]	12 [18]	200-1	160
34	Pentane-2,4-dione	123-54-6	100.12	0.699[b] -0.543[c]	200	104	1.9	1240[d] [1820][d]	12 [18]	-23	$>10^5$
35	Allyl cyanide	109-75-1	67.09	0.120	210	182	1.2	2910 [3910]	16 [22]	L	$>10^5$
36	Phenyl disulphide	882-33-7	218.34	4.4 (4.41)	0.27	0.11	2.5	2.29 [2.33]	21 [21]	58-60	4.39
37	Benzaldehyde	100-52-7	106.12	1.495 (1.48)	35	7.61	4.6	235 [392]	31 [52]	-26	1.69×10^4 (3.45×10^3)
38	2,5-Diphenylfuran	955-83-9	220.27	5.544	0.07	0.05	1.4	1.56 (0.242)	31 [4.8]	88(L)	0.086
39	α,α,α-Trifluoro-*m*-tolualdehyde	454-89-7	174.12	2.594	1.9	0.92	2.1	36.6 [71.0]	40 [77]	L	1.62×10^3
40	Butanal	123-72-8	72.11	0.834	20[e] 16	16 13.4	1.4	667 [1003]	45 [68]	-96	6.35×10^4 (6.99×10^4)
41	Pentafluorobenzaldehyde	653-37-2	196.07	2.449	2.5	1.1	2.3	56.0 [107]	51 [97]	20	2.65×10^4
42	α,α-Dichloro-*p*-xylene	623-25-6	175.06	3.866	0.14	0.039	3.6	3.35 [5.57]	86 [143]	99-101	3.17

43	1,3-Dibromopropane	109-64-8	201.90	1.987	5.3	1.79	3.0	155 [278]	87 [155]	-34	9.01×10^3
44	Acetaldehyde	75-07-0	44.05	-0.224	60	30.8	2.0	4020 [5117]	131 [166]	-125	$>10^5$ (6.67×10^5)
45	Chloroacetonitrile	107-14-2	75.50	0.219 (0.45)	(4)[f]	1.35	(3.0)[g]	2640 [3610]	1960 [2670]	L	$>10^5$

[a] See footnotes in Tables 1 and 2.
[b] Calculated for enol with correction factor of +1.00 for hydrogen bonding from enolic OH to carbonyl.
[c] Calculated for diketo tautomer
[d] Predicted LC_{50} based upon calculated log P of 0.699
[e] 12 h LC_{50} >80 mg l^{-1}
[f] 36 h LC_{50}
[g] Based upon 36 h LC_{50}

electrophile mechanism according to Scheme 2 involving an S_N2 type addition by a target nucleophilic moiety such as a sulphydryl group in a critical biochemical macromolecule. Within each series of electrophiles, the value of T_e at a particular log P value, and the log P value beyond which only narcosis is observed, will be a function of the degree of electrophilicity of the particular functional group in its own electronic environment.

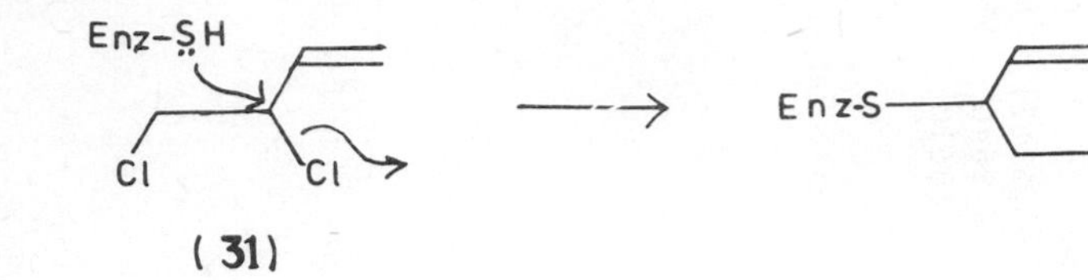

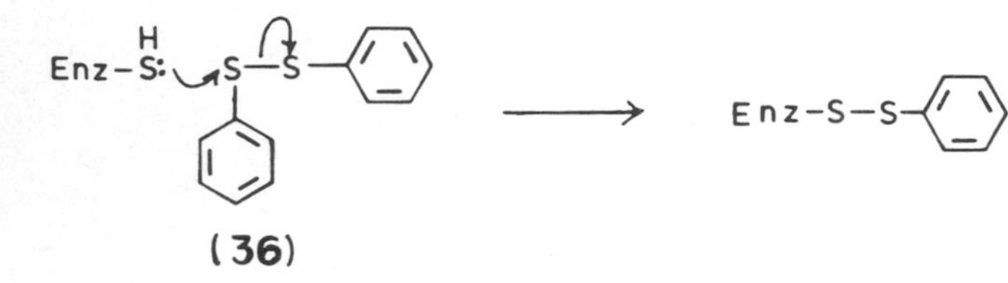

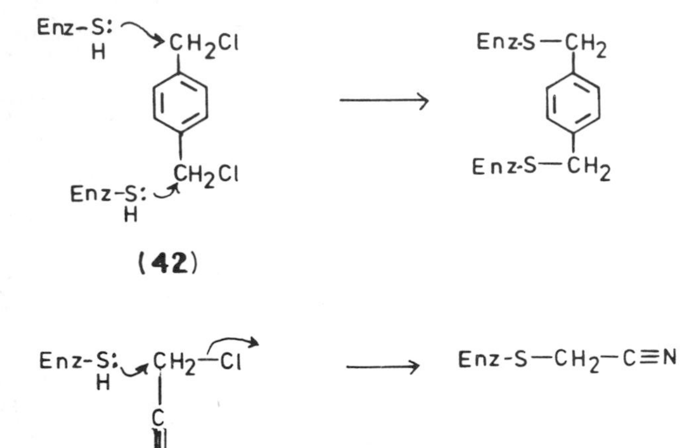

Scheme 2

The toxicity to fish of a series of simple chlorinated hydrocarbons activated for electrophilic reactivity was shown to correlate with the pseudo first-order reaction rates of the toxicants with the model nucleophile *p*-nitrobenzylpyridine (*p-NBP*).[33] Similarly, a correlation has been reported with reaction rates with the model nucleophile *p*-nitrothiophenol[34] and toxicity.

Chloroacetonitrile (45) has been shown to react with *p-NBP*, and to produce DNA strand breaks, an effect ascribed to this S_N2 mechanism upon reaction with biological target sites.[35] The chlorine in (45) is highly activated by its attachment to a carbon bearing a strongly electron-withdrawing group (nitrile) which stabilizes electronic charge development in the reaction transition state. A similar activation exists for compounds (31) and (42) where the presence of the corresponding allylic and benzylic moieties produces the same effect. In the case of (42), two of the electrophilic benzylic chloride moieties are present, doubling the effect, and allowing the possibility of cross-linking. By contrast, 1,2-dichloropropane (11), and 1,3-dichloropropane (14), in which the chlorine is not chemically activated as a leaving group, act as narcotics. The excess toxicity of 87 for 1,3-dibromopropane (43) is unexpected given the relatively low electrophilicity of bromine attached to a primary aliphatic carbon. In a metabolic study of 1,1,3,3-tetradeutero-1,3-dibromopropane, a reactive sulphonium ion (Scheme 3) was proposed as a reactive electrophile intermediate.[37]

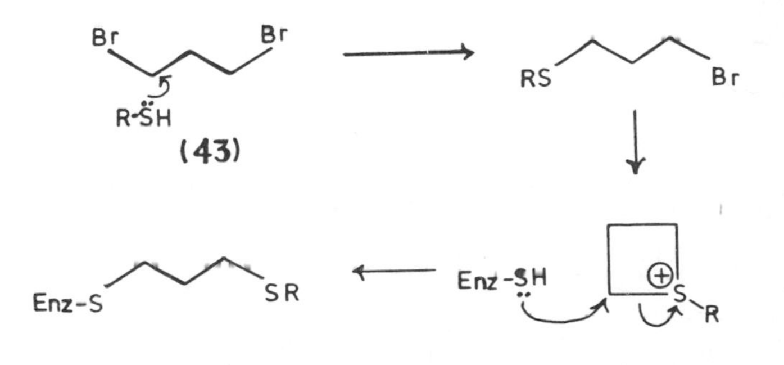

Scheme 3

Diphenyl sulphide (36) has a T_e of 21, which can be ascribed to electrophile toxicity involving attack by a sulphydryl or other biochemical nucleophile target site, and the expulsion of thiophenolate anion as a leaving group. Nucleophilic reactions of this type on sulphur have been studied in the literature,[36] and it has been concluded that such nucleophilic attack on sulphur takes place via a trigonal bipyramidal transition state (Scheme 2), with the incoming nucleophile and leaving group occupying apical positions. The site of attack is activated analogously to that of a benzyl substrate, and the second sulphur becomes a good leaving group because the derived anion is stabilized by the aromatic ring attachment. By contrast, n-butyl sulphide (10), 4,7-dithiadecane (12), and t-butyl disulphide (17) do not satisfy the requirements and LC_{50} data for these chemicals fit the narcosis model.

Given a model nucleophile with the proper 'hardness'[38] it might be possible to model the toxicities of many classes of such electrophiles on a common linear free energy scale that could be employed for correlation and predictive purposes. Electrophilic reactivity has been associated for some time with vesicant and lachrymatory action,[39] and α,α'-dichloro-*p*-xylene (42), and chloroacetonitrile (45) are cited in the Aldrich catalogue[40] as being lachrymatory. In addition, it has been demonstrated that such lachrymatory agents are toxic to fish.[41]

2.4.4 Other classes. Three compounds do not fall within these classes. Allyl cyanide can serve both as an electrophile and as a source of toxic cyanide, once the cyanide anion is released as a leaving group (Scheme 2). The T_e value of 12 calculated for 4-nitrobenzamide (33) can be interpreted in terms of a molecular mechanism involving enzymatic hydrolysis within the fish to produce ammonia and benzoic acid (Scheme 4).

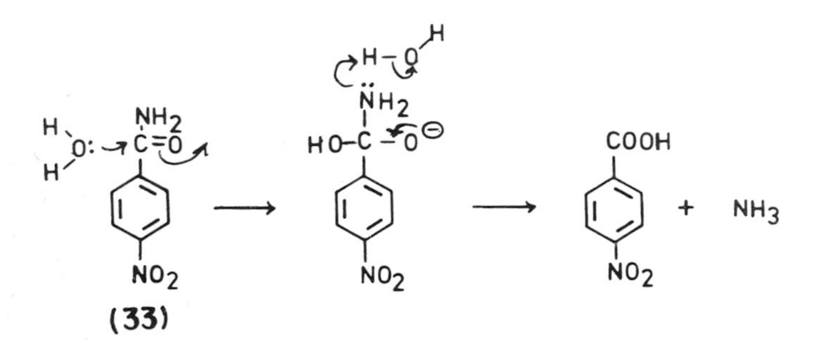

Scheme 4

The presence of the strongly electron withdrawing *p*-nitro substituent on the aromatic ring would be expected to enhance greatly the rate of this hydrolysis process. *m*-Bromobenzamide (8), and 2,4-dichlorobenzamide (21) are less strongly activated for hydrolysis and show toxicity consistent with narcosis. A similar mechanism accounts for the excess toxicity of carboxylic esters. Overton observed that esters intitially induce narcosis in tadpoles, and that this narcotic concentration follows the order predicted by the partition coefficient, but that, given sufficient test duration, other toxic effects were observed.[7,8] He attributed the latter to hydrolysis within the tadpole to the corresponding alcohol and carboxylic acid.[7,8] The excess toxicity of esters to fish has been accounted for quantitatively based upon the rate of base-catalysed hydrolysis.[42]

Compound (34), pentane-2,4-dione, contains two carbonyl groups attached to a common carbon atom bearing two hydrogens. This molecular environment leads to greatly increased acidity of these hydrogens, so that this substance will exist to a large extent as the enolized molecule. This enolization is considered to be the property in this case responsible for the observed excess toxicity. Compounds of this type, including this substance, form chelates with transition metals such as iron.[43] This chelation could interfere with the proper transport of required metals to the fish, or with the ability of the fish to dispose of toxic metals.

2.5 Interpretation of 24 Hour LC_{50}/96 Hour LC_{50} Ratio. Neely demonstrated that for certain simple non-electrolytes their solubility in water could be used to estimate the value of the ratio 24 h LC_{50}/96 h LC_{50},[44] consistent with the predictions of a first-order pharmacokinetic model. The Neely findings are consistent with the above interpretations based upon log *P* and the bilinear model, reflecting the excellent correlation between log *P* and molar water solubility for liquid solutes.[45] A value of this ratio greater than 2 is generally employed as an indicator of chronicity,[46] suggesting the need for chronic testing to assess properly the potential environmental hazard that may be posed by such a chemical.

This 24 h/96 h ratio, which will be defined by the symbol *R*, has been computed where 24 h LC_{50}s could be estimated from graphical results as discussed above.

Ten of the compounds in Table 3 have *R* values equal to or exceeding 2. However, only one of these ten, 2,6-diphenylpyridine (30), has a log *P* value of 4.5 or greater. By contrast, only two of the compounds in Table 2, hexanal (26), and chloromethylstyrene (29), have *R* values exceeding 2. Both of these chemicals can act as electrophiles (compound (29) contains a benzylic activated chloro-substituent). They have T_e values that fall within the transition between those assigned to narcosis (Table 2) and excess toxicity (Table 3). Overton ascribed the decrease in toxic concentration with time for certain organic compounds to their reactivity with cellular constituents, and accumulation at these sites within the organism.[7,8] This explanation accounts satisfactorily for the significant difference in *R* apparent in this data set between chemicals classified as narcotics (Table 2) and those acting by more specific mechanisms (Table 3).[47] Differences in *R* values for electrophiles may reflect a combination of both chemical reactivity and partition coefficient.

3 Conclusions

Quantitative structure-activity relationships are a valuable tool for systematizing toxicity data on aquatic organisms as a function of chemical structure. The application of QSAR models for predictive purposes requires some knowledge or insight into molecular mechanism of toxicity; otherwise, the toxicity of a new, untested industrial chemical may be underestimated. The narcosis model provides an estimate of baseline toxicity for non-electrolytes in general. Therefore, such a new substance can be assumed to be at least as toxic as these narcosis QSAR predictions, which are limited only by the ability to assign reliably a log *P* value. A chemical should be considered as a candidate for chronic toxicity testing if its log *P* exceeds the maximum value permitting the attainment of equilibrium partitioning in the chosen test duration, or if evidence exists for cumulative toxicity via an electrophile or other more specific toxicity mechanism.

4 References

1. R.L. Lipnick, C.S. Pritzker, and D.L. Bentley, in 'QSAR in Drug Design and Toxicology', ed. D. Hadži and B. Jerman-Blažič, Elsevier, Amsterdam, 1987, p.301.

2. R.L. Lipnick, C.S. Pritzker, and D.L. Bentley, in 'QSAR and Strategies in the Design of Bioactive Compounds', ed. J.K. Seydel, VCH, Weinheim, 1985, p.420.

3. G.D. Veith, D.T. Call, and L.T. Brooke, *Can. J. Fish. Aquat. Sci.*, 1983, **40**, 743.

4. Toxic Substances Control Act, US Public Law 94-469, October 11, 1976.

5. G.D. Veith, 'State-of-the-art of Structure Activity Methods Development', 1981, US Environmental Protection Agency, Environmental Research Laboratory, Duluth, MN, EPA-560/81-029; NTIS PB 81-187-239.

6. E. Overton, *Vierteljahrsschr. Naturforsch. Ges. Zuerich*, 1899, **41**, 383.

7. E. Overton, 'Studien über die Narkose, zugleich ein Beitrag zur allgemeiner Pharmakologie', Gustav Fischer, Jena, 1901.

8. R.L. Lipnick, *Trends Pharmacol. Sci.*, 1986, **7**, 161.

9. H. Meyer, *Arch. Exp. Pathol. Pharmakol. (Naunyn-Schmied.)*, 1899, **42**, 109.

10. F. Baum, *Arch. Exp. Pathol. Pharmakol. (Naunyn-Schmied)*, 1899, **42**, 119.

11. H. Meyer, *Arch. Exp. Pathol. Pharmakol. (Naunyn-Schmied)*, 1901, **36**, 338.

12. G.D. Veith, D.T. Call, and L.T. Brooke, in 'Aquatic Toxicology and Hazard Assessment: Sixth Symposium', ed. W.E. Bishop, ASTM STP 802, American Society for Testing and Materials, Philadelphia, 1983, p.90.

13. H. Kubinyi, *J. Med. Chem.*, 1977, **20**, 625.

14. Y.C. Martin, 'Quantitative Drug Design', Marcel Dekker, New York, 1978.

15. H. Könemann, *Toxicology*, 1981, **19**, 209.

16. R.L. Lipnick, K.R. Watson, and A.K. Strausz, *Xenobiotica*, 1987, **17**, 1011.

17. R.L. Lipnick and W.J. Dunn, III, in 'Quantitative Approaches to Drug Design', ed. J.C. Dearden, Elsevier, Amsterdam, 1983, p.265.

18. R.L. Lipnick, in 'QSAR in Toxicology and Xenobiochemistry', ed. M. Tichý, Elsevier, Amsterdam, 1985, p.39.

19. 'Acute Toxicities of Organic Chemicals to Fathead Minnows *(Pimephales promelas)*', ed. L.T. Brooke, D.J. Call, D.L. Geiger, and C.E. Northcott, Centre for Lake Superior Environmental Studies, University of Wisconsin-Superior, 1984, Vol. 1.

20. 'Acute Toxicities of Organic Chemicals to Fathead Minnows *(Pimephales promelas)*', ed. D.L. Geiger, C.E. Northcott, D.J. Call, and L.T. Brooke, Centre for Lake Superior Environmental Studies, University of Wisconsin-Superior, 1985, Vol. 2.

21. 'Acute Toxicities of Organic Chemicals to Fathead Minnows *(Pimephales promelas)*', ed. D.L. Geiger, S.H. Poirier, L.T. Brooke, and D.J. Call, Centre for Lake Superior Environmental Studies, University of Wisconsin-Superior, 1986, Vol. 3.

22. A. Leo and D. Weininger, 'Medchem Software Release 3.33', Medicinal Chemistry Project, Pomona College, Claremont, CA, 1985.

23. S.H. Yalkowsky (Editor-in-Chief), 'Arizona Data Base of Aqueous Solubility', University of Arizona, Tucson, AZ, 1987.

24. S. Banerjee, S.H. Yalkowsky, and S.S. Valvani, *Environ. Sci. Technol.*, 1980, **14**, 1227.

25. J. Hermens, H. Canton, P. Jansen, and R. DeJong, *Aquat. Toxicol.*, 1984, **5**, 143.

26. S. Broderius and M.Kahl, *Aquat. Toxicol.*, 1985, **6**, 307.

27. R.V. Thurston, T.A. Gilfoil, E.L. Meyn, R.K. Zajdel, T.I. Aoki, and G.D. Veith, *Water Res.*, 1985, **19**, 1145.

28. J. Ferguson, *Proc. R. Soc. (London), Ser. B.*, 1939, **127**, 387.

29. K. Memminki and R. Suni, *Arch. Toxicol.*, 1984, **55**, 186.

30. C. Hansch, and A. Leo, 'Substituent Constants for Correlation Analysis in Chemistry and Biology', Wiley-Interscience, New York, 1979.

31. A.C. Moffat, E.C. Horning, S.B. Matin, and M. Rowland, *J. Chromatogr.*, 1972, **66**, 255.

32. J.-C. Lhuguenot and B.F. Maume, *J. Chromatog. Sci.*, 1974, **12**, 411.

33. J. Hermens, F. Busser, P. Leeuwanch, and A. Musch, *Toxicol. Environ. Chem.*, 1985, **9**, 219.

34. A.M. Cheh and R.E. Carlson, *Anal. Chem.*, 1981, **53**, 1001.

35. F.B. Daniel, K.N. Schenck, J.K. Mattox, E.L.C. Lin, D.L. Haas, and M.A. Pereira, *Fundam. Appl. Toxicol.*, 1986, **6**, 447.

36. J.L. Kice, in 'Sulfur in Organic and Inorganic Chemistry', ed. A. Senning, Marcel Dekker, New York, 1971, Vol. 1, Chapter 6, p.153.

37. W. Onkenhout, E.J.C. Van Bergen, J.H.F. Van Der Wart, G.P. Vos, W. Buijs, and N.P.E. Vermeulen, *Xenobiotica*, 1986, **16**, 21.

38. R.G. Pearson and J. Songstad, *J. Am. Chem. Soc.*, 1967, **89**, 1827.

39. M. Dixon, *Nature (London)*, 1946, **158**, 432.

40. Aldrich Chemical Company, 'Catalog Handbook of Fine Chemicals', Milwaukee, WI, 1987.

41. R.W. Hiatt, J.J. Naughton, and D.C. Matthews, *Biol. Bull.*, 1953, **104**, 28.

42. M.J. Kamlet, R.M. Doherty, R.W. Taft, M.H. Abraham, G.D. Veith, and D.J. Abraham, *Environ. Sci. Technol.*, 1987, **21**, 149.

43. B. Ballantyne, D.E. Dodd, R.C. Myers, and D.J. Nachreiner, *Drug Chem. Toxicol.*, 1986, **9**, 133.

44. W.B. Neely, *Chemosphere*, 1984, **13**, 813.

45. C. Hansch, J.E. Quinlan, and G.L. Lawrence, *J. Org. Chem.*, 1968, **33**, 347.

46. W.J. Birge and J.A. Black, in 'Aquatic Toxicology and Hazard Assessment: Eighth Symposium', ed. R.C. Bahner and D.J. Hansen, ASTM STP891, American Society for Testing and Materials, Philadelphia, 1985, p.51.

47. R.L. Lipnick, 'A QSAR Study of Overton's Data on the Narcosis and Toxicity of Organic Chemicals to the Tadpole, *Rana Temporaria*', 'Aquatic Toxicology and Hazard Assessment: Eleventh Symposium', ed. G.W. Suter, II, and M. Lewis, American Society for Testing and Materials, Philadelphia, 1988, (in press).

48. 'Lange's Handbook of Chemistry, 13th Edn., ed. J.A. Dean, McGraw-Hill, New York, 1985.

22
Dyestuffs and the Environment — A Risk Assessment

D. Brown

IMPERIAL CHEMICAL INDUSTRIES PLC, BRIXHAM LABORATORY, OVERGANG, BRIXHAM, DEVON TQ5 8BA, UK

and R. Anliker

ECOLOGICAL AND TOXICOLOGICAL ASSOCIATION OF THE DYESTUFFS MANUFACTURING INDUSTRY, CLARASTRASSE 4/6, PO BOX 4058, CH-4005 BASEL 5, SWITZERLAND

1 Introduction

Dyestuffs may be defined as water-soluble or water-dispersible organic colorants, and the dyestuff industry has been and continues to be a fertile ground for the production of new and improved substances for use in colouring a wide variety of materials, in particular textiles.

Anliker[1] has pointed out the need for the very large number (approximately 3000) of chemically different structures which are currently used as colorants and has described how the Colour Index (C.I.) system[2] is used to classify and identify the individual products. He has also cautioned against assuming that dyestuffs belonging to a single broad class such as azo or anthraquinone dyestuffs will necessarily have similar toxicological properties. The broad classes of organic colorants and the application areas of these classes have been summarized by Clarke and Anliker[3] who also provide estimates both of world-wide production quantities in the various application areas and also of losses during manufacture and dyehouse processing operations. These production and loss data were summarized by Brown[4] as shown in Table 1.

Table 1 *World-wide dyestuff production and loss in aqueous effluents (1978)*

	Production	*Production loss*	*Processing loss*
Textile dyestuffs	360[a]	7.2	36
Paper/leather dyestuffs	90	1.8	5.4
	450	9	41.4

[a]All units are 1000 tonnes year^{-1} as active ingredient.

It can be seen from Table 1 that of the 1978 world-wide production of 0.45 million tonnes of dyestuff active ingredient, approximately 2% is estimated to be discharged in aqueous effluents during manufacture of the dyestuff and approximately 9% in dye-house operations. The actual discharge of dyestuffs to the environment is reduced by treatment of the effluent containing them and increased by any subsequent losses from the dyed article. However, most modern dyestuffs are highly substantive to the substrate to which they are applied and losses to the environment after completion of the dyeing process are considered negligible. A further conclusion from Table 1 is that it is the textile industry dyestuffs which are the major usage area and in textile dyeing the azo dyes account for 60-70% of the dyestuffs made, the anthraquinone dyestuffs being second to azo dyes in commercial importance.

Brown[4] has described the broad structure of the dyestuff industry and indicated that over 80% of dyestuffs are produced in amounts of less than 50 tonnes per year^{-1} and only 1-2% in excess of 1000 tonnes per year^{-1} (as active ingredient). Further, each individual product is sold to a large number of outlets, so that the dyestuff industry is, in general, an industry which manufactures a vast range of products, most in rather small tonnages, and sells these individual products to a large number of users, often in kilogram year^{-1} amounts.

2 Environmental Risk

'Hazard' and 'risk' have been defined by the OECD[5] and are also contained in the Glossary of Terms.

The working approach taken by most environmental toxicologists is to assume that below a certain exposure concentration (often termed the 'safe' level) the risk to environmental species will be negligible. This, of course, leaves the dilemma of how to define this 'safe' level. On the one hand there is an enormous diversity of environmental species and their different life stages which might be exposed to the substance in question. On the other hand the database of information on the properties of a chemical which might make it capable of causing adverse effects is necessarily limited.

ECETOC,[6] in the context of the EEC 6th Amendment to the Directive 67/548/EEC,[7] has addressed this general question of how limited 'base set' acute toxicity data may be used to predict an exposure level below which there is unlikely to be a significant risk. The ECETOC conclusion is that for non-bioaccumulative substances there is unlikely to be a significant risk at levels which are less than 1% of acute toxicity levels. ECETOC also points out that where exposure levels may be greater than 1% of the acute toxicity level, further sub-lethal (chronic) data (*e.g.* from level 1 and level 2 of the '6th Amendment' Directive) may be required to judge whether a particular exposure level may pose a significant risk. The US Environmental Protection Agency (EPA) also appears to follow this approach[8] and proposes a factor of 10% to relate chronic toxicity data to a predicted 'safe level'.

As will be discussed later in this chapter, it is possible to predict and in some cases to measure environmental levels of dyestuffs, and, as depicted in Figure 1, the ratio of the predicted 'safe level' to the environmental level may be considered as the 'safety margin'.

```
100 |              Measured Acute Level
   ------------------------------------------------
    |    |
    |    |
 10 |    |         Measured Sub-Acute Level
   ------------------------------------------------
    |    |    |
    |    |    |
  1 |    ↓    ↓     Predicted 'Safe Level' (A)
   ________________________________________________
    |         |
    |         |
    |         |     'Safety Margin' A/B
    |         |
    |         |
    |         ↓
    |  Measured or Predicted Environmental Level (B)
   ________________________________________________
```

Figure 1

3 Environmental Levels of Dyestuffs

3.1 Water. Although the 'point source' discharge of dyestuffs from a specific manufacturing site may pose a specific local problem, as indicated in the Introduction, the major route for the release of dyestuffs to the environment is considered to be their discharge in aqueous effluents following dye-house processing operations. For textile dyeing operations Brown[4] has indicated that the predicted environmental concentration of a dyestuff in a receiving water may be calculated from a knowledge of the daily usage of the dyestuff, the percentage uptake by the substrate being dyed (exhaustion rate), the likely degree of removal during any effluent treatment process, and the flow of the receiving river. These calculations for a medium-size textile dye-house using 30 tonnes year^{-1} dyestuff and discharging via a municipal sewage treatment works to a small receiving river (4.5 cumecs) gave total dyestuff concentrations in the river of approximately 10 μg l^{-1}. For an individual dyestuff used at 1 tonne year^{-1}, a similar calculation, but based on 100 days usage year^{-1}, gave a concentration of 1 μg l^{-1}.

There are relatively few reports of measured levels of dyestuffs in rivers. In the United Kingdom, Richardson and Waggot[9] have monitored levels of the triphenylmethane dyestuff C.I. Acid Blue 1, used as a colouring agent in toilet flush blocks. They reported levels up to 12.3 μg l^{-1} in treated sewage effluents, up to 1.7 μg l^{-1} in the Rivers Lee and Thames, and 0.6 μg l^{-1} in a reservoir used for drinking water supplies. However, the authors conclude that chlorination of levels used in normal waterworks' practice appears to destroy C.I. Acid Blue 1 and in general it was not detectable in finished water as supplied to the consumer. In this paper Richardson and Waggot also indicated some of the potential difficulties in the analysis of very low levels of dyestuffs in environmental samples. In particular they suggest that C.I. Acid Blue 1 in natural waters may be associated in some way with other organic material, *e.g.* cationic surfactants or humic or fulvic acids, and that the resulting adjunct has chromatographic properties different from those of free C.I. Acid Blue 1. The dyestuff organization ETAD has recognized the importance of obtaining reliable data on the levels of dyestuffs which may be present in natural waters, and is currently investigating appropriate analytical methods before carrying out environmental monitoring.[10]

In the USA Tincher[11] has investigated the level of six acid dyestuffs which are used in large amounts for the dyeing of nylon carpet in the Coosa River (N.W. George) basin. This area manufactures over 50% of the carpets and rugs produced in the USA and Tincher quotes the 1984 commercial dye usage as approximately 3500 tonnes. Tincher has found all six of the dyestuffs in the effluents from two waste treatment plants in amounts up to approximately 0.12 mg l^{-1}. In river samples these levels were approximately an order of magnitude lower with the maximum concentration found being that of C.I. Acid Yellow 219 at 22 μg l^{-1}, and in raw water samples (before processing for potable use) the levels were further reduced, three of the six dyes being non-detectable.

C.I. Acid Yellow 219 was, however, found in raw water samples at up to 15 μg l^{-1} and the treatment process (filtration through anthracite coal and addition of potassium permanganate) reduced this to 5 μg l^{-1}, all other dyes being below the limits of detection.

As a general comment, any appreciable quantity of dyestuff in a natural water is visible and, except in very limited areas, coloration of natural waters by dyestuffs has not been reported. Although it is not possible to give precise concentrations at which dyes would be visible, it has been suggested[4] that 0.1 mg^{-1} may be detectable, and that at 1 mg l^{-1} colour will generally be observable.

3.2 Sediments. Anliker[1] has noted that the substantive nature of dyestuffs is such that when discharged to receiving waters they are likely to be adsorbed on suspended solids and hence find their way to aquatic sediments. As will be discussed later in this chapter, the anaerobic nature of many sediments is such that dyestuffs are unlikely to remain intact in this environment, but nonetheless Tincher[12] has found dyestuff levels in the range 0.1 - 3 mg l^{-1} $^{w}/_{w}$ of dry sediment in the Coosa River basin. These sediment levels are 1-2 orders of magnitude higher than levels found by Tincher in the waters of the Coosa River basin and such sediment concentration factors are not unusual for substantive materials.

European investigations[13] on sediments taken from the Rhine at the border between Germany and Holland indicated that the major portion of the isolated coloured matter appeared to be of natural origin and no individual synthetic dyestuff could be identified to a detection limit of 0.05 mg kg^{-1}.

3.3 Agricultural Soils. Although there is no direct route by which agricultural soils may become contaminated with synthetic dyestuffs, it is possible in principle that the disposal of sewage sludge from a treatment works which receives dye-house effluents may provide an indirect route. Although there appear to be no reported data for levels of dyestuff on agricultural land, calculations using the principles elaborated by the OECD[14], give a 'worst case' level of ~1 mg kg^{-1} ($^{w}/_{w}$ of dry soil).[1]

4 Environmental Toxicity of Dyestuffs

4.1 Toxicity to Aquatic Organisms. 4.1.1 Fish. The present authors[1,4] have both referred to the standardized Safety Data Sheet[15] which ETAD member companies provide with their dyestuff products. This data set, which in respect of its ecotoxicological section gives broadly similar information to that requested in the corresponding section of the 'base set' of the EEC 6th Amendment Directive, provides ETAD with a substantial database on fish toxicity of dyestuffs.

From a survey of this database it was concluded[1] that the majority of dyestuffs are not very toxic to fish, 59% of dyestuffs have an LC_{50} in excess of a limit dose of 100 mg l^{-1} and only 2% were toxic at a level of less than 1 mg l^{-1}. Although these more toxic dyestuffs included a number of different structural types, Anliker[1] noted that the largest group were basic dyestuffs, together with some metal-complex dyestuffs.

The other industrial organization which has made a significant contribution to the available data on the toxicity of dyestuffs to fish is the American Dye Manufacturers Institute (ADMI). In the early 1970s, ADMI commissioned a range of studies on a number of dyestuffs. The reports of these studies are contained in two volumes[16] (Volume 1 has been summarized by Horning[17]) which include chapters on fish bioassay tests; the effect of dyestuffs on aerobic and anaerobic systems; the effect of 56 selected dyestuffs on the growth of the green alga *Selenastrum capricornutum*; the effect of biological treatment on the toxicity of dyestuffs to fish; the acute toxicity of 10 selected basic dyes to the fathead minnow *Pimephales promelas*; and the toxicity of aminoanthraquinone dyestuffs to fish and algae.

Of the 46 dyestuffs which were examined in the first set of fish bioassay tests 32 (70%) had 96 h LC_{50} values in excess of 100 mg l^{-1} and 3 (7%) had LC_{50} values of 1 mg l^{-1} or less. On the basis of this study ADMI noted the high toxicity of the triphenylmethane basic dyestuffs tested (C.I. Basic Violet 1 LC_{50} 0.05 mg l^{-1}) and also made some tentative structure-activity relationships for dyes containing the 1,4-diamino-anthroquinone grouping.

On this basis ADMI selected a further ten basic dyestuffs for test and, although none was as toxic as the commercial form of C.I. Basic Violet 1, all were toxic at less than 100 mg l^{-1} and the most toxic (C.I. Basic Yellow 37) had a 96 h LC_{50} of 0.8 mg l^{-1}.

ADMI also selected a further 12 aminoanthraquinone dyestuffs for a comparative study of their toxicity to fish and algae. It was found that the non-dyestuffs anthraquinone, anthraquinone-2-sulphonic acid (sodium salt), and 1-anthraquinone sulphanilic acid (sodium salt) (1-amino(4-sulphophenyl)anthraquinone) are not toxic to either fish or algae but that several of the aminoanthraquinone dyestuffs did show toxic effects to fish and/or algae. It was concluded that the type and position of substituent groups in the anthraquinone parent molecule influence toxicity, that ethanol-amino groups increase the toxicity to both fish and algae, and that cationic dissociation (basic dyestuffs) gives high toxicity. ADMI also commented that the toxic actions of aminoanthraquinone dyestuffs on fish and green algae are apparently different and that neither can be used as a prediction for the toxicity of the other. In respect of the toxic effect of dyestuffs on algae it should be pointed out that dyestuffs which absorb light at the wavelength necessary for photosynthesis may inhibit algal growth through this essentially physical effect.

No systematic long-term studies on possible sublethal chronic effects of dyestuffs on fish appear to have been reported, although Brown[4] has reported bioaccumulation studies on 30 water-soluble and 12 disperse dyestuffs up to 10 mg l^{-1} for an eight week period. Although the experiments were carried out to assess bioaccumulation, an incidental observation was that this eight week exposure of the fish, to much higher concentrations of dyestuffs than would be conceivable in the natural environment, caused no apparent adverse effects.

In regard to bioaccumulation potential of dyestuffs in fish, it would be expected that the water-soluble dyestuffs (low octanol-water partition coefficients) would not bioaccumulate. These studies reported by Brown (and others carried out by ETAD member firms) confirmed that water-soluble dyestuffs do not bioaccumulate. However, Brown also reported low bioaccumulation factors for 12 disperse dyestuffs and these materials are essentially water-insoluble and have a relatively high octanol-water partition coefficient. Anliker et al.[18-20] have discussed the reasons why disperse dyestuffs and organic pigments may show only modest accumulation factors in fish. One factor (as suggested by Opperhuizen[21]) may be a molecular size limitation which effectively prevents permeation through biological membranes, and another factor may be the relatively low solubility level of pigments in fat which effectively limits the amount of uptake.

4.1.2 Other aquatic organisms. Although the freshwater invertebrate *Daphnia magna* is included in the EEC 'base set' of data, this test was not included in the ETAD Safety Data Sheet which pre-dates the EEC '6th Amendment' Directive. Thus there is no database available to ETAD on this organism, and there appear to be no systematic published data from other sources.

The ADMI studies on anthraquinone dyestuffs and the green alga *Selenestrum capricornutum* have been discussed above but ADMI has also examined the toxic effects of 56 selected dyes to this organism.[16] The growth of the algae was assessed after 7 and 14 days in the presence of 1 and 10 mg l^{-1} of dyestuffs. Of the 56 dyestuffs tested, 15 strongly inhibited algal growth at a test concentration of 1 mg l^{-1} after 7 days of incubation. Amongst the 15 inhibitory materials, 13 were basic dyestuffs and, in fact, of the basic dyestuffs tested in this study only the C.I. commercial form of C.I. Basic Violet 10 was not inhibitory at 1 mg l^{-1}.

4.1.3 Sewage bacteria. Both ETAD[22] and ADMI[16] have studied the possible adverse effects of dyestuffs on wastewater bacteria with a view to assessing whether dyestuffs reaching a wastewater treatment plant may possibly interfere with the operation of that plant.

The screening test approach taken by ETAD was to measure the respiration rate of activated sludge which had been aerated for three hours in the presence of the test dyestuff. This test, which has now been adopted as an OECD test guideline,[23] showed that of 202 different dyestuffs tested (as their commercial formulations), only 18 showed significant inhibition at less than 100 mg l^{-1} All the inhibitory dyestuffs were basic dyes and of the 30 basic dyes tested only 12 did not show significant inhibition at a test concentration of 100 mg l^{-1}. The authors, in commenting on these results, emphasized the screening nature of the test and concluded that where significant amounts of inhibitory dyestuffs are likely to reach a sewage treatment works a more detailed assessment of the likely effects would be prudent.

ADMI has considered the possible effects of dyestuffs on both aerobic and anaerobic systems. For the aerobic study on 46 dyestuffs and a fluorescent whitening agent, the basic approach was broadly similar in principle to the ETAD method above. In the ADMI work 1500 mg l^{-1} of activated sludge was suspended in settled domestic sewage and the rate of oxygen consumption over a seven hour period measured in a control, and in a system containing 25 mg l^{-1} of dyestuff as organic carbon. Eight of the dyestuffs tested showed respiration rates which were less than 90% of the control but apart from C.I. Basic Green 4, where the rate of oxygen uptake was 49% of the control, the respiration rates for these inhibitory dyestuffs were all in excess of 70% of the control.

Attempts were also made to acclimatize activated sludge to possibly inhibitory dyestuffs by twice daily additions of dyestuff over a 14 day period. This acclimatization period had different effects with different dyestuffs, some becoming slightly less and some slightly more inhibitory. However, of 24 dyestuffs subjected to the acclimatization procedure only C.I. Vat Green 3 (58%) and C.I. Basic Blue 3 (53%) showed respiration rates of <70% of the control.

In further work ADMI showed that the inhibitory effects found at the exposure level of 25 mg l^{-1} as organic carbon disappeared at exposure levels of between 1 and 5 mg l^{-1} as organic carbon, corresponding to levels of dyestuff of between approximately two and eight times higher depending on the carbon content of the dyestuffs.

The ADMI Studies on possible effects of dyestuffs on the anaerobic digestion process were carried out by measuring a number of parameters, including the volume of gas produced in a laboratory scale sewage sludge digestion apparatus in the presence and absence of dyestuff. Of the 43 dyestuffs tested, 30 were considered to have no significant effects on anaerobic digestion when added daily at a concentration of 150 mg l^{-1}. Of the remainder, seven were dispersed using a proprietary dispersing agent 'Tamol SN' and the effects seen were ascribed to this material, four dyestuffs gave either erratic gas production or showed initial inhibition followed by recovery, and two dyestuffs, C.I. Acid Blue 45 and C.I. Disperse Blue 5, caused severe inhibition leading to process failure. Both these dyestuffs are anthraquinone dyestuffs and the suggestion was made that anthraquinones can interfere with the electron transport system of the anaerobic methanogenic bacteria, thus causing the inhibition of methane production. As far as is known there have been no systematic studies on the possible toxic effects of anthraquinone dyestuffs on the anaerobic digestion process and in particular there appears to have been no study of the levels of such dyestuffs at which significant inhibition problems might occur in actual sewage works digesters.

As an incidental to this work ADMI also noted that many dyestuffs were decolorized by the anaerobic digestion process, as also found by Brown and Laboureur[24] in work to be discussed later in this chapter.

4.2 Toxicity to Plants. ETAD[25] has organized a limited study of the possible effects of dyestuffs on plant germination and growth. For this work four dyestuffs were used:

C.I. Acid Blue 25 :a water soluble anthraquinone
C.I. 62055 dyestuff

C.I. Disperse Blue 26 :a disperse anthraquinone
C.I. 63305 dyestuff

Acid Dye C.I. 13155 :a water-soluble azo-dyestuff

C.I. Basic Blue 26 :a triarylmethane basic dyestuff
C.I. 44045 dyestuff

All four dyestuffs were incorporated into a seed compost at concentrations of 1, 10, 100, and 1000 mg l^{-1} and the germination and growth of three plant species (sorghum, sunflower, and soya) assessed. In all cases there was no effect on seed germination up to and including the maximum test concentration. In respect of growth there was no observed effect at the 100 mg l^{-1} test concentration but at the 1000 mg l^{-1} level (where the soil compost was noticeably coloured with some of the dyestuffs) there was a variable effect on growth depending on the dye and the particular plant species.

In three of the studies (C.I. Acid Blue 25, Acid Dye C.I. 13155, and C.I. Basic Blue 26) the plant foliage was analysed for the dyestuff and in the case of Acid Dye C.I. 13155 also for a possible amine metabolite. At the maximum test level (1000 mg kg^{-1}) the dyestuffs were just detectable in the foliage at concentrations up to 2 mg l^{-1} at the 1000 mg l^{-1} exposure, but since this metabolite was present as an impurity in the dyestuff it is possible that this was a direct uptake rather than an uptake of the dyestuff followed by metabolism.

5 Environmental Fate of Dyestuffs

Dyestuffs to be useful must have a high degree of chemical and photolytic stability. It is thus unlikely that in general they will give positive results in short-term tests for aerobic biodegradability, for example the 'ready' biodegradability test guidelines of the OECD.[26] Pagga and Brown[27] in a study of 87 commercial dyestuffs found that 62% of the dyestuffs tested showed significant removal of colour in static test with an appreciable concentration of activated sludge present (500 mg l^{-1}), but they concluded that it was likely that the prime mechanism for the removal of the coloured component of dyestuffs is by adsorption rather than by biodegradation. Porter and Snider,[28] in a study on the biodegradability of textile chemicals, included seven textile dyestuffs and concluded that no significant aerobic biodegradation had occurred under the conditions of their study.

However, other workers have shown evidence that certain relatively simple dye molecules may be degraded under aerobic conditions, although long adaptation periods and specialized organisms may be required.[29–31] Flege[32] has reported on the aerobic metabolites of certain disperse dyes and also commented that three fibre reactive dyes of the vinyl sulphone class were not degraded. Shaul et al.[33] have examined a number of azo dyestuffs in a pilot-scale activated sludge sewage treatment system and have found positive evidence for the biodegradation of C.I. Acid Orange 7 and C.I. Acid

Red 88, which are both rather simple azo dyestuffs with a marked similarity in structure:

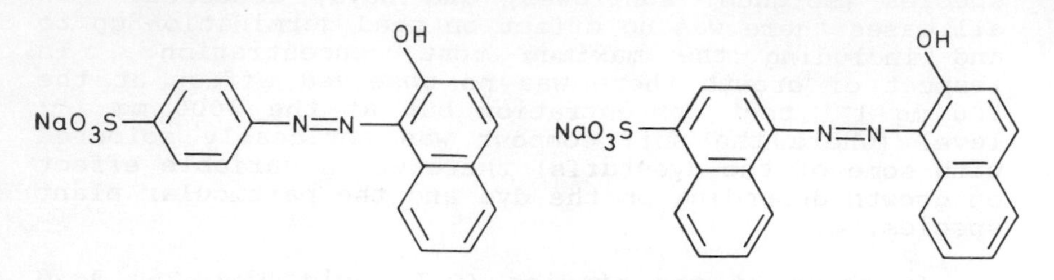

C.I. Acid Orange 7 C.I. Acid Red 88

The tendency of many dyestuffs to sorb onto sludge solids during the biological treatment of effluents containing dye-house wastes has already been mentioned. This sorptive removal during biological treatment has been termed 'bioelimination', and rather simple sludge adsorption tests to determine the likely extent of such removal have been described by Hitz et al.[34] and recently by Lieberman and Shau.[35]

<u>5.1 Anaerobic Biodegradation</u>. The reductive cleavage of azo dyestuffs under the anaerobic conditions found in sewage sludge digestion and in sediments undoubtedly occurs, and is almost certainly the major route by which this class of dyestuff is removed from the environment.

Brown and Laboureur[24] have described a rather simple method by which the decolorization of dyestuffs under anaerobic conditions may be studied and have also shown[36] that certain of the lipophilic amines likely to be produced when azo dyestuffs are reductively cleaved are biodegradable under aerobic conditions. Brown and Hamburger[37] have identified a number of the expected amine metabolites arising from the anaerobic degradation of azo dyestuffs, and have also investigated their biodegradability under both anaerobic and aerobic conditions. Under anaerobic conditions only 4,4'-diamino-3,3'-dimethoxybiphenyl was found to be degradable, but under aerobic conditions several of the sulphonated aromatic amines formed were found to degrade. These included 1-aminonaphthalene-4,8-disulphonic acid, and current ETAD studies have as their aim an investigation of the aquatic toxicity of these less degradable amines.

Wuhrmann et al.[38] have investigated the rate-determining factor in the microbial reduction of azo dyestuffs and have found that azo dyestuffs are decolorized either by the living cells or by cellular extracts.

Tincher[39] has carried out a practical demonstration that sewage sludge contaminated with dyestuffs when held under simulated land-fill conditions does not release those dyestuffs into leachate and that neither can amine metabolites, which might be expected to be produced from thee dyestuffs, be found in the leachate or interstitial water.

Very recent work by Weber[40] has shown that aromatic azo compounds are readily degraded in anaerobic sediment/water systems. Although the exact nature of the reducing agent has not been determined, the reduction process appears to be an abiotic, surface-mediated reaction.

5.2 Photolysis. Hagg and Mill[41] have examined the photostability of azo dyestuffs in water. They make the important finding that although many of these dyestuffs are photostable in pure water, in the presence of natural humic materials the photo-decomposition is strongly accelerated, probably through oxidation by singlet oxygen or oxy-radicals. These authors also comment that the formation of potentially carcinogenic aromatic amines is unlikely, since aromatic amines are highly susceptible to photolysis.

Bandlow et al.[42] have also shown that the rate of photolysis of dyestuffs is considerably increased in natural waters in comparison with distilled water.

6 Conclusions

The vast number of dyestuffs, and the many varied situations in which they are used and applied, inevitably means that an environmental risk assessment can only properly be made on an individual dyestuff and an individual dye-house basis. However, for most dyestuffs aquatic toxicity data will be available to demonstrate that provided any dyestuff discharge results in levels below those visible in a natural water, then no significant risk to water organisms in that water is likely. For those dyestuffs where this statement may not apply, it is particularly important to consider levels of discharge both in factory effluents and following biological treatment. The basic dyestuffs which are most likely to be toxic are in general very substantive to the substrate which they are dyeing and are also well removed in biological treatment processes.

The likely environmental fate of dyestuffs is by sorption either on to sewage works sludges or on to the sediments of natural waters where azo dyestuffs in particular will be degraded in an anaerobic environment. The available data indicate that the amines produced as the anaerobic metabolic products will themselves be degradable or be essentially non-toxic to aquatic life.

7 References

1. R. Anliker, 'Toxic Hazard Assessment of Chemicals', ed. M.L. Richardson, Royal Society of Chemistry, 1986, Chapter 14, 'Organic Colorants. Interpretation of Mammalian, Geno- and Eco-toxicity Data in Terms of Potential Risks'.

2. 'Colour Index', published by the Society of Dyers and Colourists, Bradford, England, and the American Association of Textile Chemists and Colorists, Research Triangle Park, NC, USA, 3rd Edn., 1971, revised 1975, 1982 and running Supplements Vol. 1-7.

3. E.A. Clarke and R. Anliker, 'The Handbook of Environmental Chemistry', ed. O. Hutzinger, Vol. 3A 'Organic Dyes and Pigments', Springer-Verlag, 1980, pp. 181-210.

4. D. Brown, *Ecotox. Environ. Safety*, 1987, 13, 139.

5. OECD, 'Chemicals Control Legislation - An International Glossary of Key Terms', OECD, Paris, 1982, p.62.

6. ECETOC Technical Report No.13, 'The EEC Sixth Amendment: A guide to risk evaluation for effects on the environment', ECETOC, 1984.

7. Directive 79/831/EEC: Council of European Communities Directive on Classification, Packaging and Labelling of Dangerous Substances, 18 September 1979; see also ref. 1, chapters 11, 12, and 23.

8. US EPA Environmental Effects Branch - Health and Environmental Review Division, 'Estimating "Concern Levels" for Concentrations of Chemical Substances in the Environment', February 1984.

9. M.L. Richardson and A. Waggott, *Ecotox. Environ. Safety*, 1981, 5, 424.

10. ETAD (Ecological and Toxicological Association of the Dyestuffs Manufacturing Industry). Current work: Monitoring dyestuffs in water. ETAD, Basle, Switzerland.

11. W.C. Tincher, 'Analysis for Acid Dyes in the Coosa River Basin', School of Textile Engineering, Georgia Institute of Technology, Atlanta, GA, Personal communication to Dr. R. Anliker.

12. W.C. Tincher, 'Survey of Coosa River Basin for Organic Contaminants from Carpet Processing', Final Report. Contract No. E-27-630-, Prot. Div., Dept of Natural Resources, State of Georgia, USA, 1978.

13. ETAD, 'Sediment Analysis', Status Report Project E3013, June 27, 1980, ETAD, Basle, Switzerland.

14. OECD, Hazard Assessment Project: Working Party on Exposure Analysis. Final Report, Berlin, 1982, p.52.

15. R. Anliker and E.A. Clarke, 'ETAD - A collaboration', *Chem. Br.*, 1982, **18**, 796.

16. ADMI, 'Dyes and the Environment - Reports on Selected Dyes and their Effects', Vol. 1, September 1973; Vol. 2, September 1974.

17. R.H. Horning, *TAPPI*, 1974, **57**, 135.

18. R. Anliker, E.A. Clarke, and P. Moser, *Chemosphere*, 1981, **10**, 263.

19. R. Anliker and P. Moser. *Ecotox. Environ. Safety*, 1987, **13**, 43.

20. R. Anliker, P. Moser, and D. Poppinger, 'Bioaccumulation of Dyestuffs and Organic Pigments in Fish, Relationship to Hydrophobicity and Steric Factors', *Chemosphere*, 1988, (in press).

21. A. v.d. Opperhuizen, E.W. Velde, F.A.P.C. Gobas, D.A.D. Liem, and v.d. J.M.D. Steen, *Chemosphere*, 1985, **14**, 1971.

22. D. Brown, H.R. Hitz and L. Schaefer, *Chemosphere*, 1981, **10**, 245.

23. OECD Guideline 209: Activated Sludge Respiration Rate Test, OECD Guideline for Testing of Chemicals, 4 April 1984.

24. D. Brown and P. Laboureur, *Chemosphere,* 1983, **12**, 397.

25. ETAD (Ecological and Toxicological Association of the Dyestuffs Manufacturing Industry) Report: Agricultural Use of Sludge Contaminated with Colorants Feb. 28, 1986, ETAD, Basle, Switzerland.

26. OECD Guidelines 301A-E. Ready Biodegradability, OECD, 12 May 1981.

27. U. Pagga and D. Brown, *Chemosphere*, 1986, 15, 479.

28. J.J. Porter and E.H. Snider, *J. Water Pollut. Control Fed.*, 1976, **48**, 2198.

29. H.G. Kulla, 'Aerobic Bacterial Degradation of Azo Dyes', in 'Microbial Degradation of Xenobiotics and Recalcitrant Compounds', FEMS Symposium No.12, ed. T. Leisinger et al. Academic Press, 1981, p.387.

30. U. Meyer, G. Overney, and A. von Wattenwyn, *Textilveredlung*, 1979, **14**, 15.

31. E. Idaka, T. Ogawa, H. Hotizu, and M. Tomoyeda, *J. Soc. Dyers Colour.*, 1978, **94**, 91.

32. R.K. Flege, Georgia Institute of Technology, 'Determination and Degraded Dyes and Auxiliary Chemicals in Effluents from Textile Dyeing Processes', Completion Report OWRP Project No. B-02-GA, 31 March 1970.

33. G.M. Schaul, R.J. Liebermann, C.R. Dempsey, and K.A. Dostal, 'Fate of Azo Dyes in the Activated Sludge Process', Paper presented at the 41st Annual Purdue Industrial Waste Conference, May 13-15, 1986.

34. H.R. Hitz, W. Huber and R.H. Reed, *J. Soc. Dyers Colour.*, 1978, **94**, 71.

35. R.J. Liebermann and G.M. Shaul, Technical Status Report for Adsorption of Azo Dyes onto Activated Sludge Solids, Contract No. 68-03-3183, April 1987, (US EPA, Cincinatti, Ohio).

36. D. Brown and P. Laboureur, *Chemosphere* 1983, **12**, 405.

37. D. Brown and B. Hamburger, *Chemosphere,* 1987, **16**, 1539.

38. K. Wuhrmann, K. Mechsner, and T. Kappeler, *Eur. J. Appl. Microbiol. Biotechnol.,* 1980, **9**, 325.

39. W.C. Tincher, Dyes in the Environment: Dyeing Wastes in Landfill, Status Report on ETAD Project, October 1987.

40. E.J. Weber, U.S. EPA Environmental Research Laboratory, Atlanta, GA, Personal communication.

41. W.R. Haag and T. Mill, *Environ. Toxicol. Chem.,* 1987, **6**, 359.

42. M. Bandlow, R. Frank, M. Herrmann, K. Hustert, D. Kotzias, H. Palar, and A. Zsolnay, 'Ausarbeitung eines abgestuften prüfungssystems für die abiotische Abbaukarkeit in wässriger Lösung', Endbericht GSF-FE-77663, 1987. Institut für Oekologische Chemie der Gesellschaft für Strahlen- und Umweltforschung (GSF–mbH, München und Battell Institut E.V., Frankfurt, FRG.

23

Selection of Substances Requiring Priority Action

C.D. Byrne

DEPARTMENT OF THE ENVIRONMENT, WATER QUALITY DIVISION, ROMNEY HOUSE, 43 MARSHAM STREET, LONDON SW1P 3PY, UK

1 Introduction

There are currently approximately one hundred thousand chemical substances registered on the European market. Some of these substances have the potential to cause harm to the aquatic environment and the organisms that it supports. The problem faced by pollution-controlling authorities is one of selecting from this large number of substances those chemicals which warrant some form of priority action. It is necessary to develop such a priority list so that the resources available both to Governments and Industry are applied in the most effective manner. The requirement for priority lists spans many areas of water pollution control. For example, in the UK there is a requirement that the Government approval of existing pesticides be periodically reviewed. This obviously entails a reassessment of the potential harm pesticides can cause aquatic life. This reassessment is likely to take a considerable amount of time and effort so it is sensible to examine those pesticides which are considered to be the most dangerous as a first-priority. Similarly, when setting national environmental quality standards, priority needs to be given to those substances posing the greatest environmental threat.

This chapter outlines a general selection scheme which can be used to identify priority aquatic pollutants in a wide range of circumstances. The scheme consists of a set of decision trees into which parameters appropriate to particular circumstances can be slotted. The identification of List I substances[1] in the context of the EEC Dangerous Substances Directive has been taken as an example of an appropriate use of the scheme, in order to illustrate its detailed workings.

2 Background

Under the Dangerous Substances Directive (76/464/EEC) Member States are required to 'eliminate pollution' by List I substances and 'reduce pollution' by List II substances.[1] It follows that the controls over the production, discharges, and uses of List I substances are intended to be more stringent than for List II. There appears, however, to be no clear view of what differentiates a List I from a List II substance excepting the rather vague notion that a List I substance is toxic, persistent, and bioaccumulative. The identification of List I substances must therefore centre on selecting those substances which require the most stringent form of control. This process involves not only an assessment of the properties of a substance but also an assessment of its current and possible future level in the environment.

A list of 129 substances or groups of substances has been identified by the Commission as being potential List I candidates (see Appendix I). Since the adoption of the Directive in 1976 agreement has been reached on the designation of less than a dozen of these 129 as confirmed List I. It is generally agreed that the process by which potential List I substances have their status confirmed needs to be speeded up. However, an essential element in this procedure is the existence of a systematic and generally accepted criterion for deciding which of the remaining 129 substances should next be considered for List I status. At the moment there appears to be no agreed criteria for determining this question of priority.

There have been many attempts at deriving methods of selection and/or ranking of hazardous or dangerous chemicals. Many of these have been developed in connection with legislation for assessing the potential environmental hazard from new chemicals. The methods fall between two extremes:

(i) At the most sophisticated end are methods which employ a mathematical model to predict or calculate the environmental concentration of a chemical. This concentration is then related to the concentration which is known to have an effect upon organisms. The closer these two concentrations are to each other, the higher the hazard ranking of the chemical. An example of this approach is the Stanford Research Institute (SRI) study carried out for the EEC.[2] The most advanced examples of models calculating the distribution of chemicals in the environment are those developed by Mackay.[3]

(ii) At the other end of the spectrum is a method of selecting chemicals on the basis of a limited number of properties such as toxicity, persistence,

and bioaccumulation. An example of this approach is given in the recent paper by Taylor et al.[4] in which any chemical with a bioconcentration factor of 1000 or more, a 96 h LC_{50} of 1 mg l^{-1} or less and a half-life of 100 days or more is considered to be a 'black' or List I substance.

There are problems with both methods:

(i) The 'model' approach is certainly the more attractive scientifically. However, the environment is extremely complex and still poorly understood. As models of the environment become more sophisticated in order to incorporate the many complexities, there is an increased probability of undetected error. Furthermore, models require a considerable quantity of reliable and accurate data concerning the physical, chemical, and toxicological properties and behaviour of substances. Regrettably this information is not usually available. Thus, though models will be useful in the future as scientific knowledge and understanding develop and data become available, they are currently unlikely to play a significant regulatory role.

(ii) The use of just three properties, as in the other extreme, can greatly over-simplify the problem. By using such a method there is a great danger of classifying as hazardous substances which do not warrant such a classification and possibly of failing to identify truly hazardous substances.

A further problem with almost all ranking and selection schemes is the assumptions which are made about the relative importance of particular properties of the chemical. For example, the toxicity of a substance may be rated on a score from one to five and the persistence on a score from one to ten. The overall ranking is obtained by adding or multiplying these scores together. However, a bias has been introduced by giving persistence a double weighting as compared to toxicity. (Even using the same weighting may be considered to be introducing bias.) The correctness of such weightings or biases is difficult to assess on scientific grounds since it is the combination of properties of a substance along with the level of input which results in it causing problems for the environment. Unfortunately in many schemes the assumptions made concerning the relative importance of various properties are not explicit.

3 Outline of UK Department of the Environment Selection Scheme

The recently developed Department of Environment Scheme has two main aims. The first is to provide a set of criteria which can be used to identify those substances that should be considered in detail for priority action. The second aim is to provide a framework within which the subsequent detailed discussions on the necessity of any particular selected substance being a priority candidate can be held. These discussions would take place in the case of List I candidates, through the normal EEC procedures, *e.g.* referral to the Group of National Experts, the Scientific Advisory Committee on Toxicology and Ecotoxicology, and the Working Group of the Council of Environment Ministers.

In devising the scheme, the limitations of available data played a large part in determining its structure. In the case of the List of 129, if toxicity, persistence, and bioaccumulation data were available in each case and the concentrations of these substances in all UK/EEC waters, biota, sediments, *etc.* were known, it would be relatively straightforward to identify those substances requiring the most stringent controls, *i.e.* those to be classed as List I. Unfortunately, not nearly enough data of this type are available, so data requirements for the selection scheme need to be adapted. It was also felt important to make the subjective parts of the scheme transparent so that they could be open to inspection and criticism and possibly modified in the future as understanding of the environment advances. The most important assumption which has been made is that no single property or characteristic warrants a substance being designated for priority action. The scheme is based on the fundamental premise that it is the combination of several of a substance's properties along with its prevalence in the aquatic environment which justifies a substance's presence on any priority list.

By analysis and extrapolation from past environmental problems caused by the introduction of hazardous substances into the aquatic environment, three broad scenarios can be drawn up which describe the sets of circumstances which have the potential to result in an 'environmental problem'. These are as follows:

i) The Short-term Scenario. This is where the concentration of the substance in the water approaches the level at which acute toxic effects may occur. The concentration in water may have reached this level either because of the combination of a number of irregular sources or the combination of a continuous source, some irregular sources, and the substance's persistence.

ii) The Long-term Scenario. This is the situation where the concentration of the substance approaches the level at which chronic toxic effects occur, due to a variety of continuous and irregular sources and the substance's persistence.

iii) The Food Chain Scenarios. This is the situation where the concentration of the substances reaches a level at which toxicity problems occur in higher organisms due to bioaccumulation through the food chain.

These scenarios set the framework for the scheme and provide the basis for defining the data requirements for candidate substances, *e.g.* toxicity, persistence, and bioaccumulation plus an indication of probable concentrations in the environment. Unfortunately, the data available on actual concentrations in the aquatic environment of many candidate substances are totally inadequate. Furthermore, as mentioned earlier, it is not currently possible to predict these concentrations accurately for a large number of substances using models. It is necessary to change the data requirements in relation to concentration in the environment by substitution with an input factor for each chemical based on known parameters such as production volume, use categories, solubility, volatility, *etc*.

Having defined the data that are needed for the selection scheme it is now possible to define the three scenarios in more precise terms. For each scenario a decision tree diagram can be drawn showing the combinations of properties and input which are considered to have the potential to lead to a serious problem. Obviously, the properties and input of substances vary in magnitude. It is therefore necessary to introduce an element of scaling into the scenarios for each property and the input factor. To achieve this, each has been divided into three rankings, high, medium, and low. This enables different combinations of varying levels of the properties and inputs to be equated. The resulting decision tree for each scenario is shown in Figures 1-3. The lines from the squares join together various rankings of the properties and input to form the combinations. Any substance whose combination of properties and input are identical to any one of those shown in the Figures may be considered a candidate for priority action. Those combinations which are not considered to warrant priority action candidature are not incorporated into the decision trees for the sake of clarity.

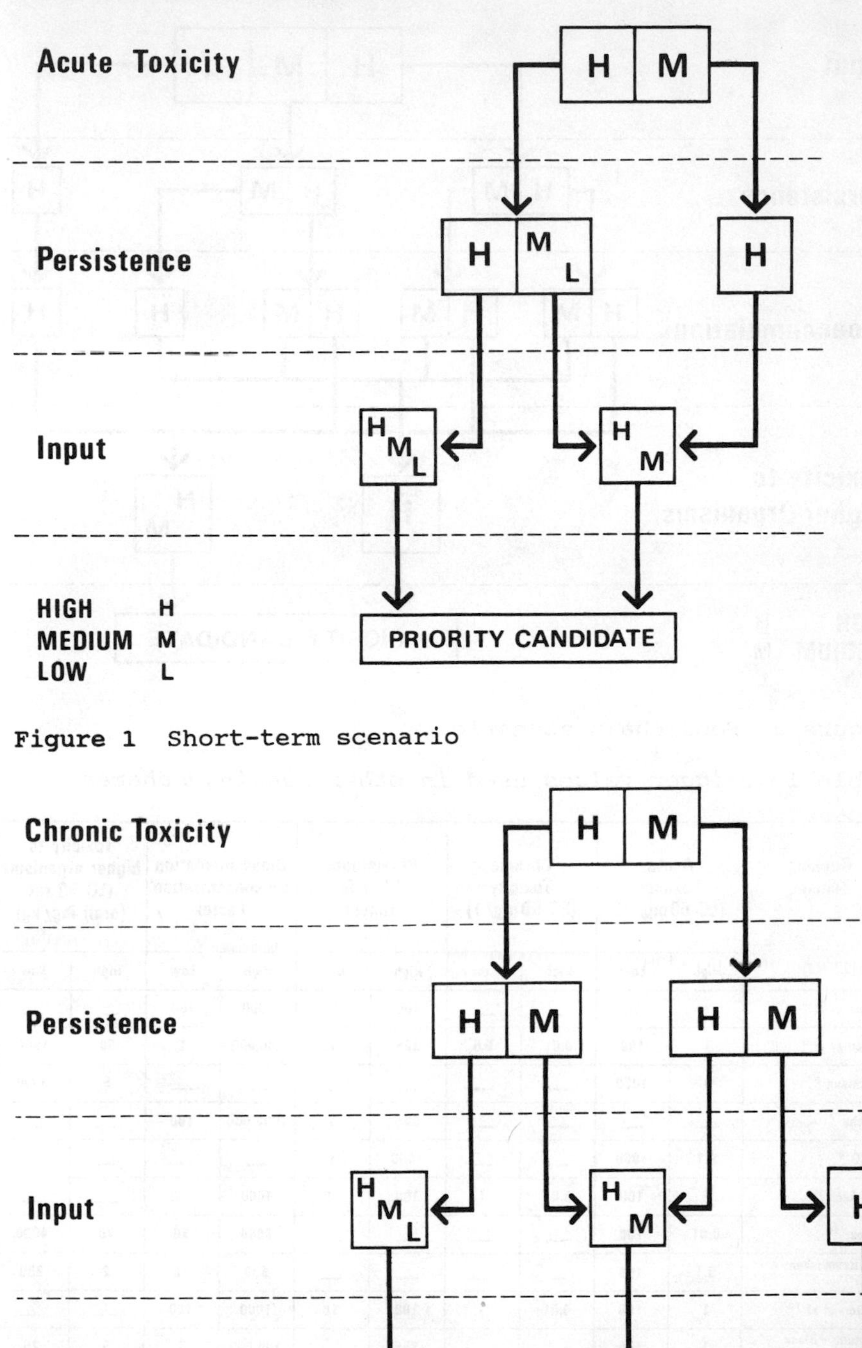

Figure 1 Short-term scenario

Figure 2 Longer-term scenario

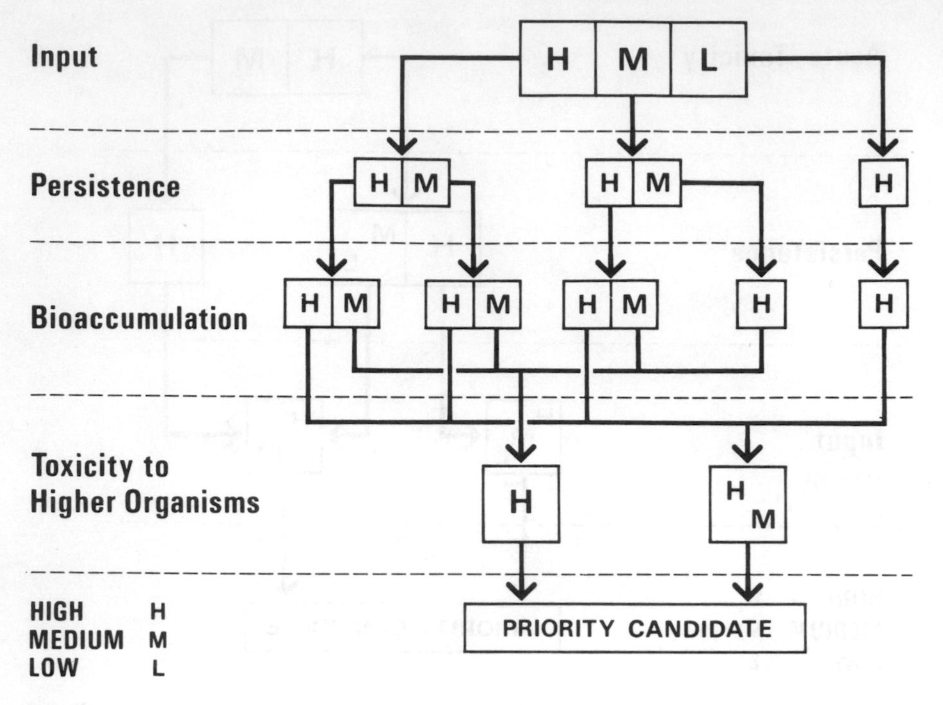

Figure 3 Food chain scenario

Table 1 *Trigger values used in other ranking schemes*

Ranking Scheme	Acute Toxicity (LC 50mg/1)		Chronic Toxicity (EC 50mg/1)		Persistence 1/2 Life (days)		Bioaccumulation Bioconcentration Factor		Toxicity to higher organisms (LD 50 rat (oral) mg/kg)	
	High	Low	High	Low	High	Low	High	Low	High	Low
Taylor et al [4]	1	10	—	—	100	10	1000	100	—	—
Welch et al [5]	1	100	0.01	0.5	365	7	100,000	1	50	1500
Portmann [6]	1	1000	—	—	—	—	—	—	5	5000
Frische [7]	—	—	—	—	365	7	1000,000	100	—	—
OECD [8]	0.1	1000	—	—	1000	1	—	—	—	—
Van Esch [9]	1	100	0.01	1	100	10	1000	100	—	—
Weber [10]	0.01	100	—	—	—	—	8000	60	20	4000
Bro Rasmussen et al [11]	0.1	100	—	—	—	—	500	10	2	200
Canton et al [12]	1	100	0.01	1	100	10	1000	100	—	—
Schmidt – Bleek et al [13]	1	100	—	—	365	—	100,000	1	5	25
Klein et al [14]	1	100	—	—	365	4	100,000	10	5	25

– no appropriate value quoted

The next more difficult step is to define the subdivisions of high, medium, and low using numeric values. Many selection schemes use some form of trigger valves, some examples are contained in Table 1. Table 2 contains the numerical values that are used in the DoE scheme as applied to List I substance selection. They are based in part on the values used in other schemes, in part from experience with confirmed List I substances and in part incorporate an element of professional judgement. The definition of the type of numerical value which replaces the high, medium, and low ranking for each property is not critical as it acts only as an indicator of that property rather than having specific environmental significance.

Table 2 *Numerical values for property rankings*

	Numerical Values				
Property Ranking	Acute Toxicity (LC 50 mg/l)	Chronic Toxicity (EC 50 mg/l)	Persistence ½ life (days)	Bioaccumulation Bioconcentration Factor	Toxicity to higher organisms (LD 50 rat (oral) mg/kg)
High	⩽ 1.0	⩽ 0.01	⩾ 100	⩾ 1,000	⩽ 50
Medium	> 1.0– < 100	> 0.01– < 1	10– < 100	999– 101	> 50– < 500
Low	⩾ 100	⩾ 1	⩽ 10	⩽ 100	⩾ 500

For example, for persistence it does not matter if you use half-lifes or time for reduction of concentration by 90% as the numerical values for this property. The important factor in determining the type of numbered values to be used is the availability of comparable data. On this basis the following have been used:

Persistence - Half-life in water in days (all scenarios)

Acute Toxicity - 96 h LC_{50} (Short-term scenario)

Chronic Toxicity - EC_{50}s or 1/100 of LC_{50}s (Long-term scenario)

Toxicity to Higher Organisms - LD_{50} Rat (oral) (Food chain scenario)

Bioaccumulation - BCFs or BCFs calculated from K_{OW}s (Food chain scenario)

(K_{OW} = octanol-water partition coefficient;
BCF = Bioaccumulation factor)

Inserting the numerical values from Table 2 into the decision trees produces a series of look-up diagrams Figures 4-6.

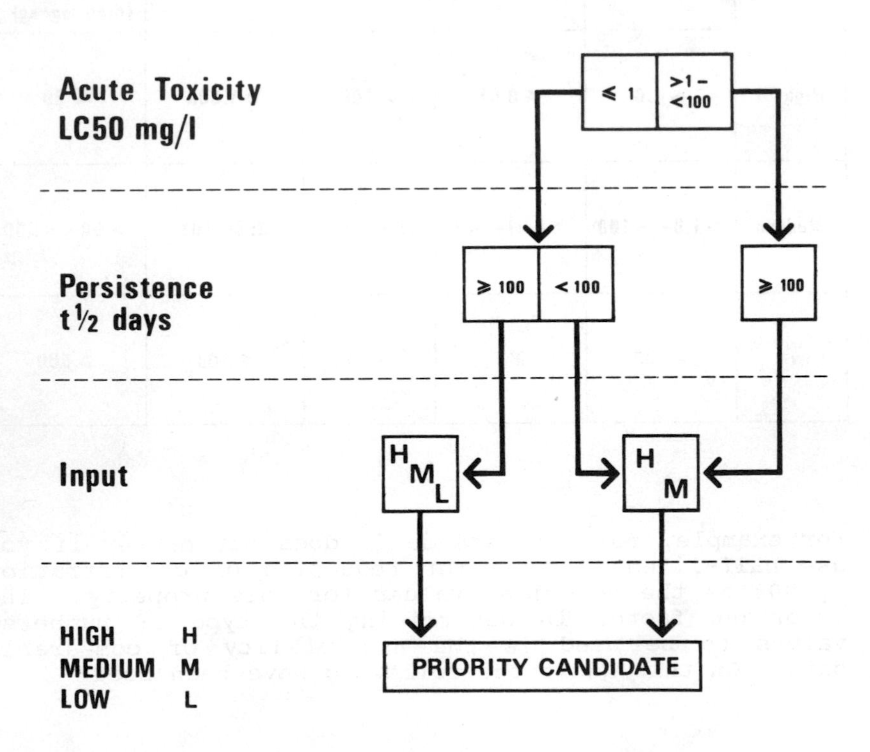

Figure 4 Short-term 'look-up' diagram

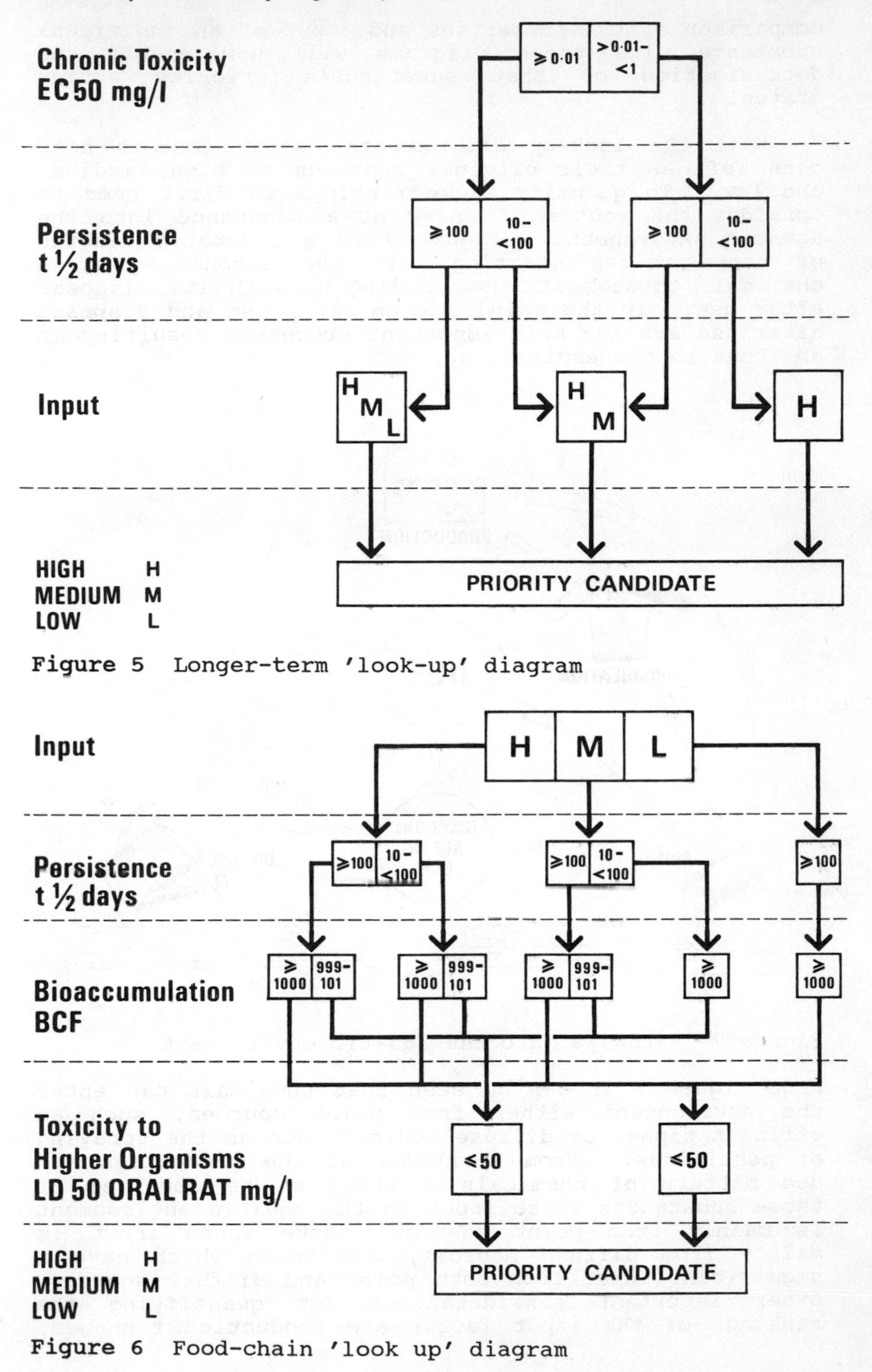

Figure 5 Longer-term 'look-up' diagram

Figure 6 Food-chain 'look up' diagram

Comparison of the properties and input of an individual substance with these diagrams will now enable the determination of that substance's priority action status.

With the look-up diagrams the input factors have been left as their original rankings of high, medium, and low. To quantify these rankings we first need to consider the routes of entry of a substance into the aquatic environment. Figure 7 is a schematic diagram of these routes starting with the production of a chemical, through its use, ending up with its disposal after use. In the main, a chemical's use and disposal after use are the more important processes resulting in an input to the environment.

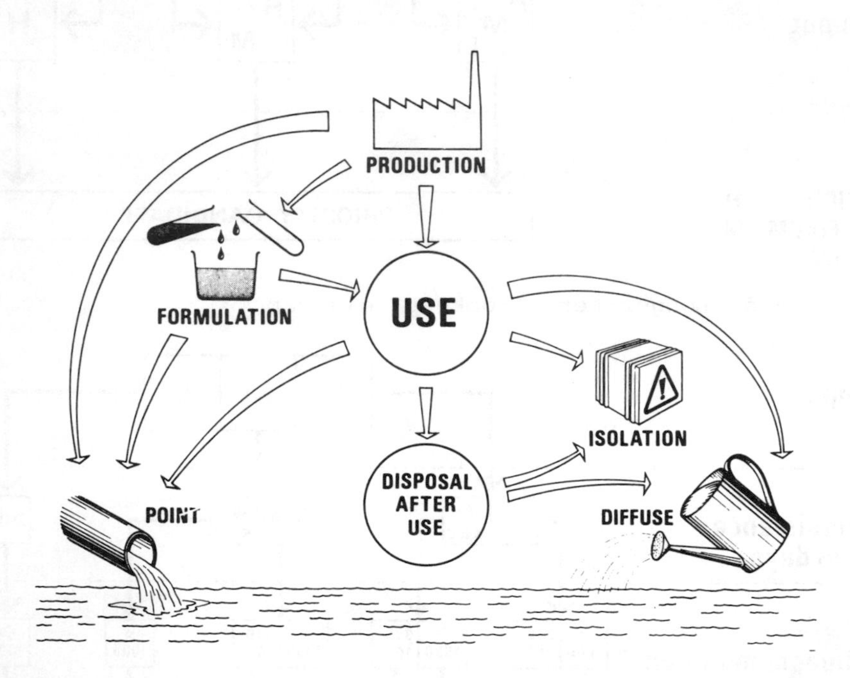

Figure 7 Pathways into the aquatic environment

From figure 7 it can be seen that chemicals can enter the environment either from point sources, such as effluent pipes, or diffuse sources such as the spraying of pesticides. From knowledge of the production and use pattern of chemicals a list can be compiled of those substances whose input to the aquatic environment is mainly from point sources, those whose input is mainly from diffuse sources, and those which have a significant input from both point and diffuse sources. Other important considerations for quantifying the rankings of the input factor are production tonnages, solubility, and volatility.

The production tonnage is important as it determines the maximum quantity of a substance that could possibly enter the environment. The solubility and volatility are important as they are indicators of how much of the production tonnage is likely to get into effluents and hence into the aquatic environment. These parameters have been combined with the source categories to form another decision tree (Figure 8). The tree is used to decide whether the input factor of a chemical is high, medium, or low. For the sake of convenience of presentation solubility and volatility have been combined to produce an 'escapability' factor. This factor is used as an indicator of the likelihood of a substance *getting into an effluent* and hence into the environment from a point source. It is calculated by dividing the solubility by the volatility. Thus a substance which has a high solubility and a low volatility will have a high 'escapability' while a substance with a low solubility and a high volatility will have a low 'escapability'.

Again, as for the various properties, by examining the literature some consensus can be obtained as to a set of numerical values used to define the parameters in Figure 8.

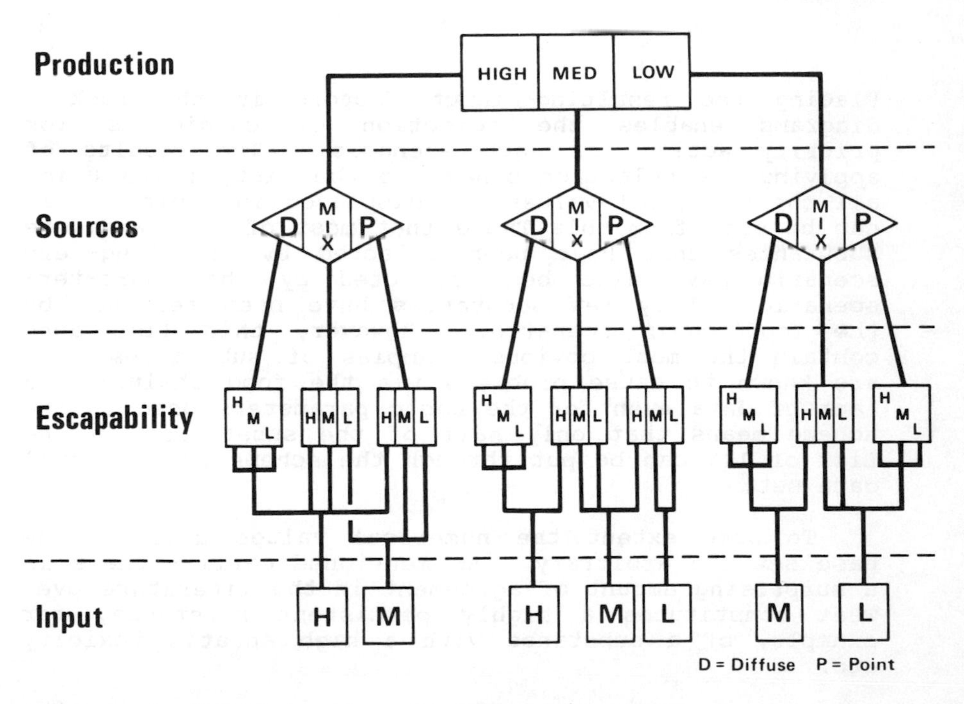

Figure 8 Input to the aquatic environment

The values used for selecting List I candidates are listed in Table 3.

Table 3 *Numerical values for input rankings*

Input Ranking	Numerical Values		
	Production (tonnes/ year)	Solubility (mg/ l)	Volatility (pa)
High	⩾10,000	⩾1,000	⩾ 0.133
Medium	<10,000 – >1,000	<1,000 – >1	< 0.133 – > 0.000133
Low	⩽1,000	⩽1	⩽ 0.000133

Placing the resulting input factors in the look-up diagrams enables the selection of candidates for priority action for each scenario. The results of applying the selection scheme to the List of 129 using all the numerical values are displayed in Table 4. It can be seen from this Table that most of the candidate substances that have been selected by the long-term scenario have also been selected by the short-term scenario. Very few substances have been selected by the food chain scenario. However, this list does contain the most obvious examples of substances that are known to cause problems *via* the food chain. The lack of data even for the basic parameters used in the scheme means that only half of the substances in the List of 129 can be put through the scheme using a full data set.

To some extent the numerical values used in the base set are arbitrary. As mentioned earlier there is a surprising amount of agreement in the literature over what constitutes a highly persistent substance, for example, of a substance with a high aquatic toxicity *etc*.

Table 4 *List I candidates using base set values*

SUBSTANCES LISTED BY VIRTUE OF AVAILABLE DATA	SHORT TERM SCENARIO	LONG TERM SCENARIO	FOOD CHAIN SCENARIO
ALDRIN	●	●	●
ANTHRACENE	●		
AZINPHOS-ETHYL	●	●	
AZINPHOS-METHYL	●	●	●
CHLOROPRENE	●	●	
3-CHLOROTOLUENE	●	●	
2,4-D	●		
DDT	●	●	●
DEMETON-O	●		
1,4-DICHLOROBENZENE	●		
1,1-DICHLOROETHYLENE	●	●	
1,3-DICHLOROPROPENE	●		
DICHLORPROP		●	
DICHLORVOS	●	●	
DIELDRIN	●	●	●
ENDOSULFAN	●		
ENDRIN	●	●	●
ETHYLBENZENE	●		
FENTHION	●	●	
GAMMA-HEXACHLORO CYCLOHEXANE (LINDANE)	●	●	
LINURON		●	
MALATHION			
MEVINPHOS	●		
MONOLINURON	●	●	
PARATHION			
PARATHION-METHYL	●		
PENTACHLOROPHENOL	●		
PYRAZON (CHLORIDAZON)	●	●	
SIMAZINE		●	
2,4,5-T	●	●	
1,2,3-TRICHLOROBENZENE	●	●	
1,2,4-TRICHLOROBENZENE	●	●	
1,1,1-TRICHLOROETHANE	●	●	
TRIFLURALIN	●	●	

However, these views are mainly based on professional judgement rather than some objective criteria. To try to understand the effect of using different base set values in the scheme has on the number and identity of the candidate substances selected, the scheme has been computerized. The results obtained from the programme runs carried out so far, varying the base set values, suggest that the scheme is fairly robust. Variation in the main parameters such as toxicity and persistence by factors of between five and ten results in a change in status for only 10% to 15% of the candidate List I substances selected by the base set. Reducing the numerical values for production by a factor of ten has a similar effect whereas adjustment of the escapability factor causes only minor changes. This form of sensitivity testing allows those substances which are on the borderline of being priority action candidates to be identified so that their data set can be re-examined and their non-selection confirmed. This is particularly useful when it comes to trying to predict likely problem substances in the future. For instance, the production of certain substances could be monitored so that at the point when they exceed a certain base set level their priority action candidate status is confirmed and appropriate control action taken to prevent an environmental problem. In the same way, as further data on the properties of the other substances become available these can be fed through the scheme, and again, the substances' priority action candidature can be reassessed.

4 Development of Scheme

At this stage the scheme is based on some rather simple concepts of what causes aquatic environmental problems. Certain more difficult areas have been left out such as carcinogenicity. It is envisaged that, as understanding of these areas improve, new scenarios will be incorporated into the scheme and the existing ones modified. The scheme can also be refined; for example, persistence could be expressed as a combination of half-lifes relating to the individual degradation pathways such as microbial, photodegradation, and hydrolysis. The limiting factor of the scheme remains the availability of data. Major effort is required in order to provide full property data for about half of the substances on the List of 129. To some extent the gaps could be filled by the use of structure-activity derived data. However, perhaps the most important parameter for which data are urgently required is that of annual production figures. The only reliable sources for such data, inevitably, are the industries producing the candidate substances.

5 Application of the Scheme

The scheme has been described with reference to the selection of List I substances. Another example of the use of the scheme is the identification of substances whose discharges should be controlled by the application of technology-based emission standards. The UK Government recently announced its intention to introduce such a system, in conjunction with environmental quality standards, for a set of priority substances to be termed the 'Red List'.[15] A scheme of the kind outlined in this chapter is likely to form the basis of identifying substances to be included on the 'Red List'. However, any control measures which are based on the results of this scheme will require a degree of stability to ensure that pollution control agencies, and those whose activities they seek to control, are not aiming at a perpetually moving target. A balance will therefore need to be struck between the need for periodic review of the properties and input data of candidate substances and the advantage of a relatively stable list of priority substances.

6 Conclusions

In summary, the proposed scheme does not attempt to model the fate of the substances in the environment, but identifies those combinations of properties and input which appear likely to have the potential to result in environmental problems. These combinations of properties and input are quantified using readily available data resulting in a set of look-up diagrams which can be easily used for deciding whether a substance should be considered for List I status. When applied to the European Community List of 129 comparison of the substances selected by the scheme with those currently classified as List I or under consideration of List I status shows a marked correlation. If nothing else this demonstrates that the results it produces accord closely with current views about priority substances based on the subjective assessment and practical experience of environmental risk. The scheme is by no means perfect and still contains a fair degree of subjectivity. It does however provide a broad framework which can be refined as knowledge and understanding of the environment progress.

7 Acknowledgements

I should like to acknowledge the help and encouragement I have received from my colleagues in the Department of the Environment and in the Water Research centre.

8 References

1. EEC Directive on pollution caused by certain dangerous substances discharged into the aquatic environment of the Community, 76/464/EEC, *Official Journal* L129.18/5/76.

2. Stanford Research Institute (1980) Elaboration of a method for evaluating the hazard to the aquatic environment caused by substances of List I of Council Directive 76/464/EEC and preparation of a list of dangerous substances to be studied by priority. Cress Report no.136.

3. D. Mackay and S. Paterson, *Environ. Sci. Technol.* 1981, **15**, 1006-1014.

4. D. Taylor, G. Diprose, and M. Duffy, *J. Com. Mark. Stud.*, 1986, **XXIV**, 225.

5. J.L. Welch and R.H. Ross, *Environ. Tox. Chem.*, 1982, **1**, 95.

6. J.E. Portmann, *Ecotox. Environ. Safety*, 1981, **5**, 56.

7. R. Frische, G. Esser, W. Schonborn, and W. Klopffer, *Ecotox. Environ. Safety*, 1982, **6**, 283.

8. OECD Environment Directorate. Water Management Group Programme on policies for water pollutants control hazard rating. ENV/WAT/77.4.

9. G.J. Van Esch, Aquatic Pollutants and their Potential Biological Effects, in 'Aquatic Pollutants Transformation and Biological Effects', ed. Hutzingen, Van Lelyveld, and Zoeteman, Pergamon Press 1978.

10. J.B. Web, *Environ. Sci. Technol.*, 1977, **11**, 756.

11. F. Bro-Rasmussen and K. Christiansen, *Ecol. Model*, 1983/4, **22**, 67.

12. J.H. Canton and W. Sloff, *Ecotox. Environ. Safety*, 1982, **3**, 126.

13. F. Schmidt-Bleek, W. Haberland, A.W. Klein, and S. Carili, *Chemosphere,* 1982, **11**, 383.

14. A.W. Klein and W. Haberland, Environmental hazard ranking of new chemicals based on European Directive 79/831/EEC Annex VII, in 'Chemicals in the Environment: Chemical Testing and Hazard Ranking - the interaction between Science and Administration', Proceedings of an International symposium, ed. K. Christiansen, B. Koch, and F. Bro-Rasmussen, 1982.

15. Lord Belstead, Parliamentary Question on Dangerous Substances, House of Lords Official Record *(Hansard)*, 19/11/87, **490**, (39), 403-404.

Appendix 1: List of substances which could belong to List I of Council Directive 76/464/EEC

Aldrin
2-Amino-4-chlorophenol
Anthracene
Arsenic and its mineral compounds
Azinphos-ethyl
Azinphos-methyl
Benzene
Benzidine
Benzyl chloride (α-chlorotoluene)
Benzylidene chloride (α,α-dichlorotoluene)
Biphenyl
Cadmium and its compounds
Carbon tetrachloride
Chloral hydrate
Chlordane
Chloroacetic acid
2-Chloroaniline
3-Chloroaniline
4-Chloroaniline
Chlorobenzene
1-Chloro-2,4-dinitrobenzene
2-Chloroethanol
Chloroform
4-Chloro-3-methylphenol
1-Chloronaphthalene
Chloronaphthalenes (technical mixture)
4-Chloro-2-nitroaniline
1-Chloro-2-nitrobenzene
1-Chloro-3-nitrobenzene
1-Chloro-4-nitrobenzene
4-Chloro-2-nitrotoluene
Chloronitrotoluenes (other than 4-Chloro-2-nitrotoluene)
2-Chlorophenol
3-Chlorophenol
4-Chlorophenol
Chloroprene (2-Chlorobuta-1,3-diene)
3-Chloropropene (Allyl chloride)
2-Chlorotoluene
3-Chlorotoluene
4-Chlorotoluene
2-Chloro-*p*-toluidine
Chlorotoluidines (other than 2-Chloro-*p*-toluidine)
Coumaphos
Cyanuric chloride (2,4,6-Trichloro-1,3,5-triazine)
2,4-D (including 2,4-D-salts and 2,4-D-esters
DDT (including metabolites DDD and DDE)
Demeton (including Demeton-o, Demeton-s, Demeton-s-methyl and Demeton-s-sulphone)
1,2-Dibromomethane
Dibutyltin dichloride
Dibutyltin oxide
Dibutyltin salts (other than Dibutyltin dichloride and Dibtyltin oxide
Dichloroanilines
1,2-Dichlorobenzene
1,3-Dichlorobenzene
1,4-Dichlorobenzene
Dichlorobenzidines
Dichlorodiisopropyl ether
1,1-Dichloroethane
1,2-Dichloroethane
1,1-Dichloroethylene (Vinylidene chloride)
1,2-Dichloroethylene
Dichloromethane
Dichloronitrobenzenes
2,4-Dichlorophenol
1,2-Dichloropropane
1,3-Dichloropropan-2-ol
1,3-Dichloropropene
2,3-Dichloropropene
Dichlorprop
Dichlorvos
Dieldrin
Diethylamine
Dimethoate
Dimethylamine
Disulfoton
Endosulfan
Endrin
Epichlorohydrin
Ethylbenzene
Fenitrothion

Appendix 1 (cont)

Fenthion
Heptachlor
(including Heptachlorepoxide)
Hexachlorobenzene
Hexachlorobutadiene
Hexachlorocyclohexane
(including all isomers and Lindane)
Hexachloroethane
Isopropylbenzene
Linuron
Melathion
MCPA
Mecoprop
Mercury and its compounds
Methamidophos
Mevinphos
Monolinuron
Naphthalene
Omethoate
Oxydemeton-methyl
PAH
(with special reference to:
3,4-Benzopyrene and
3,4-Benzofluoranthene)
Parathion
(including Parathion-methyl)
PCB
(including PCT)
Pentachlorophenol
Phoxim
Propanil
Pyrazon
Simazine
2,4,5-T
(including 2,4,5-T salts and 2,4,5-T esters)
Tetrabutyltin
1,2,4,5-Tetrachlorobenzene
1,1,2,2-Tetrachloroethane
Tetrachloroethylene
Toluene
Triazophos
Tributyl phosphate
Tributyltin oxide
Trichlorfon
Trichlorobenzene
(technical mixture)
1,2,4-Trichlorobenzene
1,1,1-Trichloroethane
1,1,2-Trichloroethane
Trichloroethylene
Trichlorophenols
1,1,2-Trichlorotrifluoro-ethane
Trifluralin
Triphenyltin acetate
(Fentin acetate)
Triphenyltin chloride
(Fentin chloride)
Triphenyltin hydroxide
(Fentin hydroxide)
Vinyl chloride
(Chloroethylene)
Xylenes
(technical mixture of isomers)

Appendix 2 Lists of Families and Groups of Substances

List I contains certain individual substances which belong to the following families and groups of substances, selected mainly on the basis of their toxicity, persistence and bioaccumulation, with the exception of those which are biologically harmless or which are rapidly converted into substances which are biologically harmless:

1. Organohalogen compounds and substances which may form such compounds in the aquatic environment
2. Organophosphorus compounds
3. Organotin compounds

4. Substances in respect of which it has been proved that they possess carcinogenic properties in or via the aquatic environment*
5. Mercury and its compound
6. Cadmium and its compounds
7 Persistent mineral oils and hydrocarbons of petroleum origin.
8. Persistent synthetic substances which may float, remain in suspension or sink and which may interfere with any use of the waters.

List II contains:

- Substances belonging to the families and groups of substances in List I for which the limit values referred to in Article 6 of the Directive have not been determined
- Certain individual substances and categories of substances belonging to the families and groups of substances listed below, and which have a deleterious effect on the aquatic environment, which can, however, be confined to a given area and which depend on the characteristics and location of the water into which they are discharged.

Families and groups of substances referred to in the second indent

1. The following metalloids and metals and their compounds:

1. Zinc	11. Tin
2. Copper	12. Barium
3. Nickel	13. Beryllium
4. Chromium	14. Boron
5. Lead	15. Uranium
6. Selenium	16. Vanadium
7. Arsenic	17. Cobalt
8. Antimony	18. Thalium
9. Molybdenum	19. Tellurium
10. Titanium	20. Silver

2. Biocides and their derivatives not appearing in List I.

3. Substances which have a deleterious effect on the taste and/or smell of the products for human consumption derived from the aquatic environment and compounds liable to give rise to such substances in water.

* When certain substances in List II are carcinogenic, they are included in category 4 of this list.

4. Toxic or persistent organic compounds of silicon, and substances which may give rise to such compounds in water, excluding those which are biologically harmless or are rapidly converted in water into harmless substances.

5. Inorganic compounds of phosphorus and elemental phosphorus

6. Non-persistent mineral oils and hydrocarbons of petroleum origin

7. Cyanides, fluorides

8. Substances which have an adverse effect on the oxygen balance, particularly: ammonia, nitrites.

Section 4: Intentional Emissions

24
Hazard and Risk Assessment and Acceptability of Chemicals in the Environment

F. Bro-Rasmussen

LABORATORY OF ENVIRONMENTAL SCIENCE AND ECOLOGY, THE TECHNICAL UNIVERSITY OF DENMARK, DK-2800 LYNGBY, DENMARK

1 Introduction

According to the storyteller, Queen Cleopatra was a bold woman. She exposed herself to many hazards. In her own capricious way she lived - and she died - by showing a high level of acceptance to many risks, such as the *courtoisie* of brutal Roman rulers. In the end, she even took what we may call a 1:1 or a 100% risk, namely when she fatally nourished a venomous snake at her bosom.

On the other hand, it is also told that she - before that event - took no chances. When she enjoyed her meals, it was the duty of her butler to predict the hazards by pretasting all her food supplies. She achieved what she thought would be a zero risk, although some of us could undoubtedly challenge that as an illusion.

However, the concept of a zero risk is also applied today, *e.g.* when we require a 100 x Safety Factor (SF) to be applied for food chemicals on top of a Non-Observed Effect Level (NOEL). Instinctively we feel uneasy, or we are even alarmed if such an interval is reduced. This may be the case when we meet certain hazardous or toxic chemicals as unacceptable contaminants in our food, such as mercury in fish, or polychlorobiphenyls in mother's milk.

Like Queen Cleopatra, we may in other cases *(e.g.* our often careless handling of household chemicals) accept lower safety margins, or under certain circumstances we may even be willing to endure certain ill-effects. This latter situation seems to be close by, when we with confidence accept that the medical doctors prescribe drugs or pharmaceutical preparations for which the profession on our behalf has been entrusted to evaluate and balance side effects against other - curative or protective - effects.

In spite of the well-established practices which can be found behind this historical analogy, it seems appropriate in this chapter to present a conceptual framework for the evaluation of hazards of and risks from environmental chemicals. This permits us to develop reasonably stringent definitions, and to exercise a more narrow range of interpretations than normally found in this context. We also recognize that it may not be possible to transfer every word which is used above to illustrate Cleopatra's endangered life directly into use in our dealing with the threatened environment as the object for our interest in risk assessments.

A few examples of environmental risk assessments will be presented, and the observation will be made that the acceptability is as important for the assessment as are the risk criteria *per se*. Our standards for risk acceptance are variable, and we will notice that our acceptance of human risks differs from - and generally is low in comparison with - the more liberal views which until today have mostly characterized the sphere of environmental risks.

2 Conceptual Framework

2.1 Chemical Hazards to the Environment. Hazard is defined in the Glossary of Terms. It can normally be described as a function of two elements, each of which characterizes the chemical in relation to the threatened target, *viz.* the exposure from the chemical, and the potential (adverse) effects resulting from such exposure.

This definition is close to the practice developed in recent years by the Organization for Economic Co-operation and Development OECD,[1] and it seems to be adopted in most evaluations of chemicals in the environment, whereas there is still a tendency in the sphere of human hazards to concentrate on the effects assessment. This is connected to a traditional interpretation which seems to dominate at least in the USA (see Chapter 6), where the exposure analysis is a part of assessment of human risks rather than attached to the evaluation of human health hazards.[2]

The hazard evaluation - as here defined - is often[3,4] described in terms of a step-wise procedure in which we may distinguish between the hazard identification (HI) and the hazard assessment (HA) (see Glossary of Terms and Figure 1). The basis for HA is an estimation of potential exposure concentrations of the chemical in the environment, in order to compare these with the concentration or dosage levels which may create (adverse) effects on specified target systems, if they are exposed.

1. <u>HAZARD IDENTIFICATION (HI)</u>

To identify hazardous properties inherent to the chemicals, *e.g.*

- persistence,
- bioaccumulative potential,
- toxicity, and
- possibly selected properties which indicate environmental mobility and reactivity.

<u>HI concludes</u> in classification or categorization of chemicals in numerical or otherwise prioritized order.

2. <u>HAZARD ASSESSMENT (HA)</u>

To assess the potential for a chemical to cause harm (adverse effects) to targets or target systems, such as human population, environmental species and/or ecosystems.

The assessment depends on information to predict

- environmental exposure (distribution or pathways - fate),
- assessment of effects with reference to identified receiving environments (targets at risk).

<u>HA concludes</u> in identification of targets at risk related to potential exposure and exposure routes.

3. <u>RISK ESTIMATION (RE)</u>

To estimate the probability that a chemical causes harm (adverse effects) as a result of a specified production, use and/or other emission into the environment.

The estimation involves a comparative study of

- environmental exposure (intensity, frequency, and duration) and
- exposure routes in the study area against data on acceptable/unacceptable effects in exposed target.

<u>RE concludes</u> in a relationship (probabilistic or quality related) between potential target-effect data and data on acceptable or unacceptable effects.

Figure 1 The logics of sequential hazard-risk assessments of chemicals

Obviously, the hazard identification and the hazard assessment procedures are both of a dominantly predictive nature. In administrative practice, they have in recent years gained considerable attention, and much effort has been given to the development of standard procedures for assessments.[5] The hazard assessments seem logical and useful tools in the communication between health authorities and environmental agencies on the one side, and industrial or commercial parties on the other, and also as a basis for information to the general public. They may serve for the development of classification schemes, for registration practices (notification), and for the clearance of new chemicals which are introduced into regulated areas of the modern, technological society.[6]

2.2 The Environmental Risk Concept. Risk is a word which is often confounded with the term hazard. It is interpreted here in its statistical context.

The exposure analysis is normally considered a prerequisite for a risk assessment by referring to specified release situations *e.g.* from a production site, from use, storage, or transport or similar sources. The risk estimation, therefore, involves identification of exposure routes and situations in which exposure (measured in intensities or concentrations, frequencies, and durations) coincides with the possibility that effects or unacceptable changes develop in individual targets at risk. The critical target-effects data may be identified at the species level (acute and long-term effects), at integrated population/ecosystem levels, and/or connected to other biological and abiotic test systems. As a matter of principle, even damages to cultural values, *e.g.* acidification effects on buildings could be included within this conceptual framework.

The risk estimation concludes in a relation (probabilistic) between these situations and the exposure-effects levels which are established as acceptable, as for example expressed in guideline levels, limit values or quality objectives, *etc.*

2.3 The Concept of Acceptability. In risk estimation as here defined, the term acceptability has become a crucial, 'new' element to be considered as a constituent part of the evaluation process. It is obvious that this concept cannot be defined solely on the basis of scientific considerations and advice, from *e.g.* natural scientists such as biologists, chemists, ecotoxicologists, *etc.* Other professional disciplines concerned with environmental protection, including social sciences, political practice, and (to some extent) public perception should be involved. On the other hand, the natural scientists involved in the analysis and the protection of the environmental qualities should not hesitate to express their concern about the need for criteria on acceptable - or rather unacceptable - effects and changes in the biological and the abiotic parts of the ecosystems.

The complexity of defining criteria for acceptable/ unacceptable effects in the environment is considerable, and it is difficult to establish operational concepts, unless we can select and utilize indicator- or key-species as well as important ecological functions for that purpose.[7] On such a basis we must attempt to distinguish between acceptable and unacceptable effects as expression of the hazards, which we impose on the environment and as measures of the risks we may take on behalf of the environment.

As humans we are definitely more willing to accept effects on other species than on Man, and adverse changes in environmental systems are often reported without sanctions or further protective reactions. In environmental contexts, therefore, the risk estimation may in such cases be said to deteriorate into an identification of 'risk qualities', *i.e.* types and intensities of effects (which are accepted), rather than being the quantified determination of frequencies and durations of unacceptable exposures, which are otherwise the main concern of risk assessments.

The 1:1 or 100% risk exemplified by Queen Cleopatra's dramatic death illustrates an exception to a rule, which we recognize in human life, namely the 'rule of zero or negligible risks'. Our question today is whether we take high risks on behalf of the environment - or rather how often do we take the high risks? And how far do we accept that as the rule? In short, is it permissible to deal with acceptance levels which are higher for the environment than those recognized for man?

3 Assessment and Acceptance of Marine Pollution

The pollution of many marine areas has developed rapidly in recent years into a stage of deterioration of coastal waters as well as to definite and unacceptable changes of the natural quality of the open sea. This has become an environmental issue of great urgency and great complexity, and it has created a situation which increasingly calls for detailed assessments in many specific scenarios. These may vary from local estuaries and fjord systems, to areas such as the Meditérranean or the Baltic Sea, or to highly exposed sections of the ocean, *e.g.* the North Sea.

Our societies are confronted with immense difficulties, and we are often hampered in our attempts to identify definitely the complex cause and effect relationships between specific chemical pollution and observed disturbances in the natural system. This is most often merely the result of lack of sound, scientific knowledge, but problems in certain areas may also in some cases be amplified or widened in the interests of professional secrecy, or by reserved attitudes for economic reasons (or even lack of will), or similar reasons. Normally, therefore, it is difficult to predict individual changes and adverse effects in the marine systems from specific pollution, and also to forecast accurately the results to be expected from pollution control when regulations are installed and enforced.

In spite of this, however, there exists today a reasonable consensus of opinion among marine and other experts in their identification of the most important elements of the chain of events, and about the major cause and effect relationships, which describe the increasing number of marine pollution situations.

3.1 Risk Assessment of Specific Chemicals. A number of directly observable effects, such as fish kills, disappearance of coastal marine vegetation, or increase in number of fish deformities can be - and often has been - directly linked to the impact of individual chemicals or heavy metals, which are candidates for a general classification as hazardous to the environment.[8] Such cases often result from emissions or are caused by effluents from point sources, *e.g.* industries, urban sewage outlets, or spills from marine vessels, and they can result from single (accidental?) discharges, or from regular or intermittent emissions.

In any such observed cases an identification of hazardous properties will normally be initiated, *e.g.* for toxicity, persistence, bioaccumulative potential, *etc.* of individually found chemicals. Similarly, the hazards of chemical mixtures or effluents have to be initially characterized by their ecotoxicological profiles.[9]

A detailed and more comprehensive hazard assessment may follow as a next step in the evaluation procedure. Potential effects are predicted or they are experimentally identified and evaluated in possible target organisms or systems. Potential concentrations are estimated for comparison under differing exposure scenarios, including judgements on worst-case situations based on prediction, on experiments or from monitoring activities.

If the hazard assessment shows that potential concentrations in certain recipients or in specific target systems will exceed critical levels for (adverse) effects, this is an indication that a risk evaluation should be initiated. Thus, the risk assessment for the specified situation is made on the basis of a preceding hazard assessment of the chemical(s), and it represents an advanced stage of the full evaluation process.

3.2 Acceptable/Unacceptable Effects of Specific Chemicals. In order to obtain results of a reasonable quality and credibility from the risk assessments, these are - or they should be - based on experience and established evidence, including information from relatively detailed exposure analyses and effects assessments (see Figure 2).

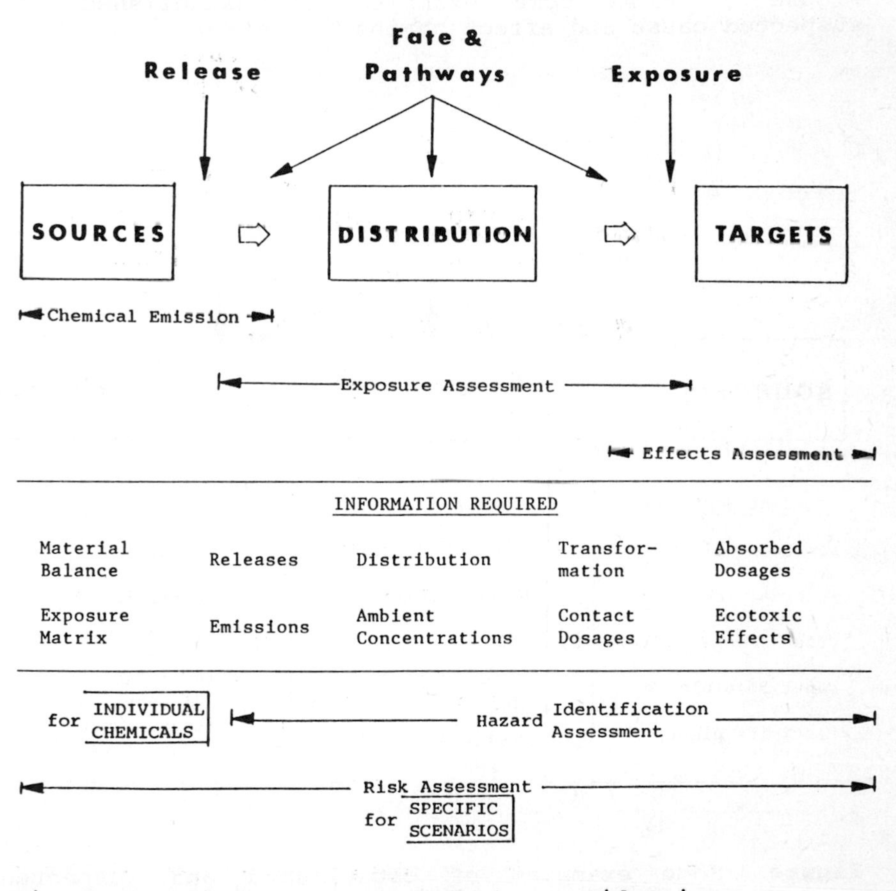

Figure 2 Elements and concepts to consider in hazard-risk assessments

Methods and technical means for performing valid risk assessments are - at least theoretically - often available for many individual priority chemicals in the environment. Reference to production and marketing information is required as the basis for release matrices and mass balance studies, to analytical techniques and computerized modelling facilities, *etc.* for exposure analyses, and to effects assessment from laboratory testings, monitoring, and surveillance data.

It is hard to point to any specific part of the combined procedure as being more (or less) critical as far as the information requirements are concerned, although it may be tempting to mention the lack of chronic ecotoxicity studies for a great number of industrial chemicals. Also disturbing is the continuous uncertainty, which is mostly attached to prediction of long-term effects at the community level or in ecosystems caused by individual chemicals. Figure 3 shows some examples of established and suspected cause and effect chains.

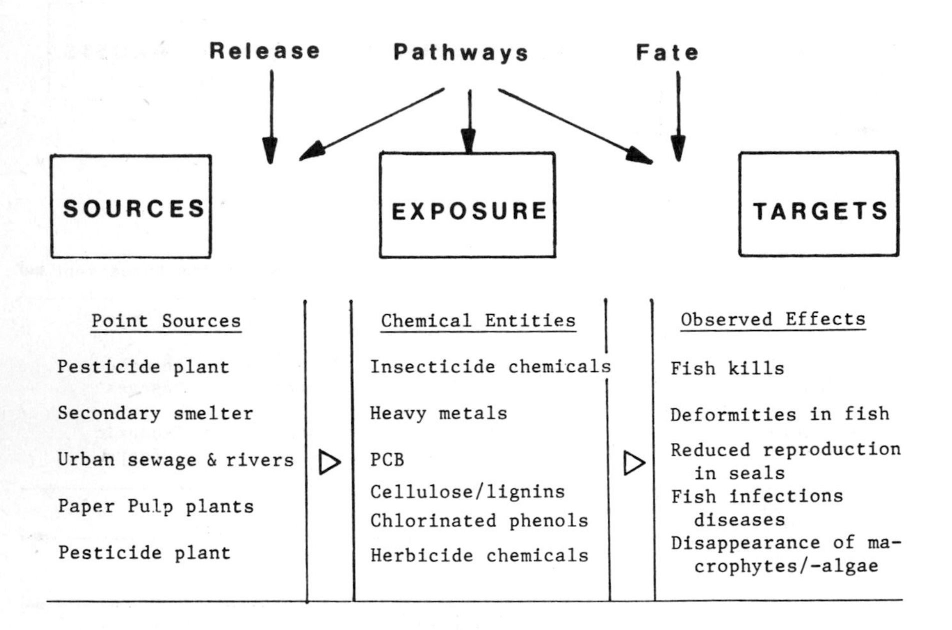

Figure 3 Some examples of established and suspected cause-effects chains connected to marine pollution

When risk evaluations are brought to their concluding steps, any critical exposure-effect data are related to corresponding levels of acceptable (or unacceptable) effects, which brings the interpretation of acceptability into focus, probably as the most critical part of the total evaluation process. An appropriate distinction between acceptable and unacceptable ecotoxic effects and changes in the ecosystem is critical for the optimum protection of the sea, but many such distinctions are debated scientifically and mostly also in the light of conflicting interests among various groups of the society.

For the time being, there seems to be a tendency in many countries to tighten up restrictions on emissions to the marine environment from point sources, but it is still the rule to accept a certain level of (adverse) effects in the environment. This can be noticed in the practice of defining (and accepting) 'mixing zones' around industrial and urban discharges instead of applying 'zero discharges' or 'best practical means' as the environmentally more appropriate strategies.

This observation is clearly connected to the introduction of 'Modified Quality Objectives' as a mode of operation in recipient quality planning in various countries, meaning that certain (adverse) effects can be defined as legally acceptable on special terms or as 'low standard quality'. Any such 'acceptable effects' will have to be defined specifically, referring to the type of effects and to the amount of effects (or the extent) which can be accepted.

The uncertainties connected to the acceptability and to the acceptance of environmental effects may also be recognized in the practice behind the establishment of Quality Objectives for individual chemicals, *e.g.* List 1 or priority chemicals within national regulations, EEC Directives, *etc.* (see also Chapter 23). In the EEC Directive on Dangerous Substances in the Aquatic Environment,[11] it is mentioned that a 'No Effect Level (NOEL) should be 'taken into account' in the setting of objectives. This observation, however, is not supported by specifications or guidance as to target organisms or biological systems to which it should be applied, nor is it prescribed how the NOEL should be determined. The acceptable (or unacceptable) effect levels are therefore left open for interpretation instead of being defined as an environmentally safe tool in the hands of the risk assessor.

3.3 Risk Assessment of Marine Eutrophication. In the description of natural processes and normal biological functions in the sea, primary importance is to be attached to the role of non-toxic, organic (biological) material, of major nutrients such as nitrogen and phosphorus, and of the available supply of oxygen in the water masses. As normal constituents (among numerous others) these substances are essential for proper structure and functions, and as nutrients or sources of energy they can hardly be categorized as intrinsically hazardous in any classification scheme for environmental chemicals.

Nonetheless, it is well established that uncontrolled inputs of nutrients and biomaterial - alone or in combination - may increase the pollution load and impose specific hazards to marine life. This is basically due to excessive primary eutrophication in the surface waters, which creates sedimentation of organic material and an increased oxygen demand - often developing into oxygen depletion - in the bottom zones as the result of the accelerated biological degradation processes.

This is a complex chain of events which has been dramatically demonstrated in recent years in an increasing number of observations of fish kills and avoidance reactions (disappearance) of fish and other species in European coastal waters and enclosed marine areas due to oxygen depletion and bottom inversions. The correlation of this development to the increased rate of nutrient inflow to the sea from landbased activities (especially agriculture) is reasonably well established, and the interest in realistic risk assessment procedures is correspondingly high.

As shown in Figure 4, marine eutrophication is characterized by a number of possible effects which vary from the increase in primary productivity of (planktonic) biomass, affecting *e.g.* the competition for specific food resources, to reductions in certain habitats (*cf.* reduction in coastal vegetation) and following changes in fish populations, as well as disruption of the biological and physical-chemical conditions at the bottom due to sedimentation and increased oxygen demand.

Among the significant effects of eutrophication is also the impact on commercial fisheries. It involves on one side increased fishing yields of certain species which may temporarily thrive from the rich nutrient supply and the productivity in pelagic zones of the open sea, *e.g.* cod and salmon. This is contrasted, however, with an increased rate of disappearance of several coastal and benthic species which are vulnerable to the fluctuating oxygen supply or depend on the threatened coastal vegetations in their reproduction and growth patterns (*cf.* eel, flounder, and lobster).

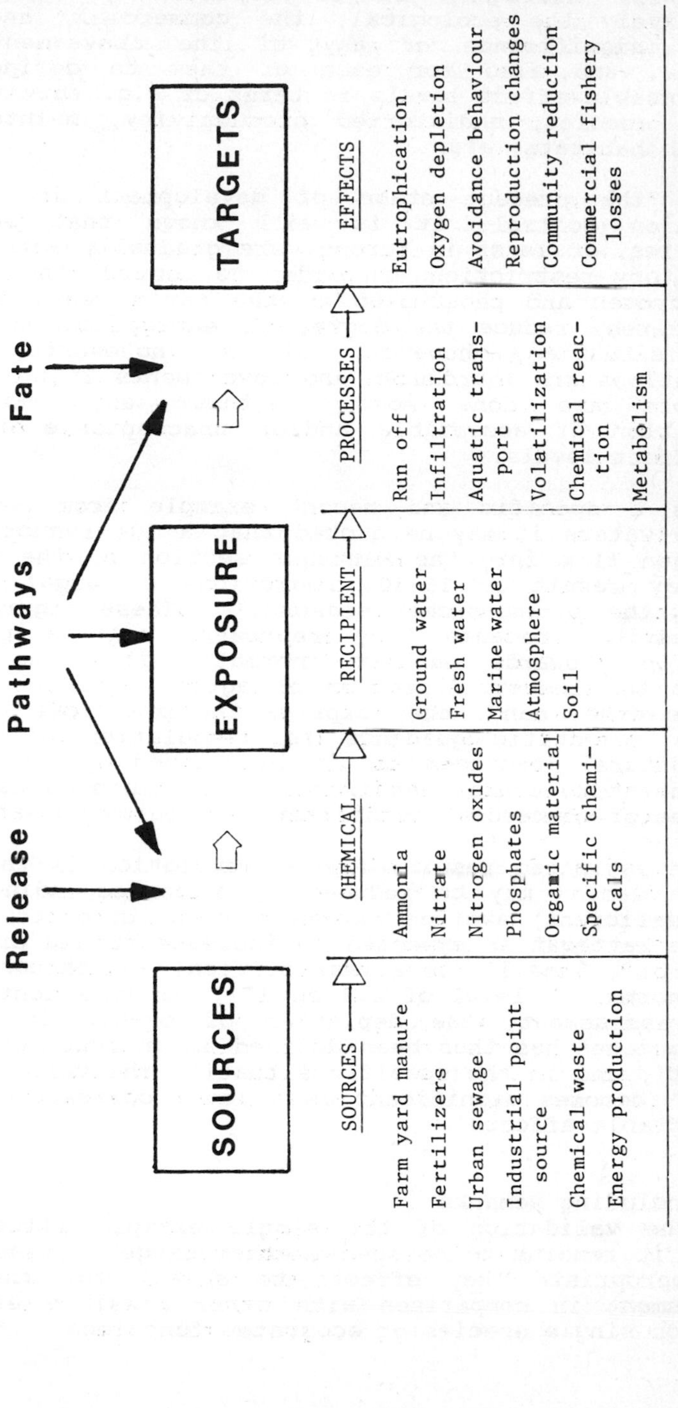

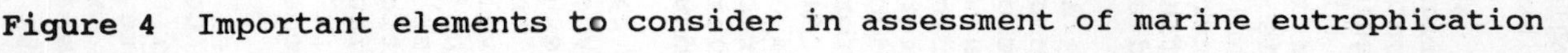

Figure 4 Important elements to consider in assessment of marine eutrophication

For a risk assessment of the marine eutrophication caused by the excessive nutrients' flow to the sea, especially nitrogen, it is important to consider selectively the ecological, the commercial, and the public significance of any of the above-mentioned effects, and also for each of them to define the unacceptable effect levels in terms of *e.g.* survival of single species, undistorted productivity, maintenance of fish habitats, *etc.*

At the present stage of development in marine pollution control, it is well known that several countries, at least in Europe, are gradually initiating regulatory restrictions in order to control the losses of nitrogen and phosphorus to the marine environment, and thereby reduce the degree of eutrophication. In this situation, however, it is noteworthy that regulations are introduced and investments in pollution controls are done mostly without any clarified definition of acceptable and/or unacceptable effects and effect levels.

As a specific and recent example from Swedish-Danish waters it may be quoted that a 50% reduction in nitrogen flow into the Kattegat section of the Baltic Sea may result in 15-30% reductions of algal blooms during the spring-summer seasons.[12] These figures are interesting, because they represent only a partial reduction towards earlier 'normal' situations, *i.e.* before the industrialization of agricultural practice. On the other hand, they express the practical targets and a scientific guidance for regulation of Danish agricultural practices, as it is planned by the Danish government to achieve environmentally improved handling and use of commercial fertilizers and farmyard manure.

An estimated result of such regulation is that the oxygen demand may be reduced by a corresponding 30%. The (deficient) average oxygen content in bottom zones of the Kattegat is expected to increase from a critical level of 2.5 mg l^{-1} to a less critical - although still not 'normal' - level of 2.8 mg l^{-1}. In the context of risk assessment, the depletion of oxygen in marine bottom zones has thus been defined as an important 'key effect', and in the specific situation the value of 2.8 mg l^{-1} becomes significant as a level corresponding to 'acceptable effects'.

4 Concluding Remarks

For the validation of the single example illustrated here, it remains to be seen whether oxygen depletion is an appropriate 'key effect' to select for the risk assessment in comparison with other possible effects, *e.g.* on single species or ecosystem functions.

Equally important, of course, is the question whether the accepted level will meet the expectations (or requirements) of not only commercial fisheries' interest, but also serve the purpose of protection of the overall quality of the marine environment.

Obviously, such questions are raised as universally important in any risk assessment. However, in the specific case the questions also express an educated doubt that the 50% reduction of nitrogen outflow into the sea may not be sufficient to ensure a restoration of the marine ecological quality. It may merely reduce the rate of a continued man-made eutrophication process.

If this judgement proves correct, we may in fact not have eliminated the risks of adverse effects caused by eutrophication. We have only restricted a process by accepting effects and effects' levels which should have been deemed unacceptable. Bearing in mind the fate of Queen Cleopatra, we have possibly changed the type and extent of potential hazard and its temporal development, but without reducing statistically the (1:1 or 100%) risk of damage to the marine ecosystem.

5 References

1. Report of the OECD Workshop on Practical Approaches to the Assessment of Environmental Exposure, Vienna, 14-18th April 1986.

2. US Environmental Protection Agency (EPA), Research Plan for Ecological Risk Assessment, Report No. CERI-86-21a, June 27th 1986.

3. L. Landner, et al., 'Systems for Testing and Hazard Evaluation of Chemicals in the Aquatic Environment', ESTHER-report No. snv pm 1631, National Swedish Environment Protection Board, 1st December 1982.

4. F. Bro-Rasmussen, 'Application of Protocols for Hazard Assessments', in 'Attitudes to Toxicology in the European Community', ed. P.L. Chambers, John Wiley & Sons, 1987.

5. 'Chemicals in the Environment', ed. K. Christiansen, B. Koch, and F. Bro-Rasmussen, Proceedings of a Symposium, 18-20th October 1982, Copenhagen.

6. 'Toxic Hazard Assessment of Chemicals', ed. M.L. Richardson, Royal Society of Chemistry, London 1986.

7. C.E. Stephan, 'Proposed Goal of Applied Aquatic Toxicology, Aquatic Toxicology and Environmental Fate: Ninth Volume', ASTM STP 921, ed. T.M. Poston and R. Purdy, American Society for Testing and Materials, Philadelphia, 1986, pp.3-10.

8. 'Ecoaccidents', ed. J. Cairns, Jr., Nato Conference Series, Series 1: Ecology, Volume 11, Plenum Press, New York, 1983.

9. NORDFORSK, 'Ecotoxicological Methods for the Aquatic Environment', Subproject No. 4, 'Characterization of Industrial Waste Water', Nordforsk, Helsinki, 1985.

10. Miljøstyrelsen, 'Guidelines for Recipient Quality Planning' (in Danish), Part 2, 'Coastal Waters', National Danish Environment Protection Agency, DK-1402, Copenhagen, Denmark.

11. EEC Directive No. 76/464, 4th May 1986 on 'Dangerous Substances for the Aquatic Environment referring to Environmental Protection Terminology', in *Official Journal of the European Communities*, 20th December 1973, No. C 112/49.

12. 'The Situation of Eutrofication in Kattegat' (in Swedish), ed. R. Rosenberg, Report No. 3272, National Swedish Environment Protection Board, November 1986.

25

Studies on the Fate of Chemicals in the Environment with Particular Reference to Pesticides

L. Somerville

SCHERING AGROCHEMICALS LIMITED, CHESTERFORD PARK RESEARCH STATION, SAFFRON WALDEN, ESSEX CB10 1XL, UK

1 Introduction

Earth has been populated by man for hundreds of thousands of years but it is only within the past ten thousand years and principally the past thousand that he has had any ecological effect upon it. Until about 8000 BC man was a hunter and a gatherer of wild plants with no domestic animals and no crops. He ate fruit and berries and helped to disperse the seeds as do birds and animals today. His waste products were probably dispersed over the land and helped to fertilize it. When he died his remains decayed and the nutrients were returned to the system. In short, man played an insignificant role within the ecosystem.

It was not until man became an agriculturalist that his effect on the environment became more significant. Settled agriculture, with the growing of some crops, has gone on in parts of the world for perhaps ten thousand years, but extensive arable farming has only existed for the past five thousand years. When wheat and other corn crops were sown broadcast, hand weeding was the only practicable method. When drilling in rows was introduced in the eighteenth century, it became much easier to keep crops clean. At the same time improved methods of separating crop from weed seeds were devised, so that sowing did not greatly contaminate the ground. Thus with clean seed and properly planted crops, cheap labour and comparatively simple horse-drawn machines kept the fields clean. At the end of the nineteenth century there was no serious weed problem for the good farmer in most parts of Britain and environmental damage resulting from agriculture was minimal.

2 Use of Pesticides

As agricultural wages rose, mechanization was introduced and some farming processes were improved, but many crops became weedier and weedier, so that different rotations had to be developed, not always with success. In recent years the advent of selective herbicides has saved the situation for the farmer. These chemicals together with insecticides, fungicides, plant growth regulators and animal health products have revolutionized agriculture but, nevertheless, problems of crop losses remain.

Despite the extensive use of pesticides, Pimentel[1] estimated that pests worldwide are destroying about 35% of all potential food crops before harvest. These losses are primarily due to insects, plant pathogens, and weeds. After the crops are harvested an additional 10-20% are destroyed by insects, micro-organisms, rodents, and birds. Thus nearly half of the potential world food material is being destroyed annually despite the extensive use of pesticides. The argument in support of the continued use and proved biocidal activity of pesticides is, therefore, strong but some constraints may well be required.

While it is difficult to obtain accurate figures for current pesticides usage, total world-wide sales of pesticides in 1986 were estimated in terms of end-user value at $17.4 billion, or in tonnage terms some 3 million tonnes.[2] Greatest usage was in North America but Western Europe was only slightly behind, mainly as a result of the increased use of fungicides in cereal crops (Table 1).

Table 1 *Pesticide sales 1986 ($ million)*

Area	*Herbicides*	*Insecticides*	*Fungicides*	*Others*	*Total*
USA	2870	1040	350	340	4600
W.Europe	1706	925	1300	420	4405
Far East	1070	1630	1010	110	3820
Latin America	530	660	230	50	1470
E.Europe	755	545	230	110	1641
Rest	615	650	130	70	1468
World total	7600	5450	3250	1100	17400

Pesticides may be expensive but they are cost effective and Pimentel et al.[3] have estimated that in the USA the average pesticide used on crops returned $4 for every $1 invested.

It can be seen, therefore, that the control of weeds, insects, and plant diseases by chemical means remains the principal option for the foreseeable future.

Paralleling the use of pesticides are, however, other potential environmental pollutants such as fertilizers, antifouling agents, wood preservatives, and industrial chemicals. The question this chapter attempts to address is whether chemicals and, in particular, pesticides, are being adequately evaluated in terms of their potential for environmental contamination prior to their introduction to the market place. Although pesticides will act as the focal point in view of their widespread use, a comparison with industrial chemicals will also be made.

3 Pesticide Research and Evaluation

Undoubtedly large sectors of the population are totally unaware of the substantial research programmes which are behind each new pesticide introduced but they may be concerned by their potential to have some effect on the environment. The reality is that pesticide research as conducted by the major multinational companies is a well-organized and intense but highly competitive industry. Each new product takes some 5-7 years to develop in programmes costing at least $15-20 million. This excludes the cost of researching products which fail to make the grade. The latter comment is particularly important since at the present time companies expect only one compound in twenty thousand to be active. In addition to the development cost, up to $10-20 million can be required to construct a manufacturing plant although occasionally an existing multi-purpose plant can be utilized. A substantial fraction of this cost is associated with environmental measures relating to the handling and disposal of wastes from the production process.

The agrochemical industry is a high-risk business, requiring substantial investment without any guarantee of an adequate return. At any time resistance to the new insecticide or fungicide may occur or a rival manufacturer may introduce a competitive product. Both situations could seriously impair the profitability of the project.

The manufacturer must take all precautions to reduce the risk associated with the investment. At all stages in its development the new pesticide is examined critically. If there are any indications of adverse effects, *e.g.* in its toxicological profile, or of long-term persistence in the environment, then termination of the project at an early stage is opted for. At all stages in its development, from its initial conception through to its ultimate manufacture, the compound continues to be critically reviewed. Even after its introduction into the market place the product is carefully monitored to ensure that its continued use does not lead to any adverse effects, to consumers or to the environment, which may necessitate its withdrawal.

Manufacturers are very demanding self-regulators of new products, particularly with regard to safety. The government regulatory bodies see only the small fraction of products which have gained the manufacturer's confidence. Equally importantly, the environment is only exposed to those products which satisfy the high standards set by industry and government regulatory agencies.

4 Environmental Assessment

As has been stated, large quantities of pesticides are used each year. For good biological efficacy, adequate coverage of the crop is essential. The latter normally requires some form of spray application either by tractor, helicopter, or fixed-wing aircraft although some chemicals are introduced directly into the soil in the form of granules or seed-dressing. Despite extensive efforts on the part of the industry to develop precision methods of application including controlled droplet spraying and electrostatic techniques, only part of the pesticide application reaches the target pest, the remainder being distributed in the non-target sectors of the agricultural ecosystem, *i.e.* the foliage, the soil, adjacent streams and hedgerows, *etc.*

It would appear therefore that numerous biological systems are at risk from pesticide use. While the soil may appear to be the ultimate site for the pesticide, consideration must also be given to the aquatic system, and to terrestrial wildlife. Consequently, numerous requirements have been proposed by regulatory authorities to evaluate not only the environmental fate of a pesticide but also its potential hazard. Current requirements for the USA and Western Europe are summarized in Table 2.

4.1. Pesticide in Soils. As can be seen in Table 2, all territories require soil degradation studies which routinely take the form of laboratory experiments. For convenience the pesticide is normally in radiolabelled form and the rate of breakdown is monitored in several different soil types under aerobic, anaerobic, and possibly sterile conditions. Information obtained would relate to the degree of mineralization of the pesticide, formation of irreversibly 'bound' residues, and identification of extractable degradation products. No breakdown under aerobic, non-sterile conditions would suggest that either the chemical was a very persistent molecule or it was toxic to the microbial population. More frequently, fairly rapid mineralization is observed and by comparison with data

Table 2 *Pesticide regulatory requirements in Western Europe and USA*

	UK	*France*	*West Germany*	*Holland*	*Belgium*	*Denmark*	*USA*
Soil degradation	Yes	Yes	Yes	Yes	Yes	Yes	Yes
Soil leaching	Yes	Yes	Yes	Yes	Yes	Yes	Yes
Soil adsorption	Yes	Yes	Yes	Yes	Yes	Yes	Yes
Soil respiration	No	No	Yes	Yes	No	Yes	No
Nitrogen transformation	No	No	No	Yes	No	Yes	No
Surface water degradation	No	No	No	Yes	No	Yes	No
Aquatic toxicology	Yes	Yes	Yes	Yes	Yes	Yes	Yes

from sterilized soil the involvement of microbial activity can be confirmed.

A typical example of soil degradation study is described by Leake et al.[4] Benazolin-ethyl (ethyl 4-chloro-2-oxobenzothiazolin-3-yl acetate) (1) is a readily translocated broad-spectrum herbicide with auxin type plant growth regulator properties, used mainly for weed control in cereals and oil seed rape. The initial laboratory study involved a formulation of benazolin-ethyl in which the active ingredient was labelled with [^{14}C] in the phenyl ring. The formulation was applied to two soils, a silt loam (Shippea Hill, pH 7.5, organic carbon 3.8%) and a sandy loam (Shelford, pH 5.8, organic carbon 3.4%), at a rate equivalent to the normal field application rate (0.6 kg a.i. ha^{-1}). The soils were incubated at 25°C and 50% moisture holding capacity (MHC) in a continuous flow-through system (Figure 1). Volatile products were trapped in a series of trapping solutions (trap 1, 0.1 M sulphuric acid, trap 2, ethane-1,2-diol, trap 3, ethanolamine).

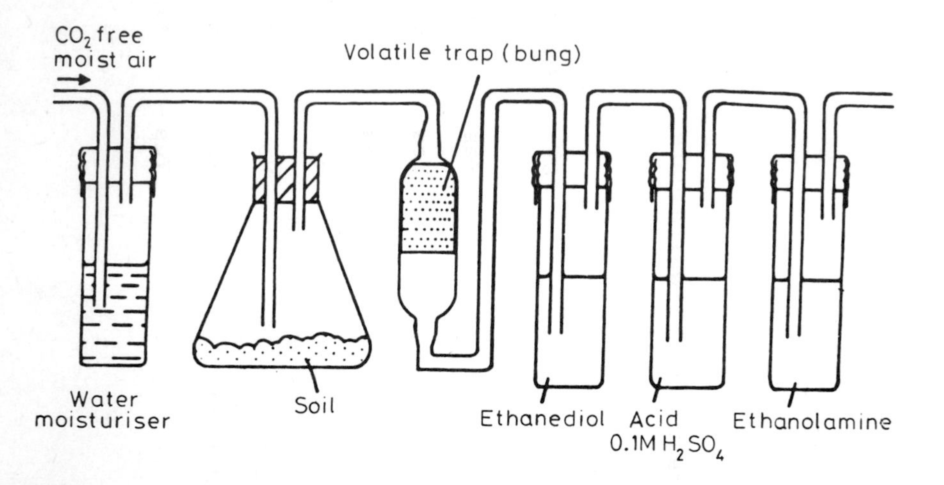

Figure 1 Flow-through system for soil degradation studies

The treated soils were analysed immediately after treatment, after 2 hours and at intervals of 1, 7, 14, 28, 42, 63, 126, 189, 252, and 365 days. At each time point, duplicate soil samples were soxhlet extracted with dichloromethane and methanol/water (9:1) or acetonitrile/water (4:1). Radioactivity in the volatile trapping solutions and solvent extracts was quantified by liquid scintillation counting (LSC). Soil extracts were concentrated by rotary evaporation under reduced pressure and aliquots analysed by thin layer chromatography. Non-extractable radioactivity (*i.e.* the 'bound' residue fraction) was quantified by combustion followed by LSC.

As can be seen in Figure 2, the primary degradation route of benazolin-ethyl under aerobic conditions was by hydrolysis of the ethyl ester to form benazolin (4-chloro-2-oxobenzothiazolin-3-ylacetic acid) (2). This initial reaction was extremely rapid with a half-life of less than one day. Degradation of benazolin then occurred, although much more slowly (half-life) 2-4 weeks), with cleavage of the acetic acid moiety to form BTS 18753 (4-chlorobenzothiazolin-2-one) (3). Further degradation via ring-opening led to mineralization of the radiolabelled carbons with up to 35% evolved as $^{14}CO_2$ in one year. Up to 66% of the applied radioactivity remained unextracted from the soil and hence was designated a 'bound' residue although there was some evidence of further mineralization of the residue to $^{14}CO_2$.

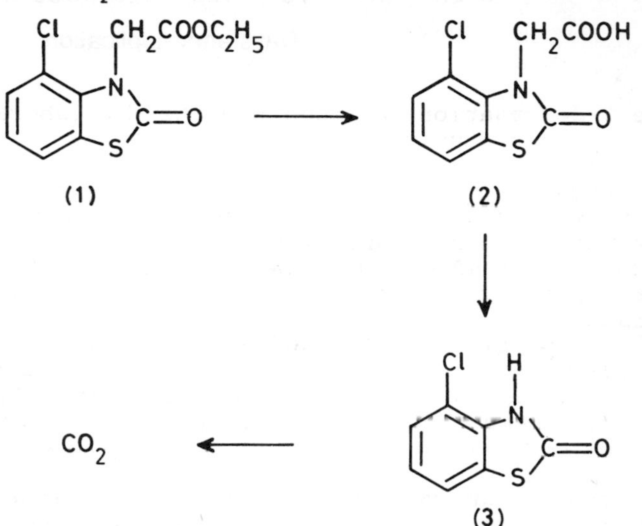

Figure 2 Degradation of benazolin-ethyl in soil

The question that must of course be answered is whether or not these laboratory data bear any relation to what would occur in the field. For benazolin-ethyl the answer from field trial data would appear to be affirmative. A second example is provided by the herbicide ethofumesate (2-ethoxy-2,3-dihydro-3,3-dimethyl-5-benzofuranyl methanesulphonate) where a good correlation between the laboratory degradation rate and field data was obtained (Figure 3). Although these two examples are obviously insufficient to prove the hypothesis, nevertheless the consensus of opinion among industrial and regulatory scientists is that laboratory degradation studies can and do provide insight into the relative persistence of a novel pesticide.

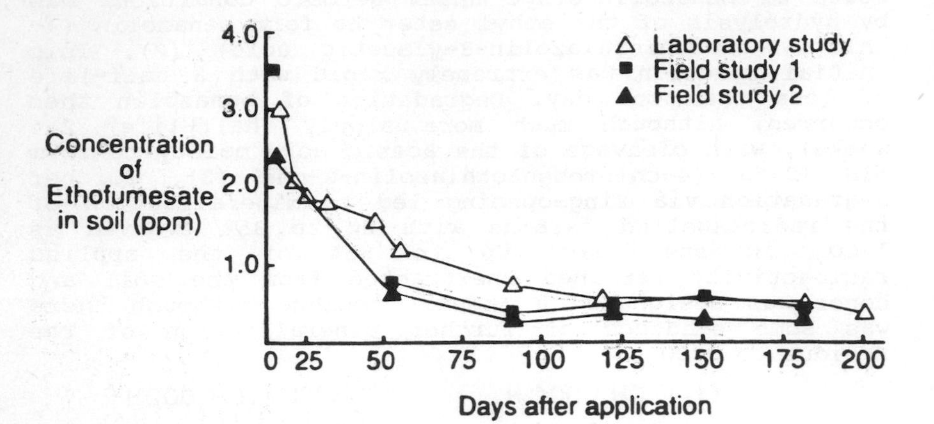

Figure 3 Degradation of ethofumesate in laboratory and field studies

Interestingly, in the past few years several regulatory authorities, and notably those of the Netherlands, Canada, and Denmark, have asked for soil degradation studies to be carried out at two temperatures approximately 10°C apart. Since the temperature for the principal study is frequently 20-22°C to comply with USA and West Germany guidelines, it follows that the only option for the second temperature is to go lower. Not surprisingly studies within the temperature range 10-15°C invariably result in longer chemical half-lives.

Of course the relative persistence in itself is not necessarily the key factor as far as environmental hazard assessment is concerned. The majority of modern pesticides are now much less persistent than their predecessors. Unfortunately a short half-life inevitably implies that the pesticide is degraded to other compounds: could any of these compounds be persistent? Modern studies not only monitor the degradation of the 'parent compound' but also the rise and subsequent decline of the breakdown products. Figure 4 shows the rate of decay of the fungicide prochloraz (*N*-propyl-*N*-[2-(2,4,6-trichlorophenoxy)-ethyl]-1*H*-imidazole-1-carboximide) in a silty clay loam.

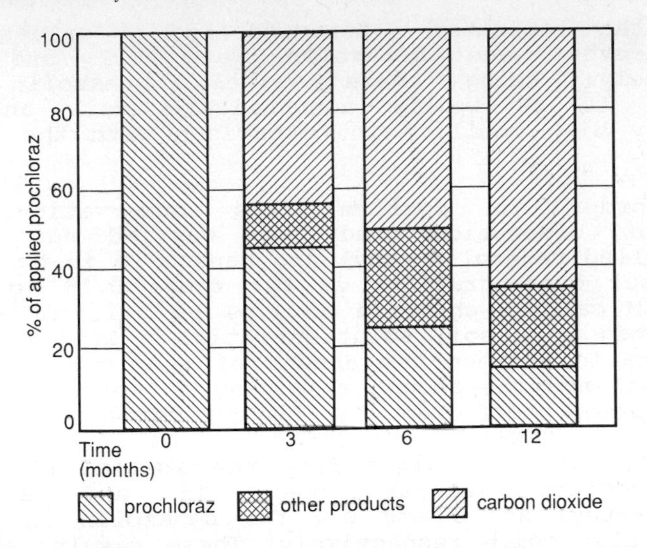

Figure 4 Degradation of prochloraz in silty clay loam soil

As can be seen, other products are formed with time but they too are steadily degraded by the action of micro-organisms, ultimately to carbon dioxide. Equally importantly, while much thought is given to methods of determining the relative mobility of the 'parent compound' in soil, an equal effort must be applied to determining the relative mobility of the degradation products.

4.2 Mobility in Soil. Mobility is an important environmental parameter since it determines whether or not a pesticide will remain on the surface of soil following application or whether it will be carried downwards by rainwater. If mobile, a compound can eventually be carried to underground sources of water (groundwater) and ultimately to lakes, ponds, rivers, *etc*. The latter of course may be teeming with aquatic life which would be immediately at risk if the pesticide or its degradation products were of sufficient toxicity.

How is soil mobility evaluated? An initial clue may be obtained by consideration of a chemical's n-octanol/water partition coefficient, since Briggs[5] suggested that the mobility of a pesticide in soil can be related to this coefficient (log P) and the soil organic content. Expressed as the degree of adsorption to soil, high log P values would suggest strong adsorption and hence low mobility, whereas low log P values would suggest the opposite. Measured log P values for benazolin-ethyl and benazolin were 2.7 and 0.8 respectively. Using these values, benazolin was predicted to be weakly adsorbed to soil and of relatively high mobility in comparison with the ethyl ester.

In order to confirm this observation, an adsorption/ desorption study was carried out using radiolabelled benazolin-ethyl and benazolin in two soil types. Four concentrations in the range 0.15 μg a.i. ml^{-1} 0.01M calcium chloride to 1.53 μg a.i. ml^{-1} were equilibrated in a soil solution ratio of 1:5. In this way the ratio between the amount of chemical adsorbed in the soil and the amount remaining in solution, *i.e.* the adsorption coefficient K_d, was determined.

The K_d values obtained from the initial slope of the Freundlich isotherm were 15 and 8 for benazolin-ethyl and 1 and 0.4 for benazolin in sandy loam and clay soils respectively. These results showed that benazolin was less readily adsorbed than benazolin-ethyl and was likely to be of relatively high mobility. The comparison of predicted (calculated) K_d values with measured values is shown in Table 3. Measured values in fact showed higher K_d values and therefore lower mobility. For most registration submissions it is also necessary to measure mobility directly by means of leaching studies. Both column leaching and thick layer chromatography studies were carried out for benazolin-ethyl.

Table 3 *Comparison of calculated K_d values with measured values for benazolin-ethyl and benazolin*

Soil type (organic matter content)		*Benazolin-ethyl*	*Benazolin*
Sandy loam (5.9%)	Calculated	7.0	0.64
	Measured	15.0	1.0
Clay loam (1.9%)	Calculated	2.3	0.21
	Measured	8.0	0.4

In the column leaching study, glass columns (internal diameter 5 cm) with the lower end drawn out to a point were filled to a height of 30 cm with 1 mm sieved air-dried soil. Columns were set up using two contrasting West German (Speyer) sandy soils (2.1 and 2.2) containing 1.8 and 4.5% organic matter. The outlets of the columns were first plugged with glass wool and the conical portions of the columns were filled with acid-washed sand. Fully deionized water was slowly applied to the tops of each column until they were saturated. Excess water was allowed to drain before radiolabelled benazolin-ethyl and benazolin were applied separately to the tops of duplicate columns. The soils were leached with 200 mm of 'rain' over a 48 h period with 24% and 4% of the applied radioactivity being recovered in the leachate from the benazolin-ethyl treated 2.1 and 2.2 soil columns respectively. Corresponding figures for the benazolin treated columns were 74% and 41%, indicating that benazolin was considerably more mobile. In the case of the benazolin-ethyl columns rapid hydrolysis to benazolin with subsequent movement of the acid occurred.

To confirm these findings the alternative technique of soil thick layer chromatography was also utilized. A schematic of the experimental system is given in Figure 5.

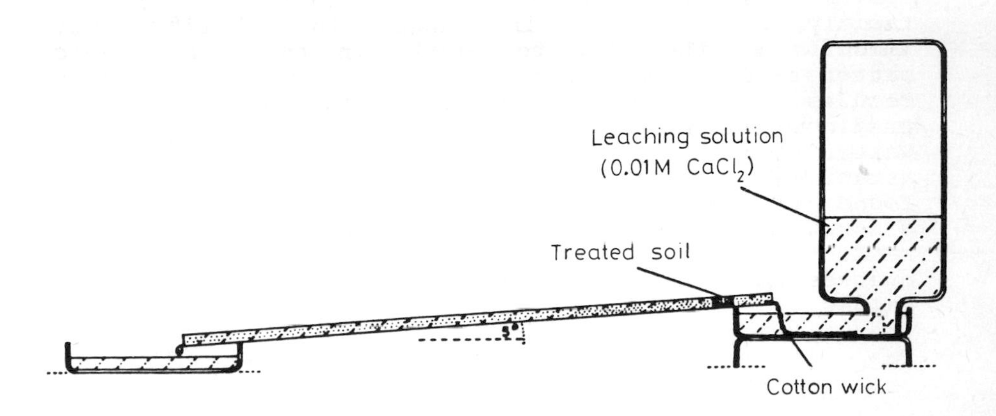

Figure 5 Apparatus for soil thick layer chromatography

The soil chromatograms were prepared on 5 cm wide, 0.5 cm thick, and 30 cm long aluminium plates. A wick made of 'J cloth' measuring 5 x 5 cm was glued to one end of the plate. The sieved soil was applied to the aluminium plates, four per soil type, by weighing each plate before and after application to ensure that approximately the same quantity of soil was added to each plate. The weight of soil is of course related to the soil type.

A 1 cm wide band was removed at 2.5 cm from the wick end of each plate by suction into a sintered glass tube and transferred to 5 ml breakers. Three portions of each soil type were then treated with acetone solutions containing approximately twice the normal field rate of benazolin-ethyl. The fourth portion of soil was treated with atrazine to act as the reference standard. The solvents were allowed to evaporate and the treated soil was replaced on the plates, ensuring good contact with the remaining soil. The plates were placed at a 5° angle to the horizontal with the wick uppermost, supported by glass petri dishes at each end in an enclosed water-saturated atmosphere. The soil chromatograms were developed using 80 ml of 0.01 M calcium chloride until the reservoir was empty. The leachate was collected for analysis while the plates were weighed and frozen. Appropriate techniques were used to analyse both the leachate and the soil plates for distribution of radioactivity and degradation products.

Helling[6] developed a system for classifying the mobility of a pesticide according to its movement on a chromatographic plate. According to this index, benazolin-ethyl was classified as immobile in sandy loam, clay, and silty loam soils and of intermediate mobility in a sand with low organic matter (0.6%). The rapidly formed benazolin ranged in mobility from immobile in the clay to mobile in the low organic matter sand. Virtually no BTS 18753 was found. These results are shown in Table 4 together with the predicted mobility index calculated from the n-octanol/ water (log *P*) partition coefficient. Radiolabelled atrazine was used as a comparative standard and was found to have the expected intermediate mobility classification (class 3) in the low organic matter sand. In general, the calculated results tended to over-estimate the mobility compared with the measured values.

Table 4 *Comparison of calculated with measured values using Helling index**

Soil type	*% Organic*	*Benazolin-ethyl calculated*	*Benazolin-ethyl measured*	*Benazolin calculated*	*Benazolin measured*
Silty clay loam	5.2	2	1	3	2
Clay	5.9	2	1	3	1
Sand	0.6	3	3	5	4
Sandy loam	2.9	2	1	4	2

*Helling index range:1 (immobile) to 5 (highly mobile).

Although mobility of benazolin in sandy soils had been demonstrated in the aforementioned laboratory studies, uncertainty remained as to whether this problem would become manifest in the field situation.

Consequently further investigations were carried out using soil lysimeters maintained outdoors.

Intact soil cores (10.5 cm i.d. x 1 m) were removed by means of heavy drilling equipment from two field sites consisting of a sandy loam soil containing 3% organic matter (o.m.) and a sand (1.7% o.m. at surface and 0.6% at 70 cm). The base of each core was fitted with a watertight buchner funnel and suspended in a large tank containing drain holes. The cores were supported in the tank by sand. Each 'lysimeter' was treated with [^{14}C]benazolin-ethyl at approximately field rate in December 1985 and leachate was collected from natural rainfall (200 mm) until May 1986.

No radioactivity was detected in the leachates over this time course. Subsequent analysis of 10 cm soil segments from each lysimeter showed that the majority of the applied radiochemical was present in the top 20 cm with no detectable penetration beyond 30 cm in the sandy loam and 60 cm in the sand respectively.

Extraction and chromatographic analysis indicated that following the treatment of the soil with benazolin-ethyl, hydrolysis to benazolin had occurred. Although mobile in soil, benazolin was not present in significant amounts below 20 cm having been further degraded to the relatively immobile product BTS 18753.

The overall conclusion from this group of studies was that although laboratory studies are extremely useful in highlighting potential problems of groundwater contamination, they should not be used in isolation from field studies where other mechanisms also affect the fate of the chemical in the soil. Recently a large-scale field study undertaken by Scheuermann et al.[7] in West Germany had confirmed this observation. Benazolin-ethyl was applied as a commercial herbicide application to a wheat field in Oldenberg province. The soil type was sand (3.3% o.m.) and the field had been equipped in the previous year with a series of 22 wells (10 cm i.d. x *ca.* 2.5 m) which had been strategically placed to reflect the groundwater flow. Throughout a nine month period, water samples were removed at two week intervals from each of the wells and analysed for residues of benazolin-ethyl, benazolin, and BTS 18753. Despite the fact that the wells were directly tapping the groundwater, no residues of any of the chemicals were detected. Equally important, regular sampling of the soil by means of cores taken to a total depth of 75 cm indicated that following the initial rapid hydrolysis of the ester to benazolin, further degradation took place. No chemicals could be detected below the top 25 cm.

4.3. Persistence in Water. While soil mobility studies are common requirements worldwide, some countries have requirements which reflect particular environmental considerations within their territory. Thus the Netherlands, since it is criss-crossed with rivers, canals, and drainage ditches, specifically requests information on the degradation of pesticides in surface water, as a potential exists for risk from spray drift. For example, the acaricide clofentezine (3,6-bis-[2-chlorophenyl]-1,2,4,5-tetrazine) had been subjected to the normal range of soil degradation and mobility studies to support registrations throughout Western Europe. To gain registration in the Netherlands, it was necessary to produce an additional evaluation of the compound's breakdown in river water with associated sediments.

Microcosms were prepared using two river sediments and their corresponding waters (Figure 6). [^{14}C]-Clofentezine was applied to the surface water at a rate equivalent to normal field rate and the microcosms were incubated at 20°C in the dark with aeration using carbon dioxide free air. Volatile products evolved from the microcosms were 'trapped' in a series of solvents as previously described.

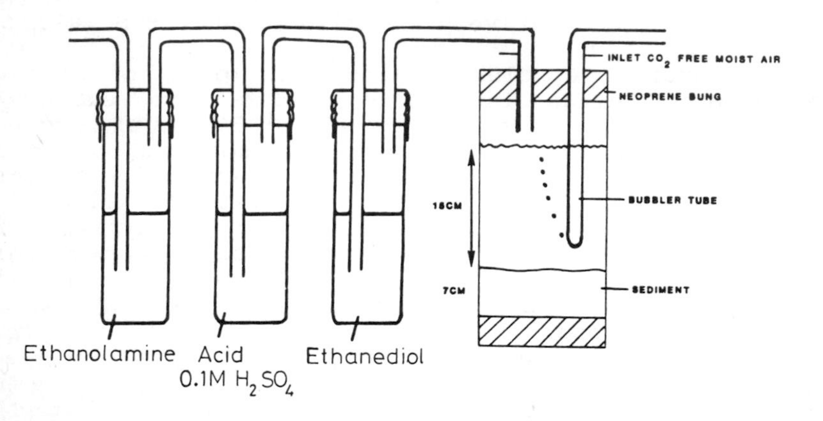

Figure 6 Flow-through apparatus for soil/sediment studies

Degradation of the acaricide in the microcosms was extremely rapid with half-lives of less than one week. In a series of related 'soil degradation' studies, longer half-lives of between four and eight weeks had been recorded according to soil type, suggesting that more effective microbial systems were present within the aquatic system. Degradation in both systems proceeded via hydrolytic cleavage of the tetrazine ring leading to the formation of a major metabolite,

2-chlorobenzoic (2-chlorobenzylidene)hydrazide. Further breakdown led to the formation of several other minor metabolites with ultimate mineralization to radiolabelled carbon dioxide.[8]

5 Ecotoxicology

It is apparent from the above studies that the microbial populations of soil and sediment systems are extremely important components of these ecosystems. Soil harbours vast numbers of bacteria, actinomycetes, fungi, and algae. Although unseen without a microscope, the micro-organisms are of prime importance not only in degrading pesticides in soil and water systems but also for breaking down plant debris into nutrients which are then available for new plant growth. Disturbance of these key biological processes may have an effect on soil fertility.

5.1 Soil Microflora. Scientists have been aware for many years of the possible adverse effects of pesticides on soil microflora. Being biologically active compounds it is inconceivable that they could control the target pest without having potential side-effects on the microflora and fauna of the soil. Indeed in the early 1970s there were indications that regulatory authorities in Europe were considering the introduction of a requirement for side-effects testing. Unfortunately the authorities were hampered by the lack of valid test regimes and it was not until 1980 that test protocols for measuring soil respiration and nitrogen transformation were adopted by several national regulatory authorities. A revision of these test protocols has appeared in a document entitled 'Recommended Laboratory Tests for Assessing the Side-Effects of Pesticides on Soil Microflora'.[9] This document was the culmination of a Workshop held in Cambridge in September 1985 which was attended by interested scientists from Universities, Government Institutions, Regulatory Authorities, and Industry.

In view of the upsurge in interest in this area, manufacturers regularly examine their products for microbial side-effects. During the early stages of development the fungicide prochloraz was evaluated against a range of soil micro-organisms since the manufacturers anticipated widescale field usage. When applied to soil samples in the laboratory, prochloraz had no significant effect on the population of bacteria, actinomycetes, and useful fungi. Following the publication of the 1980 recommendations, the fungicide was also subjected to evaluation in terms of its effects on soil respiration and nitrogen transformations.

Respiration, or the production of carbon dioxide, is a measure of the conversion of organic materials by micro-organisms and is related to the biological activity of the soil. By monitoring the carbon dioxide output of soil samples treated with prochloraz, it was shown that microbial respiration and, in turn, microbial activity were largely unaffected by the fungicide.

Nitrogen transformation is essentially the combined processes of ammonification and nitrification which are responsible for converting ammonium, released by a wide variety of micro-organisms from decaying vegetation, into nitrite and then into nitrate according to the following scheme.

Ammonification Nitrification

$$\text{Organic nitrogen} \longrightarrow NH_4^+ \underset{\textit{Nitrosomonas}}{\longrightarrow} NO_2^- \underset{\textit{Nitrobacter}}{\longrightarrow} NO_3^-$$

The latter nitrification stage is carried out by a relatively specialized group of bacteria of which *Nitrosomonas* and *Nitrobacter* are the most active. The nitrate ultimately produced is more easily adsorbed into plants than the original ammonium. When tested at field rates in a laboratory study prochloraz was shown to have no effect on this vital microbial process.

It is perhaps inappropriate to discuss further the impact of pesticides on the soil microflora as it is essentially beyond the scope of this chapter. Nevertheless, the reader should be aware that pesticides are routinely evaluated for their effects on the soil microflora.[10] A long-term decrease in microbial activity could have serious implications not only for pesticide degradation in soil and aquatic systems but also for soil fertility.

5.2 Aquatic Life. As has been suggested earlier, contamination of natural waters with a pesticide may occur as a consequence of spray drift, soil erosion, or run-off into streams. It is conceivable, therefore, that some chemicals may directly affect aquatic organisms such as fish, crustaceans, or algae as a result of their inherent toxicity or alternatively accumulate in their body tissues as a result of their lipophilicity. Standardized laboratory test methods have been devised to give information on such aspects. The simplest test for toxicity is to expose the test organism to the chemical substance for a specified time. The results are expressed as an LD_{50}, LC_{50}, or EC_{50} value. The values are a measure of the extent to which a chemical affects a test organism. Equally important is the need to test for possible bioconcentration of the chemical in fish when the latter are continuously exposed to sub-lethal concentrations.[11] One could imagine a situation whereby a pesticide is accidentally spilled into a stream. For several days the aquatic ecosystem would be continuously exposed to the chemical before fresh inflowing water would dilute it to insignificant levels. In such situations it is not only important to know whether the chemical would accumulate in the edible tissues, since fish feature regularly in the human diet, but also whether the influx of fresh water would rapidly cleanse the tissues.

Prochloraz was subjected to this series of aquatic test systems. When *Daphnia magna*, a sensitive crustacean species, were exposed to a range of concentrations of prochloraz for a period of two days, the EC_{50} was 3 mg l^{-1}. Using both static and flow-through test systems over a four day period, the toxicity to rainbow trout (*Salmo gairdneri*), a cold water fish, and bluegill sunfish (*Lepomis macrochirus*), a warm water species, was assessed at approximately 2 mg l^{-1}. Although classed as moderately toxic according to these tests, the reality of the situation is that toxic levels for fish and other aquatic organisms are never reached in practice. At normal application rates for crops the concentration of prochloraz reached in water is extremely low. Even in shallow water with a depth of 10 cm the concentration after direct overspray is less than 0.005 mg prochloraz l^{-1} water.

Since acute toxicity to fish is unlikely to be a problem with prochloraz, it follows that reassurance must also be obtained with respect to the compound's ability to concentrate in the edible tissues. Consequently, both bluegill sunfish and rainbow trout were continuously exposed to a sub-lethal concentration (0.02 mg l^{-1}) of prochloraz for 28 days. Thereafter, the fish were transferred to clean water for a further 14 days in order to assess the rate of depuration from the tissues (Figure 7).

The extent to which prochloraz accumulated in the fish can be measured by comparing the maximum residue in the edible tissue at plateau with the concentration of prochloraz in the water. This ratio is referred to as the bioconcentration factor (BCF).[11] Substances may be concentrated in the organs of the fish to different extents, depending on their chemical properties. Whereas low bioconcentration factors are preferable, high bioconcentration factors of many thousands may not be harmful to the organisms. In the current example prochloraz has an overall (whole fish) BCF of 270 in the trout and 390 in the bluegill with a BCF of 100 or less in edible tissues. In contrast to more lipophilic pesticides which have BCF values in excess of 1000, these factors are moderately low and would not lead to significant accumulation in the tissues. Equally important, the residues in the tissues were rapidly eliminated when the fish were transferred to clean water.

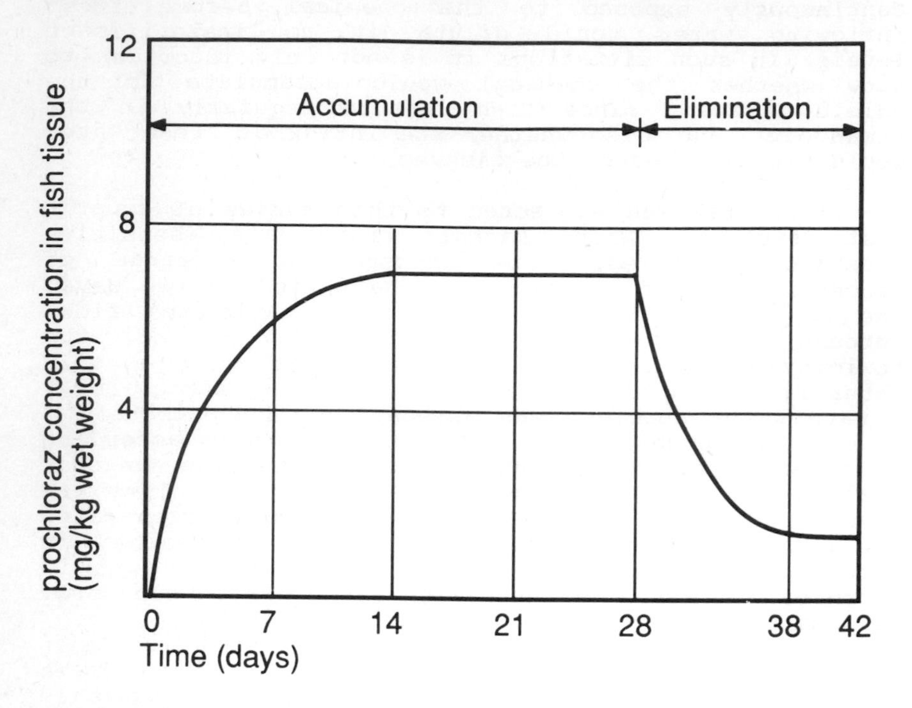

Figure 7 Effect of prochloraz on fish tissue (bluegill sunfish)

The studies mentioned above refer, of course, to the need to assess the toxicity of the new product. But what of the by-products associated with its manufacture? In the case of prochloraz (before receiving planning permission for a new production plant) the manufacturers were obliged to satisfy the Regional Water Authority with respect to the probable toxicity of effluent arising from the chemical process. The plant, which has been erected at Widnes on the River Mersey, discharges treated effluent into the river via the local sewage plant. Since parts of the River Mersey estuary are designated 'sites of special scientific interest', it was necessary to ensure that the effluent could have no direct or indirect effects on the river's ecosystem. In particular, the Water Authority wanted confirmation that the food source for migratory wading birds, in particular the bivalve *Macoma* would not be at risk.

Consequently, prior to commissioning the manufacturing plant, a substantial analytical programme was set up to monitor samples of river water and *Macoma* to ensure that neither contained any residues of prochloraz or its by-products. Equally important, after manufacturing commenced further analysis was carried out at regular intervals. The comprehensive investigation included a bioconcentration study in the bivalve using a 'model' by-product specifically radiolabelled with [^{14}C] in the phenyl ring. Since there was no evidence of bioaccumulation of prochloraz or its by-products in the *Macoma*, nor of measured residue concentrations in river water, it was possible to demonstrate that neither the bivalve nor birds using this food source were at risk from this new manufacturing plant.[12]

5.3 Avian Life. In the investigation described above, attention was focused on a somewhat specific group of birds, *i.e.* the migratory waders. While obviously no-one wishes to see these visitors placed at risk, it is equally important to ensure that the resident bird population is adequately protected. Consequently, a wide range of avian studies is routinely undertaken.[13] These include simple toxicity assessments utilizing the mallard duck (*Anas platyrhynchos*) as representative of aquatic species and the bobwhite quail (*Colinus virginianus*) as a terrestrial inhabitant. Equally relevant are studies which monitor the fertility of bird species following the ingestion of diet treated with a pesticide. Although the latter studies must of necessity be laboratory studies, they nevertheless provide valuable clues as to possible effects on wild bird populations.

When the insecticide bendiocarb (2,2-dimethylbenzo-1,3-dioxol-4-yl methylcarbamate) was being evaluated as a possible seed dressing treatment for peas and beans the fact that it was a carbamate insecticide, and therefore possibly toxic to birds, was in the forefront of investigators' minds. The compound had been used extensively in the UK as an insecticidal seed treatment on maize and sugarbeet with no significant effects on wildlife but nevertheless this new use involved a smaller seed. The risk to birds and other wildlife was therefore uncertain.

During the period March - April 1985, a field study was carried out at three locations in the East of England in which the effects of bendiocarb-treated pea seeds were monitored. The treated peas were drilled under normal agricultural conditions into bare soil on areas totalling 7.5 ha at each location. Teams of observers carried out wildlife surveillance at each site before and after drilling and at crop emergence (approximately two weeks after drilling). Surveillance consisted of dawn observations of wildlife activity followed by an extensive search of the treated areas and surrounding hedges and ditches. Wildlife casualties were collected for necropsy and residue analysis.

The period of greatest risk to wildlife, particularly birds, was immediately post-drilling when exposure to treated seeds on the soil surface was a potential hazard. Although the standard agriculture practice is to cover the seed completely at drilling, occasionally small patches of uncovered seed were observed. Only at one site however did this actually appear to have directly led to mortalities (one mallard duck and possibly two sparrows). The large variety of birds on the treated areas exhibited normal behaviour during the observation period although some evidence of seed unpalatability was seen. Even after the pea seedlings had emerged they were actively foraged by mainly pheasants, partridges, and pigeons with no adverse effects. In view of the small number of wildlife casualties which could be attributed to bendiocarb poisoning in relation to the large areas drilled with the treated peas, it was considered that the seed treatment presented a minimal risk to wildlife. The UK registration authority agreed with this view and the product was subsequently registered.[14] The product was not, however, commercialized.

5.4 Terrestrial Life. Earthworms actively participate in the turnover of nutrients in soil by removing plant debris from the soil surface and by breaking down organic matter. They also contribute to the formation of a good crumb structure in many soils and thereby play a significant role in the maintenance of soil fertility. Such valuable allies of farmers must therefore be protected.

The toxicity of prochloraz to earthworms has been evaluated in laboratory tests with the findings confirmed by means of a field study. In the laboratory, soil samples were sprayed with different concentrations of prochloraz well in excess of normal agricultural application rates. In each soil sample ten earthworms *(Lumbricus terrestris)* were incubated for a period of 14 days in darkness. After this time the effects on the earthworms were assessed. Comparisons were also made with reference standards of known toxicity.

The concentration of prochloraz which inactivated 50% of the earthworms (LC_{50} value) was 230 p.p.m. At 100 p.p.m. or below the earthworms were completely unaffected. Relating this to the field situation, the LC_{50} value was approximately 300 times the concentration which could be expected to be found in the top 10 cm of soil following a typical application.

Confirmation of this finding was obtained directly in the field situation. A number of plots 10 m^2 in area were marked out on grassland and cereals at an early stage of growth. Random plots were treated with prochloraz (as the formulated product SPORTAK) at recommended rates of application and above. Some plots remained as untreated controls. Application of the fungicide was made in June and assessments of the earthworm populations were made at six month intervals. The investigation showed that there was no difference in the numbers or distribution of earthworm species between the treated and untreated soil plots. Even application rates of prochloraz exceeding the label recommendations had no effect on earthworms.

5.5 Insect Life. Similar studies are carried out on 'beneficial' insects, since the latter are important agents in maintaining the environmental balance.[13] Many insect pests such as cereal aphids, are preyed upon as a source of food by other insects and spiders. These predators are of considerable benefit to the farmer in view of their potential to reduce crop damage. Species such as ladybirds *(Adalia bipuncta)*, houseflies *(Metasyrphus corrollae)*, staphylinid *(Tachyporus hypnorum)*, carabid beetles *(Demetrias atricapillus)*, and liniphid spiders *(Lepthyphantes tenuis)* are therefore investigated in terms of their possible toxicity to a novel pesticide.

Perhaps the most important insect from the economic point of view however is the honeybee *(Apis melliflora)*. Both the farmer and the bee-keeper gain from its efforts. The farmer values the increased pollination of his crops while the latter benefits financially from the production of honey and wax. Consequently, stringent tests are conducted to ensure that the honeybee is not at risk. Investigations start at the level of laboratory testing and progress to field studies where effects under actual use conditions are examined.

Prochloraz is an excellent example of a pesticide which might pose a risk to the honeybee as it is used extensively in cereals and oil seed rape, the latter being a major source of forage for bees in early spring.

Laboratory tests showed that the chemical was of low toxicity when applied as 1 μl droplets to the bee thorax or fed to the bees with sugar solution. Nevertheless, field evaluations were carried out to confirm these findings. In the first of two large-scale studies carried out in the UK, three large fields of flowering oil seed rape were selected. The fields were approximately 1 km apart to prevent cross-foraging by bees. One field was treated with prochloraz formulated as SPORTAK, the second received a chemical of known toxicity to bees (the toxic standard), while the third remained as the untreated control. Four beehives, fitted with dead bee traps, were placed alongside each field and one hive was also fitted with a pollen trap. The fields were treated with the agrochemicals using tractor-mounted boom sprayers during peak bee foraging activity. On the day of treatment foraging activity, numbers of dead bees, and amounts of pollen collected were measured at hourly intervals. Thereafter daily assessments were made depending upon weather conditions.

The second field study, using a different prochloraz formulation, was similar in design to the previous study except that the chemicals were applied as a low volume spray by helicopter which passed directly over the hives. In this case the bees were unable to escape the spray droplets.

The results from the studies were similar. Prochloraz had no effect on honeybees (Figure 8). Foraging activity was similar to that of the control plots resulting in normal yields of honey and wax. No effect on brood development was osbserved. Perhaps of particular importance was the fact that chemical analysis confirmed the absence of prochloraz residues in the wax.

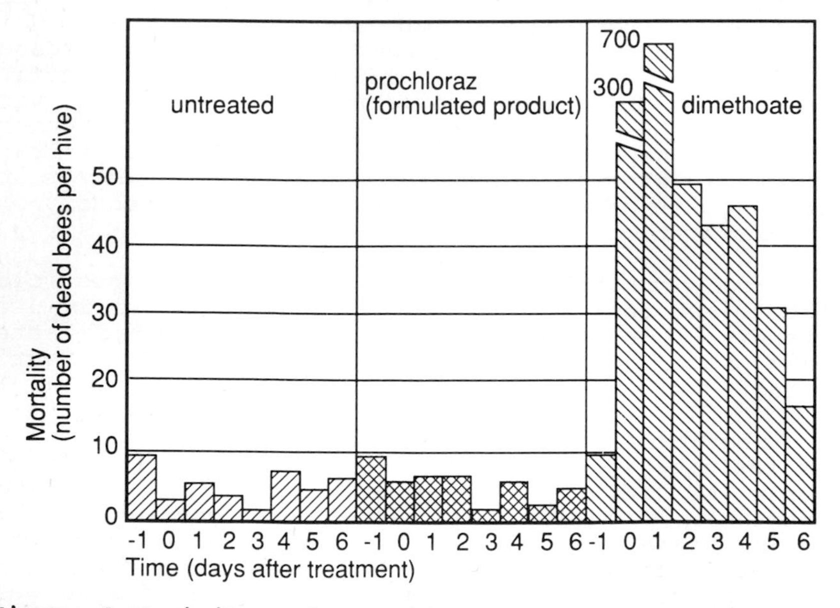

Figure 8 Toxicity of prochloraz to honeybees: field studies

It can be seen therefore that pesticides are subjected to a vigorous safety evaluation programme. All facets of an agrochemical's behaviour, and hence fate in the environment, are monitored prior to its introduction to the market place, but can the same be said for industrial chemicals?

6 Industrial Chemicals

The polychlorinated biphenyls (PCBs) are a family of stable, water-insoluble industrial chemicals that were widely used for nearly fifty years (1929-1978). By 1975 Durfee et al.[15] estimated that some half million tonnes had been produced in the USA and about 80,000 tonnes had passed into its soils, sediments, and waters. Until recently there was widespread opinion that such PCBs, particularly the more highly chlorinated congeners, were resistant to ordinary biodegradation processes and hence were extremely persistent in the environment. As late as 1978, the only known route for the environmental destruction of the more heavily chlorinated PCBs had been photolysis by solar near-ultraviolet light.[16] Modelling studies showed that solar photolysis could reduce the levels of PCBs in large lakes and oceans with half-lives of one to two years, but the major accumulations of PCBs that lie buried in aquatic sediments are clearly inaccessible to sunlight. Fortunately, dechlorination by other agents present in these environmental reservoirs removes what would otherwise be a major blockage to PCB degradation in nature.

It is worth adding that in 1984 two publications appeared[17-18] which showed that PCBs in the sediments of the upper Hudson River were undergoing a previously unreported type of decomposition. It would appear that anaerobic bacteria are responsible for this action and it may be that the full breakdown requires a two-step process consisting of dechlorination, whether by sunlight or bacteria, followed by oxidative biodegradation.[19]

Bearing in mind the carcinogenic potential of many of the PCBs, it is apparent that these chemicals came close to being environmental pollutants on a scale unprecedented this century. One accepts that at the time of their introduction the chemicals would not have been subjected to the rigorous test regimes previously described for new pesticides. If introduced today, however, would the test regimes currently applied to industrial chemicals have identified the recalcitrant nature of the PCBs?

7 Sixth Amendment Directive

In 1979 the European Communities published a Council Directive amending for the sixth time Directive 67/548/EEC relating to the classification, packaging, and labelling of dangerous substances.[20,21] This amendment, which is commonly referred to as the 'Sixth Amendment', has been incorporated into legislation by the member states and requires a manufacturer or importer of a new substance to submit 'a technical dossier supplying the information necessary for evaluating the foreseeable risks, whether immediate or delayed, which the substance may entail for man and the environment'. In addition to the dossier the manufacturer or importer must also provide

> 'A declaration concerning the unfavourable effects of the substance in terms of the various uses envisaged;
>
> the proposed classification and labelling of the substance in accordance with this Directive;
>
> Proposals for any recommended precautions relating to the safe use of the substance'.

Focusing specifically on 'foreseeable risks .... for the environment' the Directive calls upon the manufacturer or importer, *i.e.* 'the notifier', to provide information on the substance at three levels (Base set, Level 1, and Level 2) depending on the tonnage marketed. Moreover, having generated Base set data, for example, the notifier must evaluate the ecotoxicological risk to determine whether or not Level 1 studies must also be carried out. Obviously, Level 1 studies could be expanded to Level 2.

The aim of the procedure is not to prove that a substance is environmentally 'safe' but rather to indicate how a potentially hazardous chemical can be used within its proposed market with minimum risks. An outline of the data requirements for each level follows. Where the experimental procedure is similar or even identical to that described previously for the evaluation of pesticides, comment will be limited.

Base set information is required when the volume of a chemical exceeds 1 tonne year^{-1}. This relates to either a process intermediate which is being transported from one production site to another or a finished product which is being transported to the market place. When the volume reaches 10 tonnes year^{-1} data relating to Level 1 may be required by the Authorities and must be provided to them at 100 tonnes year^{-1}. At 1000 tonnes year^{-1} the threshold for Level 2 is triggered.

The Base set requirements are quite limited. The common physical-chemical properties of the substance must be supplied *e.g.* solubility, vapour pressure, n-octanol/water partition coefficient, since they provide some indication for possible distribution within the environment. In particular, the n-octanol/water partition coefficient is a useful indicator of bioaccumulation potential. As indicated previously, if the value of the coefficient is below 1000, the risk of bioconcentration in the tissues of aquatic species is low.

The main requirement in the Base set is for a measure of ready biodegradation over 28 days. Several test procedures have been proposed and they are discussed in considerable detail in ECETOC Technical Report No. 18, which attempts to introduce harmonization into the tests.[22]

The general principle of all of the ready biodegradability test procedures is the incubation of a relatively small amount of inoculum containing a variety of aerobic micro-organisms in a suitable medium (water and minerals), at neutral pH, with a concentration of dissolved oxygen above 2 mg l^{-1} and temperature between 20 and 25 °C. A limited amount of test chemical (2-100 mg l^{-1}) is added to the medium and serves as a source of carbon and energy. In the event that a chemical is shown to be readily biodegradable in the above test, then it is considered to pose little hazard to the aquatic environment. Conversely, if there is little evidence for biodegradation, then higher level requirements may be triggered.

The remaining Base set requirements relate to aquatic organisms. Acute toxicity data are required on a fish species, commonly rainbow trout (*Salmo gairdneri)* and *Daphnia magna*, to provide some measure of possible effects on food chains within the aquatic environment.

Level 1 continues to focus on the aquatic ecosystem, with the need for information on the effect of the test chemical on the growth of a unicellular algae species, *e.g. Scenedesmus pannonicus*. Once again the emphasis is on the components of the food chain. Fish and Daphnia continue to feature but now attention is placed on possible effects resulting from longer term exposure. When no plateau has been reached in the curve of concentration against time, in the LC_{50} (Base set) test on fish, a 14 day toxicity test should be carried out. For Daphnia the emphasis is now on possible effects on reproduction and consequently exposure to the test chemical is extended to 21 days. Moreover, if the n-octanol/water partition co-efficient indicated potential for bioaccumulation then it is in

Level 1 that the definitive data are required.

Level 1 acknowledges that a chemical may have the potential to reach environments other than aquatic systems. Consequently, earthworms and higher plant tests are introduced. The earthworm toxicity study is similar to that previously described whereas the plant study is somewhat novel. The Directive calls for data, expressed as an EC_{50}, on both germination and growth of a higher plant. It is not normally required for compounds which are unlikely to reach the soil in significant amounts (1 mg kg^{-1} soil) or are readily biodegradable in water. It follows, of course, that there is little need for similar requirements for a pesticide since, if it markedly affected plant germination or growth, then it would never be considered for development, except as a herbicide.

In the event that a new substance breaches the Level 2 threshold then the requirements are tailored to the chemical and biological properties of the compound and its use pattern. It is recommended therefore, that discussions should take place between the manufacturer and the competent authority to define the programme. In general, however, the following areas are of concern and must be considered.

Accumulation in fish species may have been adequately evaluated in Level 1. If, however, there was evidence of bioconcentration then a further study(s) may be required, focusing more specifically on the kinetics of uptake and depuration of the chemical. Similarly, additional studies may be required in the area of biodegradability, possibly assessing the potential of the chemical for removal under sewage treatment conditions. Fish studies may also be extended to include effects on reproduction.

New factors to be considered at this level include toxicity studies with birds and other organisms. While the latter are not specifically defined, their selection must clearly take into account the specific parts of the environment which are at risk. Thus the organisms could be anything from benthic organisms, marine organisms, soil micro-organisms, to higher plants *etc.* Finally, an assessment of the compounds mobility in soil may be required, possibly by measuring its adsorption/desorption coefficient.

It appears therefore that, since the withdrawal of the PCBs in the early 1970s, a comprehensive test regime has been introduced for the control of industrial chemicals. Potential environmental pollution is avoided by careful evaluation of a chemical prior to its introduction to the market place. This prior testing is extremely important, not exclusively for industrial chemicals, but for all possible environmental pollutants including pesticides. Once a chemical enters the environment, it may be extremely difficult to treat effectively and total elimination may be impossible.

8 DDT

DDT is an interesting example of an environmental pollutant. Following its introduction in the 1940s, the compound was used extensively and effectively worldwide. In the public health area the pesticide played an important role in the control of arthropod-borne diseases such as malaria and yellow fever. Equally significant was the compound's activity against pests of major crops including cotton, soybeans, and tobacco. Unfortunately, over a period of years a number of adverse environmental effects were also identified. These include toxicity in non-target organisms, bioaccumulation, long-term environmental persistence, and resistance in target pests. Consequently the pesticide was withdrawn in most parts of the world.

At its peak, however, huge tonnages of DDT were produced. Within the USA the chemical was first produced in 1947 at the Redstone Army Arsenal near Huntsville, Alabama. The untreated industrial waste effluent was discharged directly into a small tributary of the Tennessee River and it was not until 1965 that the manufacturer started to introduce a number of treatment schemes in an attempt to meet federal effluent requirements. Despite all efforts the treatment schemes were never completely effective and by 1970 production was terminated, with the plant demolished two years later.

Unfortunately, 14 years after the input of DDT was stopped, residue levels in the aquatic ecosystems are still extremely high. Indeed a recent investigation has shown no significant change in contamination levels over the course of a 4 year period.[23]

9 Drinking Water Directive 80/778/EEC - Parameter 55

The EEC Directive relating to the quality of water intended for human consumption has set a general limit of 0.1 $\mu g\ l^{-1}$ for any single pesticide and of 0.5 $\mu g\ l^{-1}$ for all pesticides without considering the need for the use of pesticides in modern agriculture and regardless of the toxicological properties and the assessment of risks to human health.

The pesticide industry shares the view of society that drinking water should be safe and of high quality. The industry through its association of manufacturers

(GIFAP), however, challenges the arbitrary setting of limits for a specific class of products (pesticides) independent of their toxicological properties when these have been taken into account in the evaluation of *other* product classes. The industry regards this as inconsistent and inappropriate and believes that the limits for pesticides in drinking water should be based on the toxicological profile of each individual pesticide. Use of this approach would achieve scientific credibility and would take into account the needs of agriculture whilst addressing an area of public concern. (See also Chapter 26.)

10 Conclusions

Commitment to the present legislation should prevent a recurrence of the PCB and DDT sagas but is the environment completely safe with our current test regimes?

Recently Anderson and Strand[24] reported a wildlife incident in Oklahoma, USA where 97 migratory wildfowl were found dead on an industrial site. Two separate manufacturing facilities with separate waste treatment systems were located at the site. One plant produced crystals for use in solid-state electronic devices while the other purified boron gas for a wide variety of uses. Fluoride, a by-product of the boron purification process, was discharged to waste storage lagoons for containment and treatment by evaporation. At the time of the bird-kill most available surface water in the area was frozen, which made the open waters of the waste lagoons attractive to wildlife.

Despite the use of bird-scaring devices, the waterfowl, apparently looking for open water, landed on the waste lagoon, drank the wastewater containing the fluoride, and died rather quickly from severe gastric and intestinal haemorrhaging due to acute halide poisoning. Being a responsible company, the manufacturers immediately improved the bird scaring devices and added a network of stainless steel wires to deter birds from landing in future.

This incident highlights the continued need for vigilance. Complacency must be avoided. Despite comprehensive legislation incorporating state of the art test protocols, all potential hazards cannot be foreseen. It follows , therefore, that when evaluating a new compound, albeit pesticide or industrial chemical, one must be constantly looking for that one facet of its properties which could offer a potential hazard to the environment.

11 References

1. D. Pimentel, in 'CRC Handbook of Pest Management in Agriculture', Vol. I, ed. D. Pimentel, CRC Press, Boca Raton, FL, 1981, p.3.

2. Wood MacKenzie, Agrochemical Service Edinburgh, 1987.

3. D. Pimentel, J. Krummel, D. Gallahan, J. Hough, A. Merrill, I. Schreiner, P. Wittum, F. Koziol, E. Back, D. Yen, and S. Fiance, in CRC Handbook of Pest Management in Agriculture', Vol. II, ed. D. Pimentel, CRC Press, Boca Raton, FL, 1981, p.27.

4. C.R. Leake, D.J. Arnold, S.E. Newby, and L. Somerville, Proceedings of the British Crop Protection Conference - Weeds, 1987, Vol. I, **2**, p.577.

5. G.G. Briggs, *J. Agric. Food Chem.*, 1981, **29**, 1050.

6. C.S. Helling, *Soil Sol. Am. Proc.*, 1971, **35**, 735.

7. H.-J. Scheuermann, J. Moede, and C. Wienhold, Schering AG, 1988, personal communication.

8. D.J. Arnold, C.R. Leake, and L. Somerville, Proceedings of the British Crop Protection Conference - Pests and Diseases, 1986, **2**, p.851.

9. L. Somerville, M.P. Greaves, K.H. Domsch, W. Verstraete, N.J. Poole, H. van Dijk, and J.G.E. Andersen, 'Recommended Laboratory Tests for Assessing the Side-Effects of Pesticides on Soil Microflora', 1986.

10. L. Somerville and M.P. Greaves, 'Pesticide Effects on Soil Microflora', ed. L. Somerville and M.P. Greaves, Taylor and Francis, London, 1987, pp.240.

11. J.H. Duffus 'Toxic Hazard Assessment of Chemicals', ed. M.L. Richardson, The Royal Society of Chemistry, London, 1986 p.98.

12. Proceedings of the British Crop Protection Conference Symposium 'Field Methods for the Study of Environmental Effects of Pesticides', 1988 in press.

13. D. Osborn, see ref. 11, p.247.

14. D.J. Arnold, Schering Agrochemicals Ltd., 1988, personal communication.

15. R.L. Durfee, G. Contes, F.C. Whitmore, J.D. Barden, and E.E. Hackmen, 'PCBs in the United States: Industrial Use and Environmental Distribution', PB 252012 National Technical Information Service, Springfield, VA, 1976, p.488.

16. N.J. Bunce, Y. Kumar, and B.G. Brownlee, *Chemosphere*, 1978, 7, 155.

17. R.F. Bopp, H.J. Simpson,B.I. Deck, and N. Kostryk, *Northeastern Environ. Sci.,* 1984, 3, 180.

18. J.F. Brown, R.E. Wagner, D.L. Bedard, M.J. Brennan, J.C. Carnahan, R.J. May, and T.J. Tofflenuire, *Northeastern Environ. Sci.,* 1984, 3, 167.

19. J.F. Brown, R.E. Wagner, H. Feng, D.L. Bedard, M.J. Brennan, J.C. Carnehan, and R.J. May, *Environ. Toxicol. Chem.,* 1987, 6, 579.

20. J.L. Vosser, see ref 11. pp.117.

21. H.P.A. Illing, see ref. pp.133.

22. ECETOC Technical Report No.18, Harmonization of Ready Biodegradability Tests, 1985.

23. A.R. Reich, J.L. Perkins, and G. Cutter, *Environ. Toxicol. Chem.,* 1986, 5, 725.

24. J.K. Andreasen and R.K. Stround, *Environ. Tox. Chem.*, 1987, 6, 291.

26

Pesticides in the Aquatic Environment — Data Needs for Their Control

R.J. Otter

DEPARTMENT OF THE ENVIRONMENT, WATER QUALITY DIVISION, ROMNEY HOUSE, 43 MARSHAM STREET, LONDON SW1P 3PY, UK

1 Introduction

This chapter addresses the issue of pesticides in surface and ground waters but could equally well have been concerned with other substances from other sources. The issue of pesticides has been taken up deliberately because of the increasing concern about them. The concern is increasing for three reasons.

a) Their presence is increasingly being measured and confirmed in ground and surface waters.

b) The concentrations of a range of herbicides and other biocides in source and drinking waters are giving rise to non-compliance with the European Communities Drinking Water Directive.[1] This Directive specifies that no single pesticide should be present at a concentration exceeding 100 ng l^{-1} and that their collective total should not exceed 500 ng l^{-1}. This problems is not, of course, confined to the United Kingdom but occurs throughout the European Community.

c) The selection scheme designed by the Department of the Environment (UK) and described in Chapter 23 indicates that it is pesticides which are the substances most likely to cause serious environmental problems in the aquatic environment.

All risk assessment depends ultimately on the quality of the data used to make these assessments. This chapter will demonstrate that there are many shortcomings in our databases.

2 Pesticide Usage

Because of the wide range of pesticides in use for both agricultural and non-agricultural purposes, the transient nature of many uses and the cost and difficulty of analysis, there has been little monitoring for pesticides, although the situation is now improving. What monitoring there has been has concentrated on the organochlorine insecticides, and because the use of these substances has declined the quality and quantity of monitoring is not extensive. Recent concerns about the levels of the organochlorines in birds and fish and sediments, together with the more sensitive monitoring required to demonstrate compliance with water quality standards, have revealed a number of unexpected problems with these compounds. There has been very little monitoring for the substitutes for the organochlorines and for the newer pesticides used in increasing amounts. This situation is being corrected but the increased monitoring may reveal more problems and the need for corrective action in pesticide usage (see also Chapter 25).

3 Problems to the Water Authorities and Government

The problem for water authorities and for the Government is what needs to be done? This raises the immediate problem of priorities or what needs to be done first? In an ideal world it would be possible to assess the risk of any pesticide to the organisms in the aquatic environment or to organisms such as birds or man which are exposed to the substance via the aquatic environment (or via any other route). Such assessment would enable water quality requirements to be established and priorities set for the most serious problems. They would equally well enable 'safe' chemicals to be recognized. The ideal is not achievable in practice. The environment is too complex, the number of chemicals in use is too high, and the data required for assessment often do not exist, or if they do they are not publicly available. Nevertheless governments and international bodies like the European Community, or marine pollution prevention organizations or the Paris and Oslo Commissions recognize that they have to select priorities for action so that they can use their resources most economically and provide the maximum protection for the environment they are seeking to protect. It is neither desirable nor practical to stop the use of all chemicals nor to deal with all existing chemicals simultaneously.

3.1 UK Quality Objective Approach. The United Kingdom's quality objective approach has great power to identify the more serious problems because it relates what is in the environment to what are safe or harmless concentrations of dangerous substances. As the recent European Community Directives covering discharges of organochlorine pesticides to water[2] are implemented and waters are being monitored to check compliance with the stringent quality standards set, so a number of practical and unexpected problems are being brought to light and reported by Water Authorities. Some examples can be given in a non-attributable way:

a) The importation of pesticides on fleeces and hides is leading to the release of pesticides as the fleeces and hides are processed. The companies handling these raw materials have often been unaware of the presence of pesticides in their effluents and in consequence have taken no action to control their discharges. Indeed there are no recognized methods of such control.

b) Discharges of pesticides have been discovered where the discharger did not know that the particular pesticide was present in his raw materials and did not need it in his product.

c) There have been well authenticated cases of fish kills or other damage to the ecology of rivers arising from approved uses of biocides.

   If codes of 'Good Agricultural Practice' result in unacceptable damage to waters then these codes will need to be changed.

d) Pesticides have been found to be reaching sewers from applications which do not involve aqueous processes or effluents; seed dressing or the formulation of composts at garden centres are examples. In these cases pesticides have reached the sewer after cleaning-up or after rain-fall. The consequences have been a river failing to meet an environmental standard.

e) Pesticides are used in veterinary products or in food formulations for animals or fish. The users of these products are not aware of the presence of pesticides and so they are released to the environment in uncontrolled ways.

f) Some incidents can only be explained by the unapproved use of pesticides. This should diminish as the statutory provisions of the Food and Environment Protection Act 1985 are fully implemented.

g) At least one food handling process - the washing of potatoes - has been shown to release large quantities of pesticides to sewer. Tecnazene is sprayed on to potatoes to stop them sprouting when put into store; it is released in a highly concentrated effluent when the potatoes are washed prior to marketing.

h) The substitution of pesticides such as dieldrin by-products which have turned out to be just as environmentally damaging, and in some cases more damaging to fish or invertebrates, has occurred.

i) The use of dangerous biocides such as organo-tin compounds in cooling water systems or for flushing piping networks; these chemicals must eventually be released to the aquatic environment.

3.2 Other Pesticides. There is no reason to believe that such incidents will be peculiar to organochlorine pesticides. It is likely that as other pesticides are investigated similar problems will in time be recognized, and need to be dealt with. On the other hand it would be wrong to create the impression that there is a widespread risk to all the aquatic life in all of our rivers and estuaries. That is patently not so. However, the number of unexpected problems raised by the authorities indicates that there is no room for complacency. The fact that there are rivers where the sources of measured organochlorine compounds cannot be accounted for is worrying; so is the general lack of information on concentrations of nearly all of the other pesticides.

4 Risk Assessment

4.1 Data Requirements. The risk assessment of chemicals and the setting of priorities require data. So do a lot of other questions concerned with pesticides. The manufacturer, the seller, the user, and the various controlling authorities all need different types of data because they ask different questions.

The farmer needs to know whether the pesticide will do the job and how to apply it safely.

The water authority operator faced with a report of a spillage of a chemical, known only by a number or a trade name, needs first to be able to find out the true nature of the chemical (*i.e.* by full chemical name). He may then need to know whether he should close down an abstraction source for drinking water or whether he can expect a fish kill.

A factory manager may need to know what is in his raw material and whether particular care is needed for the storage or handling of particularly toxic chemicals.

The Government needs a much greater range of information for a full environmental assessment of chemicals.

It is unlikely that a single database could even be designed to answer all possible questions and it is almost certainly much more efficient to have a series of databases for specific purposes.

4.2 Pollution/Harmful Effects. A substance can be said to be causing pollution or harm to some organism including unwanted organisms. The harm may occur to organisms distant from the point of the introduction of the substance to the environment and not only to those first exposed but also to organisms several steps along a food chain. These considerations lead to the key criteria for assessment of risk of toxicity, persistence, and bioaccumulation. In assessing toxicity proper consideration must be given to both acute and chronic toxicity and to toxicity to the various life stages of the key organisms. Persistence data are of no less importance than toxicity data because without persistence exposure times must be limited and bioaccumulation cannot be possible. In any thorough assessment of the harmful effects of a chemical the nature of degradation products or metabolites should also be considered for all of the organisms at risk. In practice, however, degradation rates and the properties of degradation products or metabolites are much less well known than are the properties of the original substance so these considerations are rarely covered fully in assessments of chemicals. Other harmful effects such as carcinogenicity, teratogenicity, mutagenicity, *etc.* also have to be considered.

No chemical harms the environment if it does not reach the environment. In any proper assessment therefore there should be a knowledge of how the chemical is released, the quantity, the area affected, and the patterns of release. Such data are generally not available.

5 Chemicals for Consideration

5.1 New Chemicals. For new chemicals there are now well established schemes for the assessment of environmental risk before marketing and use can be approved.[3] Basic environmental information on toxicity, persistence, and the actual use have to be provided to Governments. The scale and the extent of the information base increases with the quantity of material to be sold or used. These data are used in the UK to assess the risks to human health by the Health and Safety Executive and the risks to the environment by the Department of the Environment.[4,5] For the aqueous environment, and in nearly all cases it is this sector which is the most critical, assessment is based on comparing worst-case environmental concentrations with an estimated no-effect concentration (usually an LC_{50} divided by 100). This scheme ensures that there is for any new substance a modicum of the data needed for an assessment. It suffers from the fact that it does not apply to pharmaceuticals, veterinary products, or agrochemicals and although there are separate approval schemes for such substances the environmental risk assessments probably need to be reconsidered. In addition, information supplied under the scheme is confidential and limited, and once a chemical is on sale it can be used for purposes other than that for which it has been properly assessed.

5.2 Existing Chemicals. For existing chemicals the problem of knowing what priorities to adopt is much more complex. There are many potential candidates and there are serious limitations in the information publicly available on the physical and chemical properties of many substances in widespread use. The Commission of the European Communities has considered over 1500 substances as possible candidates for List I status under the Directive concerning discharges of Dangerous Substances to water (76/464/EEC) (List I quoted in Chapter 23). Even for the 129 primary candidates the consultants needed to assume data on toxicity, persistence, and bioaccumulation in order to make their selection! How many other substances for which data are even more scarce should have been on the list?

5.3 Priority Substances. The Department of the Environment considers that priority substances should be those posing the most real risks. They should be carefully selected using clearly defined procedures, and quality standards should be established as the first stage in any necessary controls. These criteria involve an implicit ability to be able to measure the substance in the environment and to know the levels at which harm or pollution occur. It is not necessary, or wise to do nothing in the absence of perfect information and all selection schemes are used to alert controlling authorities to the most likely candidates for legislative control. There has to be a fuller assessment and often a political agreement on the need for action. Agreement on priorities is easy when harm is evident. All environmental controls are based on perceived safe environmental concentrations. The necessity to control organo-tin compounds in anti-fouling paints is a good example of speedy action once a problem was recognized. The real need is the ability to anticipate the problems.

6 Selection Schemes

Although there are a number of selection schemes for dangerous substances none has universal acceptance and all have their own disadvantages. A committee of the Royal Society is studying this topic. The Department of the Environment decided that it needed a scheme both for use in the context of water pollutants in the United Kingdom and to assist in the selection of candidates for List I status in international circles. The scheme developed is that described in Chapter 23. The scheme was designed with the following criteria in mind:

It should focus on pollutants in the aquatic environment but be flexible enough to be modified for a variety of policy questions or even for other environmental sectors.

It should be simple to understand and to operate and the key decision points or assumptions should be readily identified.

It should require the minimum of information and data.

It should recognize that it is particular combinations of the properties of toxicity, persistence and bioaccumulation which result in high risk.

The scale of use and the likelihood of substances reaching the aquatic environment have to be incorporated.

Both acute and chronic toxicities and food chain accumulation should be considered.

The scheme has been shown to be practical but it has revealed a number of major problems with data. To date only published data have been used for the 129 substances in the Commission of the European Communities priority lists. For about one half of these there is an incomplete data set. The most difficult data to acquire relate to scale of use on a national or a European scale. There are major gaps in data on persistence in water and no standard way to measure it; often data relating to lifetimes in soil have had to be used. In many cases there are no chronic environmental toxicity data.

A surprising feature is the low number of substances selected on the basis of the food chain scenario. Is this because the new pesticides are safer or because we just do not have the necessary bioaccumulation data? These data would answer the question.

The future development of the scheme will be to consider a wider range of chemicals and to incorporate information from Government assessment files in a confidential way. This latter may be possible for agrochemicals but is unlikely to be possible for industrial chemicals for which assessment schemes do not exist. The scheme can be used to decide which substances are candidates for List I status or for the newly announced 'Red List' of the most toxic and persistent dangerous substances which will be subject to stringent control. In each case there will have to be a final assessment which looks at all relevant factors including teratogenicity and carcinogenicity. It has been found impossible to deal adequately with carcinogenicity because there are only two criteria:

Is a substance carcinogenic in or through the aquatic environment? To date one such substance has been so identified but it is agreed that the environmental concentrations are so low that this property cannot be exhibited. Hence the overall conclusion is that no substances are selected.

The alternative is to accept any evidence or suspicion of carcinogenicity as sufficient; in this case a large number of substances which would not qualify on grounds of persistence and toxicity should be given priority status on often very dubious evidence.

A sensible middle course does not seem to be available.

7 Environmental Distribution

Having identified priority substances the next need is an understanding of the environmental distribution and assessment against stringent quality standards so that proper judgements can be made upon the need for action. The setting of environmental quality standards is even more demanding in terms of data because they need to be based on long-term chronic tests - perhaps covering the whole life cycle of several organisms.

The main deficiencies in data for proper assessment are:

Quantities and patterns of use - so that likely hot spots can be identified and monitored and assessments made of likely environmental concentrations. There may be a need for special surveys for substances whose presence in the environment cannot be explained, or to identify previously unrecognized sources.

Better data on chronic toxicities for a range of organisms. A woefully small number of substances seem to have been studied.

Table 1 *Results of analyses of 139 samples from 70 river sites in UK (1986)*

Determinand	*A*	*B*	*C*	*D*	*E*
Eulan WA	12	5	0.006	0.05	0.005
Mitin N	2	1	0.008	0.01	0.005
Permethrin	4	2	0.023	0.06	0.005
Cypermethrin	3	1	0.01	0.015	0.005
Resmethrin	0				0.005
Cyfluthrin	0				0.005
Deltamethrin	0				0.005
Cyhalothrin	0				0.005
Fenvalerate	0				0.005
Fenpropathrin	0				0.005
Chlorobenzene	24	8	0.23	7.6	0.1
1,2-Dichlorobenzene	37	16	0.5	5.7	0.1
1,3-Dichlorobenzene	35	9	0.6	5.7	0.1
1,4-Dichlorobenzene	29	8	0.7	13	0.1
1,2,3-Trichlorobenzene	2	0	0.2	0.22	0.1
1,2,4-Trichlorobenzene	4	1	0.16	0.3	0.1
1,3,5-Trichlorobenzene	5	1	0.18	0.3	0.1
1,2,3,4-Tetrachloro-benzene	2	1	0.17	0.23	0.1
1,2,3,5-Tetrachloro-benzene					0.1
1,2,4,5-Tetrachloro-benzene	0				0.1
Pentachlorobenzene	0				0.1
Hexachlorobenzene	2	0	0.17	0.18	0.1
Hexachlorobutadiene	0				0.1
Chloroethylene	0				0.1
1,1-Dichloroethane	0				0.1
1,2-Dichloroethane	0				0.1
1,1-Dichloroethylene	8	1	1.6	4.9	0.1
1,2-Dichloroethylene	1	0	0.1	0.1	0.1
1,1,1-Trichloroethane	8	3	4.2	8.6	0.1
1,1,2-Trichloroethane	5	2	0.16	0.22	0.1
Trichloroethylene	21	9	0.6	4.3	0.1
Tetrachloroethane	6	1	0.15	3.3	0.1
Tetrachloroethylene	47	19	1.1	43.3	0.1
1,1,2-Trichloro-trifluoroethane	25	9	0.45	0.9	0.1
Atrazine	83	37	0.11	1.12	0.025
Simazine	74	33	0.12	0.82	0.025
Prometryne	9	3	0.13	0.57	0.025
Terbutryne	9	3	0.4	0.09	0.025
Propazine	11	4	0.03	0.11	0.025

A = No of positive values
B = No of sites with two positive values
C = Median of positive values ($\mu g\ l^{-1}$)
D = Maximum value ($\mu g\ l^{-1}$)
E = Limit of detection ($\mu g\ l^{-1}$)

Better data on environmental concentrations. The Department will continue to encourage water authorities to monitor for pesticides and other chemicals and will continue with its own exploratory surveys. An example of the results of the latter for a range of pesticides and industrial chemicals in spot samples is shown in Table 1. It clearly shows that some herbicides are widely distributed in our rivers and consequently the Water Research centre has been asked to prepare Quality Standards for atrazine and simazine.

Better persistence data. The water column surveys will be expanded to cover materials in biota or in sediments. If substances are not found in these media they are unlikely to be persistent. There is a need, however, for better persistence testing.

There needs to be a recognition that quality control is needed for laboratories producing data on which assessments are based. Good laboratory practice and standardized protocols need to be developed including those included in the OECD guidelines. At present accessors have to accept a lot of data of dubious quality.

8 Concluding Remarks

The Department of the Environment is starting to divert some of its limited research funds to this sort of work. However, the task of testing existing chemicals is huge and it is expected that progress nationally or internationally will be slow. Judgements have to be made on the best data available and if they are inadequate these judgements must be conservative. Much more attention must be given to improving the quality and quantity of the databases for pesticides.

9 References

1. Directive on the quality of water for human consumption, 80/778/EEC.

2. Directive on limit value and quality objectives for discharges of certain dangerous substances included in List 1 of the Annex to Directive 76/464/EEC: 86/280/EEC.

3. J.L. Vosser, *Chem. Ind. (London)*, 1988, 8-1.

4. J.L. Vosser in 'Toxic Hazard Assessment of Chemicals', ed. M.L. Richardson, Royal Society of Chemistry, London, 1986, p.117.

5. H.P.A. Illing, in ref. 4, p.133.

27
The Environmental Risks Associated with the Use and Disposal of Pharmaceuticals in Hospitals

M.G. Lee

REGIONAL QUALITY CONTROLLER, MERSEY REGIONAL HEALTH AUTHORITY, HAMILTON HOUSE, 23 PALL MALL, LIVERPOOL L3 6AL, UK

1 Introduction

Prescription-only medicines are listed in Schedule 1 of the Control of Pollution (Special Waste) Regulations 1980 as substances which are to be regarded as special waste. Since these Regulations apply to all special waste, no matter how small the quantity, the disposal of pharmaceutical waste is subject to the controls set out therein. There is a need, however, for a commonsense approach to the disposal of unwanted medicines and this need has been recognized in the joint circular 4/81 issued by the Department of Environment giving guidance on the application of the pollution regulations.[1] The circular recommends householders to dispose of small quantities of unwanted pharmaceuticals by flushing to sewer and advises Waste Disposal Authorities (WDAs) to adopt a similar *de minimis* approach in the industrial and commercial section where appropriate. Under normal circumstances, the amounts of pharmaceutical chemicals required to be disposed of by hospital pharmacy departments and indeed by retail pharmacies will be small enough to allow these chemicals to be disposed of by this route. Hence, this chapter relates solely to UK practice. The disposal of clinical waste is also covered in other UK Government publications.[2,3]

The risk associated with pharmaceutical waste which might enter the water cycle has been discussed in detail in a review by Richardson and Bowron.[4] The authors calculated that the major source of pharmaceutical chemicals in potable water would be from domestic sources including home and hospital with only a marginal contribution to the overall load by manufacturing industry. That being so, the study concluded, from analytical and biodegradation data, that few drugs were likely to survive treatment in sewage works, river retention, reservoir retention, and waterworks treatment and those drugs that did would not pose a health risk at the concentration likely to be found in water supplies.

Although the majority of pharmaceutical waste can be disposed of in this manner, for large quantities of pharmaceutical waste such as that which arises from manufacturing operations, clearance of retail pharmacies, or medicine disposal schemes, disposal via the domestic sewerage is not appropriate. Large-scale discharge of such waste into the water cycle may lead ultimately to high concentrations of drugs which, however transient, would be unacceptable. Guidelines published recently by the Pharmaceutical Society of Great Britain (PSGB) recommend pharmacists to use their professional judgement when deciding the rate or route of disposal of substances which might be particularly toxic, insidious, or persistent but up to 500 tablets or capsules (or the dosage equivalent in liquids) may be disposed of by flushing to sewer.[5]

2 Disposal of Unwanted Medicines Properly (DUMP) Campaigns

DUMP campaigns are organized publicity campaigns aimed at clearing homes of old, unused or stockpiled medicines. The schemes discourage the public from hoarding medicines and, therefore, reduce the risks of accidental poisoning which can occur particularly in young children and the elderly. Members of the public are encouraged to inspect their medicine cabinets for unused or unwanted medicines, and these are returned to a specific collection point which would normally be the local pharmacy. All returned medicines are subsequently collected at one point for disposal.

DUMP campaigns have been run throughout the country for the past fifteen years with some considerable success. However, although they reduce hazards to the public from stockpiled medicines, the accumulation of large quantities of pharmaceutical waste presents the organizing body with the problem of waste disposal.

In one such campaign in Kent, 1,364,000 tablets and 80 gallons of liquid medicines were brought in for disposal, and following a one week campaign in Nottinghamshire 3923 lb of medicines including 330,000 tablets and 100 l of liquids were returned.[6] In a recent campaign in Liverpool, 401 bags of unwanted medicines weighing 5770 lb and containing 600,000 tablets or capsules were returned. One patient returned a 10 lb bag containing 8400 tablets consisting of approximately equal numbers of just two brands of tablets.

It is possible to find highly dangerous items in the returns. For instance, during the Nottinghamshire campaign 1 kg of lead arsenate and several lots of strychnine were returned, and in Liverpool a bottle of picric acid was discarded. Incineration is the preferred method of disposal for bulk waste. It is not practical, however, to use the consignment note system laid down in the Control of Pollution Regulations for such large and diverse samples. Where the campaigns are organized with the help of a District Health Authority, a hospital incinerator may be used. It is important, however, to recognize that some pharmaceuticals, such as arsenic-containing compounds, inorganic cyanides, and chlorates, should not be incinerated. During the Liverpool DUMP campaign, pharmacists were asked to separate liquids, controlled drugs, cytotoxic drugs, and aerosols and destroy them in the usual way. In addition, explosive or potentially hazardous material was to be set aside for separate collection and disposal. These precautions isolated the potentially dangerous material and enabled the residual waste to be incinerated with minimum risk.

3 Environmental Risks in Hospitals

It is possible for some people to be affected adversely by drugs at doses which are far below the therapeutic range of the drug. The reactions that occur in persons sensitive to penicillins and phenothiazines are examples of this. Provided, however, that a person is aware of the reaction, it is possible to avoid the causative agent. There are, however, classes of drugs for which the long-term effects of continuous exposure to very low levels are not so easily definable or are not fully understood. Two such groups are cytotoxic drugs and radiopharmaceuticals.

The majority of cytotoxic drugs are carcinogenic in normal, healthy persons and possess mutagenic activity at concentrations well below their therapeutic range. Radiopharmaceuticals, which are required to be radioactive for their therapeutic and diagnostic applications, carry the risks associated with ionizing radiations. In view of the known or potential toxicity of these drug groups special precautions are needed during their preparation and handling to protect the operator and the immediate environment of the dispensing area. Care is also needed to eliminate environmental risks during disposal.

3.1 Radiopharmaceuticals. Radiopharmaceuticals are used widely in many branches of medicine and surgery. The majority of applications are for the investigation and diagnosis of disease but in some circumstances they are used for therapeutic purposes.

The applications for radiopharmaceuticals require them to be administered parenterally and so the need to maintain product sterility during dispensing must not be overlooked. Facilities for the preparation of injections of radionuclides in hospitals must meet simultaneously the requirements for good radiation and good pharmaceutical practice.

3.1.1 Environmental control within hospitals. The principles of radiation protection in the radiopharmacy are those which apply to radiation protection in all fields of work. In the UK they are based on those of the International Commission on Radiation Protection (ICRP) which are set out in its General Recommendations[7,8] and are made effective through the Ionizing Radiation Regulations 1985. These give considerable emphasis to the limitations of dose received by workers and include a detailed schedule of dose limits.

Radiopharmaceuticals are prepared for use in vertical laminar flow cabinets complying with the specification for class II safety cabinets under the British Standard for Microbiological Safety Cabinets BS 5726. They are sited in a radiopharmacy suite comprising of a changing room plus a sufficient number of clean rooms. The clean rooms receive a filtered air supply and operate under a positive pressure differential of 1.5-2.5 mm water gauge (15-25 Pa) with reference to the area outside the suite. The air passing over the working surface has been filtered by a high efficiency, sub-micron particulate air (HESPA) filter which together with the laminar air-flow ensures an aseptic working environment within the contained work station. Not all the air is recirculated within the cabinet. A proportion (approximately 30%) is exhausted into the room and so air is constantly being drawn through the front opening to compensate. This creates a negative pressure within the cabinet and, therefore, any spillages are contained within it.

All equipment used while handling radio-pharmaceuticals is shielded appropriately. Operations are carried out in a shielded tray to contain any spillage and, as far as is practical, 12 mm sheets of lead glass are used for viewing vials and vessels containing radionuclides.

3.1.2 Disposal of waste. The disposal of radio-pharmaceuticals is controlled under the Radioactive Substances Act 1960[9] and the Control of Pollution (Radioactive Waste) Regulations, 1976.[10]

Under the Radioactive Substances Act, hospital departments that handle such substances must possess and display a Certificate of Authorization for the Disposal and Accumulation of Radioactive Waste. The certificate is issued by the radiochemical inspector of the Department of the Environment and in it is specified the levels of radioactive waste that may be disposed of over a given period of time.

Table 1 lists the radionuclides that may be used in medicine together with their half-lives and examples of use. In the average District General Hospital the only radionuclide that would be used is technetium-99m (Tc-99m). Tc-99m has a short half-life allowing relatively large doses to be administered and the energy of its gamma-emission is readily detectable, ideal properties for a radiopharmaceutical. It is the daughter ion of molybdenum-99 which, because of its short half-life is normally prepared just before use by elution from a sterile generator consisting of molybdenum-99 absorbed onto alumina in a plastic column.

Table 1 *Half-lives, doses, and applications of radiopharmaceuticals currently in use*

Radionuclide	*Half-life*		*Dose/MBq*	*Application*
Calcium-47	4.5	d	0.185-0.74	Investigation of calcium therapy
Chromium-51	27.7	d	up to 370	Investigation of blood disorders
Cobalt-57	270	d	0.018-0.37	Measurement of vitamin B_{12} absorption
Gallium-67†	78	d	55-92	Detection of neoplasm
Gold-198*	65	h	9250	Radiotherapy
Indium-111†	67	h	74-185	Detection of tumours
Iodine-125 and	60	d	0.185-3.7	Investigation of thyroid function
Iodine-131†	8.06	d	3700-5500	Radiotherapy
Mercury-197*	64.4	h	3.7-11.1	Kidney imaging
Selenium-75†	120	d	7.4-11.1	Imaging of pancreas and parathyroid
Technetium-99†	6.02	h	up to 740	Bone scans; brain, liver, lung, spleen and thyroid imaging
Thallium-201†	73.5	h	55.5	Scanning myocardium
Xenon-131	5.25	d	37-185	Measurement of lung perfusion

* Use virtually ceased
† Most commonly used radionuclides

Owing to its short half-life, the environmental risks associated with the disposal of Tc-99m are slight. Waste is collected in a shielded sharps box, where it is stored for two days to allow radioactivity to decay. After checking the activity, the waste is disposed of by incineration during a designated incinerator session.

The half-life of the parent molybdenum-99 is 66 h. Unwanted generators are stored for six weeks to allow total decay of the parent ion then the solid waste is placed in sealed containers and disposed of at a tip specified in the Certificate of Authorization for Disposal (usually a Council tip).

In teaching hospitals and other major centres a wider range of radiopharmaceuticals is used although the major usage (over 90%) will still be with technetium. The more commonly used radionuclides are listed in Table 1. All these are purchased as sterile injections with a stated activity and require dilution under aseptic conditions prior to administration. The residue from these preparations can be disposed of directly into the water drainage system provided proper records are kept and subject to the limitations in the Certificate of Authorization for the disposal of liquid radioactive waste (Table 2). The Regional Radiopharmacy Unit in Mersey Regional Health Authority has no problem in complying with these restrictions.

Table 2 *Authorized levels of liquid radioactive waste that may be discharged into the drainage system from the Royal Liverpool Hospital in one calendar month*

Half-life	*Sum of total of radionuclides*
Less than 5 d	75 GBq
5 d - 1 year	40 GBq
Greater than 1 year	200 MBq

<u>3.1.3 Monitoring</u>. Monitoring of the radiation exposure of staff is organized by the Radiation Protection Adviser. Film badges provide a continuous monthly record of the exposure of the staff engaged directly in radiation work as well as others whose work may involve exposure to radiation, *e.g.* from patients receiving radioactive preparations. Also thermoluminescent dosimetry techniques are used to investigate contamination to the hands of those staff preparing radiopharmaceuticals. It has been calculated that experienced workers preparing four injections per day would receive a radiation dose to the fingers equivalent of 4 rem in a year as compared with the annual dose equivalent limit to the hands of 50 rem.[11] Thermoluminescent measurements are consistent with these estimates.

The nature of ionizing radiations enables the activity of any waste or contaminated material to be monitored, allowing environmental pollution to be quantified. Legal restrictions on the disposal of radiopharmaceuticals are based upon the medical risks associated with these products. Therefore, provided the required procedures are followed, the environmental levels will be within acceptable limits. The risks to the environment are, therefore, minimal.

3.2 Cytotoxic Drugs. The descriptive term cytotoxic is applied to those drugs used for cancer chemotherapy. Cytotoxic drugs present probably the major environmental risk in hospitals. The pharmacological properties that they require for the treatment of the various cancers result in their having both carcinogenic and other severe toxic reactions towards healthy tissue. Guidelines on the handling of cytotoxic drugs have been published by a number of professional bodies included the PSGB[12] and the Health and Safety Executive (HSE).[13] These cover mainly the risks to the operator during the handling and preparation of cytotoxic injections and make only brief references to the question of the disposal of cytotoxic waste. As medicinal products their disposal is controlled under the Control of Pollution (Special Waste) Regulations but no specific reference is made to cytotoxic drugs under these Regulations.

Perhaps the greatest problem when investigating the environmental hazards of these chemicals is that the carcinogenic and mutagenic effects of continuous exposure to microgram and sub-microgram quantities are not known. Much of the work on chromosome damage and carcinogenesis caused by antineoplastic agents is based on studies in patients to whom the drugs have been administered for the treatment of cancer, and therefore relates to relatively high doses. It has been shown, however, that chromosome damage is related to both dose and duration of therapy and is cumulative.[13] Cytotoxic drugs are also known to be teratogenic and a study in Finland indicates that there may be a causal relationship between anti-cancer drugs and fetal abnormalities in pregnant nurses who worked with these drugs.[15] These findings were based however on a relatively small number of cases and further studies are needed to confirm them. It has also been reported, following a retrospective analysis of data from seventeen hospitals during a seven year period 1973-1980, that fetal loss associated with cytotoxic agents had occurred among Finnish hospital nurses.[16]

It is essential therefore to minimize exposure to these potent carcinogens and teratogens even though the levels of absorption that may take place during handling are difficult to assess.

3.2.1 Environmental control within hospitals. Studies on several groups of workers have shown that the urine of health-care workers who routinely prepare injections of cytotoxic preparations can possess mutagenic activity. Pharmacists who reconstitute anti-cancer drugs were shown to excrete mutagenic urine over the period of exposure. When they stopped handling the drugs, the mutagenic activity fell within two days to the level of unexposed controls.[17] Detectable amounts of cyclophosphamide have also been found in the urine of nurses handling it.[18]

As with radiopharmaceuticals there is a need to achieve a balance between operator protection and product protection when dispensing cytotoxic injections. Asepsis and the maintenance of sterility is of paramount importance with this group of drugs because the immunosuppressive activity which many of them possess reduces the patient's ability to overcome infection. The recommended arrangement for dispensing cytotoxic injections is a centralized, pharmacy-based, reconstitution service. Preparations are handled in vertical flow safety cabinets isolated within a clean room. The use of a vertical flow containment cabinet has been shown to reduce the levels of mutagenic substances in the urine of pharmacy workers preparing cytotoxic drugs[19] and the cabinets also provide the required level of microbiological control.

Currently there is no British Standard for cytotoxic safety cabinets. The only national standards that specify their design, installation, and use are the Australian Standards AS 2567-1982[20] and AS 2639-1983.[21] Cytotoxic cabinets are designed to operate in the same way as microbiological safety cabinets. In addition the Australian Standard recommends that the cabinet has an activated carbon exhaust filter to trap gases or volatile effluents. AS 2639 also makes reference to a secondary barrier, which is the room housing the cabinet. Microbiological considerations require this room to be under positive pressure but optimum safety requirements are for the room pressure to be negative with reference to adjacent areas thus containing any spillages in the room. An acceptable compromise is to supply HESPA filtered air to the clean room which is then under positive pressure. In the event of a spillage, the air filtering plant is switched off, thus creating either a neutral pressure or a negative pressure especially if the cabinet exhaust is ducted outside the room.

It is widely recommended that protective clothing, including a long-sleeved gown, eye protection, facemask, and gloves should be worn when preparing cytotoxic drugs. Protection of the hands and fingers is particularly important since it has been reported that certain cytotoxic drugs are absorbed through intact skin.[18] Polyvinylchloride (PVC) and latex gloves are permeable to cytotoxic drugs, latex less so than PVC.[22,23] Recent work with carmustine has shown that it will, with time, penetrate all gloves irrespective of material and thickness.[25] The lag time before permeation is achieved is longer for thicker gloves but for some of the more commonly used sterile procedure gloves it can be as short as five minutes. There is some evidence that diffusion through the glove material is faster for hydrophobic drugs such as carmustine, thiotepa, and mustine. Nonetheless, the results raise many questions about the suitability of current recommended procedures for handling cytotoxic products. Total containment 'glove box' cabinets are being widely promoted but if it is possible for some drugs to accumulate in and permeate through the gloves incorporated in them, then hazards to the operator will be increased rather than decreased by this type of cabinet.

<u>3.2.2 Methods of disposal</u>. Protective clothing, syringes and needles, administration sets and solutions, and ampoules and vials used to prepare and administer cytotoxic injections may be contaminated and should be disposed of as such. The urine, faeces, and other waste from patients receiving antineoplastic drugs can also contain high concentrations of drug and metabolites. There is a need for a rational approach to the disposal of contaminated patient waste. Over-elaborate precautions may increase rather than decrease the risks to nursing staff. Gloves and disposable aprons used in conjunction with normal nursing procedures for the handling of blood, urine, and faeces from patients should provide adequate protection.

Cytotoxic agents in urine and faeces disposed of into the sewage system could potentially contaminate water supplies. Analysis of methotrexate levels in sewage, river, and potable water shows that, although the drug is widely used and in substantial doses (up to 22 g daily), no concentration of methotrexate in excess of 6.25 ng l^{-1} (the limit of detection) could be found in any sample of river or tap water tested.[25] Levels of about 1 $\mu g\ l^{-1}$ could be found in a sewer immediately downstream of an oncology clinic but water treatment and dispersion would reduce this level below the detectable limits. It is reasonable to deduce, therefore, that there should be no risk when patient waste containing cytotoxic chemicals is disposed of to drain.

Table 3 summarizes the information available on the deactivation and disposal of routinely used cancer drugs. Solid dose forms present relatively few problems.

Table 3

Drug	*Chemical method*	*Incineration temp./°C*
Actinomycin	-	Yes
Amsacrine	10% Sodium hypochlorite	260
5-Azacytidine	-	Yes
Azathioprine	Hot 10% Sodium carbonate[25]	Yes
Bleomycin	5% Sodium hydroxide	Yes
Busulphan	10% Sodium thiosulphate for 20 h[27]	Yes
Carmustine	10% Hydrochloric acid[21]	1000
Carboplatin	-	-
Chlorambucil	5% Sodium hydroxide[21]	Yes
Cisplatin	12% Alkaline sodium borohydride[21]	250
Cyclophosphamide	0.2M Methanolic potash for 1 h[27]	Yes
Cyclosporin A	-	Yes
Cytarabine	N/1 Hydrochloric acid[26]	-
Dacarbazine	10% Sulphuric acid for 24 h[26]	-
Daunorubicin	-	800
Doxorubicin	10% Sodium hypochlorite for 24 h[26]	700
Epirubicin	10% Sodium hypochlorite for 24 h[26]	700
Estramustine	-	Yes
Ethoglucid	10% Hydrochloric acid for 30 min [26]	Yes Yes
Etoposide	Alkaline potassium permanganate[21]	1000
Fosfestrol	-	Yes
Fluorouracil	5% Sodium hydroxide[21]	1000
Gestronol*	-	Yes
Hydroxylurea*	-	Yes
Infosfamide	-	Hospital incinerator
Lomustine*	-	Yes
Melphalan	5% Sodium hydroxide[21]	Yes
Mercaptopurine*	-	Yes
Methotrexate	-	1000
Methramycin	Large volume of trisodium phosphate[26]	300
Miltomycin C	2-5% Hydrochloric acid or sodium hydroxide for 12 h[19,26]	Hospital incinerator
Mitozantrone	-	Yes

Table 3 (continued)

Drug	*Chemical method*	*Incineration temp./°C*
Mustine	Immerse in mixture of water, 3 l: Sodium hydroxide solution (1.5 sp.gr), 1 l: Industrial methylated spirits 66 OP, 4.5 l for 48 h[26]	Incineration not recommended
Nandrolone	-	Hospital incinerator
Procarbazine*	-	Yes
Razoxane*	-	Yes
Streptozocin	10% Hydrochloric acid	Contact manufacturer
Tamoxifen*	-	Yes
Teniposide*	-	1000
Thioguamine*	-	Yes
Thiotepa	Large quantities of boiling water[21]	800
Treosulphan	10% Sodium thiosulphate for 1 h or 0.75N Sodium hydroxide for 24 h	Yes
Vinblastine	Hot water[26]	1000
Vincristine	Hot water[26]	1000
Vindesine	-	1000

* Only available as tablets or capsules

Under incineration temperature 'yes' indicates that incineration is permissible but the manufacturers are unable to recommend an incineration temperature.

The capsule shell or tablet coating protects patients and staff handling them and discarded tablets and capsules may be disposed of in small quantities by flushing to drain. The disposal of liquid waste is, however, more difficult. Chemical deactivation can be used for dealing with spillages and also for single vials containing injections residues. There are, however, practical difficulties in using such methods for the large amounts of waste generated by a centralized cytotoxic service, particularly the protective clothing and used syringes. The only method which is generally applicable to all cytotoxic waste is incineration. However, not all incinerators are suitable for this purpose. The most efficient type of incinerator is probably the gas-fired, two-stage type in which a first-stage combustion with a less than stoichiometric air:fuel ratio is followed by a second-stage with excess air, but not all hospitals possess such incinerators. In many cases manufacturers are unwilling to specify the minimum incineration temperature and those that do recommend temperatures of 700-1000°C. Few hospital incinerators can attain temperatures of 1000°C; the most efficient will reach about 800°C but a more realistic figure would be 500-600°C. Consequently the manufacturers' recommendations for cytotoxic drugs cannot be met.

The high temperatures suggested for incineration would appear to have no scientific basis or justification. All cytotoxic drugs are organic chemicals and would, therefore, be expected to be destroyed at temperatures normally attained in hospital incinerators. Daunorubicin, for example, melts with decomposition at 190°C and doxorubicin at 205°C but their recommended incineration temperatures are 700 and 800°C respectively. Clearly there is a need to investigate and validate thermal destruction methods for this group of drugs. It is possible that incinerators operate at temperatures in excess of those required to provide destruction of cytotoxic chemicals. Equally it is possible that partially degraded drug molecules may be as dangerous if not more dangerous than the parent substance.

3.2.3 Monitoring. The most sensitive method of monitoring for cytotoxic agents is mutagenicity testing; the limits of detection by these methods are significantly lower than for chemical analytical methods. Currently there is no clinical evidence of a correlation between mutagenic urine and toxic reactions in health care workers, but there is evidence of teratogenicity in nurses handling cytotoxic drugs. In view of this uncertainty, every effort should be made to minimize risk to persons handling these drugs. Mutagenicity testing offers the most sensitive method of demonstrating hazards during exposure, and hence urine samples of health care staff handling cytotoxic drugs should be tested regularly for mutagens. If a relationship between mutagenicity and toxicity can be established then a wider environmental risk assessment study should be considered.

4 Conclusions

There is no risk from the disposal of small amounts of pharmaceutical waste by flushing to drain. Accumulated waste can normally be incinerated provided hazardous materials have been removed. Itemizing such waste is impractical and unnecessary for the bulk of pharmaceuticals; hazardous substances such as cytotoxic drugs should be identified but the majority could be combined under a general heading of medicinal products.

Procedures for handling hazardous pharmaceuticals within the hospital environment should minimize risk to staff and patients. The disposal of the bulk of medicinal waste including radiopharmaceuticals is controlled and safe. However, thermal destruction methods for cytotoxic drugs require validation. Work is needed on the long-term effects of continuous exposure to microgram and sub-microgram quantities of cytotoxics.

5 Acknowledgements

The author wishes to thank Mr. P. Maltby, Principal Pharmacist, Radiopharmacy at the Royal Liverpool Hospital for helpful discussion, Mr. B. Rhodes and Mr. R. Marshall at the PSGB for their advice and assistance, and Miss S. Hemsworth for typing the manuscript.

6 References

1. Joint Circular 4/81 from the Department of Environment and Welsh Office. Control of Pollution Act 1974. Control of Pollution (Special Waste) Regulations 1980.

2. Department of the Environment, Waste Management Paper No.25, Chemical Wastes, HMSO, London, 1983.

3. Health and Safety Commission 'The Safe Disposal of Chemical Wastes', HMSO, London, 1982.

4. M.L. Richardson and J.M. Bowron, *J.Pharm. Pharmacol.,* 1985, 37, 1.

5. G.E. Appelbe, *Pharm. J.*, 1988, 240, 100.

6. M. Aslam, S. Ayling, M.A. Healy, and S.T. Garner, *Pharmacy Update*, 1987, 3, 58.

7. ICRP, 'Recommendations of the International Commission on Radiological Protection. Publication No. 26', Pergamon, Oxford, 1977, Annals of the ICRP, Vol. 1, 1.

8. ICRP, International Commission on Radiological Protection, *Phys. Med. Biol.*, 1978, 23, 1209.

9. Radioactive Substances Act 1960, HMSO, London.

10. Pollution (Radioactive Waste) Regulation, 1976, HMSO, London.

11. N.G. Trott and S.C. Lillicrap, 'Radiopharmaceuticals from Generator-produced Radionuclides', IAEA, Vienna, 1971, p.29.

12. Pharmaceutical Society Working Party report, *Pharm. J.*, 1983, **17**, 532.

13. Guidance note MS21 from the Health and Safety Executive, 'Precautions for the Safe Handling of Cytotoxic Drugs', HMSO, 1983.

14. R.G. Palmer, C.J. Dore, and A.M. Denman, *Lancet*, 1984, i, 256.

15. K. Hemminki, P. Kyyronen, and M.L. Lindbohm, *J. Epidemiol. Community Health*, 1985, **39**, 141.

16. S.G. Selevan, M.L. Lindbohm, R.W. Hornung, and K. Hemminki, *N. Eng. J. Med.*, 1985, **313**, 1173.

17. C. Macek, *J. Am. Med. Assoc.*, 1982, **247**, 11.

18. M. Hirst, S. Tse, D.G. Mills, L. Lemn, *Lancet*, 1984, i, 186.

19. R.W. Anderson, W.H. Pickett, T.V. Nguyen, J.C. Theiss, T.S. Matney, and W.J. Dana, *Am. J. Hosp. Pharm.*, 1982, **39**, 1881.

20. Australian Standard 2567-1982, Cytotoxic Drug Safety Cabinets.

21. Australian Standard 2639-1983, Cytotoxic Drug Safety Cabinets - Installation and Use.

22. T.H. Connor, J.L. Laidlaw, J.C. Theiss, R.W. Anderson, and T.S. Matney, *Am. J. Hosp. Pharm.*, 1984, **41**, 676.

23. J.L. Laidlaw, T.H. Connor, J.C. Theiss, R.W. Anderson, and T.S. Matney, *Am. J. Hosp. Pharm.*, 1984, **41**, 2618.

24. P.H. Thomas and V. Fenton-May, *Pharm. J.*, 1987, **238**, 775.

25. G.W. Aherne, J. English, and V. Marks, *Ecotoxicol. Environ. Safety*, 1985, **9**, 79.

26. S.J. Wilson, *J. Clin. Hosp. Pharm.*, 1983, **8**, 295.

28

Assessment of Cancer Risk Following Treatment by Cytostatic Drugs

N.E. Day

MRC BIOSTATISTICS UNIT, 5 SHAFTESBURY ROAD, CAMBRIDGE CB2 2BW, UK

1 Introduction

A considerable number of chemical exposures from the environment are now recognized as carcinogenic to humans.[1] Unfortunately, for only very few are quantitative data available on the relationship between dose or exposure and level of excess risk. Environmental control of potential carcinogens has often been based on quantitative data from long-term carcinogenicity experiments in animals, or even *in vitro* tests in bacterial systems. Unverified assumptions are then required to infer the potential effect in humans, and unverifiable assumptions are often made for the purposes of low-dose extrapolation. The latter posit certain forms for the dose-response relationship at low doses, beyond the realm where observational confirmation would be possible (see also Chapters 5, 8, 9, and 14). The use of experimental data for risk assessment in humans will, of course,always be necessary since, for compounds newly introduced to the environment, data on humans cannot be available. Their use, however, will certainly be sharpened if one can increase the number and range of compounds for which accurate quantitative data on humans are available. The algorithms for extrapolation will be more firmly based.

2 Cytostatic Drugs

One class of compounds, for which one can reasonably expect quantitative information to become available, is the range of cytostatic agents used in cancer chemotherapy (see also Chapter 27). The alkylating agents in particular induce acute non-lymphocytic leukaemia (ANLL), non-Hodgkins lymphoma, and possibly other cancers.[2-4] These agents are administered under controlled conditions with, usually, good records of the amount and duration of exposure. Often, the treated patients are either followed up clinically or are covered by the passive surveillance of cancer registration, so that the appearance of a second primary cancer is notified. Under these circumstances, there is a clear potential for establishing relationships between dose and time of exposure and the response.

2.1 Alkylating Agents. Study of the carcinogenicity, specifically the leukaemogenicity, of cytostatic drugs provides a model for quantitative extrapolation from both short-term tests of biological activity and long-term animal carcinogenicity assays to risk for man. It has the additional advantage that attention is focused principally on one type of compound, alkylating agents. Earlier work which attempted to relate carcinogenicity in humans and animals considered disparate groups of chemicals[5] in which a greater range of factors would influence carcinogenic expression. In this chapter, three aspects are considered: first, the dose-response for an individual compound: second, the relative dose-response of different compounds, and third, the relationship of the carcinogenic potency in humans to that seen in long-term experiments.

2.2 Dose Response. An example of the type of dose-response information currently emerging for a number of alkylating agents is given by a study carried out in the German Democratic Republic. Until recently, cyclophosphamide was the main chemotherapy agent available and, for the treatment of ovarian cancer, virtually no other cytostatic drug was used. The whole country has been covered by a high-quality cancer registration since the early 1950s, with particular attention to details of multiple primary cancers. A case-control study was undertaken of leukaemia occurring subsequent to a diagnosis of ovarian or breast cancer. Twelve such cases were identified after ovarian cancer, nine of which were acute, the remaining three being chronic myeloid. Ninety-three cases were identified after breast cancer, of which 52 were acute. The dose response for leukaemia in terms of cumulative dose of cyclophosphamide is shown in Table 1, based on the 79 cases for whom treatment information was available.[6]

Table 1 *Quantitative relationship between the relative risk of leukaemia and total dose received of cyclophosphamide*

Dose/g	*Leukaemia cases*	*Relative risk*
0	57	1.0
10	9	1.5
10-29	3	3.3
30+	5	7.3
Unknown	5	10.9

An extended study is currently in progress, focusing on leukaemia following ovarian cancer, based on cancer registries in Scandinavia, the UK, and Canada.[7] Besides sharpening the dose-response for cyclophosphamide, dose-response relationships are being determined for chlorambucil, melphalan, thio-tepa, and treosulphan. The differential leukaemogenicity of the different drugs will also be determined. In addition to the relationship between leukaemia incidence and total dose, the effect of time of administration of the drug is also being investigated. In a study of the induction by melphalan and cyclophosphamide of leukaemia and the precursor myelodysplastic syndrome, a strong indication emerged that the main effect was related to exposure within the past three, or possibly five, years.[8]

2.3 Carcinogenic Potency in Humans. With the accumulation of quantitative data on the leukaemogenicity of different alkylating agents used in chemotherapy, there is an increasing possibility to examine quantitatively the association of carcinogenic potency in humans with that in different animal species, and with activity in *in vitro* assays.

An initial attempt[4] has been made along these lines for the compounds tested above, plus busulfan[9] and methyl-CCNU (Semustine),[10] the other two cytostatic drugs which have been demonstrated to cause leukaemia. It is interesting to note that the weakness of the association is due largely to the low level of quantification possible from the animal data.

3 Conclusions

The information now being produced on the quantitative carcinogenic potency of chemotherapy drugs is of relevance in two main ways to risk assessment. First, in direct manner it provides one of the parameters on which choice of therapy should be based. With increasing success in the treatment of some malignancies, more emphasis is now being placed on reducing harmful side effects, of which leukaemia induction is among the most serious. Secondly, in a less direct, but more general manner, it provides a model for evaluating the role of data obtained from non-human experimental systems in predicting human risk.

4 References

1. 'Overall Evaluation of Carcinogenicity: An Update of IARC Monograph Volumes 1-42', International Agency for Research on Cancer, Monograph Series Supplement 7, 1988.

2. L. Penn, 'Malignancies induced by drug therapy: a review', in 'Carcinogenicity of Alkylating Cytostatic Drugs', ed. D. Schmahl and J.M. Kaldor, International Agency for Research on Cancer, Scientific Publication No. 78, Lyon, 1986.

3. D. Schmahl, 'Carcinogenicity of anti-cancer drugs and particularly alkylating agents', in 'Carcinogenicity of Alkylating Cytostatic Drugs', ed. D. Schmahl and J.M. Kaldor, International Agency for Research on Cancer, Scientific Publication No. 78, Lyon, 1986.

4. J.M. Kaldor, N.E. Day, and K. Hemminki, 'Quantification of drug carcinogenicity in humans and animals', *Eur. J. Cancer Clin. Oncol.*, 1988, (in press).

5. E. Crouch and R. Wilson, 'Interspecies comparison of carcinogenic potency', *J. Toxicol. Environ. Health*, 1979, 5, 1095-1118.

6. J.F. Haas, B. Kittelmann, W.H. Mehnert, W. Staneczek, M. Mohner, J.M. Kaldor, and N.E. Day, 'Risk of leukaemia in ovarian tumour and breast cancer patients following treatment by cyclophosphamide', *Br. J. Cancer*, 1987, 55, 213-218.

7. J.J. Kaldor, N.E. Day, P. Band, N.W. Choi, E.A. Clarke, M.P. Coleman, M. Haeama, M. Koch, F. Langmark, F.E. Neal, F. Pettersson, V. Pompe-Kirn, P. Prior, and H.H. Storm, 'Second malignancies following testicular cancer, ovarian cancer and Hodgkin's disease: an international collaborative study among cancer registries of the long-term effects of therapy', *Int. J. Cancer*, 1987, 39, 571-585.

8. J. Cuzick, S. Erskine, D. Edelman, and D.A.G. Galton, 'A comparison of the incidence of the myelodysplastic syndrome and acute myeloid leukaemia following melphalan and cyclophosphamide treatment for myelomatosis', *Br. J. Cancer*, 1987, **55**, 523-529.

9. H. Stott, W. Fox, D.J. Girling, R.J. Stephens, and D.A.G. Galton, 'Acute leukaemia after busulphan', *Br. Med. J.*, 1977, ii, 1513-1517.

10. J.D. Boice, M.H. Greene, J.Y. Killen, S.S. Ellenberg, R.J. Keehn, E. McFadden, T. Chen, and J.F. Fraumeni, 'Leukaemia and preleukaemia after adjuvant treatment of gastrointestinal cancer with semustine (methyl-CCNU', *New Engl. J. Med.*, 1983, **309**, 1079-1084.

29

The Management of Effluents/By-products of Multi-purpose Fine Chemical Manufacture

H.M. Donaldson

IMPERIAL CHEMICAL INDUSTRIES PLC, ORGANICS DIVISION, PO BOX 43, HEXAGON HOUSE, BLACKLEY, MANCHESTER M9 3DA, UK

1 Introduction

In the 19th century public concerns over hydrochloric acid pollution from the alkali industry led directly to the passing of the Alkali Act of 1863. This was the first effective environmental protection Act, eventually encompassing much of the chemical industry and establishing the Alkali Inspectorate as an enforcement authority.

The basic ingredients of this example are still with us today:

A chemical industry which manufactures for profit many products essential to society, but with inevitable by-products and effluents.

A public which wants the benefits of the products but which is concerned for its environment.

Legislation which responds to the public concern and seeks to protect the environment but accommodates the need for viable industry.

Statutory enforcement authorities which are professional, technically qualified, and with the experience to make sound judgements on the needs of industry and the risks to the environment.

Whilst these aspects will not be discussed in any detail, it is important to keep them in mind as an essential background to effluent issues in the chemical industry for which I shall use my experience with ICI in most cases.

Over the years there has been an increase in the complexity and sophistication of chemical processes and manufacturing plants. This has been reflected in the nature of the effluents from such processes and the problems of disposal. Such problems are particularly acute in multi-purpose fine chemicals plants.

It is important to differentiate between those units, typical of either petrochemical, commodity or speciality chemical manufacture which are designed with a high degree of specificity for a single product or small group of products and a multi-purpose fine chemicals plant. The description 'multi-purpose fine chemicals plant' is intended to cover a plant with multiple unit processes and unit operations which performs multi-stage chemical syntheses of a continually changing range of products.

2 Plant Types and Effluents

2.1 Olefins Plant. The first example of a plant/effluent type is a petrochemical plant as on ICI's Wilton Site, an olefins joint venture with British Petroleum. The characteristics of this plant are that it is **large, continuous and single stream.** It represents a very large investment of more than £200M and is the culmination of a long process development whereby many of the by-products or side streams are recycled into the main process streams. Clearly, recycling is necessary for economic reasons but it also minimizes the effluents. As a continuous plant it runs in a steady state for very long periods with a small number of dilute emissions such as steam and carbon dioxide and small amounts of hydrogen and hydrocarbons being burned in the flare stack.

2.2. Biocides Unit. The Biocides Unit at ICI's Huddersfield Works is an example of a **small, semi-continuous, single stream** plant designed to make a specific product. Although a much smaller investment than the olefins example, the Biocides Unit was the subject of significant investigation into improvements in efficiency, with consequent reduction in effluent. There are large variations in effluent flow and the effluents are frequently toxic, but these factors are predictable and the design incorporates recycling and effluent treatment tailored to each waste stream. A simple flow sheet is given in Figure 1. The first and second stage manufactures and associated recoveries and scrubbing systems are continuous but product isolation is batchwise. Ammonia, sulphur dioxide, and hydrochloric acid are evolved in the process and absorbed in scrubbers. Since, by their nature, biocides tend to be toxic to other organisms, filtrates and washes from the process are collected and discharged to the local sewage treatment works at a controlled rate by agreement with the Yorkshire Water Authority.

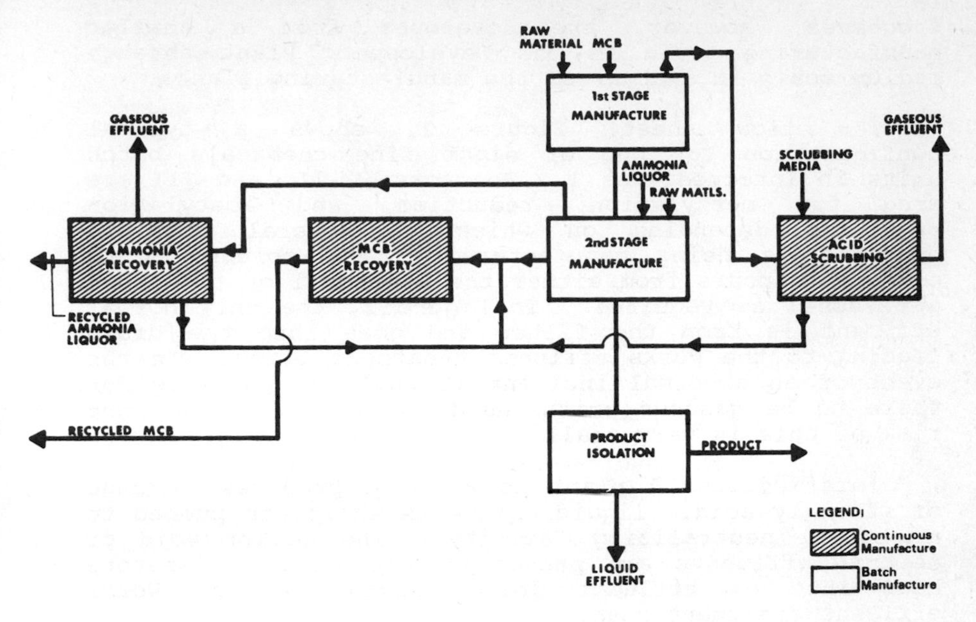

Figure 1 Biocides unit flow sheet

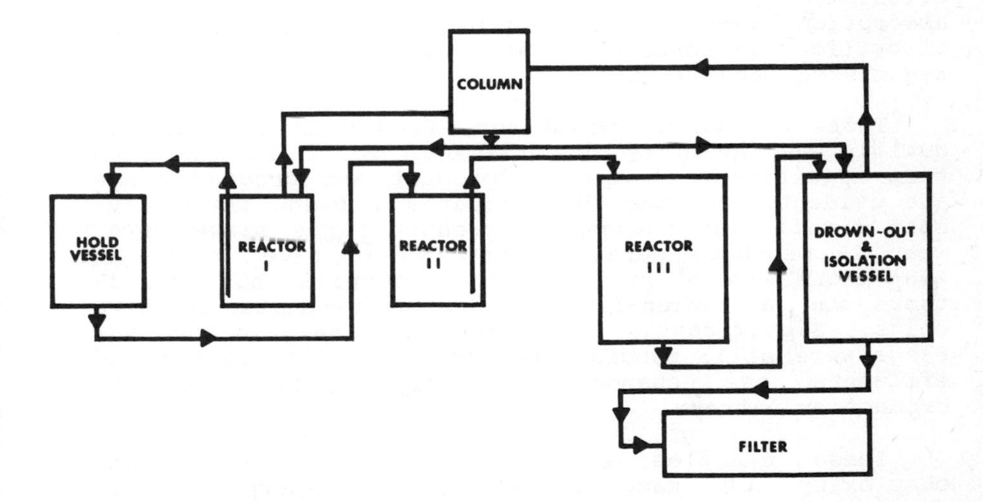

Figure 2 Intermediates unit flow sheet

<u>2.3 Intermediates Plant</u>. In contrast to the previous examples, the Intermediates 1 plant at Huddersfield is a typical modern multi-purpose fine chemicals plant. It is a medium size plant with batch operation of a wide range of processes and products. Individual batches are large, 20 - 50 t in size. Such large-scale

processes, however, are developed from a smaller manufacturing scale in the Development Plant through medium scale in another of the manufacturing plants.

The flow sheet, Figure 2, shows a typical configuration for one of eight fine chemicals batch units in Intermediates 1. Reactors I, II, and III are used for methylation, reduction, and acetylation reactions depending on which of several possible products is being manufactured. The column is to condense vapours from either the Reactor 1 or the drown out vessel as required. In Figure 2, the only normal effluent is from the filter and goes into the drain leading to the Works effluent treatment plant. In the event of an abnormal incident it would be possible for there to be gaseous emissions from the column but the risk of this is very small.

Intermediates 1 plant as a whole produces a range of strongly acidic liquid effluents which are pumped to a central neutralizing facility. The weaker acid or neutral effluents are passed through static limestone beds into an effluent drain leading to the Works effluent treatment plant.

Gaseous effluents, principally sulphur dioxide and trioxide, NO_x, and hydrogen chloride are passed through absorption towers. Some liquid effluents are disposed of by licenced contractors and some solids on the Works registered landfill site.

There are major variations, from hour to hour, in quality and quantity of effluent, while in the long term significant changes in product and process range are evident. In the Intermediates 1 Plant there have been significant changes in product range since 1974, when 29 products were manufactured in five units. In 1984 there were 52 products in eight units, but by 1987 there was a rationalization to 33 products in five units. Significantly, only nine products from the 1974 range were still manufactured in 1987. The pattern of effluents has changed similarly, reflecting the production pattern.

These examples show how the flexibility and complexity of manufacture in multi-purpose fine chemicals plant leads to more complicated effluent problems. The possibilities of by-product recovery and recycle are much reduced since either cross contamination of recycle streams between different products would be risked or duplicated recovery systems would be required often at considerable and probably uneconomical costs. The range and variation of effluents is considerable and, because much of the production is batchwise, it is more difficult to design and monitor effluent treatment equipment compared with the steady-state operation of continuous plants.

3 Mixed Chemical Works

3.1 Huddersfield. In Huddersfield Works all three types of fine chemical facility exist. There is only a small number of continuous and semicontinuous plants as described above, the large majority of plants having batch processes.

The unit processes include sulphonations, nitrations, reductions, oxidations, and diazotizations, *etc.* Unit operations include drying, solids handling, absorption, evaporation, distillation, *etc.* and the number of process stages varies from one to ten but typically involves about six. There are about 270 recovery stages in order to improve the economics and minimize effluents. The changes in product range away from the predominance of dyestuffs and pigments to agrochemicals and speciality chemicals has resulted in more difficult effluents, arising from more complex chemistry.

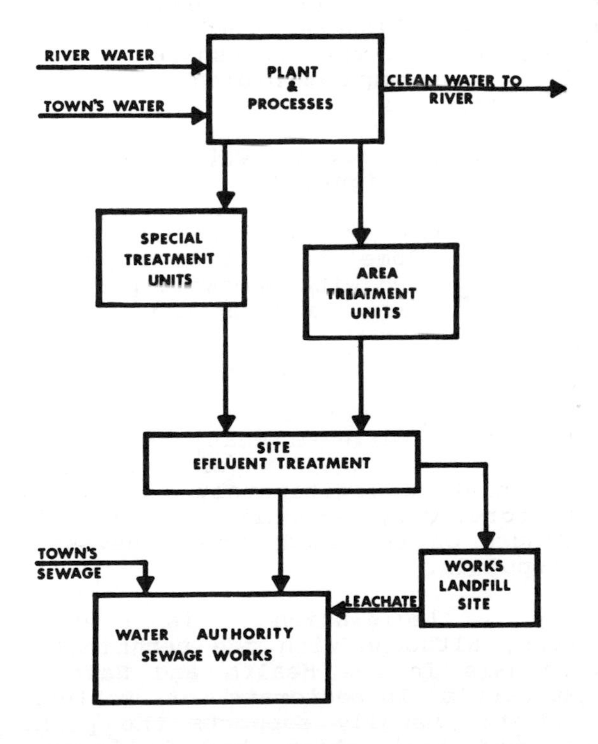

Figure 3 Water use and effluents

Such a works needs facilities for the development of product growth. Trial quantities made in laboratories have to be scaled-up through pilot plant to full scale. As the product becomes more successful an increase in scale may be necessary by using existing larger-scale plants, or by building a new plant.

This requirement encourages the building of multi-product units of various shapes and sizes, materials of construction, or specialist purposes.

3.2 Effluent Systems. The single-stream plants or those with specialist products mainly have dedicated effluent treatment systems as part of the design. The multi-purpose plant, being the major proportion of the manufacturing plants, needs the back-up of a major site treatment system. In Huddersfield's case the effluent plant adjusts the pH to approximately 9 and adds a flocculating polyelectrolyte prior to solids separation in three large settlement tanks. The settled sludge is removed and after a further thickening stage is filtered in large automatic plate and frame filter presses. The 'dry' solids are transported along with other solid waste from the works to a works operated registered landfill site near to the Works, very much a necessity for a major fine chemicals unit. The clarified overflow liquors are discharged under a consent agreement to the Yorkshire Water Authority sewage works for subsequent biological treatment along with domestic effluents.

A simple scheme showing water use and effluent treatment is shown in Figure 3.

Waste gases from the processes are absorbed in individual units. Some tars are burnt in on-site incinerators associated with specific processes. Other difficult wastes are incinerated by authorized contractors.

4 Legislation

Legislation usually reflects what the public wants to happen. While most good major companies wish to influence legislation legitimately to take account of realistic factors, they recognize that legislation is a healthy pressure on the good, and a necessary control on the unscrupulous.

Typically, legislation is drafted by administrators, although with the technical support of the professionals in the Health and Safety Executive and Her Majesty's Inspectorate of Pollution. The chemical industry usually supports the principles and motives of new legislation but believes that the legislation must work in practice; must take into account the technology which is reasonably available to most companies in the industry; and, where capital or revenue expenditure is involved, must ensure that the time of implementation is sufficient to sustain a viable industry.

Through the medium of consultative documents and discussions with the authorities, ICI seeks to ensure that its views are understood on the practicality and the potential costs of new legislation. Frequently, the experience and expertise of ICI make a positive contribution to the final form of legislation and in a number of guidance notes which interpret the legislation.

Since most legislation cannot be specific in relation to the widely varying conditions in the industry, the enforcing authorities often give their own interpretation in specific cases. It is sometimes important to ensure that these interpretations do not go significantly further than the understood intentions of the legislation, and continued professional dialogue with the enforcing authorities is an essential feature in the management of environment.

5 The Public

The general public often used to be seen as the unwashed masses who were grateful for jobs and products. The community unquestioningly saw the factory as good to be protected and supported. Enlightened management always kept moving forwards as its knowledge increased.

The skyline of the 1940s with its multiplicity of smokestacks typifies the end products of this - pea soupers and bronchitis. The public were slow to make the connections between industry and their health and environment and even then there was little expectation that anything could be done.

The Clean Air Act in the UK made a massive change in the public's perceptions: legislation could force change, industry survived and the environment was greatly improved. This trend in the public perception continued with increasing pace as prosperity allowed a realignment of their priorities. These priorities of clean air, clean rivers, and a pleasant land for their grandchildren were something for which they were prepared to pay - at least a little.

These trends have been reflected in the public's representatives, for some time in the powerful non-elected pressure groups and more recently in the elected politicians. Both have been strong influences on legislation and its interpretation by the enforcement authorities.

The discussion of the balance between the good which the chemical industry does, by manufacture of drugs, colours, crop protection chemicals, and other 'essential' products and by wealth creation, and the effects on the environment is a healthy process. Inevitably, there are groups at two extremes: people who wish to have the benefits of chemistry without the nuisance of the chemical industry, and manufacturers who still pollute in order to profit.

For the most part, the public wish to have the benefits of chemical production but not at any price. There is now an overwhelming interest in the price - the price of living with pollution and the price of putting it right. How can they make a judgement without appropriate information?

Several factors need to be considered:

The community should have a relationship of reasonable openness with its local factory and *vice-versa*. While lack of information can breed suspicion, the community can be proud of local industry. A recent open day at ICI Huddersfield Works attracted 18,000 people on a Sunday afternoon, which is some four times the number normally attending Huddersfield Town Football Club.

The community should be able to trust what the factory says. The factory cannot be secretive about incidents because of fear of adverse reaction. Above all, however, the community must see the factory improving its performance and making positive steps to avoid incidents.

The community should be able to understand what is a normal and reasonable level of effluent, particularly in the case of visible emissions.

The community must be assured that the extremes of plant failure are unlikely but anticipated and the risk of serious injury to the public is very small.

The local government authority and emergency services need to understand the factory's activities and potential problems since they are frequently the focal point for complaints. In addition, they need access to reliable help and information from the factory on a 24 hour basis.

All personnel in the factory need to be well briefed on normal activities and any incident, and motivated to help in the local community to defuse 'scaremongering' by less responsible members of the public.

6 Management of Multi-purpose Fine Chemical Sites

6.1 Policy. Before any management process can operate, a policy must be devised within which individual employees can make more detailed strategic or tactical decisions. The establishment of what is an acceptable policy is dependent on what the 'world' - neighbours, employees, legislation, public opinion, *etc.* - believes is 'right'. This 'morality' has to be weighed by the business incorporating its own perceptions and the risk/benefit balance into a company policy statement.

As an example, it is the policy of ICI to manage its activities so as to ensure that they are acceptable to the community and to reduce any adverse effects on the environment to a practicable minimum. However, any logical policy must recognize that all industrial activity has an impact on the environment, that the environment is able to absorb certain man-made effects, and that it is therefore a resource to be used as well as one to be conserved.

The use of nature as an effluent treatment mechanism is of course the basis of the argument between those who live in areas with limited absorption capability and who wish for the political solution of fixed emission standards, and those who have access to natural systems which have a much larger capability to absorb mans wastes who largely favour environmental quality standards.

In order to implement its policy, any company must aim to:

Co-operate fully with the relevant authorities in meeting its legal obligations.

Use resources with care, having particular regard for those which are scarce and non-renewable.

Participate expertly in the discussion of relevant environmental issues, with a view to ensuring that an appropriate balance is maintained between care for the environment and the benefit to society arising from the activities of the Company in each of the communities within which it operates.

Assess in advance the environmental effects of any significant or new development.

Review at regular intervals the environmental performance of existing facilities.

Have particular regard for the preservation of important habitats, avoiding adverse effects on rare or endangered species of fauna and flora.

Keep the communities in which the company operates informed of its environmental policy, and demonstrate that its conduct is both responsible and beneficial to society.

In a sense, setting a company policy is the easy part. A site-level policy has to translate the company policy into detailed procedures, taking account of the particular circumstances of the site. Within ICI's Organics Division, for instance, there are different environmental approaches for inland sites compared with coastal sites which are still compatible with the Company policy.

At Oissel Works, on the River Seine in France, a complete in house effluent plant is used to perform a total biological treatment and solids separation. At Huddersfield, another inland site only pre-treatment and solids separation are carried out in house with the biological treatment being carried out by the local authority. At Grangemouth, an estuarial site agreed liquid effluents are discharged under a monitored consent agreement with the Forth Rivers Purification Board by submerged pipeline into the tidal waters where biological treatment is carried out by the sea.

6.2 Site Organization. The policy requires a clear organization for environmental management. While accepting the line management responsibility for control of the environment, there is a need for an adequately resourced team of full time environment professionals to provide active support to the line managers.

In both the line managers and the environmental support, there is technical competence, professionalism, and a clear understanding of areas of responsibility.

6.3 Site Systems Policy. The site policy indicates the systems and procedures which must be established and maintained in order to achieve a satisfactory environmental performance. The important areas of design, modifications to plant and processes, operation of processes and equipment, and disposal of effluents must all be covered in the policy.

6.4 Site Hardware Policy. Proper operation of processes, plant, and equipment is vital in maintaining a satisfactory performance. In addition, however, it is important to have the right equipment and for it to be well maintained. Also, in the event of plant failure or breakdown, back-up plant is required. The site policy on hardware has to give a clear lead on the standards required.

6.5 Procedures and Management Systems.

6.5.1 The design of plants and processes. It is most important in the design of new processes and plants that environmental effects are allowed for at an early stage. It is at this stage that the process can be modified or the chemistry changed to achieve a more environmentally favourable process at a minimum cost. When the chemistry is optimized any outstanding environmental problems have to be resolved by plant design, perhaps with additional equipment such as scrubbers, cyclones, or liquid effluent treatment facilities. The disposal of wastes by landfill or incineration must be resolved at this stage.

It is recognized that from time to time there can be deviations from routine conditions because of plant failure or because of mistakes. These possibilities must be allowed for in the design, either to prevent them happening or to contain such incidents.

To ensure best possible results there needs to be an understood procedure for managing new projects through a project team of professionals. As part of this procedure the Division makes extensive use of Hazard and Operability Studies, a technique involving a team of experienced people examining the process and plant in a formal and structured way, to determine the safety, health, and environmental implications. For this technique to be successful an effective system for providing basic data is essential.

6.5.2 Design information. Naturally, process engineering information and engineering design standards and expertise are at the core of the design procedure. In addition, extensive information on treatment methods and on the toxic effects of effluents on flora and fauna must be obtained. In ICI's case these are provided by the Brixham Laboratory. The potential for fire and explosion and the effects of accidental emissions are also studied by the Hazards laboratories and by fire specialists. Throughout the design stage there is consultation with Her Majesty's Pollution Inspectorate (HMPI) and local authorities.

Assessment of risk to the environment from operation of plant and processes is an important part of the design process. It will be appreciated that it is more effective in technical and cost terms to identify problems at the design stage rather than when the plant is operating. Design procedures are formalised, therefore, and include several stages of hazard study. Initially the information just referred to is used to assess in broad terms the potential problems and hazards. This stage shows whether straightforward discharge of liquid effluent to the works effluent system and hence to the local authority sewage works is acceptable or whether preliminary chemical or physical treatment is necessary. Similarly the treatment of routine gaseous emissions can be assessed in terms of whether discharge to atmosphere is acceptable or whether scrubbing facilities need to be part of the design. Any problems or hazards associated with solid or tarry wastes are considered and disposal by landfill or incineration evaluated.

More detailed hazard and operability studies are carried out as the detailed flow sheets are prepared. These are examined by a formal technique using a team of managers, chemists, and engineers, when the hazard is evaluated line by line, plant item by plant item. The objective here is to look beyond the routine effluent streams to the possibility of plant or equipment failure or the likelihood of an abnormal incident involving toxic gas emissions, explosion or fire. As appropriate the risks are quantified in numerical terms as an aid to making technical and managerial judgements. This technique utilizes the mathematical analysis of fault trees and incorporates data on failure rate of equipment, computer dispersion studies *etc.*

6.5.3 Working procedures. Well designed processes and plant still need people to operate and manage them and human fallibility cannot be eliminated. It is necessary, therefore, to have comprehensive and clearly understood procedures for many activities in the chemical industry.

6.5.4 Start-up of new plants. The start-up period is particularly vulnerable in its potential for environmental problems and it is necessary, therefore, to make detailed preparations. Procedures are required for checking each plant item and associated pipework and equipment against the design intentions, for carrying out 'water' trials, for the introduction of chemicals, and for the establishment of production. Properly organized training is carried out, where possible with computer simulation and on full-size equipment.

A full record of the start-up is maintained for the benefit of plant management and for future designers.

6.5.5 Works general orders. Plant operation requires appropriately detailed operating instructions indicating the parts of the process which might give rise to a major hazard and environmental incident under abnormal conditions, and actions to take in the event of abnormality. Maintenance operations are accompanied by a rigorous 'permit to work system' to ensure that the plant is kept to design intentions. Finally, modifications to the process or plant are not undertaken without a system which ensures that the implications of the changes are fully understood and approved by professional people.

6.6 Training. The importance in preparation for start-up has been mentioned. A formal training programme is a necessary part of continued operation of the plant, not only for operators but for maintenance and support personnel and for managers.

Technical aspects are important and training naturally focuses on details of systems, plant and equipment processes, and so on. The importance of attitudes and awareness of today's environmental realities are also seen as vitally important in good performance. This is supported by training in recognizing causes of incidents, in controlling and containing them if they happen, and, in particular, the importance of rigorously reporting unusual circumstances or conditions on the plant so that incidents might be averted.

6.7 Monitoring and Auditing. A working chemical plant is a dynamic entity. Over a period of months or years personnel changes occur, sources of raw materials vary, alternative suppliers of equipment have to be used, the plant gets older, *etc.* To ensure that the design intentions and standards are preserved a system of monitoring the state of the process, plant, and procedures is necessary. It is desirable that some regular monitoring is incorporated into the various working procedures, and from time to time an audit of environmental measures on the plant is carried out. The Works liquid effluent, after treatment, is monitored continuously by flow proportional sampling and a 24 hour composite sample analysed daily. Checks on the samples are carried out separately by Brixham Laboratory. Gaseous effluents are monitored continuously in the case of the nitric acid plant, oleum plant and boiler plants, usually incorporating records and alarm systems. The gaseous effluents from batch plants are checked by intermittent sampling. An audit may be carried out in different ways, but typically a small team of people having expertise in chemistry, environmental sciences, engineering, and chemical engineering, will examine the plant, plant operations, and procedures, and recommend actions to senior management.

Auditing is an essential technique. While on a broad front it is only a snapshot check on the environmental 'health' of the management and the procedures and hardware on site, it can be followed up in considerable details if major issues are identified.

6.8 Small Companies. Clearly all of the policies, systems, techniques, and resources are defined against the backdrop of a major company with a heavily developed technical and managerial infrastructure and such systems and support would not necessarily be available as an integral part of a smaller company. All of the management systems and resources are however needed for safety and environmentally acceptable operation of multi-purpose fine chemical plants. No responsible company can embark on such a venture without them although they need not all be 'in-house' and indeed many competent contractors exist to sell such technical and professional support.

Some areas such as definition of policy, operating procedures, organisation, *etc.* are intrinsic to competent management and will of course be part of the 'in-house' capacity of the company, however, risk and hazard analysis, engineering design, toxicity testing, *etc.* can be obtained through various agencies.

The Warren Springs Laboratory, Brixham Laboratories, many reputable chemical engineering contractors and consulting engineers, the Systems Reliability Service of the UKAEA and of course discussions with H.M. Inspectors either of Factories or Pollution, can be used to supplement in house support and give the necessary complete organisation.

7 Conclusions

The relationship between chemical industry and the community is not a new problem but it has changed over the years and a new realism of a balance between risks and benefits is being established. Any chemical plant, but especially a multi-purpose fine chemicals unit, must take this into account as a prime factor in its operating plans.

There must be a clear and understood policy of operation in the works covering systems, procedures, hardware, effluent treatment, and the implications of development of new products.

For multi-purpose units, the multiplicity of new products and growth needs special attention since the norm is one of continual variation. Unit and plant systems need back up from central systems.

The public, legislators, enforcement authorities,

politicians, and pressure groups have a tremendous influence on the continuing successful operation of the chemical industry. To take account of this must be one of the first priorities of a Works Manager of a multi-purpose fine chemicals manufacturing site.

8 Acknowledgements

I would acknowledge the assistance of Dr. K. Chambers in the preparation of this chapter.

9 Reading Guide

1. Pollution: Causes, Effects and Control, Royal Society of Chemistry, London.

2. Industry and the environment in perspective: Royal Society of Chemistry, London.

3. A guide to Hazard and Operability Studies, Chemical Industries Association, London.

4. Safety Audits, Chemical Industries Association, London.

30

Trends in Risk Assessment from the Deterministic to the Probabilistic

F.R. Farmer

THE LONG WOOD, OFF LYONS LANE, APPLETON, WARRINGTON WA4 5ND, UK

The chapter considers how the growing recognition and acceptance of risk from activities having a potential to create major hazards has led to a move from a deterministic engineering approach, based on learning from accidents and derived codes of practice, to one requiring prediction of future events in a probabilistic world. This applies to nuclear and non-nuclear events. The recent growth of Probabilistic risk assessment led to a questioning of its application in relation to goals, of the relative weighting of risks of high or low probabilities in the use of the summation of frequency times consequence (f x C).

1 Introduction

Over the last twenty years or so there has been a special interest in achieving safety - or reducing risk of harm - particularly in situations where the penalty of failure could be large. From the 1st Report of the Advisory Committee on Major Hazards[1] 'It is increasingly necessary to seek to set design and operating procedures right first time. Because of their present day size and throughput, there are now many plants throughout the world where a critical first mistake can result in disaster'. (See also chapter 7.)

The early development of atomic energy was based on the assumption that good design, containment, and siting would ensure that no signficant harm would result from a serious reactor mulfunction although the potential to cause great harm had been well recognized; in 1957 WASH 750[2] estimated that 3000 might be killed as a result of an accident.

Safety was described in a deterministic way; reactors would be safe by siting, by provision of diverse and redundant safety systems, by the choice of inherently safe design, by single, double, or zero release containment. All help to ensure safety of the public against the maximum credible accident. For many years there was continued support for the concept of credible and incredible situations, reactors were safe or not safe, accidents could happen or not happen, some became incredible. Mostly, however, they became incredible because there was no design solution available, as for a major failure of a pressure vessel; pipes could fail but pressure vessels would not fail catastrophically by fast fracture if held at Nil Ductility Temperature (NDT) + 60°F - or so it was believed. This follows the conventional engineering approach, that designs would be subject to recommendations and codes of practice which convey and update relevant experience, with the implicit assumption that equipment designed to the latest codes would not fail.

In a similar way many industrial plants were considered safe by design, by operational procedures, and by virtue of a safety zone around them. In all cases the designer was required to take account of those things likely to arise by accident, wear and tear, or weather, where 'likely' was determined by codes, recommendations, and current practice.

In 1968 an enquiry into the collapse of a tall building in London led to the recommendation that 'they should be designed adequately for the maximum wind loading which they may experience'.[3]

This carries a mixture of the deterministic - they will be safe because account has been taken of all conditions including the most adverse which might occur - but is also probabilistic. The maximum, however, is not defined. Should it be 1 in 10, 1 in 100, or 1 in 1000 chance of occurrence per year?

There has always been a strong move towards codes of practice and recommendations. I was responsible for a series 'Recommendations on safety principles and practice' in the early 1960s. All too often they were deterministic and many were exhortations such as 'The possibility of failure of a circulator in the event of a single fault in any auxiliary equipment of the main circulator drive, should be reduced to a minimum', or - 'An adequate and reliable source of supply for the circulators is essential to ensure an adequate coolant flow under normal conditions'. A similar practice in the USA, 'The reactor coolant pressure boundary shall be designed, fabricated and erected and tested so as to have an extremely low probability of abnormal leakage, of rapidly propagating failure and of gross rupture'.

It is clear that at that time we were not ready openly to admit the possibility of failure, to accept a probability thereof, or to become quantitative; rather we offered exhortation and a checklist.

2 Reactor Design

A change was underway in the mid-1960s. In 1965 the US Advisory Committee on Reactor Safety[4] called attention to the possibility of a sudden large scale rupture of a water reactor pressure vessel and suggested several steps to reduce the probability and the consequence.

The growing recognition of possible failures of equipment led, on one hand, to the pursuit of an inherently safe design, as for example by the use of negative power or temperature coefficients. These can always be defeated; I have yet to see an inherently safe design of a plant having a large amount of energy and potentially dangerous material.

Another approach could be a design which could tolerate two or three coincident failures - or one having two or three protective barriers. This acceptance of possible imperfections in equipment encouraged the extensive use of redundancy and diversity in systems and components; some failures might occur but the plant would be safe. The type of failures would include those of motors, pumps, valves, instruments, *etc.*, usually in the range of 10^{-1} to 10^{-3} per year, *i.e.* those lying within current experience or record. However, the next step of how much redundancy or diversity - or how safe? - took time to evolve and engineers asked for guidance, followed best practice, and relied on case by case approach.

This was my experience, again in the mid-1960s, when asked to approve the safety of major projects in nuclear plant such as instrumentation, control systems, auxiliary power systems, *etc.* The penalty of failure could be high, even a disaster. Would the quality of design and of equipment be good enough? The fact that they may be the best available and in accordance with current codes of practice is not a satisfactory answer. We needed to know the probability of failure (or range thereof) - this was not known - and we needed to know the nature and size of the consequences and in this field a major effort had been applied for ten to twenty years.

Meanwhile, reactors were being approved in many countries, mainly on the grounds of conforming with codes or specific requirements then in force and on the grounds of being equal or better than some already approved. The move to a probabilistic approach was delayed for various reasons:

a) We do not have failure data, it would be better not to know;
b) We must rely on our judgement;
c) We cannot approve the construction of a plant which might kill people when approval itself would be seen to recognize this possibility.

This latter point became apparent to me when asked to approve projects. If the application were to be supported by a risk assessment giving failure rates and consequences, then approval of harmful consequences at estimated frequencies would become recognized and used as a basis in other submissions.

My approval in 1967 aimed to display consequences in terms of a quantity of ^{131}I released, with the intent to assess within each frequency decade whether the release was significantly greater or less than the objective presented. This was directed to the design engineer, as the requirement was independent of the site, of weather, or dose response, and within the ability of the design group to estimate.

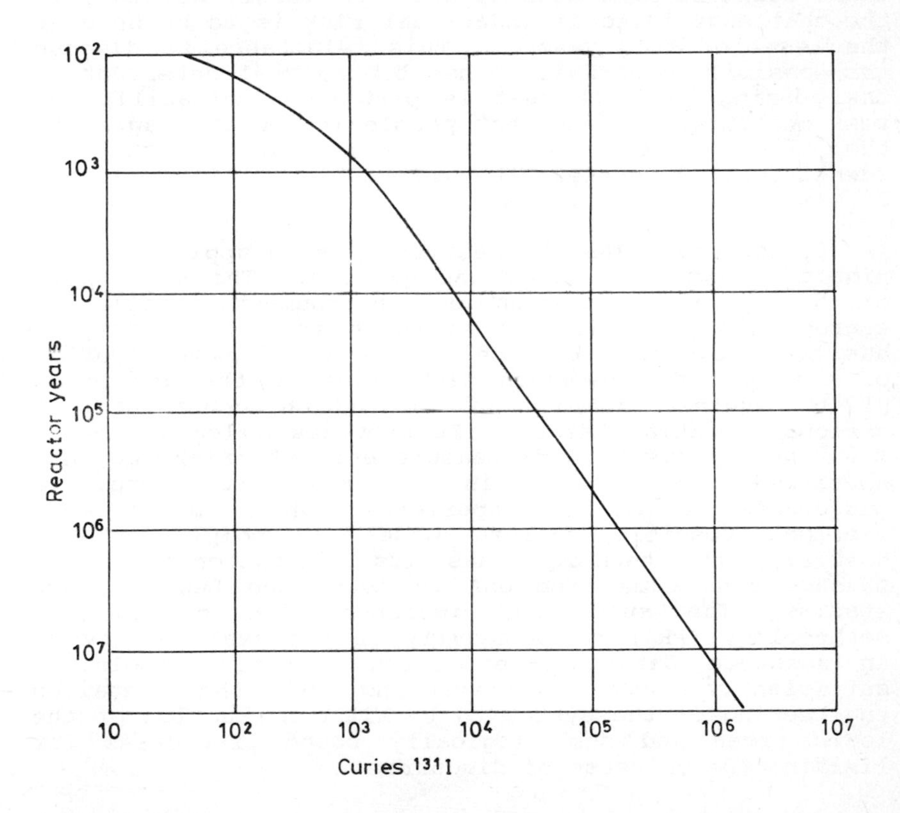

Figure 1 Proposed release criterion

In proposing this frequency/consequence (*fC*) criterion it was stated that in conforming the highest individual risk would not exceed 1×10^{-7} year^{-1} - and would additionally lead to improvements in design through the identification of failure modes or frequencies which exceed the criterion.

3 Probabilistic Risk Assessment

However, progress in probabilistic risk assessment (PRA) has not always been welcomed in the nuclear field nor has its introduction in the chemical field. Many have viewed quantified risk studies as a numbers game and have been reluctant to take analysis into the range of low probabilities; this suspicion challenges the use of prediction as against proven experience such as the move from the credible at 10^{-1} to 10^{-3} year^{-1} to the incredible of 10^{-5} to 10^{-6} year^{-1}.

After all, it is easier and more comforting to say that a plant is safe - at the 10^{-3} level - than to say there could be some harm at the 10^{-6} level; but that is the challenge faced if individual risk is to be held at the level of 10^{-7} year^{-1}. This reluctance to discuss the possibility of failure has been a real deterrent to the probing in depth that is part of a reliability of risk analysis. I find that people in industry agree to the existence of risk but wish to avoid its open identification for fear the information will be misused or misunderstood.

By the mid-1970s the need for PRA in application of highly complex technology was growing. The 6th Report of the Royal Commission on Environmental Pollution stated 'The probabilistic approach to safety analysis has been developed because of the growing realization of the need for rigorous discipline in the design of plant, whether nuclear or not which constitutes a serious potential hazard. It provides a logical basis for the analysis and measurement of risk against specified objectives, in contrast to intuitive judgements, based on experience, which might well overlook possible failure modes in complex plant. Naturally the technique has its limitations'. This message also came from USA in 1978, the Lewis Report stating 'The successful implementation of such a methodology (PRA) rests heavily on the availability of an adequate database, effective computing tools, a sufficiently deep understanding of the detailed engineering of the system to permit construction of the logic trees and some logically sound procedures for limiting the universe of discourse'.

By the mid-1970s, and in the well known Rasmussen Report WASH 1400,[7] *fC* lines had been extended in two ways - to show consequences as casualties of acute or delayed death, genetic effects, and property damage, and to extend the frequency to 10^{-9} $year^{-1}$. This opened the door for speculation as to acceptability. However, the extensive development of the *fC* presentation from PRA studies and the introduction of goals introduces a number of issues which are inter-related, such as:

The setting of goals
The slope of *fC* lines
The summation of the products $f \times C$
The extension of f from $10^{-6}/10^{-7}$ $year^{-1}$ to $10^{-9}/10^{-10}$ $year^{-1}$.

4 The Setting of Goals

There has been a reluctance to set goals, as is understandable. A starting point for both nuclear and non-nuclear has required the risk to an individual to be less than other risks to which he might be exposed. How much less will differ for various situations and as between nuclear and non-nuclear. Those discussed by Okrent 1981[8] as debated by the Advisory Committee on Reactor Safety (ACRS) include proposals for criteria for large fuel melt, early and delayed death, societal risk, for early and late effects, and cost penalties. Many of these, in some form, are being tested.

There are several fundamental issues in goal setting. Are they to be a firm requirement or a target to aim at? My *fC* line was a target to aim at in that some projects or equipments could be identified as clearly failing to meet the target and could be changed; there was insufficient data or experience to substantiate success in relation to low numbers, 10^{-6} $year^{-1}$ and lower.

The setting of a goal also has to operate in conjunction with ALARA (as low as reasonably achievable). An ALARA which exceeded the goal would not be accepted; hence ALARA should be less than or equal to the goal.

Does a goal relate to one large and identifiable project or is it determined by a number of such projects in the country - as nuclear reactors - or by the number of such projects in a single complex such as Canvey Island? This leads to consideration of societal risk, that an accident might kill 10, 100, or more, either as early or delayed casualties. Generally the highest individual risk will arise near any one installation whereas societal risk may be determined by any one of several installations and at a distance.

If the risk levels are around 10^{-4} to 10^{-5} $year^{-1}$ both criteria could be involved whereas an individual risk criterion at 10^{-7} year^{-1}, if met, could only involve consideration of societal risk of say 10 casualties by presenting a population of 10^9 people at 10^{-8} year^{-1}; as the risk at great distances to accommodate 10^9 people must be lower than the highest of 10^{-7} year^{-1}.

5 Frequency *(f)* and Time Consequences *(C)*

5.1 The Slope of *fC* lines. My first proposal had a negative slope of 1.5. I, like many others since, felt that there should be a heavy penalty for the larger consequences, since termed 'risk aversion'. However, this required a release of 10^6 Ci ^{131}I to be less than 3×10^{-8} per reactor year. I did not believe that this was capable of engineering assessment - or achievement, and was impracticable. I reverted to a slope of -1. Whatever is chosen should be meaningful and useful to design engineers. A new situation arises when we consider the summation of the product $f \times C$.

5.2 The Summation of $f \times C$. The sum of $f \times C$ has been suggested as giving a single figure to express a risk merit order, and to allow for risk aversion it has been suggested that 'to provide incentive to reduce the catastrophic potential of accidents, the risk value should be summation $f \times C^{\alpha}$, and a value of $\alpha = 1.2$ has been considered when applied to the consequence of early death.[8]

As a simple example, if we have two risk points of:
1 death at 10^{-4}
100 deaths at 10^{-6}

The summation would then be 2×10^{-4}, whereas for $\alpha = 1.2$ the sum would be 3.5×10^{-4}.

This weighting would lead to an increase in effort in design and research to reduce *f* or *C* for unlikely events such as earthquake, plane crash, *etc.* This would be costly and a diversion from the effort still required for more frequent events in pipework valves, instrumentation, *etc.* as for Three Mile Island (TMI).

I do not like any summation of $f \times C$ as it loses information and can be misused.

5.3 The Extension of *f*. *fC* diagrams were extended in the Rasmussen Report[7] up to 10^{-9} year^{-1} and in the German Light Water Report[9] to 10^{-10} year^{-1}.

This arises to respond to the pressure from those who claim that the consequences could be worse by a combination of wind, weather pattern, occupancy (football field), *etc.* which could extend an initiating event of 10^{-5} year^{-1} to a consequence at 10^{-9} year^{-1}.

This does not answer the question, what is the risk (consequence) arising from a project at a probability of 10^{-9} year^{-1}? We cannot identify and have not identified all initiators in the range 10^{-5}-10^{-9} year^{-1}, hence we cannot answer this question. I would not accept any *fC* diagram which extended beyond 10^{-6} to 10^{-7} year^{-1}.

Low numbers can arise for another reason: namely, the assumption of a linear dose/risk response with no threshold, as used in assessing a collective dose which may show 10^8 people at a risk of 10^{-8} or 10^9 people at 10^{-9} each giving one fatality. This use of equal weighting is not in accord with current social practice; in most life-saving activities more effort is deployed on those at high risk, more to reduce a risk of 10^{-3} to 1000 people than 10^{-6} to one million people, and is more likely to be successful.

The use of $\sum fC$ and the use of collective dose in determining risk gives equal weight at all frequency levels and if used in choice of alternatives in design, research, siting, *etc*. should be seen as giving equal weight. Is this what is required or intended?

6 Conclusions

Why explore low-frequency consequences below 10^{-7} year^{-1}? Is it to meet the probing of objectors? Dunster[10] has said 'It seems to be an inescapable conclusion that the concerns are generated by the source of the exposure and not the magnitude. If that is so, decreasing doses, whilst worth considering in its own right, fails to provide any real easing of concerns and fears'. (See also Chapter 7.)

I do not like the search for the low frequency, high consequence end of the spectrum. Speculative probing in that area will doubtless show that there are very small probabilities of very serious consequences for nuclear and non-nuclear events. As safety is largely judged on a comparative basis, as real or perceived, a search for a quantitative assessment at low frequency is not likely to be rewarding and can be challenged as misleading.

In conclusion, I wholeheartedly agree with the intent to set meaningful criteria, exposed to peer review, against which submissions may be judged, particularly if it encourages effort in risk assessment with the main purpose of identifying weaknesses and to modify them. I support the disciplined use of PRA to assist in design, design approval, and other safety related matters including communication between the various groups involved in decisions on major high risk plants.

7 References

1. Health and Safety Commission, First Report of the Advisory Committee on Major Hazards, HMSO, London, 1976.

2. US Atomic Energy Commission, 'Theoretical Possibilities and Consequence of Major Accidents in Large Nuclear Power Plants', WASH 750, 1957.

3. Report of the Inquiry of the Collapse of Flats at Ronan Point, Canning Town, HMSO, London, 1968.

4. Letter from ACRS to US Atomic Energy Commission, Nov. 15, 1965.

5. Royal Commission on Environmental Pollution, 6th Report 1976, para. 276.

6. H.W. Lewis et al., Risk Assessment Review Group Report to the USNRC, NUREG/CR 0400.

7. N.C. Rasmussen, 'An Assessment of Accidental Risks in US Commercial Power Plants', WASH 1400, 1975.

8. D. Okrent, 'Industrial Risks, the Assessment and Perception of Risk', Royal Society Discussion, The Royal Society, London, 1981.

9. Birkhofer, 'The German Risk Study for Nuclear Power Plants', IAEA Bulletin 22, 5/6, 1980, p.23-33.

10. J. Dunster, *The Independent,* 25 May, 1987.

Epilogue

1 Introduction

The management of risk is a matter of day-to-day decision-making for everyone, although managing risk does not necessarily mean reducing it. The hang-gliding enthusiast is choosing not to minimize personal risk, but is attempting to maximize the enjoyment and quality of life. That is to say nothing of the added risk he imposes on himself with the drink of alcohol he enjoyed as he lit his cigarette before he drove off in his car.

Thus risk management does not only mean reducing risks. To reduce a risk in the majority of cases means provision of additional resource which costs money. It sometimes costs too much money to reduce a risk further and both the resources and the finance would be far more desirably expended elsewhere. The use of most chemicals *e.g.* drugs, pesticides, colourants, all of which carry some risk, also offers a benefit. If one reduces the risk too far, we may find that the benefit has disappeared as well. Benefit, cost, and risk are always interlocked and none of these elements can be changed without affecting the others.

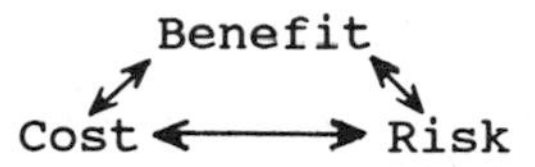

We are all exposed to chemicals; hence risk from chemicals, whether natural or man-made, is inevitable. The ways we assess these risks are important, not only to ourselves but to other animals and to the environment, to the soil, to the water, and to the air.

Public decision-making is an important part of any democracy and we all bear some responsibility for the effects of decision on risks. However, the impact of some of these decisions can be complex.

For example, if a pressure group in a developed country succeeds in having a pesticide banned worldwide it may be responsible for deaths from diseases such as malaria in less well developed and tropical countries. If nuclear power stations are not fully utilized to replace fossil fuel power stations, this may result in the exacerbation of the 'greenhouse' effect through increasing concentrations of carbon dioxide.

2 Highlights from the Preceding Chapters

In this book, an international view on these risks has been detailed in the preceding chapters.

These chapters detail, on a world-wide basis, a number of eminent scientists' views on risk assessment of chemicals in the environment. No attempt has been made to provide a chapter-by-chapter summary. However, some of the more important statements are highlighted below.

One is dealing with risks that are generally low and which can only be quantified by some assumed risk-dose relationship at very low doses. 'No observable effect level' does not guarantee that there are no effects. No one can provide an unequivocal answer to the question 'is one certain that the release of some particular chemical will harm no one or no species?' This is particularly true for man when one considers the extreme variability of response of the human system to any stimulus, and the possibilities of remote coincidence and synergisms, and weaknesses, which may render certain individuals vulnerable.

Proof of the absence of chronic effects, which may take a long time to materialize from sub-acute doses of chemicals known to have some toxic effects on animals at high concentrations, is a classical problem. Furthermore, absence of evidence is not evidence of absence. To do nothing is a decision which may have serious consequences; as positive action is necessary. Also, there are difficulties of distinguishing sharply between the effects of one xenobiotic against a general background of even lower concentrations of all xenobiotics and all natural products.

There is usually a media interest in such matters, largely due to television's insatiable appetite for visual action, in many cases for violence or disasters of all kinds and as a result it tends to distort and trivialize. This means that good and interesting material on environmental questions and on the approaches taken for safeguarding the public interest is of paramount importance. This is a task for the scientific societies and regulatory authorities, and is a never-ending task.

In itself, there is an inherent risk that scientists engaged in risk assessment in environmental matters will choose data and results which are sufficiently respectable to bolster one or other side in some political debate: in which case the outcome will often be decided on pragmatic grounds and not on risk analysis.

Such political awareness may mean one initially commences by burning cleaner coal, but it may end by burning books or witches. One should also be aware that concentration on the avoidance of small risks may blind us to the extent that we ignore the larger ones.

In Great Britain, as compared with the United States, there is much less emphasis on the quantitative assessment of chemical hazards. In other countries, for example, India, partly as a result of the Bhopal incident, the Indian Law Report has a recommendation that the import or use of any substance that is banned in the country of origin should also be banned in India. If India, or any other country, adopts this recommendation, it will be accepting certain risk management decisions of other countries and, by implication, the risk assessment on which such decisions are based.

Hence, risk assessors in all countries should perhaps devote more attention to communication than to harmonization, particularly clarifying the extent their assessment are driven by scientific considerations and to what extent by factors lying beyond the sciences.

Safety considerations related to human health and the environment may affect to a varying degree the production, availability, and international trade in many chemicals. Therefore, it is paramount that any conclusions relating to risk and safety are reached on the use of sound data and good science. The proliferation of groups undertaking toxicological evaluations and safety assessments of chemicals is today a matter of concern to both regulatory authorities and to industry since the end-points of these evaluations often diverge from one another. Furthermore, it is a poor use of a scarce resource.

There are many techniques for assessing risks associated with chemicals, from cancer research to more recent advances in immunotoxicology, epidemiology etc. But these processes themselves are complex, *e.g.* the 'one-hit' model for carcinogenicity has now been replaced by a multi-stage process in which chemicals could be implicated in a variety of different areas. *e.g.* as initiators or promoters. Furthermore, there is the question of inter- and indeed intra-species differences.

Our culture, today, produces and continues to produce a profusion of literature dealing with all concepts related to chemicals. These concepts outweigh the benefits of such unprecedented explosions of scientific and technical advances. This has in itself produced a proliferation of terminology which even the expert finds difficult to grasp, and indeed, experts from many nations will argue on the meaning of many terms. Secondly, the drive is the impelling requirement and motivation to enjoy all benefits, and at the same time without incurring undue risk. This zero-risk concept has the appearance of supporting a generation's unreasonable quest to live forever.

Today, the risk assessor needs to consider the effects of both a chemical and its metabolites as observed on a laboratory animal, fish or strains of a bacteria. This is of particular relevance when it has been clearly demonstrated that the laboratory species in question metabolizes a chemical in a totally different way to man.

These deductions are not directly relevant to the establishment of acceptable or tolerable levels in the environment (water, food, air, or soil). The transcription of levels acceptable by an animal into levels accepted in the environment is one of the most difficult tasks to be understood by scientists. This also requires an excellent interface between science, the law, and the politicians.

As a result the current methods used by toxicologists for risk assessment are best described as pragmatic and none without their flaws.

The use of several assumptions or rules, each of which over-estimates the risk by an unknown factor, can lead to an unreliable estimate of risk. Society cannot tolerate a risk assessment, which is a scientific exercise, being over or under estimated.

Today, there is an increase in the use of mathematical models in the assessment of carcinogenic risks; these models themselves can incur the risk of overshadowing the vital consideration of biological factors in risk assessment. The biological response of experimental animals in terms of number, site, frequency, and latency of cancer formation can be just as important as the numerical assessment in providing a judgement of acceptable risk.

For non-genotoxic action, when this mechanism is understood, doses below those which trigger the mechanisms are unlikely to produce a carcinogenic effect.

One question which remains is whether chemicals producing cancer in rodents, but which are not genotoxic as shown by short-term tests, pose the same risk to humans as genotoxic carcinogens do under similar circumstances. A second question is 'are chemicals which by *in vitro* tests are apparently genotoxic but do not cause cancers in rodents, artefacts?' These uncertainties must be resolved. Whilst the cell mutation theory remains viable, chemicals that show mutagenic effects, particularly in *in vivo* assays, must be considered to pose a human health risk.

Exposure is always related to quantitative toxicological factors and is an essential part of risk assessment. Figure 1 in fact reflects the immortal words of Paracelcus 'All substances are poisons, the right dose differentiates a poison from a remedy'.

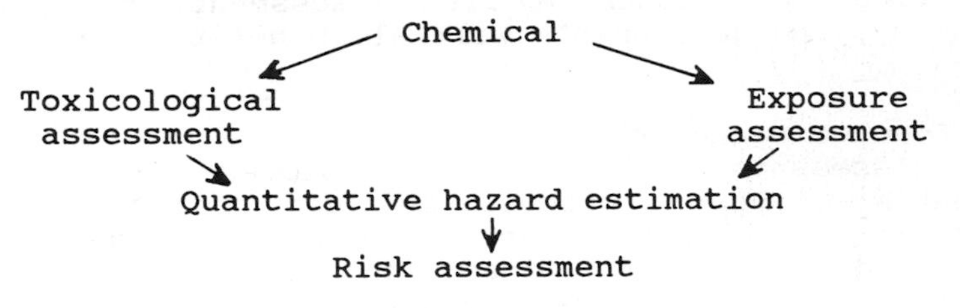

Figure 1

Risk assessment is an interactive process involving firstly, hazard identification, leading in turn to dose-response assessment and exposure assessment themselves resulting in risk characterization. These terms are contained in the Glossary of Terms at the end of this book.

People often over-estimate the frequency and seriousness of dramatic, sensational, dreaded, well-publicized causes of death, and under estimate the risk from more familiar accepted causes.

Risk managers face the necessity of making difficult decisions involving an uncertain science, thus involving the potential of grave consequences to health or to the environment, leading in turn to large economic effects on industry and consumers. Risk assessment thus provides an orderly, explicit, and consistent way to deal with scientific issues in evaluating whether a hazard exists and what is the magnitude of that hazard. Such evaluations involve significant uncertainties as the scientific data are sparse and, currently, mechanisms for adverse health impacts or environmental damage are imperfectly understood. An example relating to public perception of risk includes: hazard 'A' kills 50 people per year in a country, hazard 'B' has a one in ten probability of killing 5000 people in a neighbourhood in the next ten years. Risk assessment informs us that both have the same expected annual mortality, namely fifty. 'Outrage assessment' tells us that 'A' may be acceptable whereas 'B' is not.

These 'outrage factors' are not a distortion of the public's perception of risk, they are an intrinsic part of what risk means. They explain why people worry more about the local waste tip than geological radon, more about industrial emissions of some chemical with a very long name than say aflatoxin in peanuts.

How can scientific risk assessment be improved so that the best use of rapid scientific advances can be made? When scientific advances are used in such assessments, research is stimulated. As we succeed in improving the process of risk assessment, the public, industry, and governments will all benefit.

3 Concluding Remarks

Risk assessment is a complex process - itself a mechanism - where toxicological and ecotoxicological, chemical, biological, and exposure information is collected, refined, defined, and analysed, and note taken of all the assumptions that are necessary together with the uncertainties associated with the measurements.

Decision-makers then use such information to arrive at the appropriate strategies to address the risk associated with a chemical in an environment, whether that be the workplace, the air we breath, the water we drink, the food we, or other species, consume.

Bearing in mind the very large number of chemicals to which we have the potential to be exposed, full safety testing on all of these cannot be considered. The costs would outweigh the benefits. Furthermore, chemicals, particularly in water, are present in admixture and hence the testing of chemicals in combination would also need to be considered, an even more daunting task.

Individually, consideration can be given to the risks presented by chemicals by considering structural criteria from which conclusions may be drawn by structure-activity relationships.

There are two basic approaches:

i Formal structural analogies with chemicals known to exhibit certain risks such as carcinogenicity or immunotoxic responses, and *ii* Consideration of molecular size, shape, symmetry, reactivity, electron distribution, steric factors, and their possibility of ongoing metabolism particularly to reactive intermediates.

Such predictive techniques cannot replace the knowledge of the pharmacological or toxicological capabilities irrespective of chemical structure.

Compounds may be considered to be suspect if they induce or cause one or more of the effects indicated in Table 1.

Table 1: *Some agents or factors indicating possible adverse effects*

Alkylating agents
Aneuploidy
Antineoplastic agents
Chronic hormonal imbalance (overstimulation)
Chromosomal aberrations
Covalent binding to DNA or RNA
Degranulation of the endoplasmic reticulum
Enhanced generation of reactive oxygen species
Hepatonecrotic agents
Hydrogen bond reactors
Immune-suppressive agents
Induction and inhibition of microsomal mixed-function oxidases
Inhibitors of intercellular communication and cell membrane function
Inhibitors of mitochondrial respiration
In vitro cell transformation
Mutations
Peroxisome proliferation
Polyhalogenated compounds
Sister chromatid exchange
Spermatogenesis inhibition
Spindle poisons
Stimulators of tissue hyperplasia
Strong chelating agents
Strong surface active agents
Structural analogues of DNA bases
Teratogenic compounds
Uncouplers of oxidative phosphorylation
Unscheduled DNA synthesis

A risk assessment on a chemical can then be further considered using data generated from animal studies. But there are many problems associated with inter- and intra-species differences. For example, human susceptibility to biochemical factors affecting carcinogenicity often involves the ability to bioactivate and deactivate compounds. This varies widely in the population.

Notwithstanding the significant uncertainty in the process of predicting risk to humans, to fish, to birds, or even bacteria, or any part of the environment, toxicological research has evaluated relevance of animal model data to these areas. Obviously the most important is the application to human responses.

Much use needs to be made of the perception of risk, the use of physiologically based pharmacokinetic models, structure-activity relationships, prediction or measurement of exposure, and other factors. In this way we can start to reduce the problems engendered by human heterogeneity when data from highly homogeneous animal models are extrapolated. Major efforts must be made towards trying to understand the causes of inter- and intra-species variations so that risk assessments can be based on the enhanced understanding of the chemistry, the biology, and the fundamental mechanisms involved.

It is far from easy to undertake environmental exposure assessment and as shown in some of the chapters of this book the methodology is quite complicated.

For chemicals in general there is no standardized, authorized evaluation procedure or practice. Therefore there is every reason for concern and it should not be a complete surprise that recent industrial accidents, poisonings on a massive scale like Bhopal, spills *etc.* have markedly highlighted the awareness of these hazardous situations.

International liaison and co-operation, as is exemplified by the international support given in this book and the Conference on which it was based, has therefore become an absolute necessity. Governments should co-operate more closely with one another and give greater support to professional and learned societies and the United Nations International Programmes. Furthermore, industry should become much more involved at all levels. Indeed everyone should have a genuine interest in chemical safety; confrontation is in no-one's interest - chemical safety must be our common goal.

It is hoped that this book, with its thirty chapters written by international eminently known experts, will help all members of society to understand the realities of scientific concepts and procedures for risk assesssment, with the inherent assumption and varying degrees of uncertainties.

It must be realized within society that risks can rarely be eliminated. Risks can be minimized, controlled, and managed within defined and acceptable areas. But total risk can never be reduced to zero.

The ultimate reasons for undertaking risk assessments are to minimize both controllable and unnecessary risks, to make responsible decisions, and to take feasible and beneficial courses of corrective action.

Currently this is the best and most appropriate manner to be able to make rational decisions about the environment, the chemicals that are vital to our well-being, our lives, and our health.

The informed person will be able to understand the capabilities and limitations of science, scientists and decision-makers and will ask questions. It is via these processes of questions that all of us obtain better understanding of our world, our human way of life, and our responsibilities to each one of us in the world which depends on chemicals and our environment.

MERVYN L RICHARDSON
EDITOR

Glossary of Terms

Additional terms can be found in 'Toxic Hazard Assessment of Chemicals', ed. M.L. Richardson, The Royal Society of Chemistry, London 1986, ISBN 0-85186-897-5.

Abiotic:-Unconnected with living organisms.

Acceptable daily intake *(ADI)*:-The daily intake of a chemical that is considered without appreciable risk on the basis of all the facts known at the time it is defined.

Acute toxicity:-The adverse effect occurring within a short time of administration of a single dose of a substance or multiple doses given within 24 hours. *cf.* **Chronic toxicity** and **Subchronic toxicity.**

Additive effect:-An effect which is the result of two chemicals acting together and which is the simple sum of the effects of the chemicals acting independently. *cf.* **Antagonistic effect** and **Synergistic effect.**

Adenoma:-A tumour, usually benign, occurring in glandular tissue.

Adenocarcinoma:-A malignant tumour originating in glandular tissue.

Adverse effect:-An undesirable or harmful effect to an organism, indicated by some result such as mortality, altered food consumption, altered body and organ weights, altered enzyme levels or visible pathological change.

Adverse event:-An occurrence that produces **harm** *(q.v.)*.

Aerosol:-A compound dispersed as minute droplets or particles which allow entry to the body via the respiratory tract and widespread contamination of clothing, skin and eyes.

Aetiology:-The science of the investigation of the cause or origin of disease.

Agonist:-A chemical with a positive action in the body.

Alkylating agent:-A substance that introduces an alkyl radical into a compound in place of a hydrogen atom.

Aneuploidy:-Deviation from the normal number of chromosomes, excluding exact multiples of the normal haploid *(q.v.)* complement.

Antagonist:-A chemical that opposes an agonist.

Antagonistic effect:-The effect of a chemical in counteracting the effect of another; for example, the situation where exposure to two chemicals together has less effect than the simple sum of their independent effects; such chemicals are said to show antagonism.

Antibody:-A protein that specifically recognizes and binds to an antigen.

Antigen:-A substance that elicits a specific immune response when introduced into the tissues of an animal.

Base pairing:-The linking of the complementary pair of polynucleotide chains of nucleic acids by means of hydrogen bonds between the opposite purine and pyrimidine pairs.

Benefit:-A gain to a population. Expected benefit incorporates an estimate of the probability of achieving a gain.

Bioaccumulation:-See **Bioconcentration.**

Biochemical mechanism:-A chemical reaction or series of reactions, usually enzyme catalysed, which produces a given physiological effect in a living organism.

Biochemical Oxygen Demand (BOD):-The amount of dissolved oxygen consumed by microbiological action when a sample is incubated, usually for 5 days at 20°C and in the presence of a nitrification inhibitor, usually allyl thiourea.

Bioconcentration:-The uptake and retention of xenobiotics by organisms from their immediate environment.

Biological half-life ($t_{\frac{1}{2}}$):-The time taken for the concentration of a xenobiotic in a body fluid or tissue to fall by half by a first-order process.

Biological monitoring:-Analysis of the amounts of potentially toxic substances or their metabolites present in body tissues and fluids as a means of assessing exposure to these substances and aiding timely action to prevent adverse effects. The term is also used to mean assessment of the biological status of populations and communities of organisms at risk in order to protect them and to have an early warning of possible hazards to human health.

Biomagnification:-Bioconcentration of xenobiotics up a food chain, *e.g.* from prey to predator.

Biotransformation:-The enzyme-mediated transformation of xenobiotics via Phase 1 (*q.v.*) and Phase 2 (*q.v.*) reactions.

Cancer:-The disease which results from the development of a malignant tumour and its spread into surrounding tissues. See **Tumour**.

Carcinogenesis:-The production of cancer (see above). Any chemical which can cause cancer is said to be carcinogenic.

Carcinoma:-A malignant epithelial tumour (*q.v.*).

Ceiling value:-The airborne concentration of a potentially toxic substance which should never be exceeded in the breathing zone.

Chelation:-The trapping of a multivalent ionic species by ionic bonding to a larger water soluble molecule so as to render the ion inactive in the biological matrix and to aid excretion.

Chemical Oxygen Demand (COD):-The amount of oxygen consumed from a specified oxidizing agent in the chemical oxidation of the matter present in a sample.

Cholinesterase and pseudocholinesterase inhibitor:-A substance which inhibits the enzyme cholinesterase and thus prevents transmission of nerve impulses from one nerve cell to another or to a muscle.

Chromosomal aberration:-An abnormality of chromosome number or structure.

Chromosome:-The heredity-bearing gene carrier situated within the cell nucleus and composed of *DNA* and protein.

Chronic toxicity:-The effect of a chemical (or test substance) in a mammalian species (usually rodent) following prolonged and repeated exposure for the major part of the lifetime of the species used for the test. Chronic exposure studies over two years are often used to assess the carcinogenic potential of chemicals. *cf.* **Subchronic toxicity** and **Acute toxicity**.

Clastogens:-Agents which cause chromosome breakage.

Clone:-A large number of cells or molecules genetically identical with a single ancestral cell or molecule.

Cohort:-A group of individuals, identified by a common characteristic, who are studied over a period of time.

Competent authority (in the context of the Sixth Amendment Directive, see below):-An official government organization or group receiving and evaluating notifications of new chemicals. Such notifications are made under the provisions of national legislation implementing the European Communities Directive 79/831/EEC (The Sixth Amendment of Directive 67/548/EEC which relates to the Classification, Packaging and Labelling of Dangerous Substances).

Conjugate:-A water soluble derivative of a chemical formed by its combination with glucuronic acid, gluthathione, sulphate, acetate, glycine *etc*.

Control limit:-The limiting airborne concentration of potentially toxic substances which are judged to be 'reasonably practicable' for the whole spectrum of work activities and which must not normally be exceeded.

Cost-benefit analysis:-This is a procedure for determining whether the expected benefits from a proposed action outweigh the expected costs.

Covalent binding:-The irreversible interaction of xenobiotics or their metabolites with macromolecules such as lipids, proteins, nucleic acids.

Cytochrome P-450:-A haemprotein involved, *e.g.* in the liver, with Phase I reactions *(q.v.)* of xenobiotics.

Cytoplasm:-The ground substance of the cell in which are situated the nucleus, endoplasmic reticulum *(q.v.)*, mitochondria and other organelles *(q.v.)*.

Cytotoxic:-Causing disturbance to cellular structure or function often leading to cell death.

Damage:-A loss of inherent quality suffered by an entity. See **Harm**.

***De minimis* risk:**-A risk that is too small to be of societal concern; risks below 10^{-5} or 10^{-6} are generally viewed as *de minimis* in the United States.

Detergent:-A cleaning or wetting compound which possesses both polar and non-polar terminals or surfaces allowing interaction with non-polar molecules which renders them miscible with a polar solvent.

Detoxification:-
(1) A process, or processes, of metabolism which renders a toxic molecule less toxic by removal, alteration or masking of active functional groups.
(2) To treat patients suffering from poisoning in such a way as to reduce the probability and/or severity of harmful effects.

Detriment:-A measure of the expected harm or loss associated with an adverse event, usually in a manner chosen to facilitate meaningful addition over different events.

Disaster:-
(1) An act of nature or an act of man which is or threatens to be of sufficient severity and magnitude to warrant emergency assistance.
(2) A disruption of the human ecology which the affected community cannot absorb with its own resources.

Distribution:-Dispersal of a xenobiotic and its derivatives throughout an organism or environmental matrix, including tissue binding and localization.

Dose-effect curves:-Demonstrate the relationship between dose and the magnitude of a graded effect, either in an individual or in a population. Such curves may have a variety of forms. Within a given dose range they may be linear but more often they are not.

Dose-response assessment:-The process of characterizing the relationship between the dose of an agent administered or received and the incidence of an adverse health effect in exposed populations.

Dose-response curves:-Demonstrate the relation between dose and the proportion of individuals responding with a quantal effect *(q.v.)*. In general, dose-response curves are S-shaped (increasing), and they have upper and lower asymptotes usually but not always 100 and 0%.

Dose-response relationship:-The systematic relationship between the dose (or effective concentration) of a drug or xenobiotic and the magnitude (or intensity) of the response it elicits.

Dyestuffs:-Water-soluble or water-dispersible organic colorants.

Ecotoxicology:-The study of toxic effects of chemical and physical agents in living organisms, especially on populations and communities within defined ecosystems; it includes transfer pathways of these agents and their interaction with the environment.

Elimination:-The combination of the process of metabolism and excretion which result in the removal of a compound from the organism.

Embryo:-see Fetus.

Emission standard:-A quantitative limit on the emission or discharge of a potentially toxic substance from a particular source. The simplest system is uniform emission standard where the same limit is placed on all emissions of a particular contaminant. See **Limit value.**

Endogenous:-Arising within or derived from the body.

Endoplasmic reticulum:-A complex pattern of membranes that permeates the cytoplasmic matrix of cells.

Environmental quality objective *(EQO)*:-The quality to be aimed for in a particular aspect of the environment, for example, 'the quality of water in a river such that coarse fish can maintain healthy populations'. Unlike an environmental quality standard *(q.v.)*, the *EQO* is not usually expressed in quantitative terms.

Environmental quality standard *(EQS)*:-The concentration of a potentially toxic substance which can be allowed in an environmental component, usually air (air quality standard, or water, over a defined period. Synonym: ambient standard. See **Limit value.**

Enzymic (or enzymatic) process:-A chemical reaction or series of reactions catalysed by an enzyme or enzymes. An enzyme is a protein which acts as a highly selective catalyst permitting reactions to take place rapidly in living cells under physiological conditions.

Epidemiology:-The statistical study of categories of persons and the patterns of diseases from which they suffer in order to determine the events or circumstances causing these diseases.

Epigenetic changes:-Changes in an organism brought about by alterations in the action of genes. Epigenetic transformation refers to those processes which cause normal cells to become tumour cells without any mutations having occurred. See **Mutation, Transformation** and **Tumour.**

European Inventory of Existing Chemical Substances *(EINECS)*:-This is a list of all chemicals either alone or as components in preparations supplied to a person in a Community Member State at any time between 1st January 1971 and 18th September 1981.

Eutrophication:-A complex series of inter-related changes in the chemical and biological status of a water body most often manifested by a depletion of the oxygen content caused by decay of organic matter resulting from a high level of primary productivity and typically caused by enhanced nutrient input.

Excretion:-The process of removal of a compound or its metabolites from the body, normally via the bile or urine, but also via the lungs for volatile substances and by either minor routes such as skin, saliva and intestinal mucosa.

Exposure assessment:-The process of measuring or estimating intensity, frequency, and duration of human exposures to an agent currently present in the environment or of estimating hypothetical exposures that might arise from the release of new chemicals into the environment.

Fetus:-The young of mammals when fully developed in the womb. In human beings, this stage is reached after about 3 months of pregnancy. Prior to this, the developing mammal is at the embryo stage.

First Order process:-A chemical process where the rate of reaction is directly proportional to the amount of chemical present.

First-pass effect:-Biotransformation of a xenobiotic before it reaches the systemic circulation. The biotransformation of an intestinally absorbed xenobiotic by the liver is referred to as a hepatic first-pass effect.

Foci:-A small group of cells occurring *e.g.* in the liver, distinguishable, in appearance or histochemically, from the surrounding tissue. They are indicative of an early stage of a lesion which may lead to the formation of neoplastic nodules or hepatocellular carcinomas.

Fractile:-Let α be between 0 and 1. In a set of observations of a variable the α fractile is the number for a fraction α of the observations is less than this number. The fraction is often given in percent; the term percentile may then be used.

Frame-shift mutation:-A change in the structure of DNA such that the transcription of genetic information into RNA is completely altered because the start point for reading has been altered: *i.e.* the reading frame has been altered.

Fugacity:-
(1) The tendency for a substance to transfer from one environmental medium to another.
(2) Analogous to chemical potential as it pertains to the tendency of a chemical to escape from a phase (*e.g.* from water).

Futile cycling:-The generation of superoxide or peroxide anions through the cyclic activation of molecular oxygen by the cytochromes P-450 when these combine with a difficultly-oxygenated substrate (*e.g.* phenobarbitone) and are reduced by cytochrome P-450 reductase and NADPH. Mixed-function oxidation by the cytochromes P-450 generally reduces molecular oxygen to give the oxygenated chemical substrate plus water; in futile cycling molecular oxygen is reduced to superoxide or peroxide and the chemical substrate is left unchanged.

Genome:-All the genes carried by a cell.

Genetic toxicology:-The study of chemicals which can produce harmful heritable changes in the genetic information carried by living organisms in the form of deoxyribonucleic acid (DNA).

Genotoxic:-Able to cause harmful heritable changes in DNA.

Haploid:-The condition in which the cell contains one set of chromosomes.

Harm:-
(1) A loss to a species or individual consequent on damage.
(2) A function of the concentration to which the organism is exposed and of the time of exposure.

Hazard:-A qualitative term expressing the potential that a chemical can harm health under the conditions of exposure.

Hazard identification:-The identification of the chemical of concern, its adverse effects, target populations and conditions of exposure.

Hazard prediction/identification:-The process of recognizing a potential risk, involving both toxicity assessment and exposure assessment. Toxicity assessment determines the nature and extent of adverse health effects that the chemical in question can exert related to dose. Exposure assessment characterizes exposed populations and where possible estimates exposure levels or doses incurred. The major objective of this process is the integration of toxicity dose responses and exposure data to provide dose-response relationships.

Hepatotoxic:-harmful to the liver.

Histology:-The study of the anatomy of tissues and their cellular structure.

Histopathology:-The study of microscopic changes in tissues.

Homeostasis:-The tendency in an organism towards maintenance of physiological and psychological stability.

Hypocholesterolaemia:-A lowering of the cholesterol content of the blood.

Immediately dangerous to life or health concentration *(IDLHC)*:-The maximum exposure concentration from which one could escape within thirty minutes without any escape impairing symptoms or any irreversible health effects. This value should be referred to in respirator selection.

Immune response:-The development of specifically altered reactivity following exposure to an antigen. This may take several forms, *e.g.* antibody production, cell-mediated immunity, immunological tolerance.

Immunity:-Refers to the ability of a prophage to prevent another phage of the same type from infecting a cell or to the ability of a plasmid to prevent another of the same type from becoming established in a cell. Different mechanisms are involved in the two types of immunity. Also, transposition immunity refers to the ability of certain transposons to prevent others of the same type from transposing to the same DNA molecule.

Immunochemistry:-The study and use of antibodies as reagents and therapeutic substances, by virtue of their specific reaction with antigens.

Immunopotentiation:-An increase in the functional capacity of the immune response.

Immunosuppression:-Inhibition of the normal response of the immune system to an antigen.

Immunosurveillance:-The mechanisms by which the immune system is able to recognize and destroy malignant cells before the formation of an overt tumour.

Immunotoxic:-Harmful to the immune system.

Initiator:-An agent which starts the process of tumour formation, usually by action on the genetic material.

In Vitro:-Biological processes occurring (experimentally) in isolation from the whole organism.

In Vivo:-Within the living organism.

LC_n:-The concentration of a toxicant *(q.v.)* lethal to n% of a test population.

LD_n:-The dose of a toxicant (*q.v.*) lethal to n% of a test population.

Lesion:-A pathological disturbance such as an injury, an infection or a tumour.

Limit value *(LV)*:-The limit at or below which Member States of the European Community must set their environmental quality standards and emission standards. These limits are set by Community Directives.

Liver nodule:-A small node, or aggregation of cells within the liver.

Macrophages:-A large phagocytic cell found in connective tissues, especially in areas of inflammation.

Macroscopic (gross) pathology:-The study of tissue changes which are visible to the naked eye.

Malignancy:-A cancerous growth. (A mass of cells showing both uncontrolled growth and the tendency to invade and destroy surrounding tissues).

Malignant:-See Tumour.

Maximum allowable concentration *(MAC)*:-Exposure concentration not to be exceeded under any circumstances.

Median effective concentration *(EC_{50})*:-The concentration of toxicant or intensity of other stimulus which produces some selected response in one half of a test population.

Median effective dose *(ED_{50})*:-The statistically derived single dose of a substance that can be expected to cause a defined nonlethal effect in 50% of a given population of organisms under a defined set of experimental conditions.

Median lethal concentration *(LC_{50})*:-The concentration of a toxicant lethal to one half of a test population.

Median lethal dose *(LD_{50})*:-The statistically derived single dose of a chemical that can be expected to cause death in 50% of a given population of organisms under a defined set of experimental conditions. This figure has often been used to classify and compare toxicity among chemicals but its value for this purpose is doubtful. One commonly used classification of this kind is as follows:-

Category	LD_{50} Orally to Rat mg kg^{-1} body weight
Very toxic	<25
Toxic	>25 to 200
Harmful	>200 to 2000

Metabolic activation:-The biotransformation *(q.v.)* of relatively inert chemicals to biologically reactive metabolites.

Mixed function oxidases:-Oxidizing enzymes which are involved in the metabolism of many foreign compounds giving products of different toxicity from the parent compound.

Multigeneration study:-A toxicity test in which at least three generations of the test organism are exposed to the chemical being assessed. Exposure is usually continuous.

Mutagenesis:-The production of mutations. Any chemical which causes mutations is said to be mutagenic. Some mutagenic chemicals are also carcinogenic. See **Carcinogenesis** and **Transformation.**

Mutation:-Any relatively stable heritable change in the genetic material.

Necrosis:-Cell death (used particularly for death of cells in a focal point in a multicellular organism) due to anoxia or local toxic or microbiological action.

Neoplasm:-Any new and morbid formation of tissue, *e.g.* a malignancy.

Nephrotoxic:-Harmful to the kidney.

Non-target organisms:-Those organisms which are not the intended specific targets of a particular use of a pesticide.

No observed effect level ***(NOEL)***:-The maximum dose or ambient concentration which an organism can tolerate over a specific period of time without showing any adverse effect and above which adverse effects are detectable.

Occupational hygiene:-The applied science concerned with the recognition, evaluation and control of chemicals, physical and biological factors arising in or from the workplace which may affect the health or well-being of those at work or in the community.

Oncogene:-A retroviral gene that causes transformation of the mammalian infected cell. Oncogenes are slightly changed equivalents of normal cellular genes called proto-oncogenes. The viral version is designated by the prefix v, the cellular version by the prefix c.

Organelle:-A structure with a specialized function which forms part of a cell.

Partition coefficient:-A constant ratio that occurs when a heterogeneous system of two phases is in equilibrium; the ratio of concentrations (or strictly activities) of the same molecular species in the two phases is constant at constant temperature.

Perceived risk:-See **Risk perception.**

Permissible exposure limit *(PEL)*:-See **Threshold limit value.**

Peroxisome:-A cytoplasmic organelle *(q.v.)* present in animal and plant cells, and characterized by its content of catalase and other (peroxidase) oxidative enzymes.

Pesticides:-Defined under the UK Food & Environment Protection Act as: 'Any substance or preparation prepared or used for any of the following purposes.

a) Destroying organisms harmful to plants or to wood or other plant products,
b) Destroying undesired plants,
c) Destroying harmful creatures (that is any living organism other than a human being or a plant)',
and under the UK Control of Pesticide Regulations as:
a) 'Protecting plants or wood or other plant products from harmful organisms,
b) Regulating the growth of plants,
c) Giving protection against harmful creatures,
d) Rendering such creatures harmless,
e) Controlling organisms with harmful or unwanted effects on water systems, buildings or other structures, or on manufactured products,
f) Protecting animals against ectoparasites, as if it were a pesticide.'

Pharmacodynamics:-The study of the way in which xenobiotics exert their effects on living organisms. Also sometimes known as toxicodynamics.

Pharmacokinetics:-The study of the movement of xenobiotics within an organism. Such a study must consider absorption, distribution, biotransformation, storage and excretion. Also sometimes known as toxicokinetics.

Phase 1 reactions:-Enzymic modification of a xenobiotic by oxidation, reduction, hydrolysis, hydration, dehydrochlorination or other reactions.

Phase 2 reactions:-Enzymic modification of a xenobiotic by conjugation. See **Conjugate.**

Potentiation:-The effect of a chemical which does not itself have an adverse effect but which enhances the toxicity of another chemical.

Predicted environmental concentration:-The concentration in the environment of a chemical calculated from the available information on certain of its properties, its use and discharge patterns and the associated quantities.

Promoter (carcinogenicity):-An agent which increases tumour production by a chemical when applied after exposure to the chemical.

Protein binding:-The process by which drugs and toxins are bound to proteins other than the receptor, in the plasma or (less common) intracellularly. The bound fraction is inactive but is in equilibrium with the free fraction in the cell or plasma.

Public health impact assessment:-Applying the risk assessment to a specific target population. The size of the populations needs to be known. The end product is a quantitative statement about the number of people affected in this specific target population.

Quality assurance:-Those procedures and controls designed to monitor the conduct of toxicological studies in order to assure the quality of the data and the integrity of the study.

Quantal effect:-An effect that either happens or does not happen, *e.g.* death. Synonym: all-or-none response.

Recommended limit:-A maximum concentration of a potentially toxic substance which is suggested to be safe. Such limits often have no statutory implications and in which case a control or statutory limit should not be exceeded.

Reproductive toxicology (mammalian):-The study of the effects of chemicals on the adult reproductive and neuroendocrine systems, the embryo, fetus, neonate and prepubertal mammal.

Risk:-The likelihood of suffering a harmful effect or effects resulting from exposure to a risk factor (usually some chemical or physical or biological agent). Risk is usually expressed as the probability of occurrence of an adverse effect, *i.e.* expected ratio between the number of individuals that would experience an adverse effect in a given time and the total number of individuals exposed to the risk factor. The term absolute risk is sometimes expressed per unit dose (or exposure) or for a given dose (exposure).

Risk assessment:-The outcome of **Risk identification** and **Risk estimation** (identification and quantification of the risk resulting from a specific use or occurrence of a chemical compound including the establishment of dose-response relationships and target populations). When quantitative data on dose-response relationships for different types of population, including sensitive groups, are unavailable, such considerations may have to be expressed in more qualitative terms.

Risk control:-The type and level of control required for a specified level of risk.

Risk estimation:-The quantification of dose effect and dose response for a substance and linking exposure to the probability and nature of an effect.

Risk evaluation:-The complex process of determining the significance or value of the identified hazards and estimated risks to those concerned with or affected by the decision. It therefore includes the study of risk perception and the trade-off between perceived risks and perceived benefits.

Risk identification:-The identification of the substance of concern, its adverse effects, target populations, and conditions of exposure.

Risk management:-The managerial, decision-making and active hazard control process to deal with those environmental agents for which the risk evaluation has indicated that the risk is too high. See **Risk perception**.

Risk perception:-An integral part of 'risk evaluation'. The subjective perception of the gravity or importance of the risk based on the subject's knowledge of different risks and the moral and political judgement of their importance.

Safety (Toxicological):-Can be defined as the high probability that injury will not result from use of a substance under specific conditions of quantity and manner of use.

Short term exposure limit *(STEL)*:-The time weighted average *(TWA)* airborne concentration to which workers may be exposed for periods up to fifteen minutes, with no more than four such excursions per day and at least sixty minutes between them.

Sister chromatid exchange:-A reciprocal exchange of DNA between the two DNA molecules of a replicating chromosome.

Solvent abuse:-The intentional inhalation of volatile organic chemicals, including anaesthetic gases.

Stochastic:-Obeying the laws of probability.

Structure-activity relationship *(SAR)*:-The correlation between molecular structure and biological activity. It is usually applied to observing the effect that the systematic structural modification of a particular chemical entity has on a defined biological end-point.

Subchronic toxicity:-The adverse effects occurring as a result of the repeated daily [oral] dosing of a chemical to experimental animals for part (not exceeding 10%) of the life span. (Usually 1-3 months) *cf.* **Acute toxicity.**

Suggested no adverse response level *(SNARL)*:-The maximum dose or concentration which on the basis of current knowledge is likely to be tolerated by an organism without producing any adverse effect.

Surfactant:-A substance that lowers surface tension. *cf.* **Detergent.**

Synergistic effect:-An effect of two chemicals acting together which is greater than the simple sum of their effects when acting alone. *cf.* **Additive effect, Antagonism, Potentiation.**

Target organ dose:-The amount of a potentially toxic substance reaching the organ chiefly affected by that substance.

Teratogenesis:-Defects in embryonic and fetal development caused by a substance.

Threshold limit value-time weighted average *(TLV-TWA)*:-The time-weighted average concentration for a normal 8-hour workday and a 40-hour workweek, to which nearly all workers may be repeatedly exposed, day after day, without adverse effects

Tolerance:-The ability to experience exposure to potentially harmful amounts of a substance without showing an adverse effect.

Toxicant:-Any substance which is potentially toxic.

Toxicity:-Any harmful effect of a chemical or a drug on a target organism. See also **Acute, Chronic** and **Subchronic toxicity.**

Toxicodynamics:-See **Pharmacodynamics.**

Toxicokinetics:-See **Pharmacokinetics.**

Toxicovigilance:-The active process of identification, investigation, and evaluation of the various toxic risks in the community with a view to taking measures to reduce or eliminate these risks.

Toxin:-A toxic organic substance produced by a living organism.

Transformation:-The acquisition by a cell of new genetic markers by incorporation of added DNA. In eukaryotic cells it also refers to conversion to a state of unrestrained growth in culture resembling or identical to the tumorigenic condition.

Tumour (neoplasm):-A growth of tissue forming an abnormal mass. Cells of a benign tumour will not spread and cause cancer. Cells of a malignant tumour can spread through the body and cause cancer.

Tumorigenic:-Causing tumour formation.

Xenobiotic:-A chemical which is not a natural component of the living organism exposed to it. Synonyms: drug, foreign substance or compound, exogenous material.

Xenobiotic metabolism:-The chemical transformation of compounds foreign to an organism by various enzymes present in that organism. See also **Biotransformation** and **Xenobiotic.**

The Editor acknowledges the assistance given by the Copenhagen and Geneva offices of the World Health Organization in defining some of these terms.

Useful Addresses

Additional addresses, including those pertaining to sources of chemical information, can be found in 'Toxic Hazard Assessment of Chemicals', ed. M.L. Richardson, The Royal Society of Chemistry, London, 1986 (ISBN 0-85186-897-5)

The Royal Society of Chemistry, Burlington House, Piccadilly, London W1V 0BN, UK (from whom a list of consultants can also be obtained).

American Conference of Government Industrial Hygienists, 6500 Glenway, Building D-5, Cincinnati, Ohio 45211, USA.

Argus Research Laboratories, Inc., 935 Horsham Road, Horsham, PA 19044, USA.

Association of the British Pharmaceutical Industry, 12 Whitehall, London SW1A 2DY, UK.

Basic Information Centre for Chemistry (BASIC), CH-4002 Basle, Switzerland.

Beilstein Institute, Varentrappstrasse 40-42, D-6000, Frankfurt/M. 90, Federal Republic of Germany.

British Agrochemicals Association, 4 Lincoln Court, Lincoln Road, Peterborough PE1 2RP, UK.

British Food Manufacturing Industries Research Association, Randalls Road, Leatherhead, Surrey KT22 7RY, UK.

British Industrial Biological Research Association (BIBRA), Woodmansterne Road, Carshalton, Surrey SM5 4DS, UK.

British Toxicology Society (BTS), PO Box 10, Faversham, Kent ME13 7HL, UK.

Chemical Industries Association Ltd. (CIA), King's Building, Smith Square, London SW1P 3JJ, UK.

Chemical Inspection & Testing Institute Japan (CITI), Sumida-Ku, Tokyo, Japan.

Chemical Manufacturers Association (CMA), 2501 M Street, NW, Washington DC 20037, USA.

Chinese Academy of Preventive Medicine (CAPM), 10 Tian Tan XI Li, Beijing, China.

Department of the Environment (DOE), 2 Marsham Street, London SW1P 3EB, UK.

Department of Health & Social Security (DHSS), Hannibal House, Elephant and Castle, London SE1 6TE, UK.

European Chemical Industry Environmental and Toxicology Centre (ECETOC), Avenue Louise 250, B63, Brussels 1050, Belgium.

Environmental Agency, 1-2-2, Kasumigaseki, Chiyoda-ku, Tokyo 100, Japan.

Environmental Protection Agency (EPA), 401, M Street, SW Washington DC 20460, USA.

Food and Agriculture Organization of the United Nations (FAO), Via delle Terme di Caracalla 00100 Rome, Italy.

Gesellschaft für Strahlen- und Umweltforschung mbH München (GSF), Projekt Umweltgefährdungs-potentiale von Chemikalien (PUC), Ingolstädter Landstr. 1 - D-8042, Neuherberg, West Germany.

Health & Safety Executive (HSE), Baynards House, 1 Chepstow Place, London W2 4TF, UK.

ICI plc, Brixham Laboratory, Freshwater Quarry, Overgang, Brixham, Devon TQ5 8BA, UK.

Industry and Environment Office, United Nations Environment Programme, Tour Mirabeau, 39-43 Quai André Citröen, 75739 Paris, France.

Institute of Community Health, Department of Environmental Medicine, Odense University, J.B. Winslows Vej 19, DK-5000 Odense C, Denmark.

International Agency for Research on Cancer (IARC), 150 Cours Albert Thomas, F-69372 Lyon, Cedex 2, France.

International Labour Office (ILO), CH-1211 Geneva 22, Switzerland.

International Programme on Chemical Safety (IPCS), World Health Organization, CH-1211 Geneva 27, Switzerland.

International Register of Potentially Toxic Chemicals (IRPTC), United Nations Environment Programme, Palais des Nations, CH-1211 Geneva 10, Switzerland.

International Union of Pure and Applied Chemistry (IUPAC), Bank Court Chambers, 2-3 Pound Way, Cowley Centre, Oxford OX3 3YF, UK.

Inter-Research Council Committee on Pollution Research, c/o NERC Headquarters, Polaris House, North Star Avenue, Swindon, Wiltshire SN2 1EU, UK.

Japan Chemical Industry, Ecology, Toxicology and Information Centre (JETOC), 13-4 1-Chome, Toranomon, Minato-ku, Tokyo 105, Japan.

Laboratory of the Government Chemist, Department of Trade and Industry, Cornwall House, Waterloo Road, London SE1 8XY, UK.

Medical Research Council, Toxicology Unit, Woodmansterne Road, Carshalton, Surrey SM5 4EF, UK.

Ministry of Health and Welfare, 1-2-2, Kasumigaseki, Chiyoda-ku, Tokyo 100, Japan.

Ministry of International Trade & Industry (MITI), 1-3-1, Kasumigaseki, Chiyoda-ku, Tokyo 100, Japan.

Ministry of Labour, 1-2-2, Kasumigaseki, Chiyoda-ku, Tokyo 100, Japan.

National Academy of Sciences (NAS), Office of Public Affairs, 2101 Constitution Avenue, Washington DC 20418, USA.

National Chemicals Inspectorate, PO Box 1384, S-171 84 Solna, Sweden.

National Institutes of Health (NIH), PO Box 12233, Research Triangle Park, NC 27709, USA.

National Institute of Occupational Health, S-171 84 Solna, Sweden.

National Institute for Occupational Safety & Health (NIOSH), Robert A. Taft Laboratories, 4676 Columbia Parkway, Cincinnati, Ohio 45226, USA.

National Research Council Canada, Ottawa, Ontario KIA OR6, Canada.

Paper, Printing and Packaging Industries Research Association (PIRA), Randalls Road, Leatherhead, Surrey KT22 7RU, UK.

Regional Laboratory for Toxicology, Dudley Road Hospital, PO Box 393, Birmingham B18 7QH, UK.

Robens Institute of Industrial and Environmental Health and Safety, University of Surrey, Guildford, Surrey GU2 5XH, UK.

Warren Spring Laboratory, Department of Trade and Industry, Gunnelswood Road, Stevenage, Hertfordshire SG1 2BX, UK.

World Health Organization (WHO), CH-1211 Geneva 27, Switzerland.

Subject Index

TOXIC HAZARD ASSESSMENT OF CHEMICALS
A special feature of this publication is the extremely useful glossary of terms.
Contents
Introduction
Retrieval of Data
Interpretation of Data
Risk Assessment and Case Histories
Legislation on Chemicals
Appendix A: Glossary of Terms
Appendix B: Useful Addresses
Subject Index
Hardcover
Royal Society of Chemistry Distribution Centre
Information Services